Springer Proceedings in Advanced Robotics 15

The Springer Proceedings in Advanced Robotics (SPAR) publishes new developments and advances in the fields of robotics research, rapidly and informally but with a high quality.

The intent is to cover all the technical contents, applications, and multidisciplinary aspects of robotics, embedded in the fields of Mechanical Engineering, Computer Science, Electrical Engineering, Mechatronics, Control, and Life Sciences, as well as the methodologies behind them.

The publications within the "Springer Proceedings in Advanced Robotics" are primarily proceedings and post-proceedings of important conferences, symposia and congresses. They cover significant recent developments in the field, both of a foundational and applicable character. Also considered for publication are edited monographs, contributed volumes and lecture notes of exceptionally high quality and interest.

An important characteristic feature of the series is the short publication time and world-wide distribution. This permits a rapid and broad dissemination of research results.

More information about this series at http://www.springer.com/series/15556

spar

Jadran Lenarčič · Bruno Siciliano
Editors

Advances in Robot Kinematics 2020

Springer

Editors
Jadran Lenarčič
Jožef Stefan Institute
Ljubljana, Slovenia

Bruno Siciliano
Department of Electrical Engineering
and Information Technology
University of Naples Federico II
Naples, Italy

ISSN 2511-1256 ISSN 2511-1264 (electronic)
Springer Proceedings in Advanced Robotics
ISBN 978-3-030-50974-3 ISBN 978-3-030-50975-0 (eBook)
https://doi.org/10.1007/978-3-030-50975-0

This Springer imprint is published by the registered company Springer Nature Switzerland AG
The registered company address is: Gewerbestrasse 11, 6330 Cham, Switzerland

Foreword

At the dawn of the century's third decade, robotics is reaching an elevated level of maturity and continues to benefit from the advances and innovations in its enabling technologies. These all are contributing to an unprecedented effort to bringing robots to human environment in hospitals and homes, factories and schools; in the field for robots fighting fires, making goods and products, picking fruits and watering the farmland, saving time and lives. Robots today hold the promise for making a considerable impact in a wide range of real-world applications from industrial manufacturing to health care, transportation, and exploration of the deep space and sea. Tomorrow, robots will become pervasive and touch upon many aspects of modern life.

The *Springer Tracts in Advanced Robotics (STAR)* was launched in 2002 with the goal of bringing to the research community the latest advances in the robotics field based on their significance and quality. During the latest fifteen years, the STAR series has featured publication of both monographs and edited collections. Among the latter, the proceedings of thematic symposia devoted to excellence in robotics research, such as ISRR, ISER, FSR, and WAFR, has been regularly included in STAR.

The expansion of our field as well as the emergence of new research areas has motivated us to enlarge the pool of proceedings in the STAR series in the past few years. This has ultimately led to launching a sister series in parallel to STAR. The *Springer Proceedings in Advanced Robotics (SPAR)* is dedicated to the timely dissemination of the latest research results presented in selected symposia and workshops.

This volume of the SPAR series is dedicated to the proceedings of a special edition of ARK on Advances in Robot Kinematics. Returning to Ljubljana, Slovenia, where it was first founded in September 1988, ARK marks this year an important milestone in reaching its seventeenth gathering, establishing itself as a major anchor of research advances in robot kinematics serving the global robotics community.

The volume edited by Jadran Lenarčič and Bruno Siciliano contains 43 scientific contributions. This collection spans a wide range of research developments in robot mechanisms, kinematics, analysis, design, planning, and control.

Rich by topics and authoritative contributors, ARK brings this unique reference on the current developments and new directions in the field of kinematics. A fine addition to the SPAR series and a genuine tribute to ARK contributors, organizers, and founder!

May 2020

Bruno Siciliano
Oussama Khatib
SPAR Editors

Preface

The series of international symposia Advances in Robot Kinematics (ARK) was organized for the first time in Ljubljana in 1988. Since then, they were organized every two years, in Slovenia, Austria, Italy, France, and Spain, under the patronage of the International Federation for the Promotion of Machine Sciences (IFToMM). The first edited book was published by Springer in 1991, one year after the conference in Linz. Since 1994, a new volume has been published every two years. Each edited book is linked to a corresponding symposium, in which the participants exchange their results and opinions in a meeting that brings together the best researchers and scientists in the field of robot kinematics. The current book is the 15th, and the last three were included in the SPAR series. The book contains 43 contributions and a large team of reviewers contributed their critical and insightful recommendations to the authors.

In the 1980s, when we began organizing these symposia and publishing the books, we did not expect robot kinematics to remain at the forefront of robotics for so many years. However, in the current turbulent times of artificial intelligence, the analytical and in-depth work of kinematicians is even more important than it used to be. Kinematics remains an immense domain of topics that need to be explored if we are to continue the development of complex robot mechanisms. Industrial robots, and especially humanoid robots, open up many scientific intrigues that have not yet been answered. As an example, let us just mention that current humanoid robots are incapable of shrugging their shoulders. A movement that seems humanly childish is crucial for the reachability of the human arm, avoiding obstacles and, last but not least, for communications between people. High-performance computing and even artificial intelligence are undoubtedly tools that will help kinematicians solve the extremely difficult mathematical problems they have to deal with on a daily basis. The future of kinematics seems increasingly fascinating and unpredictable.

After 32 years, ARK is now back to where it all started. This edition is not held in the traditional June dates of the symposium because of the pandemic. As we write this preface, we still hope to be able to safely gather in Ljubljana later this year. We are grateful to the contributors of this volume for their work and enthusiasm. Some have been submitting their contributions for many years, and it is because of them that the conference maintains its high quality. Special thanks go to Tadej Petrič, Conference Publishing Chair, who provided excellent technical support, and Aleš Ude, Conference Organizing Committee Chair. We are also grateful to the Springer staff who have supported our work throughout these years.

We hope that this new ARK book will again attract scholars and researchers specializing in robot kinematics and will outline the research guidelines for many years to come.

May 2020

Jadran Lenarčič
Bruno Siciliano

Organization

Conference Chairmen

J. Lenarčič	Jožef Stefan Institute, Ljubljana, Slovenia
B. Siciliano	University of Naples Federico II, Italy

Honorary Chairman

B. Roth	Stanford University, USA

Scientific Committee

J. Angeles	McGill University, Montreal, Canada
O. Altuzarra	University of the Basque Country, Spain
M. Carricato	University of Bologna, Italy
M. Husty	University of Innsbruck, Austria
A. Kecskemethy	University of Duisburg, Germany
O. Khatib	Stanford University, USA
J. M. McCarthy	University of California at Irvine, USA
J.-P. Merlet	Inria, Sopia-Antipolis, France
V. Parenti-Castelli	University of Bologna, Italy
F. Park	Seoul National University, Korea
A. Perez	Idaho State University, USA
J. Selig	London South Bank University, UK
M. M. Stanisic	University of Notre Dame, USA
F. Thomas	Institute of Industrial Robotics, Spain
P. Wenger	Ecole Centrale de Nantes, France

Organizing Committee Chair

A. Ude — Jožef Stefan Institute, Ljubljana, Slovenia

Publications Chair

T. Petrič — Jožef Stefan Institute, Ljubljana, Slovenia

Contents

Advances in Robot Kinematics Facts and Thoughts 1
Jadran Lenarčič

Inverse Kinematics Using a Converging Paths Algorithm 7
Omid Heidari and Alba Perez Gracia

Design Parameters Influence on the Static Workspace and the Stiffness Range of a Tensegrity Mechanism 15
G. M. Cruz-Martinez, J-C Avila Vilchis, A. Vilchis Gonzalez, S. Abdelaziz, and P. Poignet

Bennett Based Balanced Butterfly Linkage, Deployable Linkage with Inherent Balance 25
Volkert van der Wijk

A Compliant Linkage for Cooperative Object Manipulation Through a Heterogeneous Mobile Multi-robot System 33
Juan Mauricio Toro Ramos, Dhruvin Doshi, Daniel Schütz, and Franz Dietrich

One Degree of Freedom 7-R Closed Loop Linkage as a Building Block of Nanorobots 41
Meysam T. Chorsi, Pouya Tavousi, Caitlyn Mundrane, Vitaliy Gorbatyuk, Horea Ilies, and Kazem Kazerounian

Modeling and Control of a Redundant Tensegrity-Based Manipulator 49
Jérémy Begey, Marc Vedrines, Pierre Renaud, and Nicolas Andreff

Motion Parameterization of Parallel Robots Used in Lower Limb Rehabilitation 57
Iosif Birlescu, Manfred Husty, Calin Vaida, Bogdan Gherman, Ionut Ulinici, Remus Bogateanu, and Doina Pisla

Robust Trajectory Planning of Under-Actuated Cable-Driven Parallel Robot with 3 Cables 65
Edoardo Idà, Sébastien Briot, and Marco Carricato

On the Plane Symmetric Bricard Mechanism 73
J. M. Selig

Exact Coupler-Curve Synthesis of Four-Bar Linkages with Fully Analytical Solutions 82
Shaoping Bai, Rui Wu, and Ruiqin Li

A General Method for Determining Algebraic Input-Output Equations for Planar and Spherical 4R Linkages 90
Mirja Rotzoll, M. John D. Hayes, Manfred L. Husty, and Martin Pfurner

The Forward Kinematics of the 4-1 Cable-Driven Parallel Robot with Non Elastic Sagging Cables 98
J-P. Merlet

The Geometrical Arrangement of Joint Constraints that Makes Natural Motion Possible: Experimental Verification on the Ankle 109
Michele Conconi, Nicola Sancisi, and Vincenzo Parenti-Castelli

Development of a Vector Geometrical Model for PKM Identification 117
J.-B. Guyon, B. Boudon, H. Chanal, and B. Blaysat

Wohlhart's Three-Loop Mechanism: An Overconstrained and Shaky Linkage 125
Andreas Müller

Invariants for Multi-twists, Screw Systems and Serial Manipulators 133
Peter Donelan

Singularity-Free Extraction of a Dual Quaternion from Orthogonal Dual Tensor 142
Daniel Condurache

Singularities in the Image-Based Visual Servoing of Five Points 150
Abhilash Nayak and Sébastien Briot

Real-Time Motion-Planning in Dynamic Environments via Enhanced Velocity Obstacle 158
Amir Shahidi, Katrin Peitsch, Stefan Klimmek, Mathias Hüsing, and Burkhard Corves

Algebraic Analysis of 3-RRC Parallel Manipulators 166
Abhilash Nayak, Martin Pfurner, Huiping Shen, and Manfred Husty

Magneto-Inertial Data Sensory Fusion Based on Jacobian Weighted-Left-Pseudoinverse 174
Janez Podobnik, Marko Munih, and Matjaž Mihelj

Evaluating the Snappability of Bar-Joint Frameworks 182
G. Nawratil

Method for Selecting Self-aligning Mechanisms Enumerated by Matroid Contractions 190
Rodrigo Luis Pereira Barreto, Andrea Piga Carboni, Roberto Simoni, and Daniel Martins

A Real-Time Capable Forward Kinematics Algorithm for Cable-Driven Parallel Robots Considering Pulley Kinematics 199
João Cavalcanti Santos and Marc Gouttefarde

Stiffness Oriented Tension Distribution Algorithm for Cable-Driven Parallel Robots 209
Etienne Picard, Stéphane Caro, Franck Plestan, and Fabien Claveau

A Forward Kinematic Code for Cable-Driven Parallel Robots Considering Cable Sagging and Pulleys 218
Marc Fabritius and Andreas Pott

Singularity and Workspace Analysis of 3-S$\underline{P}$S-U and 4-S$\underline{P}$S-U Tensegrity Mechanisms 226
Swaminath Venkateswaran and Damien Chablat

Degeneration to Infinity May Provide Information About Kinematics 234
Michel Coste and Nestor Djintelbe

Kinematic Synthesis of a Modified Jansen Leg Mechanism 242
Kevin Chen and J. Michael McCarthy

Exponential Displacement Coordinates by Means of the Adjoint Representation 250
Bertold Bongardt and John J. Uicker

A Comparative Study on 2-DOF Variable Stiffness Mechanisms 259
Christoph Stoeffler, Shivesh Kumar, and Andreas Müller

Kinematics and Orientation Workspace of a 3-DOF Parallel Robotic Wrist Actuated by Spherical Four-Bar Linkages 268
Guanglei Wu, Ning Zhang, Chuangchuang Cui, Huiping Shen, and Xuping Zhang

Analytical Determination of the Longest Cylinder Free of Gain-type Singularities Inside the Workspace of a 3-R$\underline{P}$S Spatial Manipulator 277
Argaja Deepak Shende, Bibekananda Patra, Prem Kumar Prasad, and Sandipan Bandyopadhyay

Clifford's Identity and Generalized Cayley-Menger Determinants 285
Federico Thomas and Josep M. Porta

A New Approach for Continuous Wrapping of a Thick Strand on a Surface — The Planar Case with Constant Length and Free Ends 293
Katharina Müller and Andres Kecskemethy

Higher Order Path Synthesis of Four-Bar Mechanisms Using Polynomial Continuation 303
Aravind Baskar and Mark Plecnik

Development of a Reconfigurable Four-Bar Mechanism for a Human Robot Collaborative Gripper 311
Keerthi Sagar, Vishal Ramadoss, Michal Jilich, Matteo Zoppi, Dimiter Zlatanov, and Alessandro Zanella

Kinematic Analysis of a Planar Manipulator with Anti-parallelogram Joints and Offsets 319
Philippe Wenger and Matthieu Furet

On Singularity and Instability in a Planar Parallel Continuum Mechanism 327
Oscar Altuzarra and Francisco J. Campa

Modeling and Simulation of Hybrid Soft Robots Using Finite Element Methods: Brief Overview and Benefits 335
Stanislao Grazioso, Giuseppe Di Gironimo, Luciano Rosati, and Bruno Siciliano

Exoskeleton Control Based on Network of Stable Heteroclinic Channels (SHC) Combined with Gaussian Mixture Models (GMM) 341
Tadej Petrič, Marko Jamšek, and Jan Babič

Model Predictive Controller for a Planar Tensegrity Mechanism with Decoupled Position and Stiffness Control 349
JR Jurado Realpe, Salih Abdelaziz, and Philippe Poignet

Author Index 359

Advances in Robot Kinematics Facts and Thoughts

Jadran Lenarčič(✉)

Jožef Stefan Institute, Ljubljana, Slovenia
jadran.lenarcic@ijs.si

Abstract. ARK, Advances in Robot Kinematics, has certainly left an important mark. Many top-notch robotics specialists regularly attended the symposia and contributed with their scientific papers and thoughtful discussions. The accompanying books in the ARK series, published by Kluwer Academic Publishers and later by Springer, gave the series an additional international impact. This is a brief look at how it all started and why it didn't end until today.

Keywords: Symposium on Advances in Robot Kinematics · ARK

1 Introduction

The idea of organizing an international symposium specializing in robot kinematics dates to early 1986. As a young researcher, I had just completed my Ph.D., and I sensed the need for such a conference, since there were quite a few robotic conferences and they were all general in scope. Robot kinematics was seen as a tool in other research topics, and above all, as a first stage in the mathematical modelling of robot dynamics. However, the more I delved into this area, which was not my specialty, as I am an electrical engineer by education, the more I realized that kinematics is an area that addresses specific problems that represent their own scientific field and, therefore, are not necessarily part of something else.

As a beginner in the field, I pursued two things in particular: specialized literature on robot kinematics (the idea of launching a specialized journal was also an option) and a community of researchers with whom I could collaborate, exchange ideas and results. In 1986 I started to come up with the idea of organizing an international symposium in Ljubljana (back then in socialist Yugoslavia). The main concern for me was how to attract top experts; most of them were located in the United States. I didn't worry about the funding for the symposium. This was of secondary importance to me.

I based my hope on bringing this symposium to life on already-established friendships with important personalities in robotics, among which Bernard Roth was definitely the focal point. We had met two years earlier at the Romansy symposium in Udine, where I was immediately attracted by his kindness. At the

J. Lenarčič and B. Siciliano (Eds.): ARK 2020, SPAR 15, pp. 1–6, 2021.
https://doi.org/10.1007/978-3-030-50975-0_1

same symposium, I also met some other pillars, like Oussama Khatib, Kenneth Waldron, Bruno Siciliano and David Orin.

I announced the idea of organizing a specialized symposium on robot kinematics in 1986 at the Romansy Symposium in Krakow. The interest seemed sufficiently strong, so I decided to press ahead. The title Advances in Robot Kinematics turned out very well. Some things in life do not take much time to think over. One year later, I started organizing the symposium by simply sending out invitation letters to the people I believed to be the most important in the field. It was the birth of Advances in Robot Kinematics, whose short title ARK was surprisingly well accepted by the community.

At the symposium held in Bologna in 2018, I announced that the conference in Ljubljana in 2020 would be my last in the capacity of chairman. After twenty-two years, the time has come to step down. This is not a painful or sad moment, but one full of joy, as this scientific event has certainly left an unforgettable footprint on robotics and beyond. I would like to thank everyone who contributed his or her work and enthusiasm.

2 Symposia and Related Books

Ljubljana (Slovenia, Yugoslavia) 1988 – Jadran Lenarčič
Book: Advances in Robot Kinematics, Cankarjev dom & J. Stefan Institute, 22 Contributions

Among those invited to the first ARK symposium were researchers who personally fascinated me with their scientific publications. I did not look at where they came from and whether I had met them before. Many of them I had not, but later it turned out that I was extremely lucky in my choices. I saw that a community had begun to build, which was decisive for the future work and success of this symposium. Their positive responses to my invitations were unexpected for me, beyond what I had hoped. Out of the twenty invited, only Ken Waldron apologized, and he later joined and became one of the pillars of the conference.

I had to send the invitations by regular mail and waited for at least a month or more for people to respond. There were no emails at that time. The most prominent names among the invitees were Jorge Angeles, Vladimir Lumelsky, Alberto Rovetta, Roy Featherstone, Michael McCarthy, Vincenzo Parenti-Castelli, Michael Stanisic, and substitutes were sent by Richard Paul, Kazuo Tanie, Philippe Coiffet and Andrew Goldenberg. Adding them to the aforementioned Bernard Roth, Oussama Khatib, Bruno Siciliano, Kenneth Waldron and David Orin, whose support had been the backbone of the whole project, I was fortunate enough to set up a ground-shaking group of international scientists whose research has mostly been related to robot kinematics.

The first symposium took place in Ljubljana in September, immediately after the Romansy symposium, held in Udine, Italy. Thus, some participants made use of their presence just over the border. Most importantly, I felt that I had to take care of everyone personally, which was seen as an act of hospitality and respect. With the help of sponsors, I was able to organize lunches and dinners in the cost of the conference. It was then that I came to the realization that the social events were at least as important as the technical sessions. It is crucial for the success of a conference that people come together and feel like they are important team members.

I organized the oral presentations at the symposium so that everyone had 40 min. The contributions were published in the Proceedings, which was printed in advance and available at the conference. I edited the proceedings myself and had it printed by Cankarjev Dom, the agency which was also the technical organizer of the symposium and all the other events that took place in Slovenia in the following decades. Most important, however, was the unanimous conclusion that these symposia should be continued every other year and that the symposium should be organized once in Slovenia and once somewhere else in the Alps-Adria region. The aim was to emphasize the identity of the symposium by its geographical characteristic. There was no going back after that decision.

Linz (Austria) 1990 – Sabine Stifter and Jadran Lenarčič
Book: Advances in Robot Kinematics With Enmphasis on Symbolic Computation, Springer (1991), 53 Contributions

The second symposium moved to Linz by accident. It started with a visit to Ljubljana from a political delegation of Upper Austria, who were looking for options to encourage technical and scientific cooperation between the regions. Shortly thereafter, I received a mail from Linz from Sabine Stifter proposing the co-organization of the symposium with the University of Johannes Keppler in Linz. I met her a few months later at the RISC Institute. We invited two communities to the ARK symposium: a community of kinematicians and a community of mathematicians specialising in symbolic computation. I was convinced that significant synergies could be achieved.

The symposium was held in Linz in September. Conference contributions were collected based on extended abstracts and far more submissions came than we expected. We accepted about 130 contributions. Although the conference set important new standards for the future, its focus was, in my opinion, too diffuse. The Proceedings of the abstracts were published at the conference. Based on peer reviews of these, we selected full articles that were published one year later (1991) in an edited book format by Springer. Although this symposium was quite different from the first and subsequent ones, the group of kinematicians who attended the conference, became the core of ARK, thus securing its future and long-term mission.

Ferrara (Italy) 1992 – Vincenzo Parenti-Castelli and Jadran Lenarčič
Book: 3rd International Workshop on Advances in Robot Kinematics, Antenna Verde & University of Ferrara, 44 Contributions

In 1992, ARK should have been organised in Slovenia. However, the political situation after Slovenia's independence process was not the most favourable for the organization of international events. Although there were no problems and matters were cleared up in Slovenia much earlier, I decided to make an exception and organize ARK somewhere else. I was worried about how many participants would attend an event in Slovenia. Vincenzo Parenti-Castelli solved the problem by proposing to organise ARK in Ferrara, Italy. A more ideal solution could not have been found.

Ferrara in many ways laid the foundations for today's ARK. Among other things, we first agreed with IFToMM to take over the sponsorship that ARK was awarded for all subsequent years. The link with the Linz mathematicians was abandoned, and some important new contributors from the field of robot kinematics appeared at the symposium, such as Manfred Hiller, Andres Kecskemethy, Josepf Duffy, Carlo Galletti and Jean-Pierre Merlet. The Proceedings were published and printed by the organizer in the form of full papers. New names were added to the Scientific Committee. The number of oral presentations was significantly fewer than in Linz. This proved to be appropriate, and in later years this number never exceeded 58. By doing so, the conference retained its "boutique" form. Also important was the decision to reposition future symposia to the end of June or the beginning of July.

Ljubljana (Slovenia) 1994 – Jadran Lenarčič and Bahram Ravani
Book: Advances in Robot Kinematics and Computational Geometry, Kluwer Academic Publishers, 51 Contributions

In 1994 ARK took place again in Ljubljana. In preparation for the event, my goal was how to ensure high quality and better public visibility of the proceedings. I decided to try publishing an edited book with the Dutch publishing house Kluwer Academic Publisher, which impressed me with its professionalism and expeditiousness. That's when I met (via mail) Nathalie Jacobs (Kluwer), first assistant director, later director at Kluwer Academic Publishers, Dordrecht. Working with her contributed to the creation of the impressive series of books that we know today. The work had to be conducted as efficiently as possible, as it was crucial for me to have the book printed and handed over to the conference participants at the event.

The conference was held in Ljubljana at the Club of Cankarjev dom. The organizers of the conference strived for an excellent social program in addition to an excellent scientific program. The motivation was to establish a standard for ARKs in the coming years. I trust that returning to Ljubljana was an important contribution to the identity of these symposia.

Piran/Portorož (Slovenia) 1996 – Jadran Lenarčič and Vincenzo Parenti-Castelli
Book: Recent Advances in Robot Kinematics, Kluwer Academic Publishers, 45 Contributions

We agreed to organize the ARK again in Slovenia 1996. The reason was to correct the originally planned order of organizers, which was interrupted due to the political situation in Slovenia a few years before. Piran was, after Ljubljana, the second Slovenian location and turned out to be ideal for the style of the ARK symposia. Certainly, the conference was at an important international level, but the social events were especially well organized by Cankarjev dom. I remember this event as perhaps the liveliest and with a full and friendly atmosphere, suggesting that a good conference must first create a community of people who want to participate and want to share their ideas. I do not know if the environment contributed to this, but it seems important to me that such an atmosphere was created spontaneously. During this symposium, I decided that all future books will have a different title, in combination with Advances in Robot Kinematics.

Strobl/Salzburg (Austria) 1998 – Jadran Lenarčič and Manfred L. Husty
Book: Advances in Robot Kinematics: Analysis and Control, Kluwer Academic Publishers, 58 Contributions

Piran/Portorož (Slovenia) 2000 – Jadran Lenarčič and Michael M. Stanišić
Book: Advances in Robot Kinematics, Kluwer Academic Publishers, 45 Contributions

Caldes de Malavella (Spain) 2002 – Jadran Lenarčič and Federico Thomas
Book: Advances in Robot kinematics: Theory and Applications, Kluwer Academic publishers, 51 Contributions

Sestri Levante (Italy) 2004 – Jadran Lenarčič and Carlo Galletti
Book: On Advances in Robot Kinematics, Kluwer Academic publishers, 51 Contributions

Ljubljana (Slovenia) 2006 – Jadran Lenarčič and Bernard Roth
Book: Advances in Robot Kinematics: Mechanisms in Motion, Springer, 53 Contributions

Batz-sur-Mer (France) 2008 – Jadran Lenarčič and Philippe Wenger
Book: Advances in Robot Kinematics: Analysis and Design, Springer, 48 Contributions

This time the conference did not take place in a Mediterranean or Alpine region, nevertheless it impressed us like all its predecessors.

Piran/Portorož 2010 (Slovenia) – Jadran Lenarčič and Michael M. Stanišić
Book: Advances in Robot Kinematics: Motion in Man and Machine, Springer, 58 Contributions

For the first time, articles were not included in sessions based on their topics. I distributed them using a random generator and divided the chapters in the book into neutral Part 1, Part 2, etc. The sessions at the conference were thus of greater interest.

Innsbruck 2012 (Austria) – Jadran Lenarčič and Manfred Husty
Book: Latest Advances in Robot Kinematics, Springer, 56 Contributions

Ljubljana 2014 (Slovenia) – Jadran Lenarčič and Oussama Khatib
Book: Advances in Robot Kinematics, Springer, 56 Contributions

Grasse 2016 (France) – Jadran Lenarčič and Jean-Pierre Merlet
Book: Advances in Robot Kinematics 2016, Springer – SPAAR (2018), 46 Contributions

The proceedings were first published in an electronic version by Inria Sophia Antipolis by the time of the conference in 2016. Later, the ARK Scientific Committee decided to print the book as part of the Springer SPAAR series.

Bologna 2018 (Italy) – Jadran Lenarčič and Vincenzo Parenti-Castelli Advances in Robot Kinematics 2018, Springer – SPAAR (2018), 53 Contributions

In Bologna, we celebrated the first thirty years of the conference that brought us together and has forged many new friendships.

3 Conclusion

I decided that my last conference would take place in Ljubljana in 2020, the city where it all began. I can certainly say that ARK has had a profound impact on my scientific and personal life. Future conferences are sure to follow, and a new generation of researchers is here to lead this initiative. The ARK Scientific Committee has already decided that the conference in 2022 will be held in Spain and will be organized by Oscar Altuzarra.

Robot kinematics remains an unlimited source of new scientific and practical challenges. There is scientific material for many more years of research. New technologies, such as high-performance computing and artificial intelligence, will further strengthen this research field and allow solving so far unsolvable problems. In the end, I want to thank all the ARK friends who have contributed with their professionalism and personal engagement.

Inverse Kinematics Using a Converging Paths Algorithm

Omid Heidari[1(✉)] and Alba Perez Gracia[1,2(✉)]

[1] Department of Mechanical Engineering, Idaho State University, Pocatello, USA
{heidomid,perealba}@isu.edu
[2] Remy Robotics, Barcelona, Spain

Abstract. This article presents a numerical method for the inverse kinematics of serial chains by utilizing dual quaternion formulation of the robot kinematics within a converging paths algorithm. The method is inspired by iterative techniques such as FABRIK, however adding information on the kinematics of the chain to be solved. The method has been tested with 2R, planar 4R and spatial 4R robots. Future work includes optimizing the method to compare with other fast numeric algorithms for inverse kinematics.

Keywords: Inverse kinematics · Serial manipulators · FABRIK · Iterative methods

1 Introduction

Inverse kinematics (IK) methods are widely used in robotics and computer graphics. Analytic and numerical techniques have been develop for solving the IK problem. Closed-form or analytic methods were among the first to be developed and are successful for simple chains; however, they are developed in a case-by-case strategy and they are hard to generalize. See [7] for a review. Some of the analytic methods use the description of the geometric constraints of the chain. Others use the matrix formulation to the forward kinematics equations of the chain to solve the inverse problem when this is equated to the desired position [4]. Analytic methods have the advantage of avoiding iterative solutions that in principle are more time consuming; however many times they reduce to finding the roots of a univariate polynomial [5], which carries its own numerical problems. New algorithms such as IKFast [3] allow a more systematic approach combining analytic and numerical solutions.

A good review for numerical methods, especially those applied to computer graphics, can be found in [2]. Numerical methods are grouped into those based on the Jacobian, Newton-based methods and heuristic methods. Jacobian-based methods [8] are the most popular and a variety of implementations can be found, see for instance [6]. Heuristic methods are developed as a computationally lean alternative to other numerical methods. Most of them consist of a simple approach that yields many fast iterations. Among them we highlight Forward And

J. Lenarčič and B. Siciliano (Eds.): ARK 2020, SPAR 15, pp. 7–14, 2021.
https://doi.org/10.1007/978-3-030-50975-0_2

Backward Reaching Inverse Kinematics (FABRIK) [1]. This method is an iterative approach where it tries to find the position of each joint by locating a point on line. It is not costly in terms of computation neither it produces complex solutions. However, the algorithm works best with spherical joints, mostly used for IK in computer animations and gaming.

Desired properties of IK methods are speed, numerical stability and convergence. In addition, being applicable to a broad number of kinematic chains is desirable too. In this work we present a IK algorithm that combines some of the advantages of the numerical methods with the insight of the analytic ones. The method is tested for some simple open-chain robots consisting of revolute joints. The core concept is to break the serial chain and create two legs for a given pose. Then the end-effectors (ee) of these legs move towards each other smoothly till they match and complete the chain. Moving towards a better solution at each step is insured by incorporating pseudo-inverse in Cartesian space. At each step, the solution in the trigonometric joint space is projected to the surface condition for the joint variables to make sure that the constraints are satisfied.

2 Dual Quaternion Inverse Kinematics

In this section we describe a method to state the inverse kinematics for serial chains with revolute joints. Let us consider a kinematic serial chain with n degrees of freedom; let the joint space of the chain be defined by the vector $\theta = \{\theta_1, \ldots, \theta_n\}$ and the joint axes be defined by the Plücker coordinates of the lines S_i, $i = 1, \ldots n$.

2.1 Product of Exponentials in Dual Quaternion Form

In general, for an arbitrary k-degree of freedom robot, kR, given by k joint axes $S_1 \ldots S_k$ with rotation angles $\theta_1 \ldots \theta_k$, the forward kinematics of relative displacements is given by the product of exponentials,

$$D = \mathrm{e}^{S_1 \frac{\theta_1}{2}} \mathrm{e}^{S_2 \frac{\theta_2}{2}} \ldots \mathrm{e}^{S_k \frac{\theta_k}{2}} \tag{1}$$

This can be done using any suitable algebra. In this case, we use Clifford algebra of dual quaternions which can be seen as 2^k products of cosines and sines of the angles. Select an order, for instance $(c_1c_2 \ldots c_k, s_1c_2 \ldots c_k, \ldots, s_1s_2 \ldots s_k)$. The forward kinematics can be written as a linear combination of the joint axes and Clifford products of joint axes by these coefficients,

$$[I\ S_1\ S_2 \ldots S_1S_2 \ldots S_k] \begin{Bmatrix} \cos\frac{\theta_1}{2} \cos\frac{\theta_2}{2} \ldots cos\frac{\theta_k}{2} \\ \vdots \\ \sin\frac{\theta_1}{2} \sin\frac{\theta_2}{2} \ldots \sin\frac{\theta_k}{2} \end{Bmatrix} = P, \tag{2}$$

When using this expression to solve the inverse kinematics problem, consider a desired pose P. In general, for a robot with k revolute joints and a desired position P we will always have 2^k unknowns. The forward kinematics yields 8 linear equations (6 independent). In addition to these, the trigonometric relations among the unknowns are to be considered which are bilinear and spherical relations.

2.2 Inverse Kinematics for the 2R Robot

The relative forward kinematics

$$D_{2R} = e^{S_1 \frac{\theta_1}{2}} e^{S_2 \frac{\theta_2}{2}} \tag{3}$$

is equated to the target position P. It can be collected in the products of sines and cosines. Denote $c_i = \cos \frac{\theta_i}{2}$ and $s_i = \sin \frac{\theta_i}{2}$,

$$[I \; S_1 \; S_2 \; S_1 S_2] \begin{Bmatrix} c_1 c_2 \\ s_1 c_2 \\ c_1 s_2 \\ s_1 s_2 \end{Bmatrix} = P, \tag{4}$$

Denote the $2^2 = 4$ unknowns $x_1 = c_1 c_2$, $x_2 = s_1 c_2$, $x_3 = c_1 s_2$, and $x_4 = s_1 s_2$.

To the 6 linear independent equations we need to add the bilinear condition and the spherical condition that result from the trigonometric relations, so the final system of equations is

$$I x_1 + S_1 x_2 + S_2 x_3 + S_1 S_2 x_4 = P, \quad x_1 x_4 = x_2 x_3, \quad x_1^2 + x_2^2 + x_3^2 + x_4^2 = 1 \tag{5}$$

This is a total of 8 equations, and the system is over-determined by 4 equations as expected.

Considering the rotations only, there are 4 unknowns and 5 equations, overdetermined by 1 as expected. The rotation linear system can be solved for the unknowns, which will comply with the quadratic conditions only if the rotation belongs to the workspace.

The system consists of 6 linear equations and two quadratic equations. These quadratic equations can be combined to create the surface where the angles must lie,

$$x_1^2 + x_2^2 + x_3^2 + (x_2 x_3 / x_1)^2 = 1, \tag{6}$$

which is shown in Fig. 1 below. The implicit surface admits a local parameterization at the regular points.

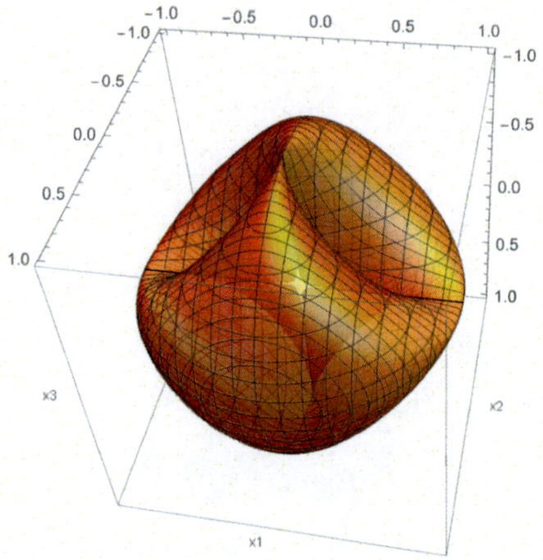

Fig. 1. Surface condition for the joint variables of the RR chain.

3 Converging Paths Algorithm

The proposed algorithm relies on breaking any serial chain into two lower-degree ones which is a well-known approach utilized in different kinematics methods. The end-effectors of these two smaller chains need to get close to each other at each step along an iterative process and reach perfectly at the final step of iteration. In this work, apart chains are called left and right leg. Left is the one having the base of the original robot and right is the one having the ee of the original manipulator as its stationary base at the given pose for IK input. Figure 2 shows a 4R serial chain which is divided into two RR chains. To explain the procedure of the algorithm and different parts of it, this example is illustrated and followed in the rest of this paper.

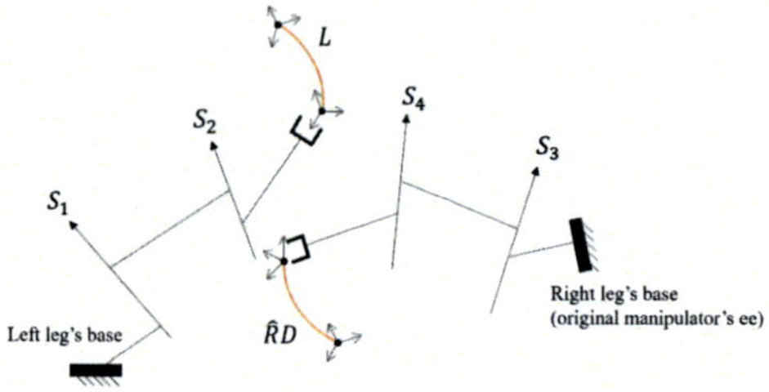

Fig. 2. 4R manipulator as two RR chains

The procedure starts with setting the left leg to a random unit dual quaternion that satisfies Study's quadric. Then, this pose is used to calculate the right leg's configuration that can be the closest. To this end, a pseudo-inverse matrix is computed which may result in a solution that does not satisfy the constraints in Eq. 5. To solve this problem, the algorithm incorporate a method of projection where solutions out of workspace are projected based on the shortest distance they can have to the constraint surface. The projected vector in the joint space is, then, used to compute the Cartesian pose of the right leg and its transformation is set equal to the left leg equations to see if it is in the workspace of the left leg.

Equation 1, for a 4R serial chain, can be converted to the following:

$$\mathrm{e}^{S_1 \frac{\theta_1}{2}} \mathrm{e}^{S_2 \frac{\theta_2}{2}} = D \mathrm{e}^{-S_4 \frac{\theta_1}{2}} \mathrm{e}^{-S_3 \frac{\theta_3}{2}} \tag{7}$$

Separating the left and right transformations of Eq. 7 by setting them to two different dual quaternion, $L = R$, results in:

$$\mathrm{e}^{S_1 \frac{\theta_1}{2}} \mathrm{e}^{S_2 \frac{\theta_2}{2}} = L, \quad \mathrm{e}^{S_3 \frac{\theta_3}{2}} \mathrm{e}^{S_4 \frac{\theta_4}{2}} = T \tag{8}$$

where hat symbol is the dual quaternion conjugate operator and $T = \hat{R}D$. As it was stated in the previous sections, Eq. 8 can be written as a linear combination of Clifford products of joint axes:

$$[\mathrm{L}]\mathbf{v} = \mathbf{l}, \quad [\mathrm{R}]\mathbf{u} = \mathbf{t} \tag{9}$$

where $\mathbf{l}$ and $\mathbf{t}$ are 8-element vectors representing dual quaternions on each side, L and T respectively. [L] and [R] are corresponding matrices containing joint axes and their Clifford products. It is noteworthy that elements of these matrices are constant depending on a reference pose and do not change by changing the configuration of the robot. Moreover, $\mathbf{v}$ and $\mathbf{u}$ are 4-element vectors including the products of sines and cosines of angels expressed in Eq. 4.

3.1 Algorithm Description

The back and forth procedure explained in the previous section between left and right leg is shown more in details by Algorithm 1. This iterative process repeats till the ee of one leg matches the ee of the other leg in its workspace. To make sure that there is a solution to the inverse kinematics problem, the forward kinematics is solved so that we have a D that is in the workspace.

Algorithm 1: Converging Paths Inverse Kinematics

Result: $\mathbf{v}$ and $\mathbf{u}$ that put the left and right leg in a configuration at which their end-effectors reach each other.

```
initialization:
L = random unit dual quaternion;
δ = 0.0001;
while constraints are not satisfied do
    l = Vector(L);
    v = [L]^{-1} l;
    v_projected = Project(v);
    l = [L] v_projected;
    L = DualQuaternion(l);
    R = L;
    r = Vector(R̂D);
    u = [R]^{-1} r;
    u_projected = Project(u);
    t = [R] u_projected;
    T = DualQuaternion(t);
    R = DT̂;
    L = R;
    if CheckConstraints(v) < δ  &&  CheckConstraints(u) < δ then
        Exit while loop;
    end
end
```

There are some functions used in this algorithm that affect its performance and speed. *Vector()* converts a dual quaternion to an 8-element vector and *DualQuaternion()* does the opposite. *CheckConstraint()* verifies if the constraints in Eq. 5 are satisfied, returning a value for each constraint that can be used to evaluate the convergence. At last, *Project()* is the function that finds a solution with the least distance between the given point and the constraint surface.

There are different approaches to accomplish this. One method uses vector calculus on the parameterization to compute the normal vector to the surface and then find the line normal to the surface and passing through the given point. This is done by finding a line whose moment is equal to the moment of the point in the direction of the line.

Another method that is used in the algorithm is through Lagrange Multipliers. The objective function to minimize is the squared Euclidean distance between the given point and points lying on the constraint surface.

The proposed algorithm is in the early stage and the authors are exploring different approaches to do the projection. However, the Lagrangian method seems to have a faster response at the current stage. Numerical results suggest convergence by iteratively approaching each leg to the other using Moore–Penrose pseudo inverse method, however convergence needs to be formally proved. At each step the pose of one leg's end effector is set equal to the other one to see if it is in its workspace or not. If not, the corresponding vector in joint space gets projected to the constraint surface to make sure that the ee is in the workspace.

3.2 Numerical Example

This section is dedicated to an example of solving the IK of a 4R serial manipulator by the proposed method. The first step is to compute matrices [L] and [R] for an arbitrary located joint axes.

$$L = \begin{bmatrix} 0 & -0.639867 & 0.0305296 & 0.732783 \\ 0 & 0.512895 & -0.510064 & 0.567498 \\ 0 & 0.572284 & 0.859595 & 0.310715 \\ 1 & 0 & 0 & -0.210788 \\ 0 & -0.13248 & 1.10913 & -0.322849 \\ 0 & -0.912657 & -0.523315 & 0.545166 \\ 0 & 0.66982 & -0.349916 & -0.138578 \\ 0 & 0 & 0 & 0.14111 \end{bmatrix} \quad R = \begin{bmatrix} 0 & -0.0689129 & 0.84182 & 0.263595 \\ 0 & -0.994039 & 0.491349 & -0.0865131 \\ 0 & -0.0844797 & -0.223418 & 0.802942 \\ 1 & 0 & 0. & 0.527558 \\ 0 & -1.12461 & 0.562694 & -0.792586 \\ 0 & 0.040865 & -0.732085 & 0.10385 \\ 0 & 0.43654 & 0.510156 & 0.0228117 \\ 0 & 0 & 0 & 0.378327 \end{bmatrix} \tag{10}$$

After 9 iterations, the algorithm finds a solution with an error of 0.001. Figure 3 shows the solution and the steps of convergence as well as the constraint surface.

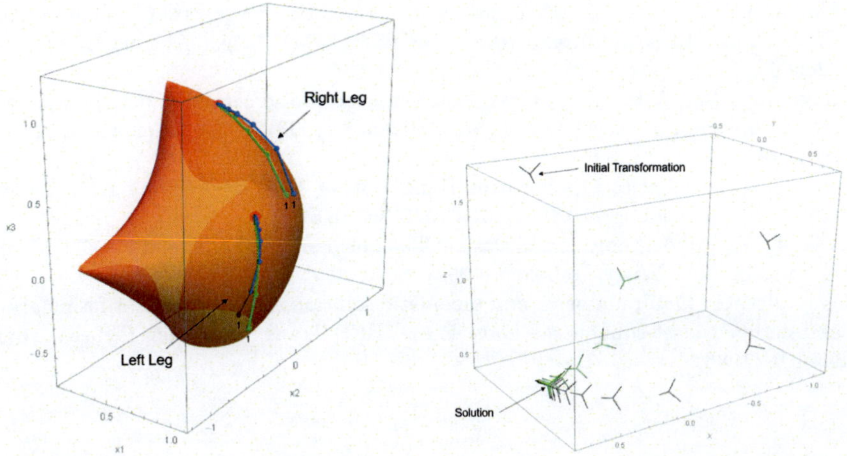

The convergence of the right and left leg to the solution in joint space in terms of x_1,x_2 and x_3. Green: $\mathbf{u}_{projected}$ and $\mathbf{v}_{projected}$, Blue: $\mathbf{u}$ and $\mathbf{v}$, Red: expected solution.

The convergence of the left (black) and right (green) leg to the solution in Cartesian space.

Fig. 3. Convergence in joint and Cartesian spaces.

The computation time is relatively high in its current version, as the steps have not been optimized for spped yet. The interest of the method is going to depend on finding efficient processes for each of the steps of the algorithm.

4 Conclusion

This work presents an inverse kinematics method based on dividing the chain and iteratively projecting the closest point on the workspace. The equations are derived from the forward kinematics and hence it can be applied systematically to serial chains. The method is tested on serial chains with 2, 3 and 4 degrees of freedom and revolute joints. In these examples, the algorithm exhibits fast convergence. Future work will include optimizing the calculations, applying the method to increasingly complex chains and comparing it to the current state of the art as well as examining its behavior at non-regular points.

References

1. Aristidou, A., Lasenby, J.: Fabrik: a fast, iterative solver for the inverse kinematics problem. Graph. Models **73**(5), 243–260 (2011). http://dblp.uni-trier.de/db/journals/cvgip/cvgip73.html
2. Aristidou, A., Lasenby, J., Chrysanthou, J., Shamir, A.: Inverse kinematics techniques in computer graphics: a survey. Comput. Graph. Forum **37**(6), 35–58 (2017)
3. Diankov, R.: Automated construction of robotic manipulation programs. Ph.D. thesis, Carnegie Mellon University (2010)
4. Manocha, D., Canny, J.F.: Efficient inverse kinematics for general 6R manipulators. IEEE Trans. Robot. Autom. **10**(5), 648–657 (1994). https://doi.org/10.1109/70.326569
5. Raghavan, M., Roth, B.: Inverse kinematics of the general 6R manipulator and related linkages. J. Mech. Des. **115**(3), 502–508 (1993). https://doi.org/10.1115/1.2919218
6. Siciliano, B., Khatib, O.: Springer Handbook of Robotics. Springer, Heidelberg (2007). https://doi.org/10.1007/978-3-540-30301-5
7. Tsai, L.W.: Robot Analysis and Design: The Mechanics of Serial and Parallel Manipulators, 1st edn. Wiley, Hoboken (1999)
8. Wampler, C.: Manipulator inverse kinematic solutions based on vector formulations and damped least-squares methods. Proc. IEEE Trans. Syst. Man Cybern. **16**(1), 93–101 (1986)

Design Parameters Influence on the Static Workspace and the Stiffness Range of a Tensegrity Mechanism

G. M. Cruz-Martinez[1(✉)], J-C Avila Vilchis[2], A. Vilchis Gonzalez[2], S. Abdelaziz[1], and P. Poignet[1]

[1] LIRMM, Université de Montpellier, CNRS, Montpellier, France
{giorgio-mackenzie.cruz-martinez,abdelaziz,poignet}@lirmm.fr
[2] Autonomous University of Mexico State, Toluca, Mexico
{avilchisg,jcavila}@uaemex.mx

Abstract. This paper deals with the impact of the design parameters on the static workspace and the stiffness range of a planar 3-DoF tensegrity mechanism. The static model is established through the energetic approach and the stiffness is derived analytically along the 3-DoF of the mechanism. The design parameters considered here are the spring stiffness and the location of the mechanism attachment points to the base. Results on the impact of these parameters are finally analyzed. This analysis constitutes a first step towards the geometric optimization of tensegrity mechanisms.

Keywords: Tensegrity mechanism · Static workspace · Stiffness range

1 Introduction

The term tensegrity was created by Richard B. Fuller as a union of 'tensional' and 'integrity' [6]. A tensegrity structure is formed entirely by a combination of rigid and flexible elements. Its configuration stands by itself and maintains its form solely because its structural members (struts) are suspended in a network of tensional elements (cables or springs) [13]. Tensegrity structures are characterized by being light, deployable and of variable stiffness [1]. Working with these structures has been of interest for engineers and researchers since tensegrity applications range from mobile robotics [5], manipulators [7,14] and robots in medical applications [3].

Developing mathematical models (a kinematic one, for instance) for tensegrity mechanisms is challenging [11,12,18] and requires to know the extension of flexible elements at all time so as to establish a relationship between the joint and the cartesian variables. If the measurement of these extensions is not available, static modeling is considered. Several methods have been proposed in the literature to determine the equilibrium configurations of tensegrity mechanisms [8,15]. Static workspace computation, using a continuous approach, has been proposed by [2] for a 2-DoF tensegrity mechanism.

J. Lenarčič and B. Siciliano (Eds.): ARK 2020, SPAR 15, pp. 15–24, 2021.
https://doi.org/10.1007/978-3-030-50975-0_3

For the tensegrity mechanism reported in this paper, a potential energy approach is considered in order to determine stable equilibrium configurations based on the stiffness analysis. This paper is an attempt to understand the influence of one geometric parameter on static workspace and the stiffness range. The geometric parameter under observation concerns the mechanism attachment points. This analysis constitutes the first step towards the geometric optimization of tensegrity mechanisms.

This paper is organized as follow. The 3-DoF planar tensegrity mechanism is described in Sect. 2. The respective static and stiffness models are presented in Subsect. 2.1. These models are synthesized using an energetic approach. The computation of the static workspace and stiffness range are presented in Subsect. 2.2. The impact of the design parameters on the static workspace and the stiffness range is finally discussed in Sect. 3.

2 Mechanism Description

Figure 1 shows a 1-bar planar tensegrity mechanism driven by 4 actuators. According to the classification proposed by Skelton [17], this mechanism is a class-1 tensegrity system. The actuators are connected to the bar, of length $2b$, using cables and springs. The springs are considered here identical of stiffnesses k. The cables, attached to the bar on p_1, are enrolled on pulleys fixed on A_1 and A_2 and are connected to springs before being enrolled on pulleys mounted on the actuators 1 and 2. The other two cables, attached to p_2, are enrolled on pulleys fixed on A_3 and A_4 and are enrolled on pulleys mounted on the actuators 3 and 4.

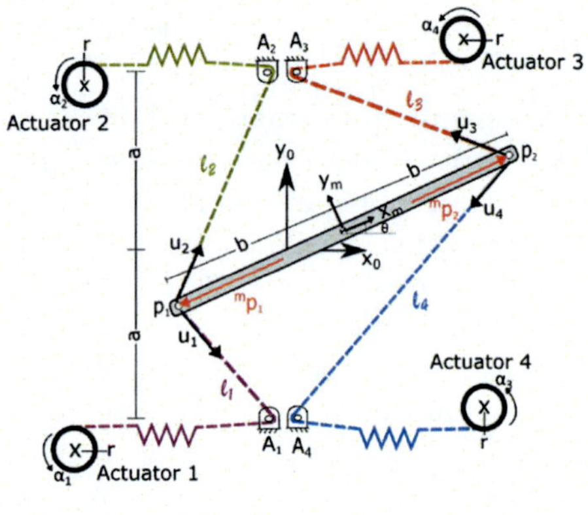

Fig. 1. Planar tensegrity mechanism

2.1 Problem Formulation

As depicted in Fig. 1, the position of the center of mass P of the bar is defined by $\mathbf{P} = [x, y]^T$ in the reference frame $\mathscr{R}_0 = (O, \mathbf{x}_0, \mathbf{y}_0)$. A mobile frame

$\mathscr{R}_m = (O_m, \mathbf{x}_m, \mathbf{y}_m)$, is attached to the bar's center of mass. The angle θ defines the bar's orientation. The coordinates of four attachment points A_1, A_2, A_3 and A_4 are defined respectively to the inertial reference frame by position vectors $\mathbf{a}_1 = [0, -a]^T, \mathbf{a}_2 = [0, a]^T, \mathbf{a}_3 = [0, na]^T, \mathbf{a}_4 = [0, -na]^T$, with n-factor allowing to change locations of attachment points A_3 and A_4 where $0.1 \leq n < 1$.

The coordinates of nodes p_1 and p_2 are defined with respect to the mobile frame $\mathscr{R}_m = (O_m, \mathbf{x}_m, \mathbf{y}_m)$ by vectors ${}^m\mathbf{p}_1 = [-b, 0]^T, {}^m\mathbf{p}_2 = [b, 0]^T$ while with respect to the inertial frame $\mathscr{R}_0 = (O, \mathbf{x}_0, \mathbf{y}_0)$ their coordinates are defined by Eq. 1 with $i = 1, 2$.

$$\mathbf{p}_i = \mathbf{P} + \mathbf{R}_m \, {}^m\mathbf{p}_i \tag{1}$$

with $\mathbf{R}_m = \begin{pmatrix} \cos\theta & -\sin\theta \\ \sin\theta & \cos\theta \end{pmatrix}$

Let's define $\ell_1 = \|\mathbf{a}_1 - \mathbf{p}_1\|$, $\ell_2 = \|\mathbf{a}_2 - \mathbf{p}_1\|$, $\ell_3 = \|\mathbf{a}_3 - \mathbf{p}_2\|$ and $\ell_4 = \|\mathbf{a}_4 - \mathbf{p}_2\|$ as the distances between the bar's attachment points and the different nodes.

The system initial configuration is defined when the position of the bar's center of mass coincides with the origin of the inertial reference frame and $\theta = 0\,rad$. The initial length in the cables/springs are determinate according with the Eq. 2 when their tensions have minima values.

$$\ell_{0g} = \ell_{0(g+1)} = \begin{cases} \sqrt{a^2 + b^2} & \text{g} = 1 \text{ for cables 1 and 2} \\ \sqrt{(na)^2 + b^2} & \text{g} = 3 \text{ for cables 3 and 4} \end{cases} \tag{2}$$

Due to the presence of springs and in order to determine the static configuration of the mechanism, an energetic approach is considered. It means that the first and second derivatives of the potential energy of the system U are used to establish the stable static equilibrium points. Equation 3 defines the total potential energy

$$U = \sum_{h=1}^{4} \frac{1}{2} k e_h^2 \tag{3}$$

where e_h is the elongation of the spring h that can be computed as in Eq. 4

$$e_h = \rho_h + \ell_h - \ell_{0h} \quad \text{for} \quad h = 1, 2, 3, 4 \tag{4}$$

with ρ_h being the displacement of the cable h. It is computed as $\rho_h = r\alpha_h$, where r is the radius of the actuator's pulley and α_h is the angular position of the actuator h.

The static equilibrium of the bar is obtained by, simultaneously, solving the Eq. 5.

$$\frac{\partial U}{\partial x} = 0, \quad \frac{\partial U}{\partial y} = 0, \quad \frac{\partial U}{\partial \theta} = 0 \tag{5}$$

A stable equilibrium configuration must satisfy inequalities in Eq. 6.

$$\frac{\partial^2 U}{\partial x^2} > 0, \quad \frac{\partial^2 U}{\partial y^2} > 0, \quad \frac{\partial^2 U}{\partial \theta^2} > 0 \tag{6}$$

2.2 Characterization of the Static Workspace and the Stiffness Range

The static workspace of a tensegrity system is defined as the set of all end-effector positions that are able to reach with a stable equilibrium configuration [2], while taking into account the unilateral nature of the cables, i.e. The tensions in all the cables must remain positive.

Each point in the workspace has a minimum and maximum stiffness value. The range of stiffness for each point is computed by subtracting the maximum and the minimum stiffness values. In order to analyze the static workspace as well as the stiffness range of a tensegrity mechanism, as a function of design parameters, the tensions limits are considered identical in all the study.

The static equilibrium of the mechanism can be expressed in a similar way as for cable-driven parallel manipulators [9,10] in accordance with Eq. 7

$$\mathbf{W}\boldsymbol{\tau} = \mathbf{f} \tag{7}$$

where $\mathbf{W}$ represents a 3×4 wrench matrix that depends on the mechanism position $\mathbf{P}$ and its orientation θ. The tensions in the cables are identified by vector $\boldsymbol{\tau} = [\tau_1, \tau_2, \tau_3, \tau_4]^T$. The vector $\mathbf{f}$ represents the wrench applied to the mechanism by means of the cables tensions. The wrench matrix $\mathbf{W}$ is computed by Eq. 8

$$\mathbf{W} = \begin{bmatrix} \mathbf{u}_1 & \mathbf{u}_2 & \mathbf{u}_3 & \mathbf{u}_4 \\ \mathbf{p}_1 \times \mathbf{u}_1 & \mathbf{p}_1 \times \mathbf{u}_2 & \mathbf{p}_2 \times \mathbf{u}_3 & \mathbf{p}_2 \times \mathbf{u}_4 \end{bmatrix} \tag{8}$$

where $\mathbf{u}_h$ represents the unit vector. Its direction is defined as the cable direction (cf. Fig. 1).

The tension in the cables must remain in between a minimum and a maximum values, respectively τ_{min} and τ_{max}. The stiffness range for a given stable equilibrium configuration is computed by varying the tensions in the cables without affecting the position and orientation of the mechanism. These tensions are computed as:

$$\boldsymbol{\tau} = \mathbf{W}^{+}\mathbf{f} + \mathbf{H}\lambda \tag{9}$$

where $\boldsymbol{W}^{+}$ is the Moore-Penrose generalized inverse of $\mathbf{W}$ [16]. $\mathbf{H}$ is a vector of dimension 4×1 whose column span the null space of $\mathbf{W}$. λ is an arbitrary scalar. To keep a stable equilibrium configuration, the wrench $\mathbf{f}$ is considered equal to a null vector. The bounds of λ can be obtained by solving the inequality 10.

$$\tau_{min} \leq \mathbf{H}\lambda \leq \tau_{max} \tag{10}$$

The bounds λ_{min} and λ_{max} allowing to compute two vector tensions solutions that enable to keep the equilibrium:

$$\begin{aligned} \boldsymbol{\tau}_{min} &= \mathbf{H}\lambda_{min} \\ \boldsymbol{\tau}_{max} &= \mathbf{H}\lambda_{max} \end{aligned} \tag{11}$$

Using the vectors $\boldsymbol{\tau}_{min}$ and $\boldsymbol{\tau}_{max}$, one can compute the minimum and maximum springs elongations. Solving the Eq. 4, can be calculated $\boldsymbol{\rho}^{min}$ and $\boldsymbol{\rho}^{max}$ both are vectors that contain the set of minimum and maximum displacements for each actuators. Now the minimum and maximum stiffness are computed using the Eq. 12 at a given pose of the mechanism ($\mathbf{P}$, θ) considering $\boldsymbol{\rho}^{min}$ and $\boldsymbol{\rho}^{max}$.

$$
\begin{aligned}
K_x^{min} &= \frac{\partial^2 U(\mathbf{P}, \theta, \boldsymbol{\rho}^{min})}{\partial x^2}, & K_y^{min} &= \frac{\partial^2 U(\mathbf{P}, \theta, \boldsymbol{\rho}^{min})}{\partial y^2}, & K_\theta^{min} &= \frac{\partial^2 U(\mathbf{P}, \theta, \boldsymbol{\rho}^{min})}{\partial \theta^2} \\
K_x^{max} &= \frac{\partial^2 U(\mathbf{P}, \theta, \boldsymbol{\rho}^{max})}{\partial x^2}, & K_y^{max} &= \frac{\partial^2 U(\mathbf{P}, \theta, \boldsymbol{\rho}^{max})}{\partial y^2}, & K_\theta^{max} &= \frac{\partial^2 U(\mathbf{P}, \theta, \boldsymbol{\rho}^{max})}{\partial \theta^2}
\end{aligned}
\tag{12}
$$

3 Discussion of the Results

This section shows the estimation of the static workspace and stiffness range. The static workspace of the system describes the pose of the end-effector in the x, y coordinates and the orientation along z axis. The parameters of the mechanism that are used were: a = 0.1 m, b = 0.1 m, $\tau_h \in [4, 10]$N and $\rho_h \in \left[\frac{4}{k}, \frac{10}{k}\right]$ m.

The Fig. 2 illustrates the relationship between the size of the workspace and the spring stiffness values k. The first row shows the static workspace of the mechanism with $n = 1$ and $k = 270$ N/m and the second row shows the static workspace when $n = 1$ and $k = 80$ N/m. The black dots represent the end-effector in a stable equilibrium configuration that the mechanism can reach, the red line shows the mechanism's boundary of static workspace with $k = 270$ N/m. In the second row the green line shows the boundary of static workspace with $k = 80$ N/m and the red line is superimposed to highlight how the workspace is affected in relationship with the modification of the stiffness springs, noting that the value of k is inversely proportional to the static workspace.

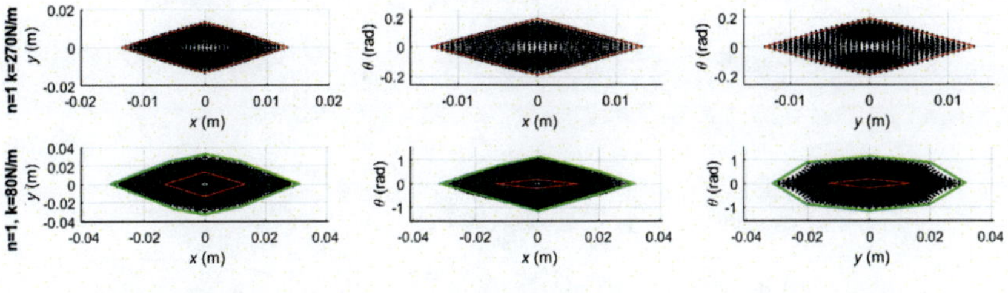

Fig. 2. Relationship between static workspace and variation of k

In order to analysis the influence of the geometrical parameters on the static workspace and stiffness range, here is presented the results when $n = 1$ (Fig. 1), $n = 0.5$ (Fig. 3a) and $n = 0.1$ (Fig. 3b). These variation of n causes that the symmetry in the y-axis be lost because the nodes A_3, A_4 reducing the distance between them.

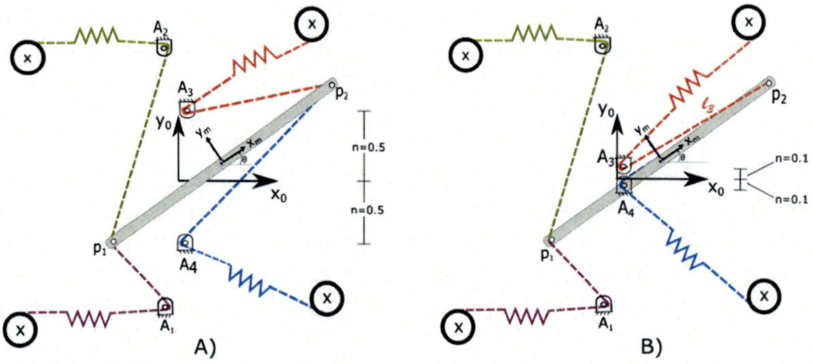

Fig. 3. A) Tensegrity mechanism with n = 0.5, B) Tensegrity mechanism with n = 0.1

The Fig. 4 illustrates the static workspace for the mechanism when $n = 1, 0.5$ and 0.1. The workspace for $n = 1$ has ranges from $x \in (-0.2, 0.2)m$, $y \in (-0.2, 0.2)m$ and $\theta \in (-1.5, 1.5)$ rad. The shape is modified throughout the changes of n, this changes are described by the projections on the planes $xy, y\theta, x\theta$. The black dots represent the center of mass in a stable equilibrium configuration that the mechanism can reach, the red line shows the mechanism's boundary of static workspace with $n = 1$ and the green line shows the boundary of static workspace for $n = 0.5$ and $n = 0.1$ respectively. As the Fig. 4 pictures, the workspace in xy is decreased as n decreases and is shifted to the other side where the attachment points joint. However the workspace $x\theta$ who is showed in row 2 increases as n decreases. The largest workspace in the plane $y\theta$ is in the configuration with $n = 0.5$.

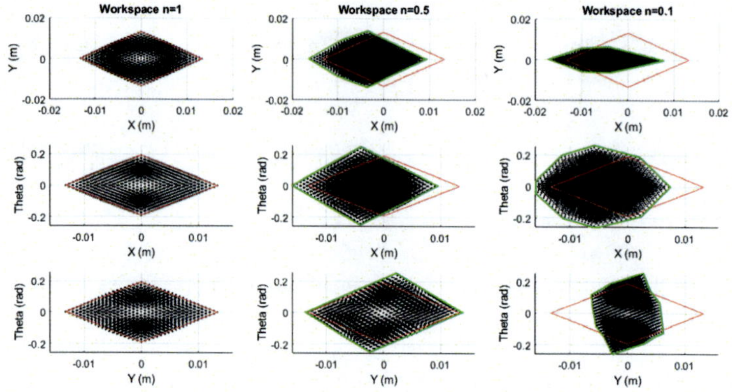

Fig. 4. Static workspace when n = 1, 0.5 and 0.1

The analysis of the stiffness range is shown below. The Fig. 5 shows the stiffness range for K_x, the Fig. 6 shows the stiffness range for K_y and the Fig. 7

shows the stiffness range for K_θ. The workspace is depicted as a volume. In order to observe the behavior of stiffness, it has been discretized in layers. The first column of each figure represents the minimum stiffness and the second column shows the stiffness range, each point in the workspace has a color according to its stiffness value. The black line represents the boundary of the workspace with $n = 1$ and it is superimposed on all the graphs to show how the workspace changes.

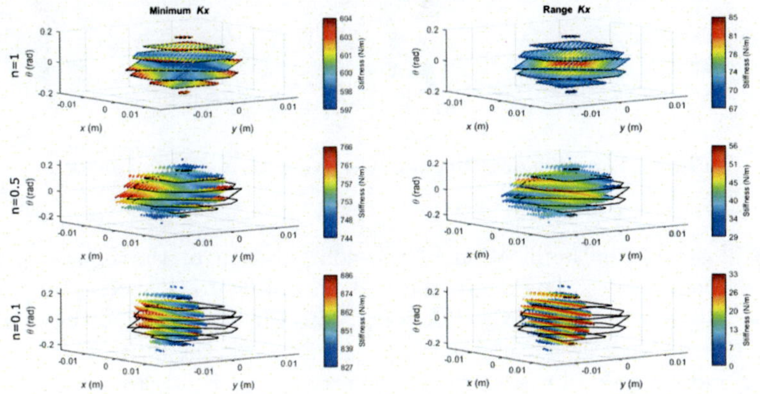

Fig. 5. The stiffness ranges K_x $n = 1,\ 0.5$ and 0.1

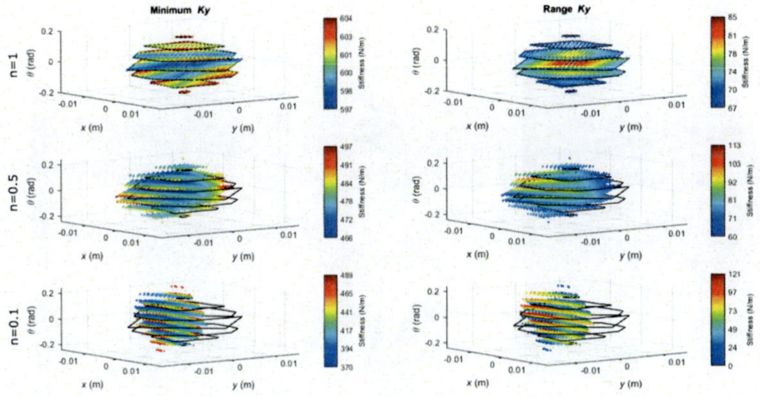

Fig. 6. The stiffness ranges K_y with $n = 1,\ 0.5$ and 0.1

This tensegrity mechanism could be used as a puncture assistance robot [3,4] because it allows stiffness modulation. The procedure requires that the effector has to be rigid when a needle is inserted in the body and in other moment it has to be soft allowing to follow the physiological movements such as breath.

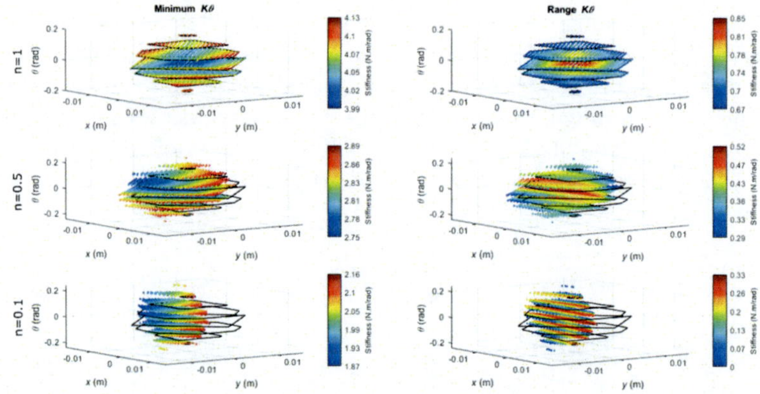

Fig. 7. The angular stiffness ranges K_θ with $n = 1,\ 0.5$ and 0.1

To illustrate the application of the computation of the Figs. 5, 6 and 7, two desired stiffness ranges are selected according to a specific application: $K_x \geq 30$ N/m and $K_y \geq 75$ N/m. The Fig. 8 illustrates the workspace that satisfies the desired stiffness range, the first row shows K_x with $n = 1, 0.5$ and 0.1 and the second row shows the workspace for K_y desired with $n = 1, 0.5$ and 0.1. The workspace for the range K_x is greater with $n = 0.5$ since it allows reach higher orientation values. The same behavior happens in the workspace for the range K_y but the increase is more notorious in the mobility orientation with $n = 0.5$. Concluding that the biggest workspace that satisfies the design conditions for the application is with the geometrical parameter $n = 0.5$.

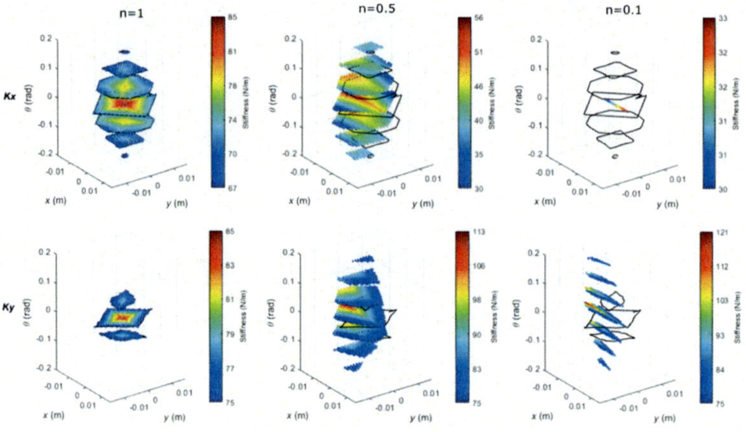

Fig. 8. Workspace using $K_x \geq 30$ N/m and $K_y \geq 75$ N/m

4 Conclusion

This paper shows the influence of the design parameters on the static workspace as well as in the stiffness range of a 3-DoF planar tensegrity mechanism. An extension of the approach analysis can be applied for tensegrity mechanism with more degrees of freedom. The analysis will help to design a geometric optimization approach that allows to define the location of the attachment points and to select the adequate springs in order to satisfy a required static workspace with a desired stiffness range.

References

1. Azadi, M., Behzadipour, S., Faulkner, G.: Variable stiffness spring using tensegrity prisms. J. Mech. Robot. ASME (2010). https://doi.org/10.1115/1.4001776
2. Boehler, Q., Charpentier, I., Vedrines, M.S., Renaud, P.: Definition and computation of tensegrity mechanism workspace. J. Mech. Robot. (2015). https://doi.org/10.1115/1.4029809
3. Boehler, Q., Zompas, A., Vedrines, M., Abdelaziz, S., Renaud, P., Poignet, P.: Experiments on a variable stiffness tensegrity mechanism for an MR-compatible needle holder. Comput./Robot Assist. Surg. (2015)
4. Bricault, I., Jauniaux, E., Zemiti, N., Fouard, C., Taillant, E., Dorandeu, F., Cinquin, P.: Light puncture robot for CT and MRI interventions. IEEE Eng. Med. Biol. Mag. **27** (2008). https://doi.org/10.1109/EMB.2007.910262
5. Friesen, J., Pogue, A., Bewley, T., De Oliveira, M., Robert, S., Sunspiral, V.: Ductt: a tensegrity robot for exploring duct systems. In: IEEE International Conference on Robotics and Automation (ICRA) (2014). https://doi.org/10.1109/IROS.2016.7759811
6. Fuller, B. (ed.): Synergetics, Explorations in the Geometry of Thinking. Collier Macmillan (1975)
7. Furet, M., Chablat, D., Fasquelle, B., Khanna, P., Chevallereau, C., Wenger, P.: Prototype of a tensegrity manipulator to mimic bird necks. In: 24ème Congrès Français de Mécanique. Brest, France (2019)
8. Furet, M., Wenger, P.: Kinetostatic analysis and actuation strategy of a planar tensegrity 2–x manipulator. J. Mech. Robot. (2019). https://doi.org/10.1115/1.4044209
9. Gagliardini, L., Gouttefarde, M., Caro, S.: Determination of a dynamic feasible workspace for cable-driven parallel robots. In: Advances in Robot Kinematics 2016, Springer Proceedings in Advanced Robotics (2017)
10. Gouttefarde, M., Merlet, J.P., Daney, D.: Wrench-feasible workspace of parallel cable-driven mechanisms. In: IEEE International Conference on Robotics and Automation. Roma, Italy (2007). https://doi.org/10.1109/ROBOT.2007.363195
11. Henrickson, J.V., Valaseky, J., Skelton, R.: Shape control of tensegrity structures. In: AIAA SPACE 2015 Conference and Exposition. Pasadena, USA (2015). https://doi.org/10.2514/6.2015-4502
12. Ji, Z., Li, T., Lin, M.: Kinematics, singularity, and workspaces of a planar 4-bar tensegrity mechanism. J. Robot. (2014). https://doi.org/10.1155/2014/967251
13. Jing, Y.Z., Ohsaki, M. (eds.): Tensegrity Structures Form, Stability and Symmetry. Springer, Heidelberg (2015). https://doi.org/10.1007/978-4-431-54813-3

14. Lessardand, S., Castro, D., Asper, W., Chopra, S.D., Baltaxe-Admony, L., Teodorescu, M., SunSpiral, V., Agogino, A.: A bio-inspired tensegrity manipulator with multi-DOF, structurally compliant joints. In: 2016 IEEE/RSJ International Conference on Intelligent Robots and Systems (IROS), pp. 5515–5520 (2016). https://doi.org/10.1109/IROS.2016.7759811
15. Manríquez Padilla, C.G., Zavala Pérez, O.A., Pérez Soto, G.I., Rodríguez Reséndiz, J., Camarillo Gómez, K.A.: Form-finding analysis of a class 2 tensegrity robot. Appl. Sci. (2019). https://doi.org/10.3390/app9152948
16. Roberts, R.G., Graham, T., Lippitt, T.: On the inverse kinematics, statics, and fault tolerance of cable-suspended robots. J. Robot. Syst. (1998). https://doi.org/10.1002/(SICI)1097-4563(199810)15:10⟨581::AID-ROB4⟩3.0.CO;2-P
17. Skelton, R., Adhikari, R., Pinaud, J.P., Chan, W.: An introduction to the mechanics of tensegrity structures. In: Conference on Decision and Control. Florida, USA (2001). https://doi.org/10.1109/.2001.98086
18. Wenger, P., Chablat, D.C.: Kinetostatic analysis and solution classification of a planar tensegrity mechanism. In: Computational Kinematics, pp. 422–431. Springer, Heidelberg (2018). https://doi.org/10.1007/978-3-319-60867-948

Bennett Based Balanced Butterfly Linkage, Deployable Linkage with Inherent Balance

Volkert van der Wijk(✉)

Department of Precision and Microsystems Engineering, Delft University of Technology, Delft, The Netherlands
v.vanderwijk@tudelft.nl

Abstract. In this paper it is shown how a 2-DoF inherently force balanced spatial deployable Butterfly Linkage is found consisting of four entangled similar Bennett linkages moving synchronously and with the common center of mass in the central joint. This linkage is derived from the Grand 4R Four-Bar Based Inherently Balanced Linkage Architecture by selecting a planar linkage with four entangled similar 4R four-bar linkages to which the Bennett conditions are applied. The inherent balance conditions are calculated, which are independent of the Bennett angles, and a CAD-model of the linkage is presented.

Keywords: Inherent force balance · Static balance · Deployable · Bennett linkage · Similar four-bar

1 Introduction

For deployable linkages, static balance - when gravity does not affect the motion of the linkage - can be of urgent importance, especially when they become large of size with significant mass. This is for instance the case in kinetic architecture where deployable structures are used as movable walls, facades, roofs and for temporary portable housing [2,4]. The use of springs in the design for balance can be very complex, dangerous in case of failure and, since in practice springs lose their stiffness properties quickly over time, not very durable [3]. Using large countermasses for balance however is not desired as well as they cause a high demand on the structural design in terms of strength and space and they also limit the portability significantly for which lightweight is essential.

The contribution of this paper is to show how the inherent balance approach can be applied to synthesize advanced deployable structures with inherent balance, which means that they are force balanced with solely the links of the linkage, not needing countermasses [6,7,9]. This is possible for specific designs of the kinematics. In this paper the synthesis of a 2-degree-of-freedom (2-DoF) inherently force balanced spatial deployable Butterfly Linkage with four entangled similar Bennett linkages moving synchronously is presented. First it is shown how this linkage is obtained from a Grand Inherently Balanced Linkage Architecture, subsequently the inherent balance conditions are derived.

J. Lenarčič and B. Siciliano (Eds.): ARK 2020, SPAR 15, pp. 25–32, 2021.
https://doi.org/10.1007/978-3-030-50975-0_4

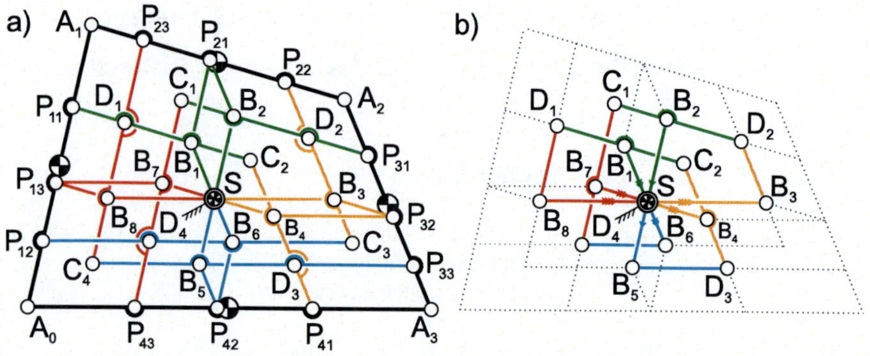

Fig. 1. a) Grand 4R Four-Bar Based Inherently Balanced Linkage Architecture from [8] where S is the common CoM of all the links for 2-DoF motions; b) 2-DoF inherently balanced linkage solution with S as the common CoM and as base pivot, obtained from the Grand Architecture by eliminating all other links.

2 Synthesis from the Grand 4R Four-Bar Based Inherently Balanced Linkage Architecture

Figure 1a shows the Grand 4R Four-Bar Based Inherently Balanced Linkage Architecture with solely mass symmetric links - when link centers of mass (CoMs) are on the line through the joints - as presented in [8]. It consists of a 4R four-bar linkage $A_0A_1A_2A_3$ with inside 20 links that are aligned with the four-bar links in a specific way based on principal vectors. The linkage architecture is 26 times overconstrained yet movable with the common CoM of all links stationary in joint S for all motions. This means that when the linkage architecture is pivoted to a base with S as fixed base pivot, the architecture is balanced with 2-DoF motion capability.

Since the architecture is highly overconstrained, a variety of links can be eliminated to obtain a normally constrained linkage solution. The choices for this do not affect the inherent balance capability, all derived solutions can be inherently balanced without the need of countermasses or additional elements. In [8] 32 inherently balanced linkages were presented which were derived from the Grand Architecture when keeping the four links of the four-bar $A_0A_1A_2A_3$. However it is also possible to find solutions without the links of the four-bar $A_0A_1A_2A_3$.

Figure 1b shows a two times overconstrained 2-DoF linkage solution that is derived from the Grand Architecture by eliminating the links of the four-bar $A_0A_1A_2A_3$ and all other internal links. It consists of the 12 links B_5SB_2, $C_1B_2D_2$, D_2B_3, B_3SB_8, B_8D_1, $D_1B_1C_2$, B_1SB_6, B_6D_4, $D_4B_7C_1$, B_7SB_4, $C_2B_4D_3$, and D_3B_5. Within the linkage, the four similar four-bars $SB_2D_2B_3 \sim B_8D_1B_1S \sim D_4B_7SB_6 \sim B_5SB_4D_3$ can be recognized. The four-bars $SB_2D_2B_3$ and $B_5SB_4D_3$ share the common link B_5SB_2, the four-bars $SB_2D_2B_3$ and $B_8D_1B_1S$ share the common link B_8SB_3, the four-bars $B_8D_1B_1S$ and $D_4B_7SB_6$ share the common link B_1SB_6, and the four-bars $D_4B_7SB_6$ and $B_5SB_4D_3$ share the common link B_7SB_4. S is the central joint where these four links have a revolute pair. Paral-

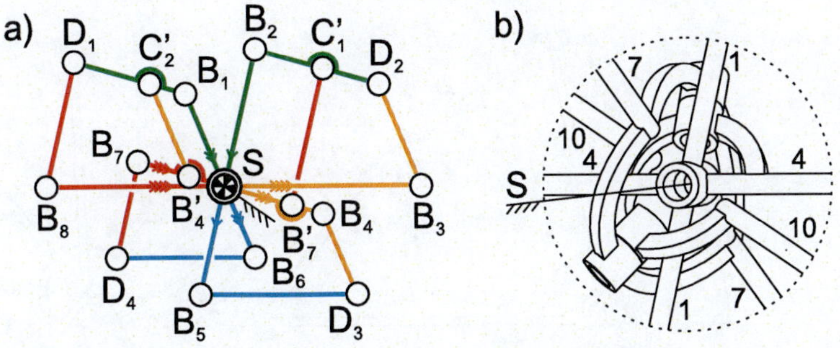

Fig. 2. a) Modification of the linkage solution in Fig. 1b where link segments C_1B_7 and C_2B_4 have become the separated and shifted links $C_1'B_7'$ and $C_2'B_4'$ to obtain a compact and normally constrained design; b) Close-up of the central joint of the spatial Butterfly Linkage in Fig. 3.

lelograms $C_1B_2SB_7$ and $C_2B_4SB_1$ constrain the four similar four-bars to move synchronously, maintaining similarity.

From the linkage in Fig. 1b it is possible to obtain a normally constrained 2-DoF linkage by removing either joint C_1 or C_2 to keep only one parallelogram. Instead it is also possible to break link $C_1B_7D_4$ into two separate links C_1B_7 and B_7D_4 and to break link $C_2B_4D_3$ into two separate links C_2B_4 and B_4D_3 to obtain a normally constrained linkage. Then it is also possible to shift the links C_1B_7 and C_2B_4 to another location without affecting the kinematics as illustrated in Fig. 2a where they are located, respectively, as $C_1'B_7'$ and $C_2'B_4'$. Practically their function is to keep links B_2D_2 and B_1D_1, respectively, in synchronous motion with link B_7SB_4.

The four similar four-bar linkages can be transformed into four similar Bennett linkages by including equal Bennett conditions [1] for each four-bar. This results in the 2-DoF spatial linkage in Fig. 3 which, because of its four deployable segments, will be referred to as the *Bennett Based Balanced Butterfly Linkage*. It is interesting to find out that the parallel links $C_1'B_7'$ and $C_2'B_4'$ in Fig. 2a are not needed to constrain the Bennetts to move similarly, which is because of the known axial overconstraints of the revolute pairs in Bennett linkages [1].

The Butterfly Linkage is shown for the Bennett conditions that opposite links have equal length ($l_1^a = l_3$, $l_2 = l_4^a$, $l_4^b = l_6$, $l_5 = l_7^a$, $l_7^b = l_9$, $l_8 = l_{10}^a$, $l_{10}^b = l_{12}$, and $l_{11} = l_1^b$, with the parameters explained in Fig. 4) and $\sin\alpha / l_1^a = \sin\beta / l_2$ where $\alpha = 40°$ and $l_1^a / l_2 = 7/9$, resulting in $\beta = 55.7°$, which are the same for all four Bennetts for similarity. The relative scaling of the four Bennetts is, in frontview, 1.0 (top right), 0.8 (top left), 0.6 (bottom left), and 0.7 (bottom right) and the angle between link 1 and link 4 is $\varphi_1 - \varphi_2 = 77°$. With respect to the balance solution of a single Bennett linkage with countermasses [5], this solution can be regarded as that the four Bennetts balance one another.

The central joint in S where four links come together with revolute pairs may be challenging to design for the spatial motions. Figure 2b shows a close-up of the design presented here. Link 1 and link 4 rotate with respect to the fixed base with revolute pairs

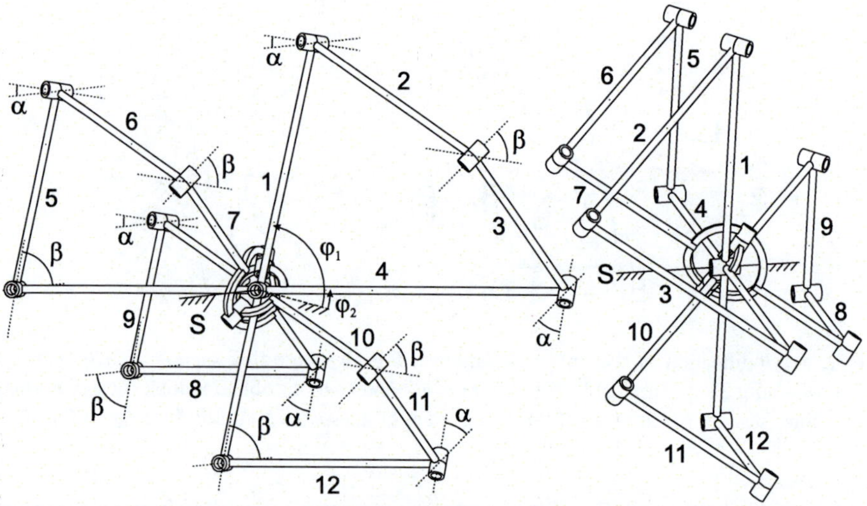

Fig. 3. The spatial deployable *Bennett Based Balanced Butterfly Linkage*, inherently balanced with four similar Bennett linkages entangled, obtained by including the Bennett conditions in the linkage of Fig. 2a. The linkage is normally constrained without the links $C_1'B_7'$ and $C_2'B_4'$ and can move with 2-DoF motions about the central pivot with all four Bennetts moving synchronously (frontview and sideview).

in S, link 7 rotates with respect to link 4 with a revolute pair in S, which is drawn with an offset to the back, and link 10 rotates with respect to link 7 with a revolute pair in S, which is drawn with an offset to the front. Because of the axial overconstraints of the revolute pairs in Bennett linkages it is also possible to replace a few revolute pairs with spherical, cylindrical, or universal joints.

3 Calculation of Parameter Values for Inherent Balance

After deriving the desired linkage solution from the Grand Architecture, the parameter values of the linkage need to be calculated for which joint S is the common CoM for all motions, i.e. for which the linkage is force balanced about S. Since the linkage solution is based on principal vectors, which are equal for planar and spatial motions, the balance conditions of the planar Butterfly Linkage in Fig. 2a are equal to the balance conditions of the spatial Butterfly Linkage in Fig. 3. In fact, the planar linkage can be seen as the single pose of the spatial mechanism when it becomes flat, i.e. assuming the Bennett angles also movable and becoming $\alpha = \beta = 0°$.

In Fig. 4 the planar Butterfly Linkage of Fig. 2a is shown with the parameters of the links and the link masses. Each link has a length l_i and a mass m_i with the link CoM located at a distance p_i from the indicated joint. Also the parallelogram links 13 and 14 are considered, located at offsets d_1 and d_2 from links 1 and 7, respectively. For the spatial Butterfly Linkage these links can be disregarded for balance by making them massless with $m_{13} = m_{14} = 0$.

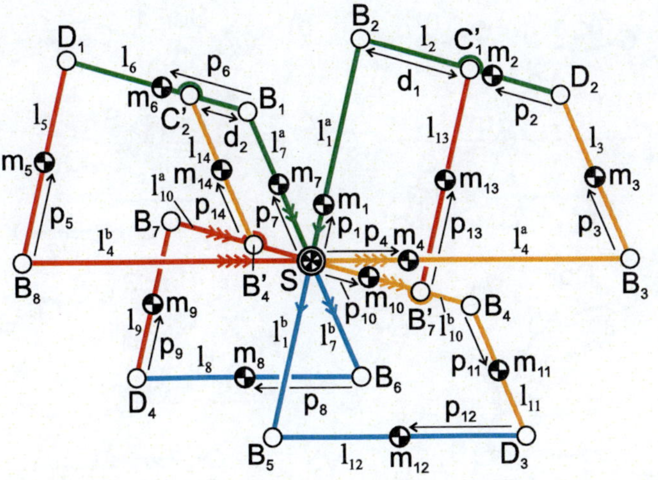

Fig. 4. Mass and link parameters of the linkage in Fig. 2a and of the spatial Butterfly Linkage in Fig. 3, which are equal.

There are two groups of inherent balance conditions to be fulfilled, the kinematic balance conditions and the mass balance conditions [8]. The kinematic balance conditions determine the similarity of the four four-bar linkages with:

$$\frac{l_1^a}{l_1^b} = \frac{l_2}{l_{10}^b} = \frac{l_3}{l_{11}} = \frac{l_4^a}{l_{12}}, \qquad \frac{l_1^a}{l_5} = \frac{l_2}{l_6} = \frac{l_3}{l_7^a} = \frac{l_4^a}{l_4^b}, \qquad \frac{l_1^a}{l_9} = \frac{l_2}{l_{10}^a} = \frac{l_3}{l_7^b} = \frac{l_4^a}{l_8} \tag{1}$$

and with $l_{13} = l_1^a$ and $l_{14} = l_7^a$. Since these properties of similarity are completely determined within the kinematics by the design of the links, they turn out to be purely geometric conditions.

The mass balance conditions determine the relations among the link mass values and the locations of the link CoMs and can be derived with a method where the linkage obtains three relative DoFs of which the linear momentum equations are written individually [6,7].

First the linkage is modeled with open loops to obtain three relative DoFs. This, for instance, is possible as illustrated in Fig. 5a where link 5 and all links in parallel, links 1, 9, and 13, are substituted with equivalent masses. Link 5 (B_8D_1) is substituted with equivalent masses m_5^a and m_5^b in B_8 and D_1, respectively, where mass equivalence is determined by $m_5^a + m_5^b = m_5$ and $m_5^a p_5 = m_5^b(l_5 - p_5)$. Similarly link 9 ($D_4B_7$) is substituted with equivalent masses m_9^a and m_9^b in D_4 and B_7, respectively, where mass equivalence is determined by $m_9^a + m_9^b = m_9$ and $m_9^a p_9 = m_9^b(l_9 - p_9)$. Link 1 ($B_5SB_2$) is substituted with equivalent masses m_1^a and m_1^b in S and B_2, respectively, where mass equivalence is determined by $m_1^a + m_1^b = m_1$ and $m_1^a p_1 = m_1^b(l_1^a - p_1)$. Link 13 ($B_7'C_1'$) is substituted with equivalent masses m_{13}^a and m_{13}^b in B_7' and C_1', respectively, where mass equivalence is determined by $m_{13}^a + m_{13}^b = m_{13}$ and $m_{13}^a p_{13} = m_{13}^b(l_{13} - p_{13})$.

The motions of the three relative DoFs of the opened linkage are illustrated in Figs. 5b-c-d. For DoF 1 in Fig. 5b link 4 is rotating about S and the links parallel to

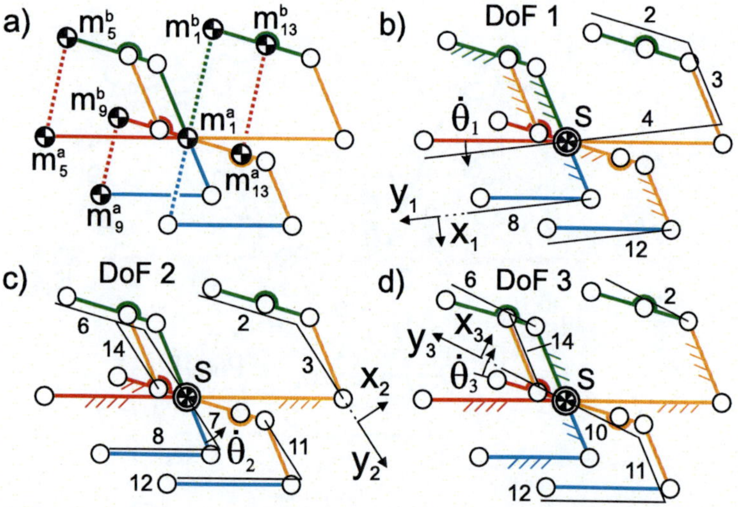

Fig. 5. a) Links 1, 5, 9, and 13 are substituted with equivalent masses to obtain an open-chain model with 3-DoF relative motion; b-c-d) Individual motion of each DoF, the mass balance conditions are derived from the linear momentum equations of each DoF individually.

link 4 (8 and 12) rotate similarly along while all other links solely translate (links 2 and 3) or remain fixed. For DoF 2 in Fig. 5c link 7 is rotating about S and the links parallel to link 7 (3, 11, and 14) rotate similarly along while all other links solely translate (links 2, 6, 8, and 12) or remain fixed. For DoF 3 in Fig. 5d link 10 is rotating about S and the links parallel to link 10 (2 and 6) rotate similarly along while all other links solely translate (links 11, 12 and 14) or remain fixed. The linear momentum equations of these three individual motions can be written with respect to the illustrated reference frames as [6]:

$$\frac{L_1}{\dot{\theta}_1} = \begin{bmatrix} -(m_1^b + m_2 + m_3 + m_{13}^b) l_4^a - m_4 p_4 + m_5^a l_4^b + \\ m_8 p_8 + m_9^a l_8 + m_{12} p_{12} \\ 0 \end{bmatrix} = \begin{bmatrix} 0 \\ 0 \end{bmatrix} \tag{2}$$

$$\frac{L_2}{\dot{\theta}_2} = \begin{bmatrix} -(m_1^b + m_2 + m_{13}^b) l_3 - m_3 p_3 - (m_5^b + m_6) l_7^a - m_7 p_7 + \\ (m_8 + m_9^a) l_7^b + m_{11} p_{11} + m_{12} l_{11} - m_{14} p_{14} \\ 0 \end{bmatrix} = \begin{bmatrix} 0 \\ 0 \end{bmatrix} \tag{3}$$

$$\frac{L_3}{\dot{\theta}_3} = \begin{bmatrix} m_1^b l_2 + m_2 p_2 + m_5^b l_6 + m_6 p_6 + m_9^b l_{10}^a - m_{10} p_{10} - \\ (m_{11} + m_{12}) l_{10}^b + m_{13}^b (l_2 - d_1) - m_{13}^a d_1 + m_{14} d_2 \\ 0 \end{bmatrix} = \begin{bmatrix} 0 \\ 0 \end{bmatrix} \tag{4}$$

with equivalent masses:

$$m_1^b = m_1\frac{p_1}{l_1^a} \qquad m_5^a = m_5(1-\frac{p_5}{l_5}) \qquad m_5^b = m_5\frac{p_5}{l_5} \tag{5}$$

$$m_9^a = m_9(1-\frac{p_9}{l_9}) \qquad m_9^b = m_9\frac{p_9}{l_9} \qquad m_{13}^a = m_{13}(1-\frac{p_{13}}{l_{13}}) \qquad m_{13}^b = m_{13}\frac{p_{13}}{l_{13}}$$

The common CoM of the linkage is in the central joint S for force balance when the linear momentum equations are equal to zero. From the linear momentum equations the three mass balance conditions of the planar and the spatial Butterfly Linkage can be directly found as:

$$-(m_1^b + m_2 + m_3 + m_{13}^b)l_4^a - m_4p_4 + m_5^a l_4^b + m_8p_8 + m_9^a l_8 + m_{12}p_{12} = 0 \tag{6}$$

$$-(m_1^b + m_2 + m_{13}^b)l_3 - m_3p_3 - (m_5^b + m_6)l_7^a - m_7p_7 + (m_8 + m_9^a)l_7^b + m_{11}p_{11} + m_{12}l_{11} - m_{14}p_{14} = 0 \tag{7}$$

$$m_1^b l_2 + m_2p_2 + m_5^b l_6 + m_6p_6 + m_9^b l_{10}^a - m_{10}p_{10} - (m_{11} + m_{12})l_{10}^b + m_{13}^b(l_2 - d_1) - m_{13}^a d_1 + m_{14}d_2 = 0 \tag{8}$$

For the spatial Butterfly Linkage these three balance conditions can be simplified by including the Bennett conditions on the link lengths $l_1^a = l_3$, $l_2 = l_4^a$, $l_4^b = l_6$, $l_5 = l_7^a$, $l_7^b = l_9$, $l_8 = l_{10}^a$, $l_{10}^b = l_{12}$, and $l_{11} = l_1^b$ and $m_{13} = m_{14} = 0$ to eliminate links 13 and 14, with which the balance conditions become:

$$-(m_1\frac{p_1}{l_1^a} + m_2 + m_3)l_2 - m_4p_4 + m_5(1-\frac{p_5}{l_5})l_6 + m_8p_8 + m_9(1-\frac{p_9}{l_9})l_8 + m_{12}p_{12} = 0 \tag{9}$$

$$-m_1p_1 - m_2l_1^a - m_3p_3 - m_5p_5 - m_6l_5 - m_7p_7 + m_8l_9 + m_9(l_9 - p_9) + m_{11}p_{11} + m_{12}l_1^b = 0 \tag{10}$$

$$m_1\frac{p_1}{l_1^a}l_2 + m_2p_2 + m_5\frac{p_5}{l_5}l_6 + m_6p_6 + m_9\frac{p_9}{l_9}l_8 - m_{10}p_{10} - (m_{11} + m_{12})l_{12} = 0 \tag{11}$$

where also the equivalent masses of Eq. 5 have been substituted. The inherent balance conditions have been verified by Matlab simulations.

4 Conclusion

In this paper the synthesis of a 2-DoF inherently force balanced spatial deployable Butterfly Linkage was presented. The linkage consists of four entangled similar Bennett linkages moving synchronously with the common center of mass in the central joint. The linkage was derived from the Grand 4R Four-Bar Based Inherently Balanced Linkage Architecture by eliminating the outer four-bar links and numerous other links to obtain a planar normally constrained linkage with four entangled similar 4R four-bar linkages. By applying the Bennett conditions the planar linkage was transformed into the spatial inherently balanced Butterfly Linkage. The balance conditions of the spatial Butterfly Linkage were derived from the planar linkage with a method based on linear momentum equations. These balance conditions are independent of the Bennett angles. A CAD-model of a realistic design of the spatial linkage was shown.

Acknowledgements. This publication was financially supported by the Netherlands Organisation for Scientific Research (NWO, 15146). The author likes to thank Gökhan Kiper for the fruitful discussions during the summer of 2019 at Izmir Institute of Technology.

References

1. Bennett, G.T.: A new mechanism. Engineering **76**, 777–778 (1903)
2. De Temmerman, N.: Design and Analysis of Deployable Bar Structures for Mobile Architectural Applications (dissertation). Vrije Universiteit Brussel, Brussels, Belgium (2007)
3. Herder, J.L.: Energy-free Systems (dissertation). TU-Delft (2001)
4. Maden, F.: Novel Design Methodologies For Transformable Doubly-Ruled Surface Structures (dissertation). Izmir Institute of Technology (2015)
5. Moore, B., Schicho, J.: Two methods for force balancing of Bennett linkages. In: Kecskeméthy, A., Müller, A. (eds.) Computational Kinematics, Proceedings of the 5th International Workshop on Computational Kinematics, pp. 241–248. Springer, Heidelberg (2009). ISBN 978-3-642-01946-3
6. Van der Wijk, V.: Methodology for analysis and synthesis of inherently force and moment-balanced mechanisms - theory and applications (dissertation). University of Twente (2014). https://doi.org/10.3990/1.9789036536301
7. Van der Wijk, V.: Design and analysis of closed-chain principal vector linkages for dynamic balance with a new method for mass equivalent modeling. Mech. Mach. Theory **107**, 283–304 (2017)
8. Van der Wijk, V.: The Grand 4R Four-Bar Based Inherently Balanced Linkage Architecture for synthesis of shaking force balanced and gravity force balanced mechanisms. Mech. Mach. Theory **150**, 103815 (2020)
9. Van der Wijk, V., Kiper, G., Yaşir, A.: Synthesis and experiments of inherently balanced umbrella canopy designs. In: Proceedings of the TrC-IFToMM Symposium on Theory of Machines and Mechanisms, p. No.23 (2015)

A Compliant Linkage for Cooperative Object Manipulation Through a Heterogeneous Mobile Multi-robot System

Juan Mauricio Toro Ramos[1(✉)], Dhruvin Doshi[2], Daniel Schütz[1], and Franz Dietrich[3]

[1] Volkswagen AG, Wolfsburg, Germany
{juan.mauricio.toro.ramos,daniel.schuetz}@volkswagen.de
[2] TU Clausthal, Clausthal-Zellerfeld, Germany
dhruvinrdoshi@gmail.com
[3] Assembly and Handling Technology, TU Berlin, Berlin, Germany
f.dietrich@tu-berlin.de

Abstract. This paper presents a compliant linkage as a standalone module that allows a mobile multi-robot system to transport objects cooperatively without explicit communication. The module is compatible with arbitrary steering kinematics and supports the cooperation among heterogeneous robots. All the required sensors and processing units are fully integrated into the module. The sensors are able to perceive robots' movements in three-dimensional space and enable them to drive on ramps with an inclination of at least 15%. In addition to the sensory capabilities, the linkage is able to compensate emerging undesired forces and simultaneously adds an extra degree of freedom to the system increasing its overall manoeuvrability. The experimental results illustrate the validity of the proposed linkage.

Keywords: Multi-robot systems · Object transport · Mechanical compliance

1 Introduction

The cooperation of multiple mobile robots enables their implementation in challenging tasks and scenarios: ranging from exploration and surveillance through foraging, coverage and object transport to sports. Nevertheless, some benefits of multi-robot systems (MRS) are often not exploited because their potentials are not taken into account during the robot design process. In object transport applications, MRSs can increase group's dexterity and flexibility because the distribution of the applied forces can dynamically change. This potential is harnessed in [2] to model a pusher/steerer system. Furthermore, by transporting the load with more than one robot, it is possible to distribute the necessary power onto all participants. They can be downsized consequently and thus increase overall system's flexibility.

Considering more than one robot transporting a rigid object, it is understandable that the mobility of each participant is constrained due to the physical connection with the other robots. Thus, even a drive down a straight path can generate undesired forces

J. Lenarčič and B. Siciliano (Eds.): ARK 2020, SPAR 15, pp. 33–40, 2021.
https://doi.org/10.1007/978-3-030-50975-0_5

which may lead to system damages. Therefore, a compliant linkage, which can decouple/damp vibrations and overcome small deviations is mandatory. On the other hand, it is important to note that depending on the robots' steering kinematic, the overall mobility of the system may also be constrained, and henceforth it may not be possible to drive through some obstacles. These two factors are the main aspects, that are addressed in this paper and have led to the development of a linkage module for mobile robots in cooperative systems. The module is easily scalable and its hardware is independent of the robot kinematic. All required sensors as well as the needed processing units to perceive three-dimensional movement of the object are fully integrated. It enables the use of indirect communication for the transport task. Simultaneously, the linkage is capable of absorbing undesirable emerging forces and add a degree of freedom through a free rotational joint.

There is a wide range of relevant work in the field of cooperative object transport, a broad review of the state-of-the-art can be found in [9]. The control algorithms are based on very diverse variables; in [4] for instance, the centroid of the participating robots and object are employed and these are estimated through an external source. In comparison, in [3] a control strategy using pheromone communication is presented. However, some works also base their algorithms on force measures and use some linkage to deal with undesired forces. In [5] the authors presented a leader-follower approach, where the leader broadcasts the desired object behaviour and each follower robot controlls its own posture based on the occurring forces, they also implemented a coupler mechanism for force compensation in two dimensions. An other leader-follower approach without explicit communication using a compliant linkage with 3 degrees of freedom is presented in [7]. Kosuge et al. [6] proposed a control algorithm for nonholonomic robots based on force feedback through a free rotational joint. In the swarm robotics field a transfer of simulation results to real robots was presented in [1], where through a compact connection module pulling forces were measured and used to perform coordinated movements through a self-organizing process instead of a leader-follower approach. Wang and Schwager [10] proposed a control algorithm for a cooperative manipulation task without direct communication using a force-feedback controller. A one degree of freedom gripper equipped with a rotatory potentiometer was used to attach the object. Both, the forces and the current velocity of the manipulated object were measured through load cells and an optical laser system respectively, to perform three different experiments. None of the works listed above consider a heterogeneous MRS, include a 3D motion perception, and integrate all the needed sensors and control units in a whole system. Their scalability is also limited due to the size and design complexity, which tends to be similar to serial robotic arms. Our approach presents a stand-alone compact module which allows heterogeneous MRS to transit through sloping surfaces.

This paper is organized as follows. Section 2 presents a methodical analysis of the emerging forces during a transport task and an evaluation of the necessity of adding an extra degree of freedom to the system. Section 3 describes the components and function of the developed module. In Sect. 4 the linkage is verified through experiments and finally Sect. 5 draws conclusion and proposes future work.

2 Force Analysis and Kinematic Considerations

Considering heterogeneous multi-robot systems where the robots present different kinematic capabilities and a direct communication is not possible, the emergence of conflict situations in which for instance, a robot tries to move in a direction that is constrained for another participant, due to the combination of holonomic and nonholonomic robots, limits the group's manoeuvrability and could lead to considerable damages. Therefore, the compensation of undesired forces and the addition of a degree of freedom between the robots and the transported object can improve the flexibility and increase the overall system performance.

2.1 Force Analysis

During cooperative object transport multiple forces and torques may arise due to various factors. From dynamics of the individual mobile robots, due to non-uniform brake paths or skidding, to unanticipated changes in the inertial moments or non synchronized driving motors. In addition to dynamics, the environment in which the MRS is deployed, is also hereby crucial. Uneven terrain or presence of slopes can also cause undesired forces and torques to occur. Lastly, delays or faulty measuring signals can lead to control deviations which are transduced into undesired effects as well.

These forces and torques can occur along any direction in the vector space xyz. In this paper x is defined as the direction of forward movement, y goes along the lateral motion and z is perpendicular to the ground. While forces include tensile and compression forces, torques contain bending and twisting moments. The direction of these depends not only on the direction of the movement but also on the relative position of the participating robots. They can arrange themselves on a single plane in infinite ways. However, there are two independent ways in which two robots can align themselves – connection along x-axis of both robots (alignment 1) and connection along y-axis of both robots (alignment 2). Based on these two arrangements, it is possible to align robots in any given configuration.

For the purpose of a qualitative analysis, a simplified model of two robots is used. They are fixed to a solid bar and connected in two different independent alignments as mentioned earlier. For each case, the occurrence of forces and torques in three-dimensional space and its effects on the MRS with respect to each of the mentioned factors were methodically analyzed and evaluated. For example, F_z can cause a torque around x-axis, which can lead to a roll-over of the mobile robot for the first arrangement. The results for the worst case scenario are summarized in form of a risk matrix in Fig. 1.

The analysis shows that especially M_x can cause large damages to the system, nevertheless the probability of occurrence is low. The force along the z-axis is responsible for the torques around any vector on the plane xy. Therefore, the damping it is important to ensure the stability of the MRS. The compensation of M_z can not only guarantee a safer transport process but notably also increase the flexibility of the system as discussed in the next section. Thus, the developed module was dimensioned to equalize all emerging forces on the plane xy and torques around z-axis directly. Additionally, the

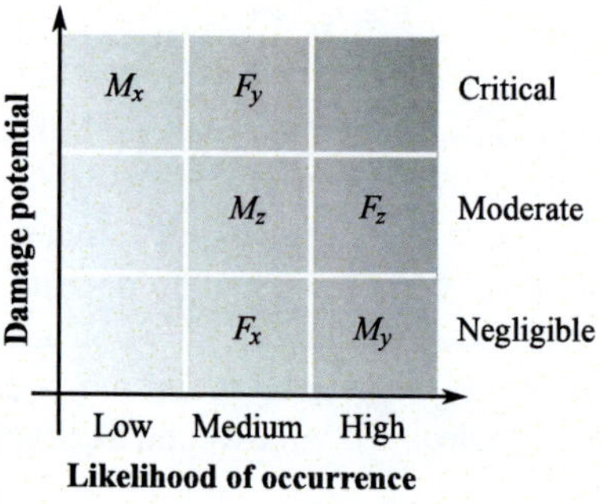

Fig. 1. Risk matrix of occurring forces and torques during cooperative object transport

remaining forces and torques can be partially dissipated through an energy absorbing mat (see Sect. 3).

2.2 Kinematic Considerations

For the design of the linkage it is important to consider the needed degree of freedom as well as the characteristics and joint space of a potential joint. While the design of a free rotational joint involves reasonable effort, the slide limits for prismatic joints are strongly limited by the mechanical design and have a direct impact on the final size of the module. In addition, the required force-compensation on the plane xy discussed above adds inherently small translational freedoms along the main axes. Hence, for pure kinematic purposes only the addition of rotational freedoms around the axes x, y or z were considered.

The addition of new rotational freedoms brings along both advantages and disadvantages. Differences in height between the robots, for example, can be compensated, ramps with a certain inclination can be overcome and the manoeuvrability can also be increased. On the other side, adding a rotational joint around any vector on the driving plane could cause stability issues and would require additional joint actoric or mechanical energy accumulators such as springs. After a qualitative analysis of the advantages and disadvantages of each arrangement, it was found that only for the motion around the vector perpendicular to the ground plane, a free rotational joint has a big positive impact. Other rotations around the remaining axes do not demand a large working range in comparison, which could be achieved through the previously introduced anti-vibrations mat. Thus, a free rotational joint with an integrated sensor unit was implemented in the design and is presented in detail in Sect. 3.

Assuming that the robots have the same height and intend to go up on a ramp, when they are arranged along an axis parallel to their motion direction while being connected to an object (see Fig. 2). The maximum compensation angle θ_c is the same as the one of the ramp (θ_r), which is 8.53° for a slope of 15%. This required partial freedom around the xy-plane, can be implemented through the combination of the force compensation mechanism and the absorbing mat (see Sect. 3).

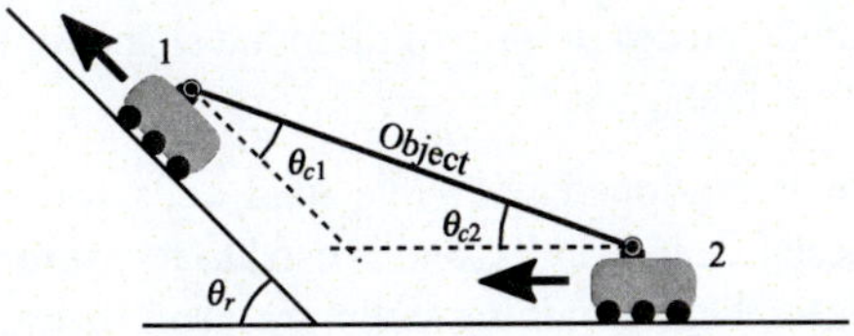

Fig. 2. Required compensation angle by critical ramp drive

3 Module Design

The presented module consists of various sensing and compensating mechanisms to enable MRSs for cooperative object transport. It was desired to keep the design as simple as possible and simultaneously accomplish all the targeted functions through custom integration of hardware. During the design process, the scalability for the future usage of the module has also been considered. This has been additively manufactured by selective laser sintering (SLS). Its final design is cube shaped with a side length of 125 mm and can be easily mounted onto mobile robots, however an external power source is necessary.

Compliant Mechanism: Purpose of the compliant mechanism is to compensate the forces in all three dimensions. For the ones along the driving plane, eight compression springs are used. These are pre-stressed from a free length of 35.5 mm to a length of 22.5 mm and are arranged as shown in Fig. 3a. The spring constant is 6.90 N/mm and their minimum length after compression is 9.5 mm. For the compensation of vertical forces, a rubber plate is fixed between the module and the robot. This damps the undesired forces and simultaneously equalizes the unevenness while driving on ramps.

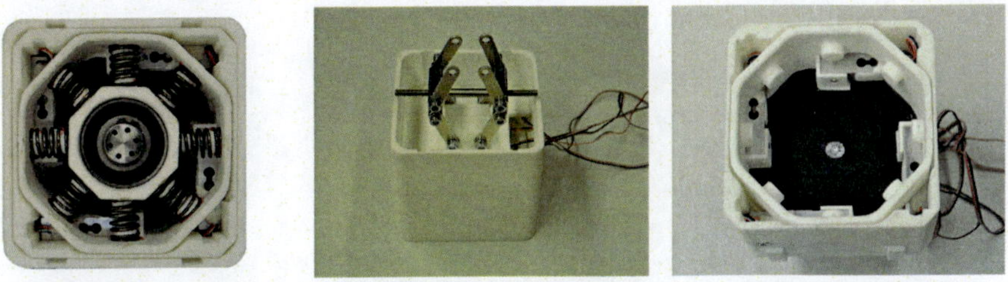

(a) Compliant mechanism (b) Height adjustment (c) Loadcell Positions

Fig. 3. Compliant linkage for mobile robots

Height Adjustment: Mobile robots in heterogeneous MRS may have different dimensions. Therefore, in order to keep the object elevation across the system uniform, a manual height adjustment for up to 100 mm with the help of a scissor jack mechanism is implemented. It consists of a threaded rod and eight metal pieces connected to two base ends. Every rotation of the threaded rod, causes a 2 mm change in the height of the module. Using eight metal pieces in form of diamond shape for the connection provides higher stability to the module.

Sensors: The module is equipped with five load cells for sensing forces in three-dimensions. Out of these five load cells, one is used to measure the weight of the object placed onto the module. The remaining cells are used to track the force applied to the object during transportation. They are positioned as shown in Fig. 3c to measure forces along the driving plane. Using two sensors in each direction, makes the sensing setup more reliable and accurate. Each load cell consists of four strain gauges, which are connected to form a Wheatstone bridge. The sensors have been calibrated with standardized weights and can operate up to 50 kg. The inclination of a ramp is measured by a inertial measuring unit (IMU) BNO055 from Bosch[1]. The Euler angles determined from accelerometer, magnetometer and gyroscope are used to estimate the inclination. The angular position of the shaft is measured with the help of a non-contact absolute magnetic Encoder from RLS[2]. It is capable of measuring $360° \pm 0.1°$ and is equipped with multi-turn capabilities. It uses asynchronous serial communication and is self-calibrated, when the system is booted. For the signal processing a microcontroller STM32F4 was integrated.

4 Experiments

To validate the correct functioning of the proposed module, five experiments were performed (see Fig. 4). A KUKA LBR iiwa[3] was used to transmit a certain force and thus generate a displacement of the module axle. Four experiments were conducted by applying forces in different directions along the xy-plane. Additionally, a torque around the y-axis was applied in order to reproduce a ramp-drive and test the module's capabilities. For each experiment the initial and final position of the robot's TCP were registered and used to calculate the applied displacement which were considered as reference.

(a) (b) (c) (d)

Fig. 4. Experimental setup: a) overview of the four conducted experiments on the plane xy. b) x-deviation (trial 1). c) diagonal deviation (trials 2, 3, 4). d) ramp drive simulation (trial 5).

[1] www.bosch.com.

[2] www.rls.si.

[3] www.kuka.com.

To determine the repeatability R_p of the module, the procedure presented in [8] was used. For each experiment ten trials i were performed. The difference between all measured values of each trial and their mean was calculated for every sensor j and is represented by l_{ij}. Sequentially, mean $\overline{l_j}$ and standard deviation σ_j of l_{ij} were determined. Thus, the repeatability of the module can be calculated as follows

$$R_p = \overline{l_j} + 3\sigma_j \tag{1}$$

For the determination of the accuracy A_c of the proposed linkage, the differences between the calculated angles based on the measured forces of each trial and the reference values from the KUKA robot were calculated. For both repeatability and accuracy the worst value was taken into consideration and are summarized in the Table 1.

Table 1. Repeatability and accuracy of the module.

	Experiment 1	Experiment 2	Experiment 3	Experiment 4
R_p	1.4°	1.1°	2.4°	5.6°
A_c	17.9°	12.1°	16.1°	16.1°

The results can be considered plausible if the elasticity of the mat and tolerance errors due to the manufacture process (SLS) are taken into account and bearing in mind the fact that the springs are capable of partially rotating around the *xy*-plane. However, the obtained accuracy can potentially be improved by fusing the data from the IMU, the load cell in z direction and the *xy* cells.

In Sect. 2.2 was shown that considering a ramp with 15% slope in a critical scenario, a maximum angle of 8.5° is required. In the conducted experiment (see Fig. 4d), a height difference of 28 mm was realized, which corresponds to an angle of 9.1°.

5 Conclusion and Future Work

A methodical analysis of the criticality and necessity of force compensation in form of a risk matrix together with a qualitative evaluation of additionally required degrees of freedom for heterogeneous mobile multi-robot systems were presented. On this basis a compliant linkage as a stand-alone module for compensation of forces in three-dimensions was proposed. The capability of the linkage to compensate undesired forces and torques along with the results of the integrated sensors were validated through different experiments. Moreover, the repeatability and accuracy of the module were calculated. In future work, control algorithms will be presented and their implementation with the obtained sensor accuracy for different robot kinematic combinations has to be validated. Additionally, a sensor-fusion strategy for a ramp drive is in development.

References

1. Baldassarre, G., Trianni, V., Bonani, M., Mondada, F., Dorigo, M., Nolfi, S.: Self-organized coordinated motion in groups of physically connected robots. IEEE Trans. Syst. Man Cybern. Part B (Cybern.) **37**(1), 224–239 (2007). https://doi.org/10.1109/TSMCB.2006.881299
2. Brown, R.G., Jennings, J.S.: A pusher/steerer model for strongly cooperative mobile robot manipulation. In: Proceedings 1995 IEEE/RSJ International Conference on Intelligent Robots and Systems. Human Robot Interaction and Cooperative Robots, vol. 3, pp. 562–568 (1995). https://doi.org/10.1109/IROS.1995.525941
3. Fujisawa, R., Imamura, H., Matsuno, F.: Cooperative transportation by swarm robots using pheromone communication. In: Martinoli, A., Mondada, F., Correll, N., Mermoud, G., Egerstedt, M., Hsieh, M.A., Parker, L.E., Støy, K. (eds.) Distributed Autonomous Robotic Systems: The 10th International Symposium, pp. 559–570. Springer, Heidelberg (2013). https://doi.org/10.1007/978-3-642-32723-0_40
4. Habibi, G., Kingston, Z., Xie, W., Jellins, M., McLurkin, J. (eds.): Distributed centroid estimation and motion controllers for collective transport by multi-robot systems. In: 2015 IEEE International Conference on Robotics and Automation (ICRA) (2015). https://doi.org/10.1109/ICRA.2015.7139356
5. Hashimoto, M., Oba, F., Zenitani, S. (eds.): Coordinative object-transportation by multiple industrial mobile robots using coupler with mechanical compliance. In: Proceedings of IECON 1993 - 19th Annual Conference of IEEE Industrial Electronics (1993). https://doi.org/10.1109/IECON.1993.339306
6. Kosuge, K., Oosumi, T., Satou, M., Chiba, K., Takeo, K. (eds.): Transportation of a single object by two decentralized-controlled nonholonomic mobile robots. In: Proceedings of 1998 IEEE International Conference on Robotics and Automation (Cat. No. 98CH36146), vol. 4 (1998). https://doi.org/10.1109/ROBOT.1998.680884
7. Ota, J., Buei, Y., Arai, T., Osumi, H., Suyama, K.: Transferring control by cooperation of two mobile robots. J. Robot. Soc. Japan **14**(2), 263–270 (1996). https://doi.org/10.7210/jrsj.14.263
8. Schütz, D.: Aufgabenspezifische Einrichtung von Parallelstrukturen mit Binäraktoren: Zugl.: Braunschweig, Techn. Univ., Diss. 2013. Schriftenreihe des Instituts für Werkzeugmaschinen und Fertigungstechnik der TU Braunschweig. Vulkan-Verl., Essen (2013)
9. Tuci, E., Alkilabi, M.H.M., Akanyeti, O.: Cooperative object transport in multi-robot systems: a review of the state-of-the-art. Front. Robot. AI **5**, 59 (2018). https://doi.org/10.3389/frobt.2018.00059
10. Wang, Z., Schwager, M.: Kinematic multi-robot manipulation with no communication using force feedback. In: 2016 IEEE International Conference on Robotics and Automation (ICRA), pp. 427–432 (2016)

One Degree of Freedom 7-R Closed Loop Linkage as a Building Block of Nanorobots

Meysam T. Chorsi[1], Pouya Tavousi[1], Caitlyn Mundrane[2], Vitaliy Gorbatyuk[3], Horea Ilies[1(✉)], and Kazem Kazerounian[1,2(✉)]

[1] Department of Mechanical Engineering, University of Connecticut, Storrs, CT 06269, USA
{horea.ilies,kazem.kazerounian}@uconn.edu
[2] Department of Biomedical Engineering, University of Connecticut, Storrs, CT 06269, USA
[3] Department of Chemistry, University of Connecticut, Storrs, CT 06269, USA

Abstract. A robotic manipulator is a series of links connected with kinematic pairs designed so that its configuration (three dimensional pose – particularly at the end effector) is controllable. The control is achieved through a series of one degree of freedom revolute or prismatic joints. Nanorobots are linkage mechanisms at nanoscale. These robots offer considerable promise in diverse applications in nanotechnology, biotechnology and medicine. Similar to their macroscale siblings, a nanorobot must have a controllable kinematic structure to perform tasks. However, at nanometer levels creation of controllable single degree of freedom kinematic pairs are unlikely. Therefore controllability of the nanorobot remains a technological challenge. In this paper, we present a peptide-based linkage composed of protein building-block molecules designed as a one degree-of-freedom closed-loop 7R spatial mechanism. Traditional spatial kinematic notation and formulations are used to solve the loop closure equations implicitly. Two different kinematic branches of this molecular linkage are identified. The designed molecule is fabricated and its motion is observed at distinct configurations using nuclear magnetic resonance (NMR) spectroscopy. These results validate our formulation, and indicate that 7R spatial closed loop linkages are a suitable building block for nanorobots and nanomachines.

Keywords: Nanorobot · Nanomachines · Peptide · Kinematic closure

1 Introduction

The molecular machinery responsible for carrying out essentially all critical functions in living things has been developed through an evolutionary process that started almost 4 billion years ago. Unsurprisingly, designing and building artificial nanomachines with prescribed functions for performing vital tasks in medical, environmental, and industrial arenas remains a monumental challenge, partly because the design principles developed by nature through evolution are not well understood and, therefore, not immediately applicable to engineered nanomachines. The level of complexity associated with the

M. T. Chorsi and P. Tavousi—Contributed equally to this work.

J. Lenarčič and B. Siciliano (Eds.): ARK 2020, SPAR 15, pp. 41–48, 2021.
https://doi.org/10.1007/978-3-030-50975-0_6

natural protein machinery makes it extremely difficult to mimic the nature by reverse engineering the natural process. But, much as the manufacturing revolution was fueled by the concept of interchangeable parts designed to precise specifications, the ability to custom design, control, and fabricate nanomachines with prescribed functions can bring about a new revolution in engineering, medicine, and biotechnology. So far, however, in the absence of a systematic method for designing and controlling artificial nanomachines, the only known attempts to start this revolution have followed ad hoc trial and error design processes [4, 7].

Having the right set of nanoscale building blocks is perhaps the most critical factor in designing nanomachines. Fortunately, important insights from across multiple disciplines suggest a practical approach to addressing these questions. The first, from molecular biology, is that the precise sequence in which the amino acids are linked together is often sufficient to determine the three-dimensional conformation(s) the resulting polypeptide ultimately folds into or unfolds from. In other words, a specific sequence of amino acids determines the physicochemical properties of the resulting protein that define its function and interaction with other molecules. Another fundamental attribute of any protein molecule is that its folding into one or more stable conformations, as well as the transition (that is, the motion) between these conformations, which determine the function of the protein, are both controlled by the environment in which the protein exists.

Another insight is that hundreds of years of research into mechanism design show that the one-degree-of-freedom (DOF) machines are the simplest, most effective, robust, and widely used designs because of their completely predictable, deterministic, and repeatable motion, which is fully determined by a single parameter. In fact, the vast majority of macroscale mechanisms that have been engineered throughout human history are either fundamentally one-DOF mechanisms or use them as building blocks [2]. This, in turn, suggests the designers of artificial nanomachines should perhaps explore first the design space generated by molecules (for example, amino acids; DNA) that can produce a one-degree-of-freedom molecular structure. Furthermore, in robotics, manipulator pose is controlled through a series of one degree of freedom joints. However, at nanometer levels creation of controllable single degree of freedom kinematic pairs are unlikely. Therefore creating controllable one degree of freedom nanomechanisms is critical for realization of nanorobotics.

In this paper, we present a peptide-based linkage composed of protein building block molecules designed as a controllable one degree-of-freedom closed-loop 7R spatial mechanism.

2 Kinematic Structure

Amino acids (N_i-C_{α_i}-C_i) are building blocks of proteins consecutively linked by a covalent chemical bond known as a peptide bond (C_i-N_{i+1}). Due to the secondary resonance stabilization [13], peptide bonds reside in a single plane called the peptide plane (Fig. 1a). All of the atoms in a peptide plane are fixed with bond lengths and bond angles nearly the same in all amino acids. The peptide planes are therefore considered to be rigid groups of atoms linked in the chain by covalent bonds at the alpha carbons

(C_{α_i}). These planes rotate about the N-$C\alpha$ and $C\alpha$-C bonds, known as dihedral angles. Each amino acid has two torsional angles, ϕ and ψ. Proline is an exception to this rule with only one dihedral angle, ψ. Consequently, a protein molecule can be kinematically defined as a set of rigid links serially connected by revolute joints, hence making a large body of knowledge in kinematics available for their analysis [5, 8, 10].

2.1 Mobility Analysis

Mobility analysis of spatial kinematic structures (Grubler–Kutzbach criterion) indicates that the simplest one degree of freedom closed loop linkage is a single loop system of seven rigid links connected by seven one degree of freedom joints such as revolute joints. Using amino acids as links in this closed loop linkage, a proposed arrangement would be the polypeptide chain GPGPP (G for glycine, and P for proline). A glycine amino acid has two joints, and proline has one. Therefore the GPGPP linkage is a single closed loop 7R system resulting in one degree of freedom.

2.2 Zero Reference Position Description

Zero position notation and the corresponding formulation [6, 9] developed for robotic manipulators offers an excellent methodology for kinematic analysis of protein molecules. The following is a very brief overview of this kinematic notation and formulation. The chains are originally modeled as an open loop (serial) chain rigid links connected by revolute joints. A suitable (although arbitrary) configuration of the chain is designated as the zero reference position and all of the joint variables are set to zero at this configuration. At this configuration the unit vectors u_{0i} (along joint i) and body vectors b_{0i} (from the center of ith joint to the center of $(i+1)$th joint) are recorded as the chain parameters. The set of u_{0i} and b_{0i} ($i = 1$ to n) completely defines the kinematic structure of a chain of n amino acids where all joint variables (dihedral angles) are set to zero (0 index shows the zero position). At zero position, the end-effector position is defined as (Fig. 1b):

$$[P_{0H}] = \sum_{i=1}^{7} [b_{0i}] \tag{1}$$

At any other configuration where the joint values are known, the updated joint and body vectors as well as the position and orientation of the last link (end effector) are calculated as follows:

$$\begin{aligned} [u_n] &= [R_i]\,[u_{0n}] \\ [u_t] &= [R_i]\,[u_{0t}] \\ [u_a] &= [R_i]\,[u_{0a}] \\ [P_H] &= \sum_{i=1}^{7} [R_i]\,[b_{0i}] \end{aligned} \tag{2}$$

where,

$$[R_i] = \prod_{i=1}^{7} R(\theta_i, u_{0i}) \tag{3}$$

and $R(\theta_i, u_{0i})$ is a rotation matrix about axis u_{0i} [12]. The unit vectors (u_i), body vectors (b_i), end-effector position (P) and end-effector orientation unit vectors (u_a axial and u_t transverse) completely define the kinematic structure of the chain at any configuration.

The GPGPP nanorobot presented in this paper is modeled as a kinematic serial open loop linkage shown in Figs. 1c, and 1d. The details about the formulation and indexing scheme is presented in [14].

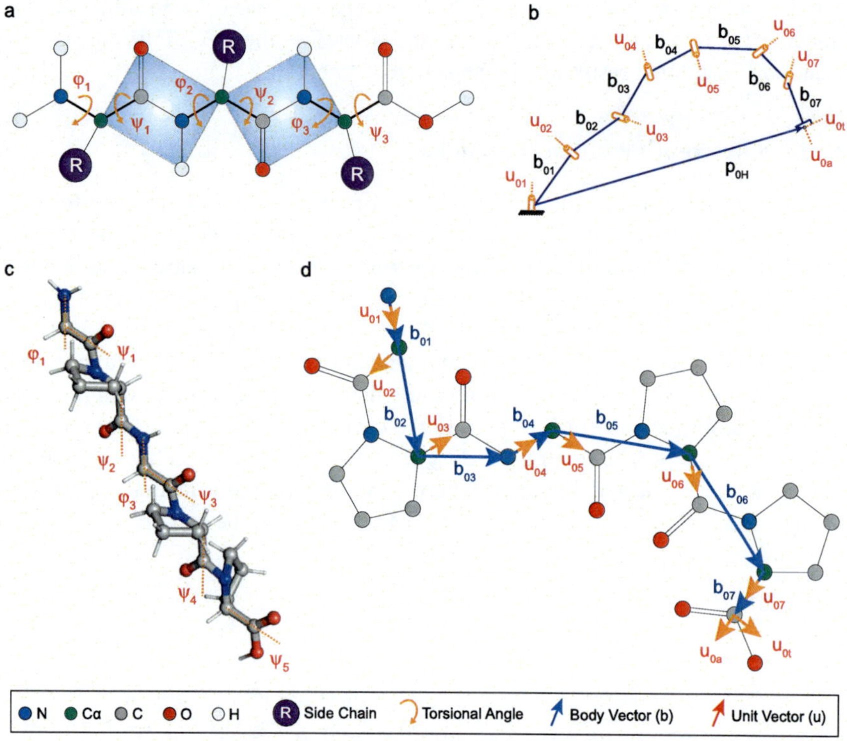

Fig. 1. The GPGPP chain is modeled as a kinematic linkage. a, Peptide planes and torsional angles in a peptide chain. b, 7-DOF robotic manipulator at zero position (0 index shows the zero position). c, Open chain polypeptide chain of GPGPP and the corresponding dihedral angles (joints). d, Open chain of the GPGPP peptide at zero reference position and the corresponding body and unit vectors.

2.3 Loop Closure Equations

To convert the open loop GPGPP chain to a closed loop 7R linkage, the last link (end effector) must be positioned and oriented at the base of the first link. Analytically this results in a system of nonlinear equations referred to as the loop closure equations. Zero position formulation results in 12 equations (Eq. 2) with 7 unknowns (7 joint angle variables). The 12 equations represent 3 independent positional equations and 9 equations entailing the orientation (with only 3 of them independent). The closed form solution of these equations is not easily obtainable.

To solve this system of nonlinear equations we utilized MATLAB unconstrained nonlinear system solver, fminunc. This solver is based on quasi-Newton method. It is well known that there are multiple sets of solutions for nonlinear systems. Therefore in numerical methods, the particular set that the solution converges to is highly sensitive to the initial guess. In our numerical experiment we repeated the procedure with numerous initial points striving to obtain all physically possible configurations of the subject peptide linkage.

Figure 2a depicts two solution sets (solid and dashed curves) derived for the GPGPP peptide. Here, we randomly selected one of the seven variable angles as the control angle (ψ_1), and plotted the variation of the other six variable angles against the control angle. The two solution curves meet each other at the singular positions. These input-related singularities are referred to as dead center positions, where the control angle is no longer able to move the linkage and the mechanical advantage of the mechanism goes to zero [15]. The region between the two singular positions is defined as the range of motion [1].

3 Experimental Observation

The GPGPP cyclic peptide was chemically synthesized by a biotech company, GenScript, and we performed experimental structural analysis to validate the synthesized chain exhibits the conformational changes predicted by the one-degree-of-freedom model.

We investigated the observed structural conformation via nuclear magnetic resonance (NMR) spectroscopy of the peptide. The GPGPP compound was dissolved in DMSO solvent and a series of routine NMR experiments were conducted. The data collected and interpreted from these experiments are used with ARIA software package [11]) to produce a list of at most 20 candidate structures with the lowest energy levels.

By aligning the 20 candidate structures output by ARIA from the NMR data, we see structural differences, illustrated in Fig. 2b, that can be attributed to either (1) expected angle variations in the seven variable angles, or (2) changes in the distances between atoms that are assumed to be constant. To further investigate the dominant factor in the observed conformational differences illustrated in Fig. 2b, we selected one of the 20 candidate structures at random (it happened to be number 18) and investigated the possible range of motion with long-established kinematic simulations. In other words, in our simulation, we obtained the range of conformational changes across which the structural integrity of ring molecule was preserved and only the value of seven dihedral angles associated with the variable joints was altered.

The 14 curves in Fig. 2a (seven corresponding to the first solution and seven corresponding to the second solution) represent the theoretical predictions of these variations based on the kinematic model, while the dots correspond to the dihedral angle values obtained from NMR data for the seven variable joints of the 20 candidate structures. This figure shows excellent comparison between the predicted and measured results and further supports the conjecture that the dominant factor for the structural differences displayed in Fig. 2b are the variations in the seven variable angles predicted by our model.

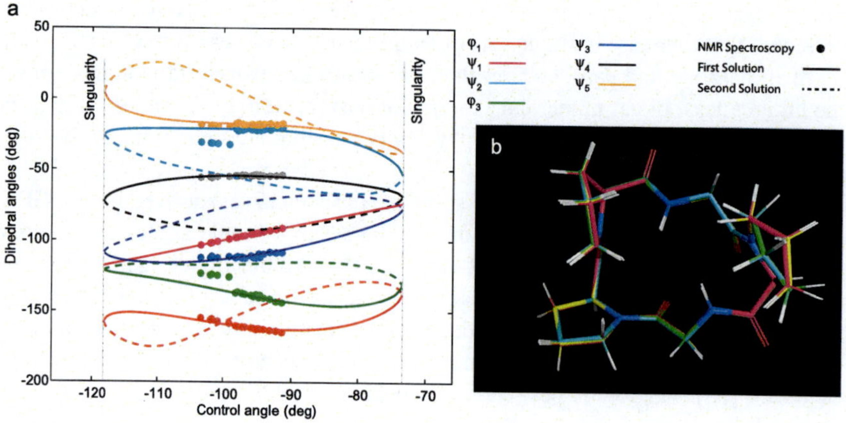

Fig. 2. a, Changes in variable dihedral angles as functions of change in control variable, ψ_1 (solid and dashed curves correspond to the first and second solution sets, respectively). Through simulation, seven curves were produced that reflected the predicted range of conformational changes across which the structural integrity of ring molecule was preserved and only the value of seven dihedral angles associated with the variable joints was altered. Dots, on the other hand, correspond to the dihedral angle values obtained from NMR data. Correlation between the predicted and measured variations is observed, supporting the hypothesis of one-degree-of-freedom motion for the synthesized machine. b, Alignment of some of the 20 candidate structures. Structural differences can be clearly seen among these candidates.

4 Motion Control and Functionalization

The common one-degree-of-freedom macroscale mechanisms, such as the ubiquitous four-bar linkage [3], are always functionalized to obtain a mechanism that performs desired tasks. The pervasive crank-slider, double wishbone suspension, and Watt's linkages, for example, are all fundamentally four-bar mechanisms–that is, they have one degree of freedom. Similarly, the availability of the simplest possible movable one-DOF protein molecule opens the door to numerous functionalization strategies for obtaining practical nanomachines that perform various tasks at the nanoscale.

Our strategy for building functional molecular machines is to design a simple mechanism with controllable motion, followed by integrating molecular components for fractionalizing the produced motion. Current work is investigating the possibilities for adding molecular components to the designed one-DOF mechanisms to produce functional mechanisms capable of various tasks. Figure 3 demonstrates such an attempt.

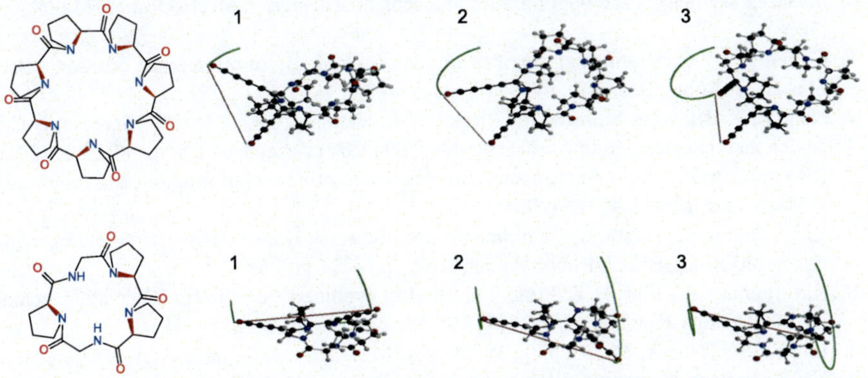

Fig. 3. We are investigating the integration of rigid molecular components with the links of the designed mechanisms to exploit their controllable motions for generating functions. At left are the structural formulae for two molecules. At right are the possible curve generations, highlighted in green, through multiple steps.

5 Conclusions and Outlook

We have presented an overview of a novel method capable of synthesizing the one-DOF seven link 7R spatial linkages composed of protein (peptide) building-block molecules. We have used traditional advanced kinematic formulations to model the linkage and evaluate output angles as a functional of input angle. The designed molecule is fabricated and its motion is observed at distinct configurations using nuclear magnetic resonance (NMR) spectroscopy. These results validate our formulation, and indicate that 7R spatial closed loop linkages are a suitable building block for nanorobots and nanomachines. Because of the unique controllability of the one degree of freedom systems, in our view this discovery could represent a major breakthrough in biotechnology in general, and in the design of nanomachines in particular.

References

1. Almestiri, S.M., Murray, A.P., Myszka, D.H., Wampler, C.W.: Singularity traces of single degree-of-freedom planar linkages that include prismatic and revolute joints. J. Mech. Robot. **8**(5), 051,003 (2016)

2. Artobolevsky, I.I.: Mechanisms in Modern Engineering Design: A Handbook for Engineers, Designers and Inventors. Mir, Moscow (1975)
3. Artobolevsky, I.I.: Mechanisms in Modern Engineering Design; a Handbook for Engineers, Designers and Inventors, vol. 4. Cam and Friction Mechanisms; Flexible-link Mechanisms. Mir (1977)
4. Bath, J., Turberfield, A.J.: DNA nanomachines. Nat. Nanotechnol. **2**(5), 275 (2007)
5. Chirikjian, G.S., Kazerounian, K., Mavroidis, C.: Analysis and design of protein based nanodevices: challenges and opportunities in mechanical design. J. Mech. Des. **127**(4), 695–698 (2005)
6. Gupta, K.C.: Kinematic analysis of manipulators using the zero reference position description. Int. J. Robot. Res. **5**(2), 5–13 (1986)
7. Hao, Y., Kristiansen, M., Sha, R., Birktoft, J.J., Hernandez, C., Mao, C., Seeman, N.C.: A device that operates within a self-assembled 3D DNA crystal. Nat. Chem. **9**(8), 824 (2017)
8. Kazerounian, K.: From mechanisms and robotics to protein conformation and drug design. J. Mech. Des. **126**(1), 40–45 (2004)
9. Kazerounian, K., Gupta, K.: Manipulator dynamics using the extended zero reference position description. IEEE J. Robot. Autom. **2**(4), 221–224 (1986)
10. Kazerounian, K., Ilieş, H.T.: Protein molecules: evolution's design for kinematic machines. In: 21st Century Kinematics, pp. 217–244. Springer, Heidelberg (2013)
11. Linge, J.P., Habeck, M., Rieping, W., Nilges, M.: ARIA: automated NOE assignment and NMR structure calculation. Bioinformatics **19**(2), 315–316 (2003)
12. Murray, G.: Rotation about an arbitrary axis in 3 dimensions (2013). http://inside.mines.edu
13. Nelson, D., Cox, M.: Lehninger Principles of Biochemistry. Macmillan Worth Publishers (2005)
14. Tavousi, P., Behandish, M., Ilieş, H.T., Kazerounian, K.: Protofold ii: enhanced model and implementation for kinetostatic protein folding. J. Nanotechnol. Eng. Med. **6**(3), 034,601 (2015)
15. Yan, H.S., Wu, L.l.: On the dead-center positions of planar linkage mechanisms. J. Mech. Transmissions Autom. Des. **111**(1), 40–46 (1989)

Modeling and Control of a Redundant Tensegrity-Based Manipulator

Jérémy Begey[1,2(✉)], Marc Vedrines[1], Pierre Renaud[1], and Nicolas Andreff[2]

[1] ICube, Strasbourg, France
jeremy.begey@etu.unistra.fr
[2] Femto-ST, Besançon, France

Abstract. Tensegrity-based mechanisms draw attention in particular for their deployability and compliance. However, task-based design and control of such systems are still open topics. In a previous work, a tensegrity-based manipulator was designed to respect the remote center of motion constraint. Only the workspace was then analyzed. Here, we develop the kinematic model of this manipulator and then exploit it for the control. Such a manipulator is redundant. The use of redundancy is discussed and evaluated in simulation with two control schemes using Jacobian based controllers.

Keywords: Tensegrity mechanisms · Redundant manipulator · Kinematic control

1 Introduction

Tensegrities are pre-stressed structures composed of bars and cables, firstly exploited [10] for their high resistance. Several authors focused on actuating these structures to design tensegrity mechanisms. If some structural elements are chosen elastic, these mechanisms can be compliant with controllable stiffness. Therefore, these mechanisms are of high interest when interactions with the environment can occur, particularly when contact management is critical. For instance, it was used for grasping tasks [12], design of flexible mobile robot [7] or to carry out medical procedures [4]. In [4], a tensegrity-based manipulator has been proposed to manage a Remote Center of Motion (RCM) constraint often met in medical procedures. It is introduced in Fig. 1, and presented in detail in Sect. 2. The device is built for quasi-static positioning. Only preliminary analysis of the robot workspace was yet achieved. Further development is hampered by the absence of analytical Forward Kinematic (FK) model and the control strategy. In this paper, kinematics are developed and used to investigate manipulator control.

Solutions to the FK problem correspond to stable configurations of the pre-stressed mechanism. Amongst others, energetic approaches have been proposed [6]. Depending on the type of elastic elements, solving the FK problem remains complex [3,13] and it is often treated numerically [3–5]. In this paper, we show using an energetic approach that an analytical solution can be determined for the considered manipulator. This constitutes the first contribution of the paper.

J. Lenarčič and B. Siciliano (Eds.): ARK 2020, SPAR 15, pp. 49–56, 2021.
https://doi.org/10.1007/978-3-030-50975-0_7

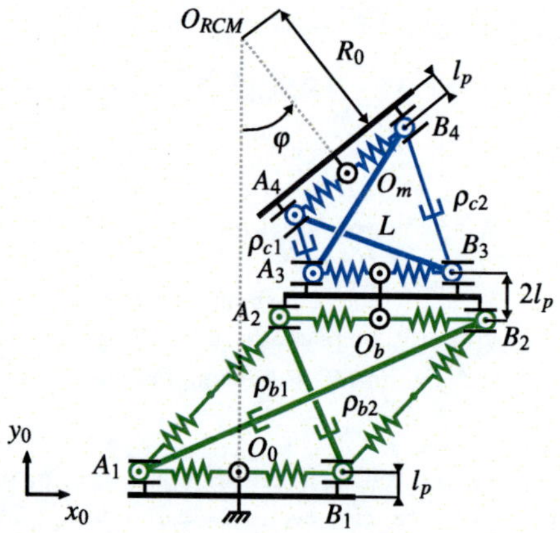

Fig. 1. The considered tensegrity-based manipulator composed of two tensegrity mechanisms: a bar-actuated one in green and a cable-actuated one in blue.

The proposed manipulator is considered for providing a RCM in the plane. Therefore, this four actuator mechanism is redundant with respect to this task. Control of redundant tensegrities has been considered to achieve shape control while optimizing a given criterion and get for instance time-optimal or energy-optimal control [1,11]. In conventional robotics, control of redundant systems is making use of the kinematic Jacobian [9]. Unlike most tensegrity-based robots, the Jacobian matrix of the manipulator proposed here can be derived analytically. The aforementioned control methods are thus well suited and can be investigated for this class of robots. Control schemes exploiting the redundancy are then proposed and evaluated in simulation as a preliminary study. It is the second contribution.

The manipulator is described and modeled in Sect. 2. The control schemes and the obtained results in simulation are then discussed in Sect. 3. Conclusions are finally drawn in Sect. 4.

2 Description and Modeling of the Manipulator

The considered mechanism is planar and composed of two serially assembled X-shaped tensegrities called Snelson Crosses (SC) [10]. A bar-actuated SC, displayed at the bottom of Fig. 1, is mounted on the base. It is actuated by modifying the lengths ρ_{b1} and ρ_{b2} of the bars attached to nodes A_1, B_2 and B_1, A_2 respectively. The SC on top is cable-actuated with control of the lengths ρ_{c1} and ρ_{c2} of non-elastic cables attached to nodes A_3, A_4 and B_3, B_4 respectively. The bars of this SC have a length L. Following [4], analysis is conducted while neglecting gravity forces. Thanks to the intermediate component placed between the SCs, the static models of the SCs are decoupled, and the manipulator can be assessed by independently analyzing each SC. For sake of symmetry, two elementary springs are placed on the top and bottom sides of each SC. Two serially connected springs are then similarly placed between the nodes A_1, A_2 and B_1,

B_2, with all springs of stiffness $2K$ and free-length $l_{0c}/2$ (resp. $l_{0b}/2$) for the cable- (resp. bar-) actuated SC. We will refer in the following to a spring as the equivalent elastic element of the two serially attached springs on each side. The end-effector of the manipulator is the platform on top of the mechanism (Fig. 1). For given position of the remote center of motion O_{RCM} and orientation φ, it is simple to determine the corresponding coordinates of O_m. The FK problem consists then to determine from the joint coordinates $\mathbf{q} = [\rho_{b1}, \rho_{b2}, \rho_{c1}, \rho_{c2}]^T$ the end-effector pose $\mathbf{x} = [x,\ y,\ \varphi]^T$, with (x, y) the position of O_m the median point of segment A_4B_4, and φ the orientation.

In [4], it was observed with numerical analysis that the bar-actuated SC shape is a parallelogram for all reachable configurations. We could also observe that these configurations seem diamond-shaped. Thus, all springs would be of same length which would be expressed in Eq. (1), with $L_{A_iB_j}$ the length between nodes A_i and B_j. In [2], the stable configuration of a cable-actuated SC was analyzed and Eq. (2) was in addition demonstrated for zero free-length springs.

$$L_{A_1B_1} = \frac{1}{2}\sqrt{\rho_{b1}^2 + \rho_{b2}^2} \tag{1}$$

$$L_{A_3B_3} = \sqrt{L^2 - \rho_{c1}\rho_{c2}} \tag{2}$$

When springs with non-zero free-lengths are used, closed-form solutions of these configurations were not found due to the increase of computation complexity induced by the free-length [3, 13]. Here, we propose to use the conditions of tensegrity mechanism equilibrium configuration [5] to derive analytically the validity conditions of Eq. (1) and Eq. (2) for non-zero free-lengths. Let $\mathbf{q}_1 = [\rho_{b1}, \rho_{b2}]^T$ and $\mathbf{q}_2 = [\rho_{c1}, \rho_{c2}]^T$ be the sets of joint coordinates of the bar- and the cable-actuated SCs. The SCs are compliant so $\mathbf{q}_{ui}$ designates sets of unconstrained joint coordinates. Existence of a stable static equilibrium can then be expressed from the mechanism potential energy $U_i(\mathbf{q}_i, \mathbf{q}_{ui})$ (Eq. (3)), verifying also that the tensions in the elements in traction are positive.

$$\frac{\partial U_i(\mathbf{q}_i, \mathbf{q}_{ui})}{\partial \mathbf{q}_{ui}} = 0 \quad \frac{\partial^2 U_i(\mathbf{q}_i, \mathbf{q}_{ui})}{\partial \mathbf{q}_{ui}^2} > 0 \quad i = \{1,\ 2\} \tag{3}$$

For both SC, the potential energy is written with $\mathbf{q}_{u1} = [L_{A_1B_1},\ L_{A_1A_2},\ L_{B_1B_2}]^T$ for the bar-actuated SC and $q_{u2} = L_{A_3B_3}$ for the cable-actuated SC, having respectively three and one unconstrained coordinates [3]. It gives:

$$\begin{aligned} U_1 &= \frac{1}{2}K\left((L_{A_2B_2}(\mathbf{q}_1, \mathbf{q}_{u1}) - l_{0b})^2 + \sum_{j=1}^{j=3}(\mathbf{q}_{u1}(j) - l_{0b})^2\right) \\ U_2 &= \frac{1}{2}K\left((L_{A_4B_4}(\mathbf{q}_2, q_{u2}) - l_{0c})^2 + (q_{u2} - l_{0c})^2\right) \end{aligned} \tag{4}$$

The first and second derivatives are then derived and the lengths in $\mathbf{q}_{ui}$ are replaced by their proposed expressions (1) or (2). It then leads that all the first derivatives are equal to zero. The expressions of the second derivatives are not detailed here for paper clarity, but are strictly positive if the components of $\mathbf{q}_i$, the elongation and the stiffness of the springs are strictly positive, which is always verified. Thereby, we have analytically demonstrated that the observed configurations correspond to tensegrity configurations even with non-zero free-lengths. The springs of the bar- (resp. cable-) actuated SC present a free-length l_{0b} (resp. l_{0c}) and a maximal elongation δ_{eb} (resp. δ_{ec}). To ensure admissible tensions for springs and cables in a practical implementation, the lengths of the springs must be included in $[l_{0b},\ l_{0b}+\delta_{eb}]$ (resp. $[l_{0c},\ l_{0c}+\delta_{ec}]$) for the bar- (resp. cable-) actuated SC.

Stability analyses showed that the configuration of the mechanism is independent of the spring stiffness and free-length (Eq. (1) and (2)). Modeling of the manipulator is then reduced to the manipulation of geometrical relationships and the forward static model can be derived analytically. The pose of the end-effector is expressed as:

$$\mathbf{x} = \begin{bmatrix} O_m \\ \varphi_m \end{bmatrix} \quad \text{with} \quad O_m = O_b + \begin{bmatrix} 0 \\ 2l_p \end{bmatrix} + \begin{bmatrix} x_c \\ y_c \end{bmatrix}, \quad O_b = O_0 + \begin{bmatrix} x_b \\ y_b \end{bmatrix} \tag{5}$$

$$\begin{cases} x_b = \dfrac{\rho_{b1}^2 - \rho_{b2}^2}{4L_{A_1B_1}} \\ y_b = \dfrac{\rho_{b1}\rho_{b2}}{2L_{A_1B_1}} \end{cases} \qquad \begin{cases} x_c = \dfrac{\rho_{c1}^2 - \rho_{c2}^2}{4L_{A_3B_3}} \\ y_c = \left(\dfrac{\rho_{c1}+\rho_{c2}}{2}\right)\sqrt{1+\dfrac{(\rho_{c1}\rho_{c2}-\rho_{c1}^2)^2}{4\rho_{c1}^2\left(\rho_{c1}\rho_{c2}-L^2\right)}} \end{cases} \tag{6}$$

and

$$\varphi_m = 2\arccos\left(\frac{\rho_{c1}-\rho_{c2}}{2L_{A_3B_3}}\right) - \pi \tag{7}$$

The kinematic Jacobian matrix $\mathbf{J}$ of the manipulator maps the Cartesian velocities of the end-effector $\dot{\mathbf{x}}$ with respect to the joint velocities $\dot{\mathbf{q}}$: $\dot{\mathbf{x}} = \mathbf{J}\dot{\mathbf{q}}$. The pose of the end-effector is defined by three coordinates while four actuators are used. The system is then redundant and $\mathbf{J}$ is a 3×4 matrix. The kinematic Jacobian matrices $\mathbf{J}_b$ and $\mathbf{J}_c$ of the bar- and cable-actuated SC respectively are given in (8) and (9).

$$\mathbf{J}_b = \left[\frac{\partial \mathbf{x}}{\partial \mathbf{q}_1}\right] = \begin{bmatrix} \dfrac{\partial O_m}{\partial \mathbf{q}_1} \\ 0 \end{bmatrix} = \frac{1}{8L_{A_1B_1}^3} \begin{bmatrix} \dfrac{\rho_{b1}\left(\rho_{b1}^2+3\rho_{b2}^2\right)}{2} & -\dfrac{\rho_{b2}\left(3\rho_{b1}^2+\rho_{b2}^2\right)}{2} \\ \rho_{b2}^3 & \rho_{b1}^3 \\ 0 & 0 \end{bmatrix} \tag{8}$$

$$\mathbf{J}_c = \left[\frac{\partial \mathbf{x}}{\partial \mathbf{q}_2}\right] = \begin{bmatrix} \dfrac{\partial O_m}{\partial \mathbf{q}_2} \\ \dfrac{\partial \varphi_m}{\partial \mathbf{q}_2} \end{bmatrix} = \frac{1}{8L_{A_3B_3}^3} \begin{bmatrix} E_1(\rho_{c1},\rho_{c2}) & -E_1(\rho_{c2},\rho_{c1}) \\ \dfrac{E_2(\rho_{c1},\rho_{c2})}{\sqrt{4L^2-(\rho_{c1}+\rho_{c2})^2}} & \dfrac{E_2(\rho_{c2},\rho_{c1})}{\sqrt{4L^2-(\rho_{c1}+\rho_{c2})^2}} \\ -E_3(\rho_{c1},\rho_{c2}) & E_3(\rho_{c2},\rho_{c1}) \end{bmatrix} \tag{9}$$

with

$$\begin{aligned} E_1(\rho_i,\rho_j) &= 4L^2\rho_i - 3\rho_j\rho_i^2 - \rho_j^3 \\ E_2(\rho_i,\rho_j) &= 8L^4 - 4\rho_i L^2(\rho_i+3\rho_j) + \rho_j(3\rho_i-\rho_j)(\rho_i+\rho_j)^2 \\ E_3(\rho_i,\rho_j) &= 8L_{A_3B_3}^2\left(1-\rho_i(\rho_i-\rho_j)/2L_{A_3B_3}^2\right)\left(\sqrt{1-(\rho_i-\rho_j)^2/4L_{A_3B_3}^2}\right)^{-1} \end{aligned} \tag{10}$$

which leads to

$$\dot{\mathbf{x}} = \mathbf{J}_b \begin{bmatrix} \dot{\rho}_{b1} \\ \dot{\rho}_{b2} \end{bmatrix} + \mathbf{J}_c \begin{bmatrix} \dot{\rho}_{c1} \\ \dot{\rho}_{c2} \end{bmatrix} \tag{11}$$

and the kinematic Jacobian matrix is finally expressed as

$$\mathbf{J} = [\mathbf{J}_b,\ \mathbf{J}_c] \tag{12}$$

3 Control Loop and Simulation

Kinematic Jacobian-based control for redundancy management can now be investigated. To carry out a positioning task in the Cartesian space with a redundant manipulator, the joint velocities are computed as:

$$\dot{\mathbf{q}} = \mathbf{P}\mathbf{J}^{\dagger}(\mathbf{q})(\mathbf{x} - \mathbf{x}^{*}) \tag{13}$$

with $\mathbf{J}^{\dagger}$ the Moore-Penrose pseudo-inverse Jacobian and $\mathbf{P}$ a negative diagonal matrix. Using Eq. (13), the redundancy is not exploited and the pseudo-inverse then minimizes the Euclidean norm of joint velocities. To add secondary tasks, the gradient projection method [9] can be used to take advantage of the null-space of $\mathbf{J}$ and optimize a given cost function $h(\mathbf{q})$. The joint velocities are then expressed as:

$$\dot{\mathbf{q}} = \mathbf{P}\left(\mathbf{J}^{\dagger}(\mathbf{q})(\mathbf{x} - \mathbf{x}^{*}) + k\left[\mathbf{I} - \mathbf{J}^{\dagger}(\mathbf{q})\mathbf{J}(\mathbf{q})\right]\left(\frac{\partial h(\mathbf{q})}{\partial \mathbf{q}}\right)^{T}\right) \tag{14}$$

with k a positive scalar. With the control scheme presented in (13), the trajectory is controlled in the Cartesian space but not in the joint space. However, if the actuator limit positions are reached, trajectory distortions will be induced. The RCM constraint can thus be violated, which is not acceptable. Also, the control of a tensegrity mechanism requires that the cables and springs are in tension at all time. To respect this constraint, actuator strokes are chosen accordingly. Moreover, as shown in Eq. (1) and Eq. (2), an increase (resp. decrease) of the joint position implies an increase (resp. decrease) of the spring lengths. Therefore, maximizing the distance between the joint positions and their limits will also maximize the distance between the spring length and the lengths at rest and at maximal elongation. This approach is then also interesting to avoid situations with low or high tensions in the springs and cables. As a consequence, a cost function $h(\mathbf{q})$ is proposed to minimize the distance between the measured joint positions and the mean value of the reachable positions following [8]. With ρ_{bmin}, ρ_{bmax} and ρ_{cmin}, ρ_{cmax} the actuator limits of the bar-actuated and the cable-actuated SC respectively, the cost function is written as

$$\begin{aligned} h_a &= \frac{1}{4}\sum_{i=1}^{2} p_1\left(\frac{\rho_{bi} - \alpha_{bi}}{\rho_{bmax} - \alpha_{bi}}\right)^2 + p_2\left(\frac{\rho_{ci} - \alpha_{ci}}{\rho_{cmax} - \alpha_{ci}}\right)^2 \\ \alpha_{bi} &= (\rho_{bmax} + \rho_{bmin})/2 \qquad \alpha_{ci} = (\rho_{cmax} + \rho_{cmin})/2 \end{aligned} \tag{15}$$

with p_1 and p_2 some coefficients to be tuned.

The obtained control loop is displayed in Fig. 2. A pose feedback $\mathbf{x}_m$ is considered. It can be implemented by measuring the joint coordinates and using the forward kinematic model or using an exteroceptive measurement. This choice is not discussed in this paper. We also consider that the motions are slow enough to avoid actuator velocity saturation and the transfer function of the actuators is considered to be unitary.

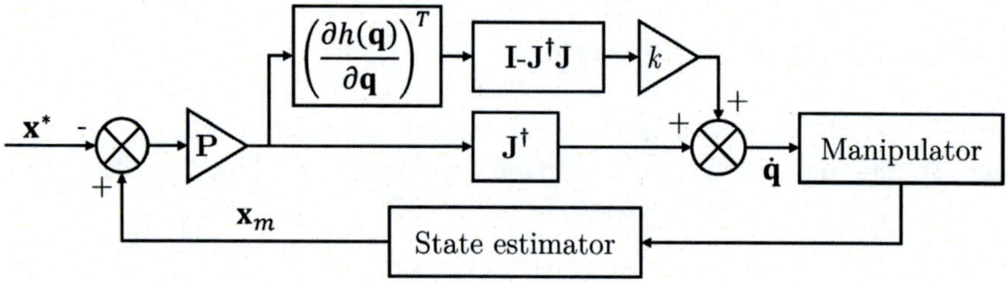

Fig. 2. Proposed control loop.

In order to evaluate the proposed control schemes, dynamic simulation is implemented with the Simscape™ Toolbox of Matlab Software (MathWorks, USA). This simulation is thereby independent from the models used for control. Weightless springs and cables are considered while the bars have a mass m_b, the end-effector and the decoupling element a mass m_p. Parameters are reported in Table 1. In practical implementation there are inevitably dissipative forces. We here simulate the latter with dampers placed in parallel to the springs and in the prismatic joints with a damping coefficient of 20 N.s/m. It is adjusted by a trial-and-error process.

A comparison with ($k \neq 0$) and without ($k = 0$) optimization is performed. The end-effector trajectory is displayed in Fig. 3. During the whole process, the manipulator respects the RCM constraint. End-effector path is controlled by modifying the radius value R_0 (Fig. 1). The manipulator end-effector follows the sequence of points ABCDEA. Along CD, $R_0 = 200$ mm and along BE, $R_0 = 235$ mm. The angle φ varies between -35 and 35°. The evaluation is carried out with an end-effector velocity of 5 mm/s and $\mathbf{P} = -diag(2,2,0.1)$. For the cost function minimization we choose $p_1 = p_2 = 1$ and $k = 10^6$. The latter is chosen through a trial-and-error method.

The obtained Cartesian trajectories and errors on the radius and the angle are displayed in Fig. 3. The end-effector follows the desired trajectory with average path tracking errors of 0.88 (resp. 0.28) mm on the radius and 0.22 (resp. 0.19) ° on the angle without (resp. with) optimization. In Fig. 3, a distortion of the Cartesian trajectory can be observed when the simple pseudo-inverse is used. It is due to the saturation of the joint positions. More specifically, it is caused by the actuators of the bar-actuated SC which reach their minimal bounds and the ones of the cable-actuated SC which reach their maximal bounds. When the control scheme with optimization is used some errors caused by the dynamics can still be observed. However, joint position saturation and tension limits are avoided.

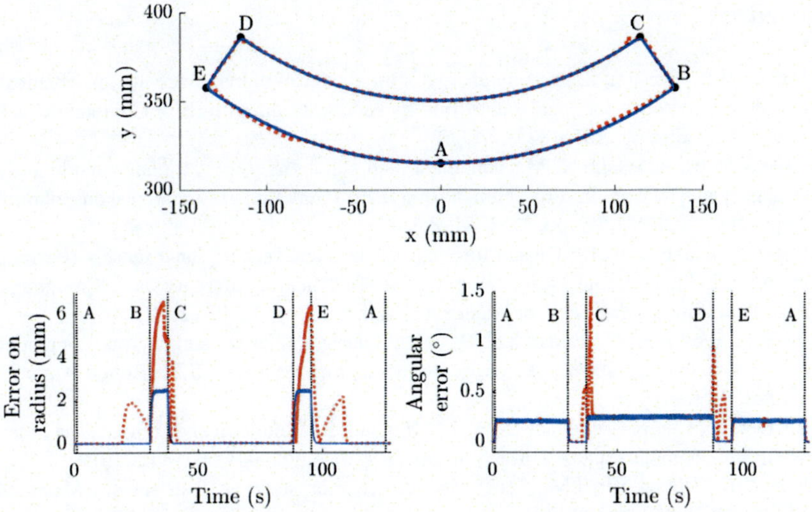

Fig. 3. Top: Cartesian trajectories without (dotted red line) and with (solid blue line) optimization. Bottom: Errors on the RCM constraint. In dotted red and in solid blue the curves obtained using the control schemes proposed in (13) and in (14) respectively.

Table 1. Parameters for simulation of the proposed manipulator.

Parameter	Value	Unit	Parameter	Value	Unit
L	200	(mm)	O_p	[0 500]	(mm)
K	0.25	(N/mm)	m_b	0.25	(kg)
l_{0b}, δ_{eb}	140	(mm)	m_p	0.25	(kg)
l_{0c}, δ_{ec}	100	(mm)	$[\rho_{bmin}, \rho_{bmax}]$	[200, 500]	(mm)
l_p	20	(mm)	$[\rho_{cmin}, \rho_{cmax}]$	[20, 170]	(mm)

4 Conclusions and Perspectives

In this paper, the kinematic model of a tensegrity-based manipulator with RCM constraint was derived. It was then used to develop and achieve early-stage tests of Jacobian-based control schemes. Optimization of the joint positions was proposed to address actuator saturation and tension issues taking advantage of the tensegrity mechanism redundancy. Promising results are obtained. Implementation of other control strategies for saturation avoidance, such as set-based control, is a first perspective of this work. The use of redundancy to minimize other criteria such as the potential energy to improve energy consumption and safety will also be considered.

Acknowledgements. This work is supported by the Investissements d'Avenir program (Robotex ANR-10-EQPX-44, Labex CAMI ANR-11- LABX-0004), the French National Research Agency (ANR Multiflag, ANR-16-CE33-0019) and by the Grand Prix Scientifique 2018, Fondation Charles Defforey, Institut de France.

References

1. Aldrich, J.B., Skelton, R.E.: Time-energy optimal control of hyper-actuated mechanical systems with geometric path constraints. In: Proceedings of the 44th IEEE Conference on Decision and Control, pp. 8246–8253 (2005)
2. Arsenault, M., Gosselin, C.M.: Kinematic and static analysis of a planar modular 2-DoF tensegrity mechanism. In: Proceedings 2006 IEEE International Conference on Robotics and Automation. ICRA 2006, pp. 4193–4198 (2006)
3. Bayat, J., Crane III, C.D.: Closed-form equilibrium analysis of planar tensegrity structures. In: ASME. International Design Engineering Technical Conferences and Computers and Information in Engineering Conference, vol. 8, no. 31, pp. 13–23 (2007)
4. Begey, J., Vedrines, M., Renaud, P.: Design of tensegrity-based manipulators: comparison of two approaches to respect a remote center of motion constraint. IEEE Robot. Autom. Lett. 1788–1795 (2020)
5. Boehler, Q., Charpentier, I., Vedrines, M., Renaud, P.: Definition and computation of tensegrity mechanism workspace. J. Mech. Robot. **7**(4), 044,502–044,502–4 (2015)
6. Juan, S.H., Mirats Tur, J.M.: Tensegrity frameworks: static analysis review. Mech. Mach. Theory **43**(7), 859–881 (2008)
7. Bakker, L.D., Matsuura, D., Takeda, Y., Herder, J.: Design of an environmentally interactive continuum manipulator. In: The 14th IFToMM World Congress (2015)
8. Liégeois, A.: Automatic supervisory control of the configuration and behavior of multibody mechanisms. IEEE Trans. Syst. Man Cybern. **7**(12), 868–871 (1977)
9. Siciliano, B.: Kinematic control of redundant robot manipulators: a tutorial. J. Intell. Robot. Syst. **3**(3), 201–212 (1990)
10. Snelson, K.: Continuous tension, discontinuous compression structures. US Patent US3169611A (1965)
11. Sultan, C.: Tensegrity deployment using infinitesimal mechanisms. Int. J. Solids Struct. **51**(21), 3653–3668 (2014)
12. Sumi, S., Böhm, V., Schale, F., Zimmermann, K.: Compliant gripper based on a multistable tensegrity structure. In: New Trends in Mechanism and Machine Science, pp. 143–151. Springer, Cham (2017)
13. Wenger, P., Chablat, D.: Kinetostatic analysis and solution classification of a planar tensegrity mechanism. In: Zeghloul, S., Romdhane, L., Laribi, M.A. (eds.) Computational Kinematics, Mechanisms and Machine Science, pp. 422–431. Springer, Heidelberg (2018)

Motion Parameterization of Parallel Robots Used in Lower Limb Rehabilitation

Iosif Birlescu[1], Manfred Husty[2,3], Calin Vaida[1], Bogdan Gherman[1], Ionut Ulinici[1], Remus Bogateanu[4], and Doina Pisla[1](✉)

[1] CESTER, Technical University of Cluj-Napoca, Cluj-Napoca, Romania
{iosif.birlescu, calin.vaida, bogdan.gherman, ionut.ulinici, doina.pisla}@mep.utcluj.ro
[2] Unit Geometry and CAD, University of Innsbruck, Innsbruck, Austria
manfred.husty@uibk.ac.at
[3] Robotics Institute, Joanneum Research, Klagenfurt, Austria
[4] S.C. Alvi Technik S.R.L., Cluj-Napoca, Romania
office@alvi-technik.com

Abstract. In this paper the constraint equations for two parallel robots for lower limb rehabilitation are derived. A parameterization method was used where the mobile platform coordinates (written with the Study parameters) were eliminated and the constraints were computed with Groebner bases. For the two parallel robots (RAISE, RECOVER ankle module) the dependencies between the active joints of the robot and the anatomic joints were obtained. For RAISE the results agreed with previous ones, whereas for RECOVER ankle module the results showed simple equations which may be used for the complete analysis of the mechanism and for the control development.

Keywords: Motion parameterization · Parallel robot · Rehabilitation

1 Introduction

Stroke is one of the leading causes of motor disability with its occurrence estimated to increase in the future [1]. To regain motor capabilities, stroke patients must undergo a rehabilitation process where the patients' limbs are mobilized either by human personnel (e.g. kinetotherapists) or by specialized devices (e.g. rehabilitation robots). A specific type of lower limb rehabilitation robots are the ones that mobilize the limb using a free motion kinematic chain (acting like an exoskeleton) guided by input chains [2, 3]. A motion parameterization approach for such robots was introduced in [2] were the dependency between specific revolute joints (adjacent to the anatomic joints of the lower limb) and the robot actuators was studied achieving constraint equations which describe the rehabilitation exercises in a straightforward manner. Moreover the constraints derived with the method can be used for a singularity analysis and operational workspace analysis which in turn provides valuable information for the interaction between the robot and the patient (facilitating the safe operation of the robot). In this paper, the motion parameterization approach described in [2], is used on two rehabilitation parallel robots, however, the constraints are obtained using algebraic

J. Lenarčič and B. Siciliano (Eds.): ARK 2020, SPAR 15, pp. 57–64, 2021.
https://doi.org/10.1007/978-3-030-50975-0_8

techniques (e.g. Groebner bases) for two main reasons: i) the constraints will be computed faster without the need of the algorithm described in [2]; ii) the method may be exploited for more complex mechanisms. In Sect. 3, this approach is exemplified on the RAISE parallel robot to show that the results agree with previous work. In Sect. 4, the constraints of the ankle module of the RECOVER parallel robot [3] are determined, showing that the method is usable for more complex mechanisms. Section 5 presents results and a discussion and finally the conclusions are presented in Sect. 6.

2 Mathematical Background

Every Euclidian displacement which describes the relative position between a mobile coordinate frame and a fixed one may be described using a point Q from the projective space P^7 [4]. The point $Q(x_0{:}x_1{:}x_2{:}x_3{:}y_0{:}y_1{:}y_2{:}y_3)$ is based on a dual quaternion and the entries of the quaternion are also called the Study parameters of SE(3) which must fulfill the following relations [4]:

$$x_0^2+x_1^2+x_2^2+x_3^2 \neq 0, \; x_0y_0+x_1y_1+x_2y_2+x_3y_3=0, \tag{1}$$

were Eq. (1) represents a normalizing condition and the Study quadric respectively [4]. In order to describe a parallel mechanism with the Study parameters, each kinematic chain within the mechanism is mathematically described by multiplying each quaternion which describes the mechanism components (links and joints) starting from the fixed coordinate frame and ending at the mobile coordinate frame. The free motion parameters can be eliminated by means of algebraic techniques (e.g. Groebner bases to compute elimination ideals [5]). All the equations describing all the kinematic chains of the parallel mechanism must be fulfilled which physically means that the chains are coupled at the origin of the mobile coordinate frame. Generally speaking, this approach describes the displacement of the end-effector relative to the robot base. In [2, 3], the authors investigated parallel robots for rehabilitation that act like exoskeletons with a kinematic chain that is attached to the limb such that the revolute joints of the chain are adjacent to the anatomic joints. Consequently, the dependency between the robot actuators and the mechanical joints adjacent to the anatomic ones is more important than the relative displacement between the robot end-effector and its base. One method for the motion parameterization of such robotic systems is presented in [2]. The method uses the Study parameters to describe each kinematic chain of the parallel robotic systems, and eliminates the Study parameters by subtracting sets of equations. This process relates various motion parameters of the parallel mechanism and the elimination of the unwanted parameters is achieved by solving equations and using back substitution. Another approach to eliminate the unwanted motion parameters is to compute elimination ideals (as detailed in [5]) over the robot joint space, which makes the parameterization method usable for more complex mechanisms (as presented in Sect. 4).

3 Motion Parameterization of the RAISE Parallel Robot

A case study is presented in this section showing how to derive the constraints of the RAISE parallel robot (Fig. 1) [6] for lower limb rehabilitation, with a straightforward and arguably more elegant method (using Groebner bases). RAISE consists of four kinematic chains [2]: one free motion kinematic chain of UR type (which guides the lower limb) with the revolute joints R_{h1}, R_{h2} (located adjacent to the hip joint) and the R_k revolute joint (adjacent to the knee joint); two RPPR input kinematic chains which guide the free motion chains via the active joint q_1 and q_2 respectively allowing the hip flexion/extension and the knee flexion/extension; one kinematic chain of PRR type which guides the free motion chain in a rotation around the R_{h1} revolute joint axis, allowing the hip abduction/adduction motion.

The parameterization of each kinematic chain of the RAISE parallel robotic system was already presented (see Eqs. 16–18 in [2] were the free motion kinematic chain is denoted M_{C0} and the first two input chains are denoted M_{C1} and M_{C2} respectively). The third kinematic chain is omitted (by setting t_1, s_1, v_1 to zero in the original equations – which represent the hip abduction/adduction motion) since it does not affect the motion of M_{C1} and M_{C2}, i.e. changing the angle α_{h1} (see Fig. 1) does not affect the configuration of the rest of the mechanism (fact that was also shown in [2] see observation 2).

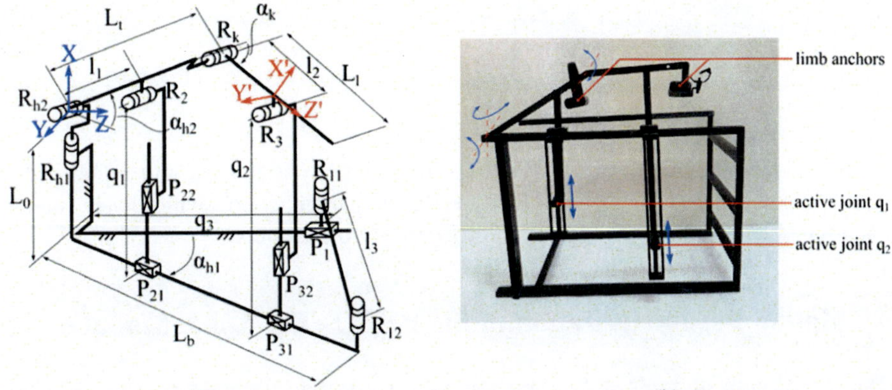

(a) (b)

Fig. 1. RAISE parallel robotic system; a) kinematic scheme [2], b) 3D scaled mockup (1:5).

To obtain the constraints for the RAISE parallel robot two sets of polynomials are defined: $J_1 = M_{C0} - M_{C1}$ (Eq. 2) and $J_2 = M_{C0} - M_{C2}$ (Eq. 3) respectively:

$$
\begin{aligned}
J_1 := \langle & 2\rho_0(t_2t_3+1) - 2\rho_1(s_2s_3+1), 2\rho_0(-t_3+t_2) - 2\rho_1(-s_3+s_2), \rho_0((-t_2-t_3)L_t \\
& -l_2(-t_3+t_2)) - \rho_1(((L_0-q_1)s_3 - L_t + l_1 - l_2 + l_{f1})s_2 + (-L_t + l_1 + l_2 - l_{f1})s_3 + L_0 \\
& -q_1), \rho_0((-l_2+L_t)t_3t_2 - l_2 - L_t) - \rho_1(((L_t - l_1 - l_2 - l_{f1})s_3 - L_0 - q_1)s_2 + (-L_0 \\
& +q_1)s_3 - L_t + l_1 - l_2 - l_{f1})\rangle
\end{aligned}
\qquad (2)
$$

$$J_2 := \langle 2\rho_0(t_2t_3+1) - 2\rho_2, 2\rho_0(-t_3+t_2) + 2\rho_2 v_2, \rho_0((-t_2-t_3)L_t - l_2(-t_3+t_2)) \\ -\rho_2(l_{f2}v_2 - L_0 - q_3), \rho_0((-l_2+L_t)t_3t_2 - l_2 - L_t) - \rho_2((-L_0+q_3)v_2 - l_{f2})\rangle, \tag{3}$$

where the tangent of half angle substitution is used: t_2 describing the R_{h2} joint, t_3 the R_k joint (within M_{C1}), s_2 describing R_2, s_3 describing R_k (within M_{C2}), and v_2 describing R_3. The homogenizing parameters ρ_0, ρ_1, and ρ_2 (which satisfy the strong normalizing condition $x_0^2 + x_2^2 = 1$) are also introduced in Eqs. (3–4) and they are:

$$\rho_0 = \pm\frac{1}{2\sqrt{(t_2^2+1)(t_3^2+1)}} \quad \rho_1 = \pm\frac{1}{2\sqrt{(s_2^2+1)(s_3^2+1)}} \quad \rho_2 = \pm\frac{1}{2\sqrt{v_2^2+1}} \tag{4}$$

The first two polynomials from Eq. (2) together with the values for ρ_0 and ρ_1 (Eq. 4) lead to a simple solution ($t_2 = s_2$, $t_3 = s_3$, $\rho_0 = \rho_1$) [2] which is also obvious from Fig. 1. This result will be substituted in Eq. (2) to facilitate the next step in the computation. Equations (2) and (3) generate ideals over the joint space of the RAISE parallel robotic system (denoted J_1 and J_2). The free motion parameters are eliminated (using Groebner bases) by computing the following elimination ideals (G_1 from J_1 and G_2 from J_2):

$$J_{1\_0} := \langle L_0t_2^2 - q_1t_2^2 + 2l_1t_2 + L_0 - q_1\rangle \tag{5}$$

$$J_{2\_0} := \langle (L_0t_3^2 - q_2t_3^2 + 2l_2t_3 + L_0 - q_2)t_2^2 + (L_tt_3^2 - l_2t_3^2 + L_t + l_2)2t_2 \\ + L_0t_3^2 - q_2t_3^2 - 2l_2t_3 + L_0 - q_2\rangle \tag{6}$$

Equations (5) and (6) are without any free motion parameters and represent the input output constraint equations for the RAISE parallel robot, which were already determined in [2] using the algorithm proposed by the authors.

4 Motion Parameterization of the RECOVER Ankle Module

The second case study illustrates how the parameterization method works for the ankle module of the RECOVER parallel robot, which is based on mechanism that is more complex than RAISE. The RECOVER [7] robotic system is designed for the motor rehabilitation of the lower limb and it consists of 2 DOF a planar module for the hip and knee (flexion/extension) guidance and a 2 DOF spatial module for the ankle (flexion extension and abduction/adduction) guidance [3]. The ankle rehabilitation module is illustrated in Fig. 2 [3] and it consists of three kinematic chains: one free motion chain of RR type consisting of the revolute joints R_{a1} and R_{a2} having orthogonal axes which intersect in the origin of both coordinate systems (the fixed one *XYZ* and the mobile one *X'Y'Z'*); two input chains of PSS type actuated by q_3 and q_4 respectively, both chains ending at the origin of the coordinate systems consisting of the spherical joints $S_{1,3}$, $S_{2,4}$ and the links l_{a1}, l_{a2}, L_0 and L_a. In the technical design the center of rotation of the ankle joint should coincide with the origin of the coordinate systems, allowing a direct description of the rehabilitation exercise.

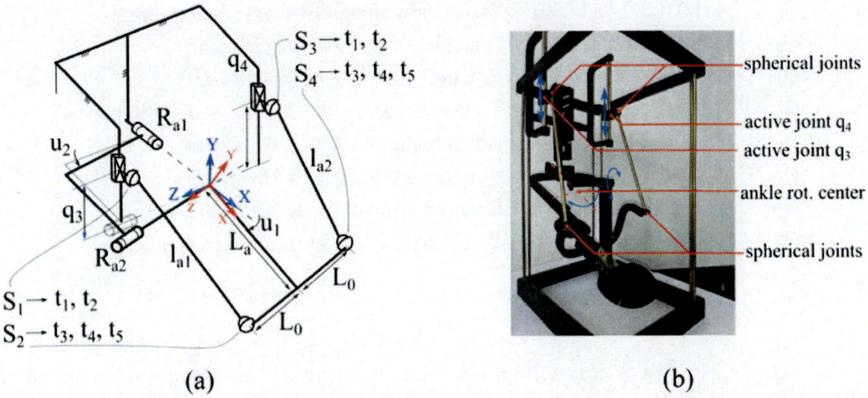

Fig. 2. RECOVER ankle module; a) kinematic scheme [3], b) 3D scaled mockup (1:1).

The Study parameters of the three kinematic chains are obtained by multiplying the dual quaternions of each mechanical element (links and joints) starting from the fixed coordinate frame and ending at the mobile coordinate frame as follows:

$$C_0 = Q_{r1} \cdot Q_{r2} \tag{7}$$

$$C_{1,2} = Q_{\pm L0} \cdot Q_{q3,4} \cdot Q_{Z1} \cdot Q_{Y1} \cdot Q_{a1} \cdot Q_{X2} \cdot Q_{Z2} \cdot Q_{Y2} \cdot Q_{\mp L0} \cdot Q_{-La} \tag{8}$$

were the dual quaternions of the right hand side of Eqs. (7) and (8) are detailed in Table 1 for the free motion chain and in Table 2 for the input chains respectively. The tangent of half angle is used for the rotation parameters (u_i for the free motion chain and t_i for the first and second input chain respectively).

Note: the only differences in the parameterizations of the chains are the active joints (q_3 and q_4 for C_1 and C_2 respectively) and the geometric displacement L_0 (along Z axis) which has opposite signs for the two input chains. To allow the motion of the mobile platform only 5 rotations (of the spherical joints $S_{1,3}$, $S_{2,4}$) t_i $i = 1..5$ are needed.

Table 1. Dual quaternions used for the free motion chain param.

	Dual quaternion	Description
Q_{r1}	$[1{:}0{:}0{:}u_1{:}0{:}0{:}0{:}0]$	Rotation around the Z axis (free joint)
Q_{r1}	$[1{:}u_2{:}0{:}0{:}0{:}0{:}0{:}0]$	Rotation around the Y axis (free joint)

Table 2. Dual quaternions used for the input chain param.

	Dual quaternion	Description
$Q_{\pm L0}$	$[1{:}0{:}0{:}0{:}0{:}0{:}0{:}\mp L_0/2]$	Translation along the Z axis (geometric param.)
$Q_{q3,4}$	$[1{:}0{:}0{:}0{:}0{:}0{:}\text{-}q_{3,4}/2{:}0]$	Translation along the Y axis (active joint)
Q_{Z1}	$[1{:}0{:}0{:}t_1{:}0{:}0{:}0{:}0]$	Rotation around Z axis (free joint)
Q_{Y1}	$[1{:}0{:}t_2{:}0{:}0{:}0{:}0{:}0]$	Rotation around Y axis (free joint)
$Q_{la1,2}$	$[1{:}0{:}0{:}0{:}\text{-}l_{a1,2}/2{:}0{:}0{:}0]$	Translation along the X axis (geometric param.)
Q_{X2}	$[1{:}t_3{:}0{:}0{:}0{:}0{:}0{:}0]$	Rotation around X axis (free joint)
Q_{Z2}	$[1{:}0{:}0{:}t_4{:}0{:}0{:}0{:}0]$	Rotation around Z axis (free joint)
Q_{Y2}	$[1{:}0{:}t_5{:}0{:}0{:}0{:}0{:}0]$	Rotation around Y axis (free joint)
$Q_{\mp L0}$	$[1{:}0{:}0{:}0{:}0{:}0{:}0{:} \pm L_0/2]$	Translation along the Z axis (geometric param.)
$Q_{\text{-}La}$	$[1{:}0{:}0{:}0{:}0{:}L_a/2{:}0{:}0]$	Translation along the X axis (geometric param.)

In Eqs. (7)–(8) two polynomial ideals are formed over the joint space of the RECOVER ankle module, $J_1 = \rho_0 C_0 - \rho_1 C_1$, $J_2 = \rho_0 C_0 - \rho_2 C_2$, where ρ_i are homogenizing parameters which could be computed according to [2]. As it turns out their computation is not necessary since the end result (the constraint equations) will be independent of them (they will factor out). Moreover, to facilitate the computation, the geometric parameters are substituted with the following numerical values: $L_0 = 1$, $l_{a1} = 5$, $l_{a2} = 5$, $L_a = 4$; these represent the ratios between the links in the original mechanism design and not dimension in mm (ratios would be 35 mm to obtain the original dimensions). The two resulting polynomial ideals $J_{1,2}$ are:

$$\begin{aligned}
J_{1,2} := \langle & \rho_0 - 2\rho_{1,2}((t_4t_5 + t_3)t_1t_2 + (t_1 - t_4)t_3t_5 - (t_2t_3 + t_1)t_4 - t_2t_5 + 1), \rho_0 u_2 - 2\rho_{1,2}((t_1t_4 \\
& + 1)t_2t_3t_5 + (t_1t_3 - t_2 + t_5)t_4 + (t_5 - t_2)t_1 + t_3), \rho_0 u_1 u_2 - 2\rho_{1,2}((t_4 - t_1)t_2t_3t_5 + (t_2 + t_5)t_1t_4 \\
& + (t_1 - t_4)t_3 - t_2 - t_5), \rho_0 u_1 - 2\rho_{1,2}((t_2 - t_5)t_1t_3t_4 + (t_1t_5 + t_4t_5 + t_3)t_2 - t_3t_5 + t_4 + t_1), \\
& \rho_{1,2}((-q_{3,4}t_1 + q_{3,4}t_4 + t_1t_4 - 9)t_2t_3t_5 + (q_{3,4}t_2 + q_{3,4}t_5 - 9t_3)t_1t_4 + (q_{3,4}t_3 - 9t_5 - t_2)t_1 - \\
& (q_{3,4}t_3 - 9t_2 - t_5)t_4 - (t_2 + t_5)q_{3,4} + t_3), \rho_{1,2}((-q_{3,4}t_3t_4 \mp 2t_3t_5 - q_{3,4}t_5 - t_4t_5 - t_3 \pm t_4)t_1t_2 \\
& + (q_{3,4}(t_1t_3 - t_2) \pm 2t_2t_3 \pm 2t_1 + t_3)t_4t_5 + (q_{3,5} + 9t_1)t_3t_5 - (q_{3,4} + 9t_4)t_2t_3 - (q_{3,4} \mp 2t_3 + \\
& 9t_4)t_1 - (9t_5 \pm 1)t_2 - (q_{3,4} \pm t_3)t_4 \mp 2t_5 - 1), \rho_{1,2}((-q_{3,4}t_4t_5 \mp t_3t_4t_5 - q_{3,4}t_3 + 9t_3t_4 + 9t_5 \\
& \pm 2)t_1t_2 + (q_{3,4}t_3 + t_1t_3 - t_2 \mp 2)t_4t_5 - (q_{3,4}t_1 \pm t_2 + 9)t_3t_5 + (q_{3,4}t_4 - 1)t_2t_3 + (q_{3,4}t_4 \mp 2t_3t_4 \\
& \mp 2t_5 - 1)t_1 + (q_{3,4}t_5 \pm 2t_4)t_2 - q_{3,4} \mp 2t_3 + 9t_4), \rho_{1,2}((q_{3,4}t_1t_4 + q_{3,4} + 9t_1 + t_4)t_2t_3t_5 + \\
& (q_{3,4}t_3 - 9t_2 + t_5)t_1t_4 + (q_{3,4}t_5 - q_{3,4}t_2 + t_3)t_1 + (q_{3,4}t_5 - q_{3,4}t_2 + 9t_3)t_4 + q_{3,4}t_3 - t_2 + 9t_5)\rangle
\end{aligned} \tag{9}$$

The elimination ideals (J_{1_0}, and J_{2_0}) are determined next by computing Groebner bases: G_1 from the J_1 using lexdeg monomial ordering ($\rho_1 < \rho_0 < t_1 < t_2 < t_3 < t_4 < t_5 < u_1 < u_2 < q_3$), G_2 from J_2 using lexdeg monomial ordering ($\rho_2 < \rho_0 < t_1 < t_2 < t_3 < t_4 < t_5 < u_1 < u_2 < q_4$). The elimination ideals J_{1_0}, and J_{2_0} contain only one equation each (independent of the free motion parameters t_i):

$$\begin{aligned}
J_{1_0} := \langle & \rho_0((u_1^2u_2^2 + u_1^2 + u_2^2 + 1)q_3^2 - (4u_1^2u_2 + 16u_1u_2^2 + 16u_1 - 4u_2)q_3 \\
& -5u_1^2u_2^2 - 9u_1^2 - 5u_2^2 - 9)\rangle
\end{aligned} \tag{10}$$

$$J_{2\_0} := \langle \rho_0((u_1^2 u_2^2 + u_1^2 + u_2^2 + 1)q_4^2 + (4u_1^2 u_2 - 16u_1 u_2^2 - 16u_1 - 4u_2)q_4 \\ -5u_1^2 u_2^2 - 9u_1^2 - 5u_2^2 - 9)\rangle \quad (11)$$

5 Results and Discussion

Equations (10)–(11) represent the input-output constraint equations for the RECOVER ankle module. To illustrate an example for the inverse kinematics the ankle flexion angular value is chosen to be 15° (u_1 = 0.1316525) and the adduction angular value is set to 10° (u_2 = 0.08748866). Solving the constraints for q_3 and q_4 reveal 4 solutions detailed in Table 3 (where only the first solution is possible in the mechanism design). To illustrate an example for the forward kinematics the first solution from Table 3 is substituted into the constraints, yielding 8 solutions (for u_1, u_2) out of which 4 are real valued and are presented in Table 4.

Table 3. Inverse kinematics numerical solutions for the ankle module.

Sol. No.	Active joint q_3 [times 35 mm]	Active joint q_4 [times 35 mm]
1	3.985597181	4.430520804
2	3.985597181	-2.024505900
3	-2.250507289	4.430520804
4	-2.250507289	-2.024505900

Table 4. Forward kinematics numerical solutions for the ankle module.

Sol. No.	Ankle flexion/extension (u_1)	Ankle abduction/adduction (u_2)
1	0.1316525000	0.087488660
2	0.1953587215	12.89128695
3	7.595753973	-0.08748866011
4	5.118788617	-12.89128695

Although the parameterization of the ankle module was rather complex to begin with (see Eq. 9), the input output equations (Eqs. 10–11) achieved with the proposed method are relatively simple. Furthermore, the kinematics (for velocities and accelerations) and the singularities can be analyzed using Eqs. 10–11 (with classical vector methods) which are essential studies for the further development of the rehabilitation module (e.g. for the control and to determine appropriate ranges of motion to facilitate patient safety). The 4 solutions for the inverse kinematics are straight-forward to visualize since the platform composed of L_0 and L_a may be kept in the same configuration with both positive and negative values for the active joints. For the forward kinematics, some solutions (according to the authors' opinion) represent configurations where the links l_{a1} and l_{a2} are "crossed" and the platform is flipped. A similar behavior was found in [8] for the 2SPRR-U mechanism used for the ankle of a humanoid robot (where the authors reported 6 solutions for the forward kinematics). However, by imposing appropriate motion ranges for the joints (through mechanical design and control) only one solution for the forward kinematics is achievable (for the RECOVER ankle module the solution no 1 in Table 4).

6 Conclusion

Based on a parameterization method where the coordinates of the mobile platform are eliminated, the input-output constraint equations for two parallel robots (used in lower limb rehabilitation) were determined. The equations were obtained in both cases using algebraic techniques (i.e. Groebner bases) showing that: i) for the RAISE parallel robot the results agreed with previous work; ii) for a more complex mechanism (RECOVER ankle module), the method achieved simple equations which will be used for further work regarding the development of the rehabilitation module (e.g. for the control). Furthermore, the additional expense in deriving the polynomial constraint equations pays off, because the obtained equations can be used for the complete analysis to determine the safe operation of the investigated rehabilitation robot.

Acknowledgements. The paper presents results from the research activities of the project ID 37_215, MySMIS code 103415 "Innovative approaches regarding the rehabilitation and assistive robotics for healthy ageing" co-financed by the European Regional Development Fund through the Competitiveness Operational Programme 2014-2020, Priority Axis 1, Action 1.1.4, through the financing contract 20/01.09.2016, between the Technical University of Cluj-Napoca and ANCSI as Intermediary Organism in the name and for the Ministry of European Funds.

References

1. Multi Annual Roadmap, euRobotics AISBL (2017)
2. Husty, M., Birlescu, I., Tucan, P., Vaida, C., Pisla, D.: An algebraic parameterization approach for parallel robots analysis. Mech. Mach. Theory **140**, 245257 (2019)
3. Gherman, B., Birlescu, I., Plitea, N., Carbone, G., Tarnita, D., Pisla, D.: On the singularity-free workspace of a parallel robot for lower-limb rehabilitation. Proc. Rom. Acad, Ser. A **20** (4), 383–391 (2019)
4. Husty, M., Pfurner, M., Schrocker, H.P., Brunnthaler, K.: Algebraic methods in mechanism analysis and synthesis. Robotica **25**(6), 661–675 (2007)
5. Cox, D., Little, J., O'Shea, D.: Ideals, Varieties, and Algorithms. An Introduction to Computational Algebraic Geometry and Commutative Algebra, pp. 276–277. Springer, Heidelberg (2007)
6. Pisla, D., Birlescu, I., Vaida, C., Gherman, B., Tucan, P., Carbone, G., Plitea, N.: Parallel robot for lower limb rehabilitation. Patent Pending A/00334/04.06.2019
7. Pisla, D., Gherman, B., Nadas, I., Pop, N., Craciun, F., Tucan, P., Vaida, C., Carbone, G.: Innovative paralel robot for lower limb rehabilitation. Patent Pending, A00391/27.06.2019
8. Kumar, S., Nayak, A., Peters, H., Schulz, C., Mueller, A., Kirchner, F.: Kinematic analysis of a novel parallel 2SPRR+1U ankle mechanism in humanoid robot. In: ARK 2018. Springer Proceedings in Advanced Robotics, vol. 8., pp. 431–439. Springer, Cham (2018)

Robust Trajectory Planning of Under-Actuated Cable-Driven Parallel Robot with 3 Cables

Edoardo Idà[1(✉)], Sébastien Briot[2], and Marco Carricato[1]

[1] Department of Industrial Engineering, University of Bologna, Bologna, Italy
{edoardo.ida2,marco.carricato}@unibo.it

[2] CNRS, Laboratoire des Sciences du Numérique de Nantes (LS2N), Nantes, France
Sebastien.Briot@ls2n.fr

Abstract. Cable-driven parallel robots (*CDPRs*) are under-actuated if they use a number of cables smaller than the degrees of freedom (*DoF*) of the end-effector (*EE*). For these robots, the constraint deficiency on the *EE* may lead to undesirable *EE* oscillations along the path that it is supposed to track. This paper proposes a trajectory-planning method for underactuated *CDPRs* which is robust against dynamic-model uncertainties or parameter variation, aiming at minimizing *EE* oscillations along a prescribed path. Oscillation reduction and robustness are achieved by means of Zero-Vibration Multi-Mode Input Shaping and Dynamic Scaling of a reference trajectory. Simulation results show the effectiveness of the method on a 3-cable 6-*DoF* robot.

Keywords: Cable-driven parallel robot · Under-actuated systems · Trajectory planning · Input shaping · Multi-mode systems · Dynamic scaling

1 Introduction

Cable-driven parallel robots (*CDPRs*) employ cables in place of rigid-body links in order to control the end-effector (*EE*) pose. A *CDPR* is underactuated if the number of cables is smaller than the number of the *EE* degrees of freedom (*DoFs*). This means that only a sub-set of the generalized coordinates of the *EE* can be controlled, while the others are determined by the system dynamics, possibly leading to undesirable *EE* oscillations. The use of *CDPRs* equipped with a limited number of cables is justified in several applications, in which the task to be performed requires the control of a limited number of *EE DoFs* or a limitation of mobility is acceptable to enhance workspace accessibility or decrease system cost. In recent years, increasing effort has been devoted to investigate the kinematics [1,3] and dynamics [5,6,10] of these manipulators.

The trajectory-planning problem for point-to-point motions of under-actuated *CDPRs* has proven to be a major challenge, seldom studied. Idàà et al. [6] presented a method to design rest-to-rest motions with prescribed path and motion time, for a generic under-actuated *CDPR* whose dynamic model is perfectly known. Hwang et al. [5] presented a method based on Multi-Mode Zero-Vibration Input-Shaping

J. Lenarčič and B. Siciliano (Eds.): ARK 2020, SPAR 15, pp. 65–72, 2021.
https://doi.org/10.1007/978-3-030-50975-0_9

(*MMZVIS*) to conveniently shape the input motion laws of a 3-cable *CDPR* performing trajectories with a limited variation of the robot natural frequencies. Both theoretical and experimental results showed effectiveness of either methods, but robustness to model uncertainties was never considered.

In this paper, we propose a trajectory-planning method for point-to-point motions of an under-actuated 3-cable *CDPRs* that is robust against the variation of configuration-dependent parameters in the robot dynamic model, which affects the robot natural frequencies. Robustness is achieved by combining *MMZVIS* and Dynamic Scaling (*DS*) of the reference trajectory. Section 2 introduces the robot model, and a methodology to compute its natural frequencies based on [5]. Section 3 describes the novel trajectory-planning method. Section 4 reports the results of numerical simulation, and Sect. 5 draws conclusions.

2 *CDPR* Modelling

Consider a 6-*DoF CDPR* composed of an *EE* coupled to the base by 3 actuated cables. $Oxyz$ is an inertial frame, $Px'y'z'$ is a mobile frame attached to the *EE* (Fig. 1a). The platform pose is described by $\mathbf{p}$, the position vector of P, and $\mathbf{R}$, a rotation matrix parametrized by the Euler angles $\boldsymbol{\epsilon} = [\phi, \theta, \chi]^T$ according to the *Tait-Bryan xyz* convention. The platform generalized coordinates are $\mathbf{q} = [\mathbf{p}, \boldsymbol{\epsilon}]^T$.

The i-th cable, modelled as massless and inextensible, is guided into the workspace from point B_i, described by $\mathbf{b}_i$ in $Oxyz$, by a swivel pulley (see [6] for modelling details) and it is attached to the platform at point A_i (Fig. 1a). G is the platform center of mass. The coordinates of G and A_i are described in the mobile frame by vectors ${}^P\mathbf{r}'$ and ${}^P\mathbf{a}'_i$, and in the inertial frame by $\mathbf{r} = \mathbf{p} + \mathbf{r}' = \mathbf{p} + \mathbf{R}\,{}^P\mathbf{r}'$ and $\mathbf{a}_i = \mathbf{p} + \mathbf{a}'_i = \mathbf{p} + \mathbf{R}\,{}^P\mathbf{a}'_i$. Vector $\boldsymbol{\rho}_i = \mathbf{a}_i - \mathbf{b}_i$ is tangent to the pulley, and the geometrical constraint imposed by the i-th cable on the platform is:

$$\boldsymbol{\rho}_i^T \boldsymbol{\rho}_i - [l_i - \widehat{B_i D_i}]^2 = 0 \tag{1}$$

where l_i is the total cable length, comprising the rectilinear length $\|\boldsymbol{\rho}_i\|$ and the arc $\widehat{B_i D_i}$ wrapped onto the pulley. The time derivatives of Eq. (1) leads, after some computation to [6]:

$$\mathbf{J}_i \dot{\mathbf{q}} = \dot{l}_i, \qquad \dot{\mathbf{J}}_i \dot{\mathbf{q}} + \mathbf{J}_i \ddot{\mathbf{q}} = \ddot{l}_i \tag{2}$$

where $\mathbf{J}_i$ is the i-th row of the analytic Jacobian of the system:

$$\mathbf{J}_i = [\hat{\boldsymbol{\rho}}_i^T \;\; -\hat{\boldsymbol{\rho}}_i^T \tilde{\mathbf{a}}'_i \mathbf{H}], \qquad \mathbf{H} = \begin{bmatrix} 1 & 0 & s_\theta \\ 0 & c_\phi & -s_\phi c_\theta \\ 0 & s_\phi & c_\phi c_\theta \end{bmatrix} \tag{3}$$

$\hat{(\cdot)}$ denotes a unit vector, $\tilde{(\cdot)}$ denotes the skew-symmetric representation of the vector product.

The dynamic model of the *CDPR* emerges from the platform dynamics, subject to the cable constraints. m is the *EE* mass, ${}^P\mathbf{I}_G$ its constant inertia matrix about G in

$Px'y'z'$, $\mathbf{I}_G = \mathbf{R}\,{}^P\mathbf{I}_G\mathbf{R}^T$ the inertia matrix in $Oxyz$, and $\mathbf{I}_P = \mathbf{I}_G - m\tilde{\mathbf{r}}'\tilde{\mathbf{r}}'$ its representation about P. If $\mathbf{f}_G$ and $\mathbf{m}_G$ are the resultant external force and moment about G, the platform equilibrium about P leads to [6]:

$$\mathbf{M}\ddot{\mathbf{q}} - \mathbf{s} + \mathbf{J}^T\boldsymbol{\tau} = \mathbf{0}_6, \qquad \mathbf{J}^T = [\mathbf{J}_1^T\,\mathbf{J}_2^T\,\mathbf{J}_3^T] \tag{4}$$

$$\mathbf{M} = \begin{bmatrix} m\mathbf{I}_3 & -m\tilde{\mathbf{r}}'\mathbf{H} \\ -m\mathbf{H}^T\tilde{\mathbf{r}}'^T & \mathbf{H}^T\mathbf{I}_P\mathbf{H} \end{bmatrix}, \qquad \mathbf{s} = \begin{bmatrix} m(\tilde{\mathbf{r}}'\dot{\mathbf{H}} + \tilde{\boldsymbol{\omega}}\tilde{\mathbf{r}}'\mathbf{H})\dot{\boldsymbol{\epsilon}} + \mathbf{f}_G \\ \mathbf{H}^T[-(\mathbf{I}_P\dot{\mathbf{H}} + \tilde{\boldsymbol{\omega}}\mathbf{I}_P\mathbf{H})\dot{\boldsymbol{\epsilon}} + \mathbf{m}_G + \tilde{\mathbf{r}}'\mathbf{f}_G] \end{bmatrix} \tag{5}$$

where $\boldsymbol{\tau} = [\tau_1, \tau_2, \tau_3]^T$ contains the cable tensions, and $\mathbf{I}_3$ and $\mathbf{0}_6$ are the (3×3) identity matrix and the (6×1) null vector, respectively. The generalized coordinates can be partitioned in 3 *actuated* coordinates $\mathbf{q}_a$ and 3 *unactuated* coordinates $\mathbf{q}_u$, whose time evolution is determined by the system dynamics driven by the actuated variables. According to this partition, Eq. (4) can be written as:

$$\begin{bmatrix} \mathbf{M}_{aa} & \mathbf{M}_{au} \\ \mathbf{M}_{ua} & \mathbf{M}_{uu} \end{bmatrix} \begin{bmatrix} \ddot{\mathbf{q}}_a \\ \ddot{\mathbf{q}}_u \end{bmatrix} - \begin{bmatrix} \mathbf{s}_a \\ \mathbf{s}_u \end{bmatrix} - \begin{bmatrix} \mathbf{J}_a \\ \mathbf{J}_u \end{bmatrix} \boldsymbol{\tau} = \mathbf{0}_6 \tag{6}$$

Given the under-actuated nature of the system, cable tensions can be algebraically eliminated from Eq. (6), leading to the system internal-dynamics equation:

$$\mathbf{A}\ddot{\mathbf{q}}_a + \mathbf{U}\ddot{\mathbf{q}}_u - \mathbf{s}' = \mathbf{0}_3 \tag{7}$$

where:

$$\mathbf{A} = \mathbf{M}_{ua} - \mathbf{J}_u\mathbf{J}_a^{-1}\mathbf{M}_{aa}, \qquad \mathbf{U} = \mathbf{M}_{uu} - \mathbf{J}_u\mathbf{J}_a^{-1}\mathbf{M}_{au}, \qquad \mathbf{s}' = \mathbf{s}_u - \mathbf{J}_u\mathbf{J}_a^{-1}\mathbf{s}_a \tag{8}$$

If $\mathbf{q}_a$ and its time derivatives are assigned, $\mathbf{q}_u$ can be determined by solving Eq. (7) with an assigned initial condition. In case the initial condition is a stable equilibrium pose, it may be evaluated as in [3]. Equation (7) may also be used to determine the *EE* oscillation natural frequencies around equilibrium configurations. To do so, $\dot{\mathbf{q}}_a$ and $\ddot{\mathbf{q}}_a$ can be isolated in Eq. (2) and the result substituted into Eq. (7), which can then be linearized about an equilibrium configuration $\mathbf{q}_0$:

$$\mathbf{M}_u(\mathbf{q}_0)\Delta\ddot{\mathbf{q}}_u + \mathbf{K}_u(\mathbf{q}_0)\Delta\mathbf{q}_u = \mathbf{0}_3 \tag{9}$$

where $\Delta\mathbf{q}_u = \mathbf{q}_u - \mathbf{q}_{u,0}$ is the deviation from the equilibrium of the unactuated variables, $\mathbf{M}_u(\mathbf{q}_0)$ and $\mathbf{K}_u(\mathbf{q}_0)$ are (3×3) symmetric positive-definite matrices (constant for an assigned $\mathbf{q}_0$) whose derivation can be found in [5]. Natural frequencies may be determined as the eigenvalues of matrix $\mathbf{M}_u(\mathbf{q}_0)^{-1}\mathbf{K}_u(\mathbf{q}_0)$, and it is clear that they change as $\mathbf{q}_0$ varies through the workspace (see Fig. 1b, corresponding to the path $\mathbf{q}_0(u)$ followed by the robot described in Sect. 4).

(a) 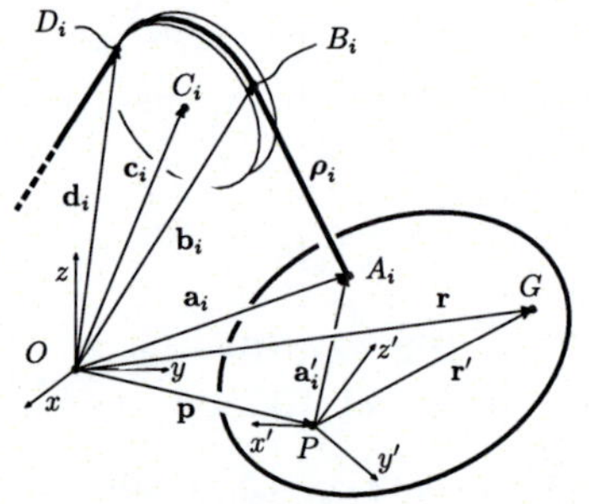

(b) 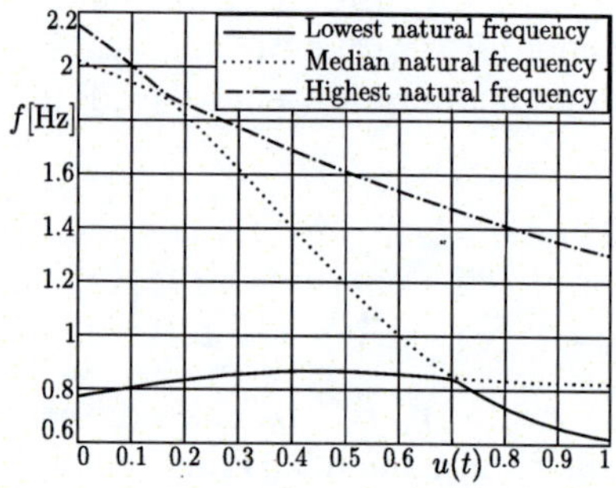

Fig. 1. *CDPR* properties: (a) Geometric model, (b) natural frequencies along a path

3 Trajectory Planning by Input Shaping and Dynamic Scaling

When planning a trajectory of duration T of the actuated coordinates $\mathbf{q}_a$, it may be convenient to separately design its geometric *path*, namely $\mathbf{q}_a = \mathbf{q}_a(u)$, and the *motion law* $u(t)$, with $u(0) = 0$ and $u(T) = 1$ [2]. If the path is assigned, we need to determine the motion law only. In the case of line-segment paths, which are the ones considered in this paper for the sake of simplicity, the trajectory connecting two set-points $\mathbf{q}_{a,s}$ and $\mathbf{q}_{a,f}$ is:

$$\mathbf{q}_a(t) = \mathbf{q}_{a,s} + (\mathbf{q}_{a,f} - \mathbf{q}_{a,s})u(t) \tag{10}$$

A method for the design of $u(t)$ allowing the residual oscillations of $\mathbf{q}_u$ to be eliminated, when both the motion time T and the path geometry are prescribed, was presented in [6], which suffers though from two drawbacks: the motion-law computation time is a priori unbounded (thus, it may be large), and it depends on the perfect knowledge of system parameters. Multi-Mode Zero-Vibration Input Shaping (*MMZVIS*) was instead used in [5] to shape actuated-coordinate trajectories; however, in order to reduce the delay time introduced by *IS*, a non-robust version of the latter was employed, leading to satisfactory results only in case of motions in a horizontal plane, where the natural frequencies of the robot at hand were only slightly variable.

IS is an approach dedicated to the reduction of oscillations of second-order linear dynamic systems, and it is well-known for its simplicity [7]. Its implementation requires the convolution of a series $S(t)$ of impulses, called the *input shaper*, with a reference signal. By denoting the convolution operator with $*$, the input shaping of the trajectory in Eq. (10) is given by:

$$\mathbf{q}_a(t) * S(t) = \mathbf{q}_{a,s} + (\mathbf{q}_{a,f} - \mathbf{q}_{a,s})\,(u(t) * S(t)), \quad S(t) = \sum_{i=1}^{k} A_i \delta_i(t - t_i) \tag{11}$$

where $\delta_i(t = t_i) = 1$, $\delta_i(t \neq t_i) = 0$, A_i is the impulse amplitude, t_i is the time at which it occurs, and k is the number of impulses. The time delay added by the *IS* is simply

$t_d = t_k$, the time location of the k-th impulse. The pairs (A_i, t_i) can be determined by setting to zero the amplitude $A_{\%}(f)$ of the Fourier transform of $S(t)$, for an assigned frequency f and k impulses (*Zero-Vibration* or *ZV IS* [7], Fig. 2a):

$$A_{\%}(f) = \sqrt{\left(\sum_{i=1}^{k} A_i \cos(2\pi f t_i)\right)^2 + \left(\sum_{i=1}^{k} A_i \sin(2\pi f t_i)\right)^2} = 0 \Longrightarrow (A_i, t_i) \qquad (12)$$

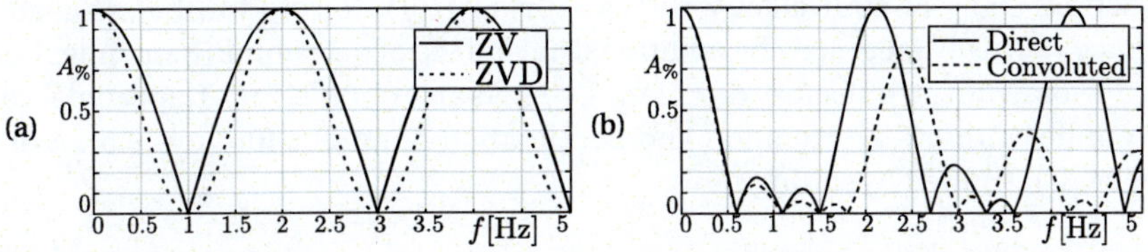

Fig. 2. (a) ZV and ZVD *IS* for $f = 1$ Hz, (b) Direct and Convoluted *IS* for $f_j = 0.6, 1.1, 1.5$ Hz

By definition, an input shaper is able to eliminate oscillations at the given frequency f (and some multiples of it), and to reduce the amplitude of oscillation associated with every other frequency (see to the graph ZV in Fig. 2a). If the shaper is supposed to eliminate the oscillations at μ frequencies f_j ($j = 1,\ldots,\mu$), two techniques can be employed, which result in a different number of impulses and time delays [8]. The so-called *direct method* always uses the minimum number of impulses $k_{dir} = 1 + \mu$, whereas the so-called *convolved method* leads to $k_{conv} = 2^{\mu}$. Usually, $t_{d,dir} < t_{d,conv} = \sum_{j=1}^{\mu} 1/(2f_j)$ but $A_{\%}$ is slightly higher when it is not zero [8] (see Fig. 2b for $f_j = 0.6, 1.1, 1.5$ Hz, $\mu = 3$). On the practical side, a convolved *IS* is easy to determine, because it results from the convolution of μ *ZV* shapers (each one with 2 impulses), which can be computed analytically, whereas a direct *IS* has to be numerically computed by imposing Eq. (12) to be satisfied simultaneously at $f_1,\ldots,f_{\mu}$, which requires $1+\mu$ impulses only (one impulse more for each additional frequency in the shaper). It should be noted that the minimum number of impulses of a direct shaper is always preferable for real-time implementation, since the amount of time required for the calculation of $u(t) * S(t)$, for each t, and thus trajectory computational complexity, increases with the number of impulses in the shaper.

The natural frequencies of a system are not always precisely known, due to either uncertainty in the dynamic-model parameters or a variation of its internal configuration (both cases occur to a robot). Loosely speaking, a trajectory is robust against natural-frequency uncertainty or variation if the amplitude spectrum of $\mathbf{q}_a(t) * S(t)$ is limited over some frequency range. In line-segment paths, for assigned set-points, only $u(t) * S(t)$ is responsible for the location of maximum and minimum values of this amplitude spectrum. If $U(f)$ is the amplitude of the Fourier trasform of $u(t)$, the amplitude spectrum of $u(t) * S(t)$ is $U(f)A_{\%}(f)$.

Limiting the amplitude of $U(f)A_{\%}(f)$ over a frequency range around a given frequency f_0 can be achieved by *Robust IS*, which uses additional impulses, and thus variables (A_i, t_i), to either set $A_{\%}(f_0) = \partial A_{\%}(f_0)/\partial f = ... = \partial^h A_{\%}(f_0)/\partial f^h = 0$ (*ZV-Derivative ZVD IS* [7], Fig. 2a), or set $A_{\%}(f) = 0$ in frequencies neighboring f_0 (*Extra-Insensitive EI IS* [9]). In either case, the amplitude of $U(f) * A_{\%}(f)$ is flattened, and thus limited, around f_0. The addition of impulses, though, augments the time delay of the shaper, thus increasing the trajectory duration, and its computational complexity. As an alternative way to obtain similar robustness results on $U(f)A_{\%}(f)$ without the need to modify the *IS*, we propose to use *Dynamic Scaling* (*DS*) of the reference trajectory [4]. An optimal value of T (as well as other trajectory parameters), for a fixed motion law profile (trapezoidal velocity, polynomial, etc.), can be determined by setting $U(f)$, or equivalently $\ddot{U}(f) = -(2\pi f)^2 U(f)$, to zero for the assigned frequency f. This strategy has the additional advantage of determining an upper bound for the total robust-trajectory duration, $T + t_d$. As an example, for a trapezoidal velocity profile, with αT acceleration and deceleration durations ($0 < \alpha \le 0.5$), α and T can be determined by setting:

$$\ddot{U}(f) = \frac{2}{(1-\alpha)\alpha(\pi T)^2 f} \| \sin[(1-\alpha)\pi f T] \sin[\alpha \pi f T] \| = 0 \tag{13}$$

By considering two frequencies f_1 and f_0, such that $f_1 \ge f_0$, and setting the arguments of the two sine functions to π, we obtain $\alpha_{opt} = f_0/(f_1 + f_0)$ and $T_{opt} = (f_1 + f_0)/(f_0 f_1)$, where T_{opt} is strictly decreasing with f_1. It is interesting to notice that, for $f_1 \to f_0$, $\alpha_{opt,s} = 0.5$ and $T_{opt,s} = 2/f_0$, while, for $f_1 \to +\infty$, $\alpha_{opt,l} = 0$ and $T_{opt,l} = 1/f_0$. This in turn means that the optimal time is bounded by the lowest frequency and, for any f, α is always well defined. Moreover, total robust trajectory duration is bounded by: $T_{opt} + t_d \le T_{opt,s} + t_{d,conv}$.

4 Numerical Example

To verify the effectiveness of the combination of *IS* and *DS*, we consider a linear trajectory of a 6-*DoF* 3-cable *CDPR* with a trapezoidal velocity motion law. Model parameters are as in [6] and are not reported here due to space limitation. Natural frequencies along the path between $\mathbf{q}_s = [0.793, 1.180, -0.208, 0.058, -0.641, 0.042]$

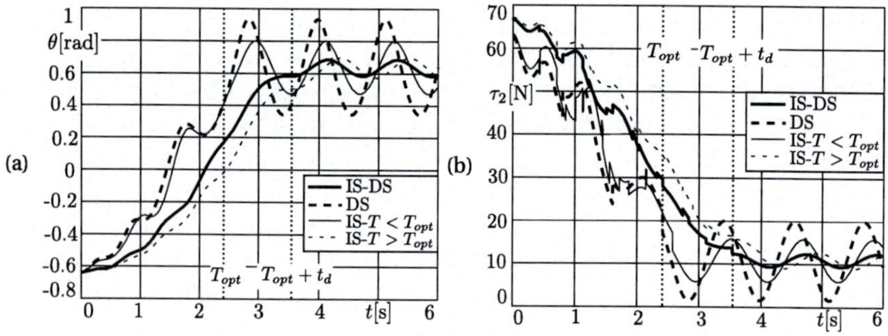

Fig. 3. Computed variables with critical oscillatory behaviors: (a) θ, (b) τ_2

and $\mathbf{q}_f = [-0.826, 1.104, -1.424, -0.041, 0.634, 0.002]$ (units in m and rad) vary as in Fig. 1b. The equilibrium value of τ_2 in $\mathbf{q}_f$ is very small, and potential *EE* oscillation in this configuration could lead to cable slackness. Since the system has 3 frequency spectra, an *IS* with $\mu = 3$ is used, as in [5], but a direct method (with 4 impulses) is employed, and the corresponding frequencies are heuristically selected as the minimum, maximum, and median ones from Fig. 1b, i.e. $f = 0.621, 1.247, 2.154\,\text{Hz}$. In addition, since $U(f)$ is decreasing with f and only 2 frequencies can be employed for the determination of α_{opt} and T_{opt}, the minimum and median ones are considered for *DS*. These choices lead to:

$$IS: \begin{array}{c} A_1 = A_4 = 0.2965,\ A_2 = A_3 = 0.2035 \\ t_1 = 0\,\text{s},\ t_1 = 0.42\,\text{s},\ t_3 = 0.705\,\text{s},\ t_4 = 1.125\,\text{s} \end{array} \qquad DS: \begin{array}{c} \alpha_{opt} = 0.332 \\ T_{opt} = 2.413\,\text{s} \end{array} \tag{14}$$

The parameters in Eq. (14) are used to compute trajectory in Eq. (11), where the actuated variables are selected as the position $\mathbf{p}$ of the *EE* reference point ($\mathbf{q}_{a,s}$ and $\mathbf{q}_{a,f}$ are then the first 3 elements of $\mathbf{q}_s$ and $\mathbf{q}_f$, respectively). Equation (7) is numerically solved with assigned initial conditions $\mathbf{q}_u = [0.058, -0.641, 0.042]\,\text{rad}$ and $\dot{\mathbf{q}}_u = [0,0,0]\,\text{rad/s}$, in order to determine the evolution of the unactuated variables along the assigned trajectory. This step allows one to assess oscillations, but also to check for cable slackness. In case these specifications are not satisfactorily met, an intuitive solution may be to robustify *IS*: oscillations are naturally reduced and the increase in time delay bounds cable tensions near their static equilibrium values. Most critical oscillatory variables are reported in Fig. 3 for four trajectories: the first one is planned according to *DS* only, the second one is an *IS* version of the first one (*IS-DS*), the third and fourth ones are *IS* versions of trapezoidal motion laws with $\alpha = 1/3$ (commonly employed value) and $T = 0.6T_{opt} < T_{opt}$ (*IS*-$T < T_{opt}$) and $T = 1.2T_{opt} > T_{opt}$ (*IS*-$T > T_{opt}$), respectively. It is clear that the use of *IS* leads to a more limited oscillatory behavior, and smaller oscillations occur for an increasing T (since the amplitude of $U(f)$ is decreasing with T). However, on the practical side, T should be as small as possible, and *IS* may not lead to satisfactory results if T is too small. Basically, *DS* helps in the choice of the best value of T (and α, for a trapezoidal velocity profile) that optimizes the transition time and limits the oscillatory behavior. Additionally, cable slackness due to oscillations is avoided with *IS-DS* (see Fig. 3b, where $\boldsymbol{\tau}$ is computed from Eq. (6)). In the end, if critically low oscillations are required, total transition time can always be increased by means of robust *IS*.

5 Conclusion

In this paper, we proposed the combined use of *IS* and *DS* for the robust trajectory planning of under-actuated *CDPRs*. Simulation results were presented for a 6-*DoF* 3-cable *CDPR* moving along a linear path with a trapezoidal velocity motion law. Results are satisfactory, as oscillations are appreciably reduced, when compared to a trajectory planned with *DS* only. In the future, smoother motion laws with a larger number of parameters and generic geometric paths will be considered, in order to avoid cable tension discontinuities.

Acknowledgements. The authors thank V. Mattioni for the help in preparing the illustrations.

References

1. Berti, A., Merlet, J.P., Carricato, M.: Solving the direct geometrico-static problem of underconstrained cable-driven parallel robots by interval analysis. Int. J. Robot. Res. **35**(6), 723–739 (2016)
2. Biagiotti, L., Melchiorri, C.: Trajectory planning for automatic machines and robots. Springer, Heidelberg (2008)
3. Carricato, M., Merlet, J.P.: Stability analysis of underconstrained cable-driven parallel robots. IEEE Trans. Robot. **29**(1), 288–296 (2013)
4. Hollerbach, J.M.: Dynamic scaling of manipulator trajectories. J. Dyn. Syst. Measur. Control **106**(1), 102–106 (1984)
5. Hwang, S.W., Bak, J.H., Yoon, J., Park, J.H.: Oscillation reduction and frequency analysis of under-constrained cable-driven parallel robot with three cables. Robotica 1–21 (2019)
6. Idà, E., Bruckmann, T., Carricato, M.: Rest-to-rest trajectory planning for underactuated cable-driven parallel robots. IEEE Trans. Robot. **35**(6), 1338–1351 (2019)
7. Singer, N.C., Seering, W.P.: Preshaping command inputs to reduce system vibration. J. Dyn. Syst. Measur. Control **112**(1), 76–82 (1990)
8. Singhose, W., Crain, E., Seering, W.: Convolved and simultaneous two-mode input shapers. IEE Proc. - Control Theory Appl. **144**(6), 515–520 (1997)
9. Singhose, W.E., Porter, L.J., Tuttle, T.D., Singer, N.C.: Vibration reduction using multi-hump input shapers. J. Dyn. Syst. Measur. Control **119**(2), 320–326 (1997)
10. Zarei, M., Aflakian, A., Kalhor, A., Masouleh, M.T.: Oscillation damping of nonlinear control systems based on the phase trajectory length concept: an experimental case study on a cable-driven parallel robot. Mech. Mach. Theory **126**, 377–396 (2018)

On the Plane Symmetric Bricard Mechanism

J. M. Selig(✉)

London South Bank University, London SE1 0AA, UK
seligjm@lsbu.ac.uk

Abstract. In this paper the motion of the plane symmetric Bricard 6R mechanism is studied. A simple method based on the dimensions of intersecting varieties in the Study quadric is used to show that the mechanism is mobile. The degree of the motion of the third link (the one adjacent to the plane of symmetry) relative to the plane of symmetry is found. The degree and genus of the motion of the third link relative to the first link is also found. This curve in the Study quadric is given as the intersection of the variety generated by the RR dyad formed by second and third joints with the variety of displacements that keep the fourth joint axis in the special linear line complex whose axis is the axis of the first joint. Finally, the motion of the symmetry plane when the second link is fixed is considered. The symmetry planes comprise the common tangent planes to a pair of circularly symmetric hyperboloids.

Keywords: Plane symmetry · Overconstrained mechanisms · Elliptic curves

1 Introduction

At the end of the 19th century Bricard discovered several types of mobile 6R mechanisms, [2]. In this work one of these types, the plane symmetric 6R, will be studied in some detail. Many other workers have looked at this mechanism previously, see for example [3] and [7] and references therein. The focus in this work is a little different. The idea is to study the mechanism using simple geometrical techniques and then use knowledge from Algebraic geometry to give very general results concerning the degree and genus of various curves associated to the mechanism.

The mechanism under consideration consists of six revolute joints arranged in a closed loop. The axes of the first and fourth joints lie in a plane Π, the fifth and sixth joint axes are the reflections in Π of the third and second joint axes respectively, see Fig. 1.

Notice that, with this arrangement the axes of the second and sixth joint will meet at a point on Π, as will the axes of the third and fifth joint. The line joining these two intersection point must lie in Π and hence will meet (or be coplanar with) the axes of first and fourth joints. Hence, all six joints lie in a special line complex and are thus linearly dependant.

J. Lenarčič and B. Siciliano (Eds.): ARK 2020, SPAR 15, pp. 73–81, 2021.
https://doi.org/10.1007/978-3-030-50975-0_10

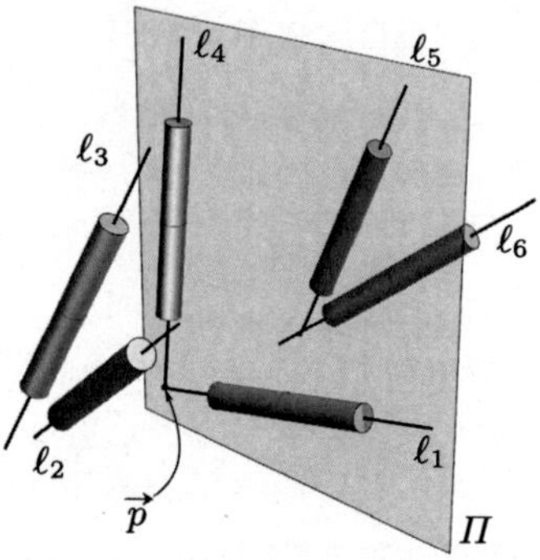

Fig. 1. The arrangement of joints in the plane symmetric bricard mechanism. Note that, for clarity, the links are not shown here.

2 Line in a Plane

The first task is to show that the mechanism is indeed mobile. Suppose we fix the plane Π and the axis of the first joint in Π. If the mechanism is mobile then the fourth joint will move but can only move in the plane Π. So we are led to consider the set of rigid-body displacements that can move a line in such a way that it remains in a fixed plane. This displacement subvariety will be a four dimensional subvariety of Study quadric, call it X for the moment. It may be parameterised as a rotation about the line, composed with planar displacements in Π. This can be thought of as a Segre variety $X = \mathbb{P}^1 \times \mathbb{P}^3$. To be specific, let Π be the xz-plane and assume that the fourth joint axis is the x-axis. That is, $\ell_4 = i$. The rigid-body motions that maintain ℓ_4 in the plane Π can be parameterised using dual quatrnions as,

$$
\begin{aligned}
g = (\alpha_0 + \alpha_1 j + \alpha_2 \varepsilon i + \alpha_3 \varepsilon k)(c + si) \\
= c\alpha_0 + s\alpha_0 i + c\alpha_1 j - s\alpha_1 k - s\alpha_2 \varepsilon + c\alpha_2 \varepsilon i + s\alpha_3 \varepsilon j + c\alpha_3 \varepsilon k
\end{aligned}
$$

This can be written as,

$$
\begin{aligned}
a_0 &= c\alpha_0, & c_0 &= -s\alpha_2, \\
a_1 &= s\alpha_0, & c_1 &= c\alpha_2, \\
a_2 &= c\alpha_1, & c_2 &= s\alpha_3, \\
a_3 &= -s\alpha_1, & c_3 &= c\alpha_3.
\end{aligned}
$$

In general, a $\mathbb{P}^3 \times \mathbb{P}^1$ Segre variety is defined by 6 degree 2 equations. These equations represent quadric hypersurfaces in $\mathbb{P}^7$ and can be expressed as requiring the following matrix to have rank 1,

$$
\text{Rank} \begin{pmatrix} a_0 & a_2 & c_1 & c_3 \\ a_1 & -a_3 & -c_0 & c_2 \end{pmatrix} = 1.
$$

The six quadrics are given by the vanishing of the 2×2 sub-determinants of the matrix,

$$\begin{aligned} Q_1 &= a_0a_3 + a_1a_2 = 0, & Q_4 &= a_2c_0 - a_3c_1 = 0, \\ Q_2 &= a_0c_0 + a_1c_1 = 0, & Q_5 &= a_2c_2 + a_3c_3 = 0, \\ Q_3 &= a_0c_2 - a_1c_3 = 0, & Q_6 &= c_1c_2 + c_0c_3 = 0. \end{aligned}$$

Notice that, the Study quadric lies in the linear system formed by these quadrics, $Q_S = Q_2 + Q_5$. The degree of this Segre variety is known to be $\binom{3+1}{1} = 4$, [6].

Intersecting this subvariety with the 3-dimensional subvariety $\mathbb{P}^1 \times \mathbb{P}^1 \times \mathbb{P}^1$ generated by the first three joints of the mechanism will produce a one-dimensional variety in the Study quadric. This curve represents a rigid-body motion that can be followed by the final link of the open loop chain formed from the mechanism's first three joints and links. Since the rigid motions keeps the axis of the fourth joint in the plane Π, the reflection of the first four joints in Π will produce the same motion of ℓ_4.

The above argument demonstrates the mobility of the mechanism since the intersection is a curve. The degree of the motion of the third link as a curve in the Study quadric can also be found. In [10] it was shown that the set of rigid-body displacements that maintain a given line in a fixed linear line complex is the intersection of the Study quadric with another quadric hypersurface in $\mathbb{P}^7$. Now consider a pair of lines ℓ_a and ℓ_b, both in Π and meeting at a point $\vec{p}$. The displacement variety of group elements that maintain ℓ_4 in the special linear line complex with axis ℓ_a will be denoted Q_a and similarly Q_b for the variety of displacements that maintain ℓ_4 in the special line complex determined by ℓ_b. Recall, that a special linear line complex is the set of all lines meeting or parallel to the axis of the complex. Now the intersection of these three quadrics $Q_S \cap Q_a \cap Q_b$, is clearly a four dimensional variety that contains the variety we are interested in $X \subset Q_S \cap Q_a \cap Q_b$. As a four-dimensional subvariety of $\mathbb{P}^7$, $Q_S \cap Q_a \cap Q_b$ is a complete intersection and hence will have degree 8. However, as a subvariety of the Study quadric it has homology class $4\sigma_4$, where σ_4 is the generator for homology in dimension 4, see [11]. Now X is only one component of this variety, there is another component. The other component of $Q_S \cap Q_a \cap Q_b$ consists of the group elements that maintain the incidence between ℓ_4 and $\vec{p}$, the intersection point of ℓ_a and ℓ_b. This component can also be seen to be isomorphic to the Segre variety $\mathbb{P}^3 \times \mathbb{P}^1$, call it Y say. So $Q_S \cap Q_a \cap Q_b = X \cup Y$, both X and Y have degree 4 and hence their homology class in Q_s must both be $2\sigma_4$.

Next look again at the first three joints, they form a $\mathbb{P}^1 \times \mathbb{P}^1 \times \mathbb{P}^1$ Segre variety. This variety has homology class $2\sigma_A + 4\sigma_B$ in the Study quadric, see [9]. So if we intersect the two varieties X and $\mathbb{P}^1 \times \mathbb{P}^1 \times \mathbb{P}^1$, the homology of the resulting curve will be,

$$2\sigma_4 \cap (2\sigma_A + 4\sigma_B) = (2 \times 2 + 2 \times 4)\sigma_1 = 12\sigma_1.$$

In dimension 1, the homology class σ_1 coincides with the degree and so we can conclude that the degree of this motion will be 12 in general.

There is a special case to consider. Suppose the first three joins of the mechanism, (and hence the final three also), have the design parameters of a Bennett linkage, see [9]. In this case the variety of displacements that the third link can perform is given by the complete intersection of the Study quadric with another quadric hypersurface and a $\mathbb{P}^5$. The homology of this subvariety is thus $2\sigma_A + 2\sigma_B$ and so the degree of the motion followed by the third link, relative to the symmetry plane is,

$$2\sigma_4 \cap (2\sigma_A + 2\sigma_B) = (2 \times 2 + 2 \times 2)\sigma_1 = 8\sigma_1.$$

That is, the motion is represented by a degree 8 curve in the Study quadric.

Finally here, notice that the varieties X and Y introduced above can be generated by mechanical linkages, a spherical joint and a prismatic joint for Y and a planar (E) joint linked to a revolute joint for X. For the last case here, the axis of the revolute joint must be parallel to the plane of the planar joint.

3 Fixed First Link

In the previous section the motion of the third link of the mechanism was found relative to the fixed plane Π. It is more usual to look at the relative motion of one link relative to another link, most commonly, the motion of a link relative to the opposite link in the loop. Here, the motion of the third link relative to the first will be investigated.

Fixing the first link means fixing the first and second joints. The plane Π can now rotate about the first joint. The fourth joint must still remain in Π but now as Π moves we can see that ℓ_4 will remain in the special line complex with axis ℓ_1. Note that this motion can be mechanically generated by a UPU linkage, see [8]. As mentioned in the previous section, the variety of rigid-body motions that keep a line in a fixed linear line complex is given by the intersection of the Study quadric with another quadric hypersurface.

On the other hand, the motion of the third link must also lie in the displacement subvariety generated by the second and third joints. This is an RR dyad and from [11], the group elements produced by such a linkage are given by the intersection of the Study quadric with a $\mathbb{P}^3$. So the motion of the third link relative to the first is given by the intersection of a pair of quadrics with a $\mathbb{P}^3$. Intersecting each quadric with the $\mathbb{P}^3$ gives a 2-dimensional quadric lying in the $\mathbb{P}^3$. The intersection of two quadrics in a $\mathbb{P}^3$ is, in general, an elliptic quartic curve, [6].

To investigate this more closely, the lines $\ell_1, \dots \ell_4$ will be written as dual quaternions as,

$$\ell_i = \omega_i + \varepsilon v_i = (\omega_{ix} i + \omega_{iy} j + \omega_{iz} k) + \varepsilon(v_{ix} i + v_{iy} j + v_{iz} k).$$

As 8-component vectors the lines will be written in partitioned form as,

$$\boldsymbol{\ell}_i = \begin{pmatrix} 0 \\ \vec{\omega}_i \\ 0 \\ \vec{v}_i \end{pmatrix}.$$

Now from [10], the quadric defining the displacement variety that keeps ℓ_4 in the special linear line complex defined by ℓ_1 can be written as,

$$\mathbf{g}^T Q \mathbf{g} = 0$$

where $\mathbf{g} = (a_0, a_1, a_2, a_3, c_0, c_1, c_2, c_3)^T$ is the column vector of homogeneous coordinates in $\mathbb{P}^7$. In partitioned form the 8×8 symmetric matrix Q can be written as,

$$Q = \begin{pmatrix} \Xi & \Upsilon \\ \Upsilon & 0 \end{pmatrix}$$

where,

$$\Xi = \begin{pmatrix} 0 & (\vec{\omega}_1 \times \vec{v}_4 + \vec{v}_1 \times \vec{\omega}_4)^T \\ (\vec{\omega}_1 \times \vec{v}_4 + \vec{v}_1 \times \vec{\omega}_4) & V_4\Omega_1 + \Omega_1 V_4 + \Omega_4 V_1 + V_1 \Omega_4 \end{pmatrix},$$

and

$$\Upsilon = \begin{pmatrix} 0 & (\vec{\omega}_1 \times \vec{\omega}_4)^T \\ (\vec{\omega}_1 \times \vec{\omega}_4) & \Omega_1\Omega_4 + \Omega_4\Omega_1 \end{pmatrix},$$

where, Ω_i is the 3×3 anti-symmetric matrix corresponding to the vector $\vec{\omega}_i$ and V_i corresponds to the vector $\vec{v}_i$, see [10].

The motion of the final link of the RR dyad can be parametrised by the sines and cosines of the joint angles

$$g = (\cos\frac{\theta_2}{2} + \sin\frac{\theta_2}{2}\ell_2)(\cos\frac{\theta_3}{2} + \sin\frac{\theta_3}{2}\ell_3).$$

If we abbreviate $\cos\frac{\theta_i}{2}$ to c_i and $\sin\frac{\theta_i}{2}$ to s_i, the above equation can be expanded to,

$$g = c_2c_3 + c_2s_3\ell_3 + s_2c_3\ell_2 + s_2s_3\ell_2\ell_3.$$

The parametrisation can be written in partitioned 8-vector form as,

$$\mathbf{g} = \begin{pmatrix} c_2c_3 - s_2s_3(\vec{\omega}_2 \cdot \vec{\omega}_3) \\ s_2c_3\vec{\omega}_2 + c_2s_3\vec{\omega}_3 + s_2s_3\vec{\omega}_2 \times \vec{\omega}_3 \\ s_2s_3(\vec{\omega}_2 \cdot \vec{v}_3 + \vec{v}_2 \cdot \vec{\omega}_3) \\ s_2c_3\vec{v}_2 + c_2s_3\vec{v}_3 + s_2s_2(\vec{\omega}_2 \times \vec{v}_3 + \vec{v}_2 \times \vec{\omega}_3) \end{pmatrix}$$

The variety determined by the parameterisation satisfies the equation for the Study quadric and the 3-plane mentioned above. So substituting this into the quadric above, $\mathbf{g}^T Q \mathbf{g} = 0$, will give an equation for the elliptic quartic curve:

$$\begin{aligned}(2 1^T Q\boldsymbol{\ell}_3 c_3 s_3 &+ \boldsymbol{\ell}_3^T Q \boldsymbol{\ell}_3 s_3^2)c_2^2 \\ &+2(\mathbf{1}^T Q\boldsymbol{\ell}_2 c_3^2 + (\mathbf{1}^T Q\boldsymbol{\ell}_2\boldsymbol{\ell}_3 + \boldsymbol{\ell}_2^T Q\boldsymbol{\ell}_3)c_3s_3 + \boldsymbol{\ell}_3^T Q\boldsymbol{\ell}_2\boldsymbol{\ell}_3 s_3^2)c_2s_2 \\ &(\boldsymbol{\ell}_2^T Q\boldsymbol{\ell}_2 c_3^2 + 2\boldsymbol{\ell}_2^T Q\boldsymbol{\ell}_2\boldsymbol{\ell}_3 c_3s_3 + (\boldsymbol{\ell}_2\boldsymbol{\ell}_3)^T Q\boldsymbol{\ell}_2\boldsymbol{\ell}_3 s_3^2)s_2^2 = 0,\end{aligned}$$

where $\mathbf{1}$ is the 8-vector corresponding to the dual quaternion 1. Notice that, $\mathbf{1}^T Q\mathbf{1} = 0$. Now if we divide this equation through by $s_2^2 s_3^2$ and then replace c_2/s_2 by λ and c_3/s_3 by μ, the above equation can be written in the form,

$$a(\mu)\lambda^2 + 2b(\mu)\lambda + c(\mu) = 0,$$

where

$$a(\mu) = 2\mathbf{1}^T Q\boldsymbol{\ell}_3\mu + \boldsymbol{\ell}_3^T Q\boldsymbol{\ell}_3,$$
$$b(\mu) = \mathbf{1}^T Q\boldsymbol{\ell}_2\mu^2 + (\mathbf{1}^T Q\boldsymbol{\ell}_2\boldsymbol{\ell}_3 + \boldsymbol{\ell}_2^T Q\boldsymbol{\ell}_3)\mu + \boldsymbol{\ell}_3^T Q\boldsymbol{\ell}_2\boldsymbol{\ell}_3,$$
$$c(\mu) = \boldsymbol{\ell}_2^T Q\boldsymbol{\ell}_2\mu^2 + 2\boldsymbol{\ell}_2^T Q\boldsymbol{\ell}_2\boldsymbol{\ell}_3\mu + (\boldsymbol{\ell}_2\boldsymbol{\ell}_3)^T Q\boldsymbol{\ell}_2\boldsymbol{\ell}_3.$$

In general, it is possible to solve for λ in terms of μ by using the familiar formula for solving quadratic equations. The appearance of a square root in the solution confirms that the curve is not rational in general. However, if the discriminant of the quadratic vanishes identically, then we do obtain a rational parameterisation of the curve. A rational quartic curve in $\mathbb{P}^3$ must have a singularity, hence the condition for the curve to be rational is the same as the condition for it to have a singularity. This is a very similar situation to the Grashof condition for planar 4-bar mechanisms, see [4]. The discriminant of the quadratic is a quartic in μ,

$$b(\mu)^2 - a(\mu)c(\mu) = \Delta_4\mu^4 + \Delta_3\mu^3 + \Delta_2\mu^2 + \Delta_1\mu + \Delta_0,$$

The coefficients are functions of the design parameters of the mechanism,

$$\Delta_4 = \left(\mathbf{1}^T Q\boldsymbol{\ell}_2\right)^2,$$

$$\Delta_3 = 2\left(\mathbf{1}^T Q\boldsymbol{\ell}_2\right)\left(\mathbf{1}^T Q\boldsymbol{\ell}_2\boldsymbol{\ell}_3\right) + 2\left(\mathbf{1}^T Q\boldsymbol{\ell}_2\right)\left(\boldsymbol{\ell}_2^T Q\boldsymbol{\ell}_3\right) - 2\left(\mathbf{1}^T Q\boldsymbol{\ell}_3\right)\left(\boldsymbol{\ell}_2^T Q\boldsymbol{\ell}_2\right),$$

$$\Delta_2 = 2\left(\mathbf{1}^T Q\boldsymbol{\ell}_2\right)\left(\boldsymbol{\ell}_3^T Q\boldsymbol{\ell}_2\boldsymbol{\ell}_3\right) + \left(\mathbf{1}^T Q\boldsymbol{\ell}_2\boldsymbol{\ell}_3\right)^2 + 2\left(\mathbf{1}^T Q\boldsymbol{\ell}_2\boldsymbol{\ell}_3\right)\left(\boldsymbol{\ell}_2^T Q\boldsymbol{\ell}_3\right)$$
$$+ \left(\boldsymbol{\ell}_2^T Q\boldsymbol{\ell}_3\right)^2 - 4\left(\mathbf{1}^T Q\boldsymbol{\ell}_3\right)\left(\boldsymbol{\ell}_2^T Q\boldsymbol{\ell}_2\boldsymbol{\ell}_3\right) - \left(\boldsymbol{\ell}_2^T Q\boldsymbol{\ell}_2\right)\left(\boldsymbol{\ell}_3^T Q\boldsymbol{\ell}_3\right),$$

$$\Delta_1 = 2\left(\mathbf{1}^T Q\boldsymbol{\ell}_2\boldsymbol{\ell}_3\right)\left(\boldsymbol{\ell}_3^T Q\boldsymbol{\ell}_2\boldsymbol{\ell}_3\right) + 2\left(\boldsymbol{\ell}_2^T Q\boldsymbol{\ell}_3\right)\left(\boldsymbol{\ell}_3^T Q\boldsymbol{\ell}_2\boldsymbol{\ell}_3\right)$$
$$- 2\left(\mathbf{1}^T Q\boldsymbol{\ell}_3\right)\left((\boldsymbol{\ell}_2\boldsymbol{\ell}_3)^T Q\boldsymbol{\ell}_2\boldsymbol{\ell}_3\right) - 2\left(\boldsymbol{\ell}_3^T Q\boldsymbol{\ell}_3\right)\left(\boldsymbol{\ell}_2^T Q\boldsymbol{\ell}_2\boldsymbol{\ell}_3\right),$$

$$\Delta_0 = \left(\boldsymbol{\ell}_3^T Q\boldsymbol{\ell}_2\boldsymbol{\ell}_3\right)^2 - \left(\boldsymbol{\ell}_3^T Q\boldsymbol{\ell}_3\right)\left((\boldsymbol{\ell}_2\boldsymbol{\ell}_3)^T Q\boldsymbol{\ell}_2\boldsymbol{\ell}_3\right).$$

Where, for example, $\mathbf{1}^T Q\boldsymbol{\ell}_2 = \vec{\omega}_2 \cdot (\vec{\omega}_1 \times \vec{v}_4) + \vec{\omega}_2 \cdot (\vec{v}_1 \times \vec{\omega}_4) + \vec{v}_2 \cdot (\vec{\omega}_1 \times \vec{\omega}_4)$. The geometric meaning of this term vanishing can be found by considering coordinates with origin at $\vec{p}$, the common point of the two lines ℓ_1 and ℓ_4 in Π. By choosing this point as origin both $\vec{v}_1 = \vec{v}_4 = \vec{0}$ and the condition becomes, $\vec{v}_2 \cdot (\vec{\omega}_1 \times \vec{\omega}_4) = 0$. Now, since $\vec{\omega}_1 \times \vec{\omega}_4$ is normal to Π we can see that the condition implies that the plane defined by the line ℓ_2 and $\vec{p}$ must be normal to Π.

Space precludes a full analysis of the five conditions above. Note, however, that singularity is a precondition for the elliptic quartic to degenerate into two or more rational components. The possible degenerations of this mechanism have been recently studied in [3] and [7].

A little more can be said about the case where the discriminant does not vanish identically. In general, the discriminant is a quartic in the variable μ. Such a quartic can have 0, 2 or 4 real roots. Solving for λ in terms of μ gives a map from the configuration curve of the mechanism to θ_2, or more precisely $\cot(\theta_2/2)$. Away from the roots of the discriminant, this map is clearly 2-to-1. Recalling that the configuration curve is an elliptic quartic, Harnack's theorem indicates that the curve can have up to 2 real ovals. That is, the curve can have one or two disjoint real components. In terms of the plane symmetric mechanism this means that it may have one or two disjoint modes of operation. Since the modes are disjoint, to get from one mode to the other the mechanism will have to be disassembled and reassembled in the other mode. Hence these are often refered to as assembly modes of the mechanism. This phenomenon occurs in planar and spherical 4-bar mechanism where both 1 assembly mode and 2 assembly mode mechanisms are possible, see [4]. If the discriminant has four real roots then the configuration curve must have two real ovals and moreover the second joint, directed along ℓ_2 will not rotate fully in either assembly mode, such a joint is usually called a rocker. When the discriminant has only 2 real roots the configuration curve has just one assembly mode and the second joint is still a rocker. If the discriminant doesn't have any real roots, then the second joint must be a crank, meaning that it is fully rotatable. However, in this case extra information is required to tell whether the curve has one or two assembly modes. More details on these ideas can be found in [5], but essentially the roots of the discriminant correspond to ramification, or branch points of the map from the configuration curve to the $\mathbb{P}^1$ determined by the second joint angle. Note also, the intersection of quadric surfaces in space is also of interest in Computer Graphics and CAD and the above analysis has also been considered in [12].

4 Second Link Fixed

Although the mechanism is plane symmetric so far the possibility that one of the links performs a plane symmetric motion has not been considered. Fixing the second link of the mechanism it is clear that fifth link will perform a plane symmetric motion, the reflection of the second link in Π, see [1, ch. IX, § 8]. In this section the motion of Π as the mechanism moves will be considered.

If the second link is fixed then the second and third joint axes are fixed. So the axes of the first joint ℓ_1 will rotate about the axis of the second joint to form a regulus of a one-sheeted hyperboloid of revolution. Likewise, ℓ_4 forms such a regulus about ℓ_3. Now, as the mechanism moves the symmetry plane Π will contain both joint axes ℓ_1 and ℓ_4. If a plane contains a line from a regulus of a hyperbolid then it must also contain a line from the other regulus on the hyperboloid, so that the intersection has degree 2. Hence, such a plane must be tangent to the hyperboloid at some point. So we can see that, as the mechanism moves the plane Π is always a common tangent to the two hyperboloids formed by the lines ℓ_1 and ℓ_4. To study planes in $\mathbb{P}^3$ it is usual to consider plane as point in a dual $\mathbb{P}^3$. The tangent planes to a quadric surface form a quadric surface in the dual $\mathbb{P}^3$. If the quadric surface is given by an equation of the form,

$$\vec{r}^T M \vec{r} = 0,$$

where M is a symmetric 4×4 matrix and $\vec{r}^T = (x, y, z, w)$ are the homogeneous coordinates of $\mathbb{P}^3$. Then the equation satisfied by the "points" in the dual $\mathbb{P}^3$ will be,

$$\left(\vec{q}^*\right)^T \mathrm{Adj}(M)\vec{q}^* = 0,$$

where $\vec{q}^*$ are the coordinates of a plane in the dual space and $\mathrm{Adj}(M)$; the adjugate matrix of M.

The common tangent planes to a pair of hyperboloids can be seen to be represented by the intersection of a pair of quadric surfaces in the dual $\mathbb{P}^3$. As above, this is, in general an elliptic quartic curve.

5 Conclusion

The above is of necessity, rather sketchy. Nevertheless the work contains some observations which would appear to be novel. In particular, the identification of the configuration curve of the mechanism as an elliptic quartic curve and the determination of the motion of the symmetry plane as an elliptic quartic in the dual $\mathbb{P}^3$. These ideas prompt several questions about these mechanisms; are there designs of plane symmetric Bricard 6Rs with 2 assembly modes? The 1-parameter family of planes determined by Π as the mechanism moves will be the envelope of a plane curve, what is the degree of this curve and how does it lie in relation to the mechanism?

References

1. Bottema, O., Roth, B.: Theoretical Kinematics. Dover Publications, New York (1990)
2. Bricard, R.: Mémoire sur la théorie de l'octaèdre articulé. Math. Pures Appl. **3**(3), 113–148 (1897)
3. Feng, H., Chen, Y., Dai, J.S., Gogu, G.: Kinematic study of the general plane-symmetric Bricard linkage and its bifurcation variations. Mech. Mach. Theory **116**, 89–104 (2017). https://doi.org/10.1016/j.mechmachtheory.2017.05.019
4. Gibson, C.G., Newstead, P.E.: On the geometry of the planar 4-bar mechanism. Acta Appl. Math. **7**, 113–135 (1986). https://doi.org/10.1007/BF00051348
5. Gibson, C.G., Marsh, D.: Concerning cranks and rockers. Mech. Mach. Theory **23**(5), 355–360 (1988). https://doi.org/10.1016/0094-114X(88)90049-3
6. Harris, J.: Algebraic Geometry a First Course. Springer, New York (1992)
7. López-Custodio, P.C., Dai, J., Rico, J.: Branch reconfiguration of Bricard linkages based on toroids intersections: Plane-symmetric case. Trans. ASME J. Mech. Robot. **10**(3), 031002. https://doi.org/10.1115/1.4039002
8. Selig, J.M.: Some mobile overconstrained parallel mechanisms. In: Lenarčič, J., Merlet, J.P. (eds.) Advances in Robot Kinematics. Springer Proceedings in Advanced Robotics, vol. 4, pp. 139–147. Springer, Cham (2016)
9. Selig, J.M.: Some remarks on the *RRR* linkage. In: Lenarčič, J., Khatib, O. (eds.) Advances in Robot Kinematics, pp. 77–85. Springer, Heidelberg (2014). https://doi.org/10.1007/978-3-319-06698-1_9

10. Selig, J.M.: On the geometry of the homogeneous representation for the group of proper rigid-body displacements. Romanian J. Tech. Sci. - Appl. Mech. (Spec. Issue New Trends Adv. Robot.) **58**(1–2), 27–50 (2013). http://www.imsar.ro/MEC-APPL_2013_1_2/2013_58_1_a1_Selig.pdf
11. Selig, J.M.: Geometric Fundamentals of Robotics. Springer, New York (2005)
12. Wang, W., Goldman, R., Tu, C.: Enhancing Levin's method for computing quadric-surface intersections. Comput. Aided Geom. Des. **20**(7), 401–422 (2003). https://doi.org/10.1016/S0167-8396(03)00081-5

Exact Coupler-Curve Synthesis of Four-Bar Linkages with Fully Analytical Solutions

Shaoping Bai[1(✉)], Rui Wu[2], and Ruiqin Li[2]

[1] Department of Materials and Production, Aalborg University, Aalborg, Denmark
shb@mp.aau.dk
[2] School of Mechanical Engineering, North University of China, Taiyuan, China

Abstract. Path synthesis from an algebraic coupler curve, or coupler-curve synthesis, is a classic problem, in which all linkage parameters are to be determined for a given special sextic polynomial. In previous works, it is shown that exact solutions exist for a given coupler curve, which can be found through a combined analytical and geometric approach. In this paper, a method is developed to get all solutions analytically. An example is included to demonstrate the new method.

Keywords: Four-bar linkages · Exact path synthesis · Coupler-curve synthesis · Polynomials

1 Introduction

Path synthesis is a classic kinematics problem, which is to find dimensions of a mechanism that is able to trace a path with a number of prescribed locations [1–5]. Traditionally, the path synthesis is carried out by solving a set of synthesis equations, which consists of a number of multi-variable higher-order polynomial equations, for which it is difficult to obtain solutions. Approach of polynomial continuation was applied to solve 9-point synthesis problem [6]. With the huge amount of possible solutions, identifying physically meaningful linkages becomes another problem. An alternative way for the path problem is to search a library that is pre-built based on kinematic analysis to match a curve [7]. This approach avoids the challenging equation solving. However, both the library building and path curve matching are time consuming. Solving the synthesis problem with optimization techniques has been extensively explored [8–10]. While optimization methods allow us to specify more points (10–20 or more) to describe a desired curve, solution convergence is not always guaranteed.

Another approach of path synthesis is to take advantage of coupler curve equations, a sextic bivariate polynomial with fifteen coefficients [11,12]. Once all coefficients are determined, a couple curve is uniquely obtained. While this approach sounds interesting, it was not really implemented. A major problem

J. Lenarčič and B. Siciliano (Eds.): ARK 2020, SPAR 15, pp. 82–89, 2021.
https://doi.org/10.1007/978-3-030-50975-0_11

of parameter determination was the overdeterminacy, namely, there are fifteen coefficient equations that can be established, while the problem has only nine unknowns. Until recently, a progress was made with this approach, as reported in [13,14], where the overdeterminacy was eliminated through a new formulation. A determined system of coefficient equations was derived, which allows us to find exact solutions. Furthermore, solutions were obtained with a method combining analytical and geometric approaches. With the method, three linkages, namely, Cognate linkages, were found for a given algebraic coupler curve.

In this paper, a fully analytical method is further developed. The method, as demonstrated by example, is able to find exact solutions of linkage parameters for a given couple curve equation. This is achieved by algebraic manipulation of a system of coefficient equations. Moreover, the method yields a small system of equations which is readily solved for all solutions, namely, solutions to three cognate linkages. Such a method is more efficient and able to apply to more synthesis cases.

2 Problem Formulation

We present briefly in this section the formulation of a determined system of coefficient equations. More details can be found in [14].

A four-bar linkage is illustrated in Fig. 1, whose coupler curve is Γ that is visited by a landmark point P of the body. The linkage is generally defined with nine parameters, $\mathbf{x} = [m, h, b_1, b_2, d_1, d_2, l_2, l_3, l_4]^T$, where the link lengths $l_2, l_3, l_4 > 0$. The length l_1 of the base link can be calculated from the coordinates of two points B and D.

The problem is to find the nine parameters, which allow the linkage tracing Γ given in the form of a sextic polynomial. This problem is essentially the inverse problem to the linkage analysis, in which the coupler curve of a known linkage is determined and analyzed.

The coupler-curve equation of four-bar linkages is given as following:

$$\begin{aligned} f(x,y) = K_1(x^2+y^2)^3 + (K_2x+K_3y)(x^2+y^2)^2 + (K_4x^2+K_5xy+K_6y^2)(x^2+y^2) \\ + K_7x^3 + K_8x^2y + K_9xy^2 + K_{10}y^3 + K_{11}x^2 + K_{12}xy + K_{13}y^2 \\ + K_{14}x + K_{15}y + K_{16} = 0 \end{aligned} \tag{1}$$

where coefficients $K_i, i = 1, \ldots, 16$ are functions of linkage parameters, for which the detailed expressions are not included due to space limit. By introducing $k_i = K_{i+1}/K_1, i = 1, \ldots, 15$, Eq. (1) becomes

$$\begin{aligned} f(x,y) &= (x^2+y^2)^3 + (k_1x+k_2y)(x^2+y^2)^2 + (k_3x^2+k_4xy+k_5y^2)(x^2+y^2) \\ &+ k_6x^3 + k_7x^2y + k_8xy^2 + k_9y^3 + k_{10}x^2 + k_{11}xy + k_{12}y^2 \\ &+ k_{13}x + k_{14}y + k_{15} = 0 \end{aligned} \tag{2}$$

which is a polynomial equation of exact degree 6. In total, we have 15 coefficients $\{k_i\}_1^{15}$, all functions of nine parameters.

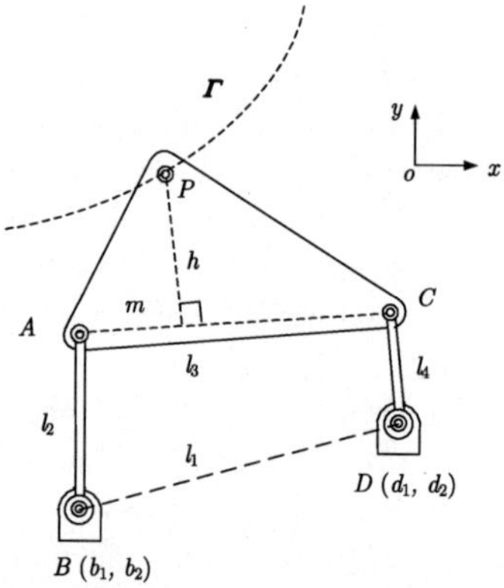

Fig. 1. A generic four-bar linkage with nine design parameters, including link lengths, l_2–l_4, the position of point P in the coupler link located by (m, h), and the positions of two joint centers $B(b_1, b_2)$ and $D(d_1, d_2)$

With any given algebraic equation of a four-bar linkage coupler curve, a system of coupler-curve coefficient (C^3) equations with all nine linkage parameters can be established

$$k_i(\mathbf{x}) - k_i^* = 0; \quad i = 1, \ldots, 15 \tag{3}$$

where $\{k_i^*\}_1^{15}$ are the known coefficients of the given coupler curve.

Equation (3), with more equations than variables, stands for an overdetermined system. However, it has shown in Bai's work that the system is determinant. A determinant system has been formulated, as described briefly here, which admits exact solutions.

Recall that $K_1 = -{l_3}^2/4$, we introduce a parameter $r = l_3/2$, which implies

$$k_i = -K_{i+1}(\mathbf{x})/r^2, \quad i = 1, \ldots, 15 \tag{4}$$

Substituting $l_3 = 2r$ into all coefficients k_i yields new expressions of coefficients in terms of l_2, l_4, m, h and the coordinates $(b_1, b_2), (d_1, d_2)$ and parameter r.

For the coefficient functions, they have the following properties:

Property 1 Coefficients k_1, k_2 and k_4 are independent of l_2 and l_4

Property 2 Coefficients k_3 and k_5 are linear in l_2^2 and l_4^2; moreover, the coefficient of l_2^2 in k_3 is identical to its coefficient in k_5. The same holds for the term l_4^2

Property 3 Coefficient pairs (k_6, k_8) and (k_7, k_9) have the same features as the pair (k_3, k_5)

Property 4 Coefficients $k_{10}, \ldots, k_{15}$ contain terms of l_2^4, $l_2^2 l_4^2$, l_4^4, l_2^2, l_4^2 only.

Upon these properties, a system of equations with a reduced number of unknowns, namely, a seven-dimensional array of design variables defined as $\mathbf{y} = [r, m, h, b_1, b_2, d_1, d_2]^T$, is established.

As k_1, k_2 and k_4 are independent of l_2 and l_4, their corresponding equations in (3) can be expressed in terms of variable $\mathbf{y}$ as

$$k_j(\mathbf{y}) - k_j^* = 0, \quad j = 1, 2, 4 \tag{5a}$$

Moreover, by subtracting one from the other in each pair of the coefficient equations of (k_3, k_5), (k_6, k_8) and (k_7, k_9), terms with l_2 as well as l_4 are cancelled, which leads to three new equations

$$k_i - k_j - (k_i^* - k_j^*) = 0, \quad (i, j) \in \{(3, 5), (6, 8), (7, 9)\} \tag{5b}$$

all independent of l_2 and l_4.

Lastly, six coefficient equations for $k_{10}, \ldots, k_{15}$, containing terms including l_2^4, $l_2^2 l_4^2$, l_4^4, l_2^2, l_4^2 only, are written as:

$$t_{i,1} l_2^4 + t_{i,2} l_2^2 l_4^2 + t_{i,3} l_4^4 + t_{i,4} l_2^2 + t_{i,5} l_4^2 + t_{i,6} = 0; \quad i = 1, \ldots, 6 \tag{5c}$$

which are cast in a compact form as

$$\mathbf{T}\mathbf{q} = \mathbf{0} \tag{5d}$$

where $\mathbf{q} = [l_2^4,\ l_2^2 l_4^2,\ l_4^4,\ l_2^2,\ l_4^2,\ 1]^T$, and $\mathbf{T}$ is a 6×6 matrix whose entries include the terms of Eq. (5c). Equation (5d) implies that

$$g(\mathbf{y}) = \det(\mathbf{T}) = 0 \tag{5e}$$

Equations (5a), (5b) and (5e) are in total seven equations, containing seven unknowns. They compose a determined system, which admits exact solutions.

3 Analytical Solution

The determined system of seven equations for seven unknowns in last section was solved utilizing both analytical and geometric methods [13, 14]. As we will find out, the system can be simplified to an even smaller system, namely, a system of four equations for four unknowns, which yields readily analytical solutions, as described presently. Equations (5a), (5b), after expansion and simplification, become

$$hb_2 - hd_2 - mb_1 + m\,d_1 + 4\,rb_1 + 2\,rd_1 + r\,k_1^* = 0 \tag{6a}$$
$$-hb_1 + hd_1 - mb_2 + md_2 + 4\,rb_2 + 2\,rd_2 + r\,k_2^* = 0 \tag{6b}$$
$$2\,hb_1{}^2 - 2\,hb_2{}^2 - 2\,hd_1{}^2 + 2\,hd_2{}^2 + 4\,mb_1b_2 - 4\,md_1d_2$$
$$-8\,rb_1b_2 - 8\,rb_1d_2 - 8\,rb_2d_1 + rk_4^* = 0 \tag{6c}$$
$$4\,hb_1b_2 - 4\,hd_1d_2 - 2\,mb_1{}^2 + 2\,mb_2{}^2 + 2\,md_1{}^2$$
$$-2\,md_2{}^2 + 4\,rb_1{}^2 + 8\,rb_1d_1 - 4\,rb_2{}^2 - 8\,rb_2d_2 - rk_3^* + rk_5^* = 0 \tag{6d}$$
$$-4\,hb_1{}^2d_2 - 8\,hb_1b_2d_1 + 8\,hb_1d_1d_2 + 4\,hb_2{}^2d_2 + 4\,hb_2d_1{}^2$$
$$-4\,hb_2d_2{}^2 + 4\,mb_1{}^2d_1 - 8\,mb_1b_2d_2 - 4\,mb_1d_1{}^2 + 4\,mb_1d_2{}^2$$
$$-4\,mb_2{}^2d_1 + 8\,mb_2d_1d_2 - 8\,rb_1{}^2d_1 + 16\,rb_1b_2d_2$$
$$+8\,rb_2{}^2d_1 - rk_6^* + rk_8^* = 0 \tag{6e}$$
$$4\,hb_1{}^2d_1 - 8\,hb_1b_2d_2 - 4\,hb_1d_1{}^2 + 4\,hb_1d_2{}^2 - 4\,hb_2{}^2d_1$$
$$+8\,hb_2d_1d_2 + 4\,mb_1{}^2d_2 + 8\,mb_1b_2d_1 - 8\,mb_1d_1d_2 - 4\,mb_2{}^2d_2$$
$$-4\,mb_2d_1{}^2 + 4\,mb_2d_2{}^2 - 8\,rb_1{}^2d_2 - 16\,rb_1b_2d_1$$
$$+8\,rb_2{}^2d_2 - rk_7^* + rk_9^* = 0 \tag{6f}$$

From Eqs. (6a) and (6b), we can find

$$m = \frac{r}{L_1}(k_1^*\,b_1 - k_1^*\,d_1 + k_2^*\,b_2 - k_2^*\,d_2$$
$$+ 4(b_1{}^2 + b_2{}^2) - 2(b_1d_1 + b_2d_2 + d_1{}^2 + d_2{}^2)) \tag{7}$$
$$h = -\frac{r}{L_1}(k_1^*\,b_2 - k_1^*\,d_2 - k_2^*\,b_1 + k_2^*\,d_1 - 6\,b_1d_2 + 6\,b_2d_1) \tag{8}$$

where

$$L_1 = l_1^2 = b_1{}^2 - 2\,b_1d_1 + b_2{}^2 - 2\,b_2d_2 + d_1{}^2 + d_2{}^2 \tag{9}$$

Substituting m and h into the remaining equations and simplifying,

$$2\,k_1^*\,b_2 + 2\,k_1^*\,d_2 + 2\,k_2^*\,b_1 + 2\,k_2^*\,d_1 + 8\,b_1b_2$$
$$+4\,b_1d_2 + 4\,b_2d_1 + 8\,d_1d_2 + k_4^* = 0 \tag{10a}$$
$$-2\,k_1^*\,b_1 - 2\,k_1^*\,d_1 + 2\,k_2^*\,b_2 + 2\,k_2^*\,d_2 - 4\,b_1{}^2 - 4\,b_1d_1 + 4\,b_2{}^2$$
$$+4\,b_2d_2 - 4\,d_1{}^2 + 4\,d_2{}^2 - k_3^* + k_5^* = 0 \tag{10b}$$
$$4\,k_1^*\,b_1d_1 - 4\,k_1^*\,b_2d_2 - 4\,k_2^*\,b_1d_2 - 4\,k_2^*\,b_2d_1 + 8\,b_1{}^2d_1 - 16\,b_1b_2d_2$$
$$+8\,b_1d_1{}^2 - 8\,b_1d_2{}^2 - 8\,b_2{}^2d_1 - 16\,b_2d_1d_2 - k_6^* + k_8^* = 0 \tag{10c}$$
$$4\,k_1^*\,b_1d_2 + 4\,k_1^*\,b_2d_1 + 4\,k_2^*\,b_1d_1 - 4\,k_2^*\,b_2d_2 + 8\,b_1{}^2d_2 + 16\,b_1b_2d_1$$
$$+16\,b_1d_1d_2 - 8\,b_2{}^2d_2 + 8\,b_2d_1{}^2 - 8\,b_2d_2{}^2 - k_7^* + k_9^* = 0 \tag{10d}$$

In the above simplified four equations, r is coincidentally cancelled out. The system of equations contains variables of b_1, b_2, d_1, and d_2 only. The equation can be further simplified as:

$$- 4\,k_1^*\,{b_1}^2 + 4\,k_1^*\,{b_2}^2 + 8\,k_2^*\,b_1 b_2 - 8\,{b_1}^3 + 24\,b_1{b_2}^2 - 2\,k_3^*\,b_1$$
$$+2\,k_4^*\,b_2 + 2\,k_5^*\,b_1 - k_6^* + k_8^* = 0 \qquad \text{(11a)}$$
$$-8\,k_1\,b_1 b_2 - 4\,k_2^*\,{b_1}^2 + 4\,k_2^*\,{b_2}^2 - 24\,{b_1}^2 b_2 + 8\,{b_2}^3 - 2\,k_3^*\,b_2$$
$$-2\,k_4^*\,b_1 + 2\,k_5^*\,b_2 - k_7^* + k_9^* = 0 \qquad \text{(11b)}$$
$$-4\,k_1^*\,{d_1}^2 + 4\,k_1^*\,{d_2}^2 + 8\,k_2^*\,d_1 d_2 - 8\,{d_1}^3 + 24\,d_1{d_2}^2 - 2\,k_3^*\,d_1$$
$$+2\,k_4^*\,d_2 + 2\,k_5^*\,d_1 - k_6^* + k_8^* = 0 \qquad \text{(11c)}$$
$$-8\,k_1^*\,d_1 d_2 - 4\,k_2^*\,{d_1}^2 + 4\,k_2^*\,{d_2}^2 - 24\,{d_1}^2 d_2 + 8\,{d_2}^3 - 2\,k_3^*\,d_2$$
$$-2\,k_4^*\,d_1 + 2\,k_5^*\,d_2 - k_7^* + k_9^* = 0 \qquad \text{(11d)}$$

The above equations stand for a system of four unknowns of the pivoting points. The first two equations are about the coordinates of point B, while the last two about point D. A close examination of the two groups of equations find that they are identical. This can be readily explained, as a four-bar linkage has three cognate linkages, which are built with three pivoting points on the grounded link. In this light, we need only to solve one set of equations, for example, Eq. (11a) and (11b). Three sets of real solutions will be found for b_1 and b_2. For each set of solutions, d_1 and d_2 can take values from two other sets correspondingly, thus ending up total six sets of real solutions.

Upon solutions of b_1, b_2, d_1, and d_2 obtained, by substituting their solutions, along with Eqs. (7) and (8), into Eq. (5e), a polynomial equation of r is obtained. Its positive solutions yield the solution of r. Upon the solutions of r and locations of points B and D, values of m and h are calculated.

With the solutions for $\mathbf{y}$, l_2 and l_4 can be uniquely determined due to Property 2 and 3. For example, we can take coefficient equations about k_5 and k_6 to find corresponding solutions for l_2 and l_4, thereby finding all unknowns.

4 Example

We include an example to demonstrate the developed method. The coupler curve is given in the form of a sextic as

$$\begin{aligned} f =& x^6 + 3\,x^4y^2 + 3\,x^2y^4 + y^6 + 0.1875\,x^5 + 0.5375\,x^4y + 0.375\,x^3y^2 + 1.075\,x^2y^3 + 0.1875\,xy^4 \\ &+ 0.5375\,y^5 - 0.125859\,x^4 + 0.16625\,x^3y - 0.017968\,x^2y^2 + 0.16625\,xy^3 + 0.10789\,y^4 \\ &- 0.010304\,x^3 - 0.007859\,x^2y + 0.006695\,xy^2 - 0.035859\,y^3 + 0.0112623\,x^2 \\ &+ 0.006071\,xy - 0.007966\,y^2 + 0.00255779\,x + 0.00141746\,y + 0.00017165 = 0 \end{aligned} \qquad (12)$$

$$0.017 + 0.4675\,b_1 + 0.3325\,b_2 + 0.75\,{b_2}^2 - 8.0\,{b_1}^3 - 0.75\,{b_1}^2$$
$$+24.0\,{b_2}^2 b_1 + 4.3\,b_1 b_2 = 0 \qquad \text{(13a)}$$
$$-0.028 - 0.3325\,b_1 + 0.4675\,b_2 + 2.15\,{b_2}^2 + 8.0\,{b_2}^3 - 2.15\,{b_1}^2$$
$$-24.0\,{b_1}^2 b_2 - 1.5\,b_1 b_2 = 0 \qquad \text{(13b)}$$

Table 1. Linkage synthesis results for the example (unit: m)

No	$[b_1, b_2]$	$[d_1, d_2]$	r	m	h	l_2	l_3	l_4	Note
1	[−0.25, −0.1]	[0.2, −0.2]	0.2	0.15	0.15	0.15	0.4	0.35	M1
2	[−0.25, −0.1]	[−0.0437, 0.0312]	0.0397	0.1060	−0.1060	0.2121	0.0795	0.1856	M3
3	[0.2, −0.2]	[−0.25, −0.1]	0.2	0.25	−0.15	0.35	0.4	0.15	M1
4	[0.2, −0.2]	[−0.0437, 0.0312]	0.1275	0.3001	0.18	0.2915	0.2551	0.1093	M2
5	[−0.0437, 0.03125]	[0.2, −0.20]	0.1275	−0.045	−0.18	0.1093	0.2551	0.2915	M2
6	[−0.0437, 0.03125]	[−0.25, −0.10]	0.0397	−0.0265	0.106	0.1856	0.0795	0.2121	M3

Using the foregoing method, a small system of two equations for two unknowns was obtained for point B, as given in Eqs. (13a)–(13b). The system admits 9 solutions in total. Of them, only three solutions are real. While they stand for the coordinates of pivoting point B, they are also the coordinates of D, as discussed before. The coordinates of point D are thus simply given by assigning these values to point D as well, ending up with six solutions in total. From the pivoting point coordinates, other unknowns are found accordingly.

The entire linkage parameters obtained for this example are listed in Table 1. The six solutions stand for three linkages, noted by M1 to M3, which are cognate linkages of the given couple curve. There are two solutions for each linkage, representing two modes of assembly. Figure 2 displays the three linkages, along with the coupler curve. In the figure, we show only the linkages in one branch.

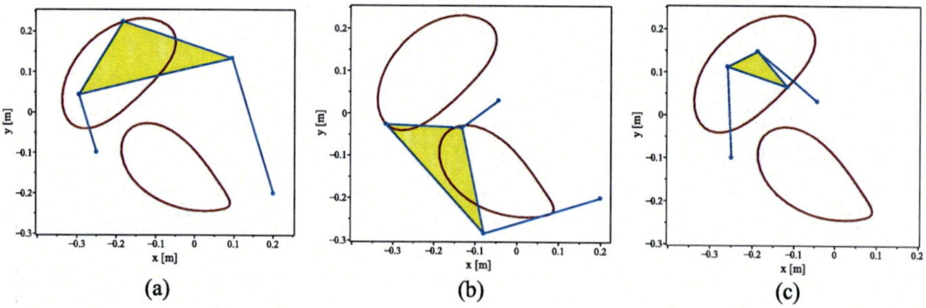

Fig. 2. Three cognate linkages showing together with the coupler curve, (a) Linkage M1, (b) Linkage M2, (c) Linkage M3

5 Conclusions

The paper reports a progress in exact path synthesis problem. A method to find all parameters analytically is developed on the basis of a determined system of coefficient equations obtained from previous works. A small system of four equations of four variables was derived, which allows parameter determination directly and efficiently.

This work, combined with the work reported in [13, 14], solves a fundamental problem in linkage synthesis, namely, exact path synthesis from a given algebraic curve. The solution has several implications on both continuous and discrete path synthesis, for example, the nine-point discrete path synthesis problem, among others.

References

1. Burmester, L.: Lehrbuch der Kinematik. Arthur Felix-Verlag, Leipzig (1888)
2. Bottema, O., Roth, B.: Theoretical Kinematics. North-Holland Pub. Co., New York (1979)
3. Subbian, T., Flugrad Jr., D.R.: Four-bar path generation synthesis by a continuation method. ASME J. Mech. Des. **113**(1), 63–69 (1991)
4. Sun, J., Liu, W., Chu, J.: Dimensional synthesis of open path generator of four-bar mechanisms using the Haar Wavelet. ASME J. Mech. Des. **137**(8) (2015)
5. Bai, S.: Geometric analysis of coupler-link mobility and circuits for planar four-bar linkages. Mech. Mach. Theory **118**, 53–64 (2017)
6. Wampler, C.W., Morgan, A.P., Sommese, A.J.: Numerical continuation methods for solving polynomial systems arising in kinematics. ASME J. Mech. Des. **112**(1), 59–68 (1990)
7. Kota, S.: Automatic selection of mechanism designs from a three-dimensional design map. J. Mech. Des. **114**(2), 359–367 (1992)
8. Sancibrian, R., Viadero, F., García, P., Fernández, A.: Gradient-based optimization of path synthesis problems in planar mechanisms. Mech. Mach. Theory **39**(8), 839–856 (2004)
9. Cabrera, J.A., Simon, A., Prado, M.: Optimal synthesis of mechanisms with genetic algorithms. Mech. Mach. Theory **37**(10), 1165–1177 (2002)
10. Bulatović, R., Djordjevic, S.: Optimal synthesis of a four-bar linkage by method of controlled deviation. Theor. Appl. Mech. **31**, 265–280 (2004)
11. Blechschmidt Jr., J.L., Uicker, J.J.: Linkage synthesis using algebraic curves. ASME J. of Mech. Des. **108**(4), 543–548 (1986)
12. Ananthasuresh,G.K., Kota, S.: Synthesis of four-bar linkages for path generation via the coupler curve equation. In: Proceedings of Third National Applied Mechanisms & Robotics Conference, Cincinnati, OH, USA, 8–10 November 1993. paper 83
13. Bai, S.: Determination of Linkage Parameters from Coupler Curve Equations, pp. 49–57. Springer International Publishing, Cham (2015)
14. Bai, S., Angeles, J.: Coupler-curve synthesis of four-bar linkages via a novel formulation. Mech. Mach. Theory **94**, 177–187 (2015)

Check for updates

A General Method for Determining Algebraic Input-Output Equations for Planar and Spherical 4R Linkages

Mirja Rotzoll[1](✉), M. John D. Hayes[1], Manfred L. Husty[2], and Martin Pfurner[2]

[1] Department of Mechanical and Aerospace Engineering, Carleton University, Ottawa, ON, Canada
mirjarotzoll@cmail.carleton.ca

[2] University of Innsbruck, Unit Geometry and Surveying, Innsbruck, Austria

Abstract. A new and completely general method for determining the algebraic input-output (IO) equations for planar and spherical 4R linkages is presented in this paper. First, the forward kinematics transformation matrix of an arbitrary planar or spherical open 4R kinematic chain is computed in terms of its Denavit-Hartenberg parameters, where the link twist and joint angles are converted to their tangent half-angle parameters. This transformation matrix is mapped to its corresponding eight Study coordinates. The serial kinematic chain is conceptually closed by equating the forward kinematics transformation to the identity matrix. Equating the two corresponding Study arrays yields four equations in terms of the four revolute joint angle parameters. Gröbner bases are then used to eliminate the two intermediate joint angle parameters leaving an algebraic polynomial in terms of the input and output joint angle parameters and the four twist angle or link length parameters. In the limit, as the sphere radius becomes infinite and the link twist angle parameters are expressed as ratios of arc length and sphere radius in the general spherical algebraic IO equation, the only terms that remain are those in the planar 4R IO equation.

Keywords: Algebraic input-output equation · Planar and spherical four-bar linkage · Study coordinates · Kinematic mapping

1 Introduction

Four-bar linkages, consisting of four rigid bodies connected by revolute (R) joints have fascinated mathematicians and engineers for centuries. One of the greatest successes was the establishment of an input-output (IO) equation by F. Freudenstein, which correlates the driver input angle ψ to the follower output angle φ according to a function $\varphi = f(\psi)$ [5]. The IO equation developed in [5] is trigonometric, whereas in [6] an algebraic version is derived by mapping the constraint equations of the driver and follower into Study's kinematic image space [1, 11].

J. Lenarčič and B. Siciliano (Eds.): ARK 2020, SPAR 15, pp. 90–97, 2021.
https://doi.org/10.1007/978-3-030-50975-0_12

Let a, b, c, d be the link lengths of the four-bar mechanism, and ψ and φ the respective input and output angles, then the movement of the mechanism is governed by the following *IO* equation

$$Au^2v^2 + Bu^2 + Cv^2 - 8abuv + D = 0 \tag{1}$$

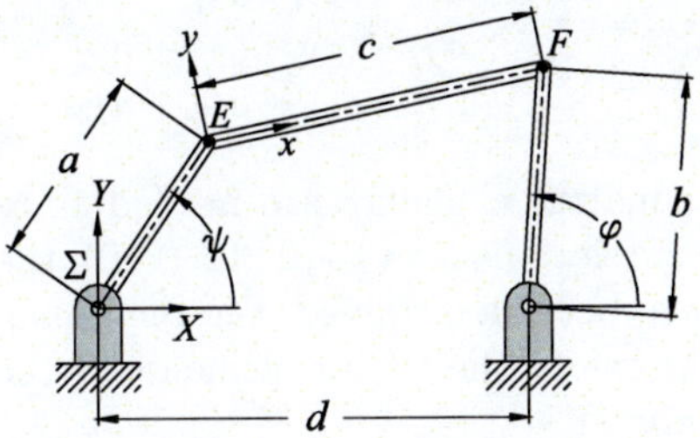

Fig. 1. Planar 4R function generator.

where

$$\begin{aligned}
A &= (a-b-c+d)(a-b+c+d),\\
B &= (a+b-c+d)(a+b+c+d),\\
C &= (a+b-c-d)(a+b+c-d),\\
D &= (a-b+c-d)(a-b-c-d),\\
u &= \tan\frac{\psi}{2},\\
v &= \tan\frac{\varphi}{2}.
\end{aligned}$$

In addition, it was shown in [10] that Eq. (1) is not only valid for planar four-bar linkages containing revolute joints, but also for planar four-bar linkages containing up to two prismatic joints. In this paper we will describe a method that can be applied to both planar and spherical four-bar linkages to derive the respective *IO* equations. Moreover, we will show that planar linkages can be interpreted as special cases of spherical linkages, thus, expanding the generality of the *IO* equation obtained in [6].

2 Study's Kinematic Mapping

Consider a coordinate system Σ_2 that moves with a rigid body relative to a stationary reference frame Σ_1. The Euclidean displacement group $\mathscr{D} \in SO(3)$ can be represented by

$$\mathbf{p}' = \mathbf{A}\mathbf{p} + \mathbf{t} \tag{2}$$

where $\mathbf{p}$ is a 3×1 position vector in Σ_2, $\mathbf{A}$ is a proper orthogonal 3×3 rotation matrix, $\mathbf{t}$ is a 3×1 position vector of the origin of Σ_2 expressed in Σ_1, and $\mathbf{p}'$ is the 3×1 position vector of $\mathbf{p}$ expressed in Σ_1 [7, 9].

Displacements of kinematic chains are often parametrised using the Denavit-Hartenberg (DH) convention [3]. The four associated DH parameters are the link lengths a_i, link twist angles τ_i, joint angles θ_i, and link offsets d_i. According to this convention the coordinate transformation from the coordinate system for joint i relative to the coordinate system of the previous joint $i-1$ is given by

$$ {}^{i-1}_{\ \ i}\mathbf{T} = \left[\begin{array}{ccc|c} \cos\theta_i & -\sin\theta_i\cos\tau_i & \sin\theta_i\sin\tau_i & a_i\cos\theta_i \\ \sin\theta_i & \cos\theta_i\cos\tau_i & -\cos\theta_i\sin\tau_i & a_i\sin\theta_i \\ 0 & \sin\tau_i & \cos\tau_i & d_i \\ \hline 0 & 0 & 0 & 1 \end{array}\right] = \left[\begin{array}{ccc|c} & \mathbf{A} & & \mathbf{t} \\ & & & \\ \hline 0 & 0 & 0 & 1 \end{array}\right]. \tag{3}$$

With Study's kinematic mapping distinct Euclidean displacements can be represented as distinct points $x = [x_0 : x_1 : x_2 : x_3 : y_0 : y_1 : y_2 : y_3]^{\mathrm{T}} \in P^7$ where the first four entries are obtained using the matrix elements a_{ij} of $\mathbf{A}$

$$x_0 : x_1 : x_2 : x_3 = \begin{cases} 1 + a_{11} + a_{22} + a_{33} : a_{32} - a_{23} : a_{13} - a_{31} : a_{21} - a_{12}, \\ a_{32} - a_{23} : 1 + a_{11} - a_{22} - a_{33} : a_{12} + a_{21} : a_{31} + a_{13}, \\ a_{13} - a_{31} : a_{12} + a_{21} : 1 - a_{11} + a_{22} - a_{33} : a_{23} + a_{32}, \\ a_{21} - a_{12} : a_{31} + a_{13} : a_{23} + a_{32} : 1 - a_{11} - a_{22} + a_{33}. \end{cases} \quad (4)$$

Four different combinations of the rotation matrix elements are needed since certain displacements make one or more of the relations lead to $(0 : 0 : 0 : 0)$, the exceptional generator in P^7. Once the x_i have been determined, the remaining four entries are computed as linear combinations of the vector elements of the translation $\mathbf{t}$ and the x_i determined above, giving

$$\begin{aligned} y_0 &= \tfrac{1}{2}(t_3x_3 + t_2x_2 + t_1x_1), & y_1 &= \tfrac{1}{2}(t_3x_2 - t_2x_3 - t_1x_0), \\ y_2 &= \tfrac{1}{2}(-t_3x_1 + t_1x_3 - t_2x_0), & y_3 &= \tfrac{1}{2}(-t_3x_0 + t_2x_1 - t_1x_2). \end{aligned} \quad (5)$$

These eight Study parameters must fulfill the Study condition in order to represent a Euclidean displacement, meaning the eight ratios represent a point on Study's seven dimensional quadric $\mathscr{S}$

$$x_0y_0 + x_1y_1 + x_2y_2 + x_3y_3 = 0 \quad (6)$$

excluding the exceptional generator $\mathscr{E}$

$$(x_0 : x_1 : x_2 : x_3) = (0 : 0 : 0 : 0). \quad (7)$$

3 Planar Four-Bar Linkage

To derive the algebraic *IO* equation for planar four-bar mechanisms using the DH convention [3] and Study's kinematic mapping [11], we first consider the four-bar mechanism to be an open kinematic chain connected by four rotational joints as shown in Fig. 2. The respective DH parameters are listed in Table 1. Note that for planar mechanisms all link twists and all link offsets are identically zero. This simplifies the overall transformation matrix ${}^0_4\mathbf{T}$, which maps the coordinates of points described in the end-link coordinate frame to those of the base frame:

$${}^0_4\mathbf{T} = {}^0_1\mathbf{T}\,{}^1_2\mathbf{T}\,{}^2_3\mathbf{T}\,{}^3_4\mathbf{T}, \quad (8)$$

where the transformation matrices ${}^0_1\mathbf{T}$, ${}^1_2\mathbf{T}$, ${}^2_3\mathbf{T}$ and ${}^3_4\mathbf{T}$ are evaluated according to Eq. (3). The computed transformation matrix can be mapped onto Study's quadric using Eqs. (4, 5) resulting in a Study array with zero entries for x_1, x_2, y_0 and y_3. After normalizing, the remaining four Study parameters become

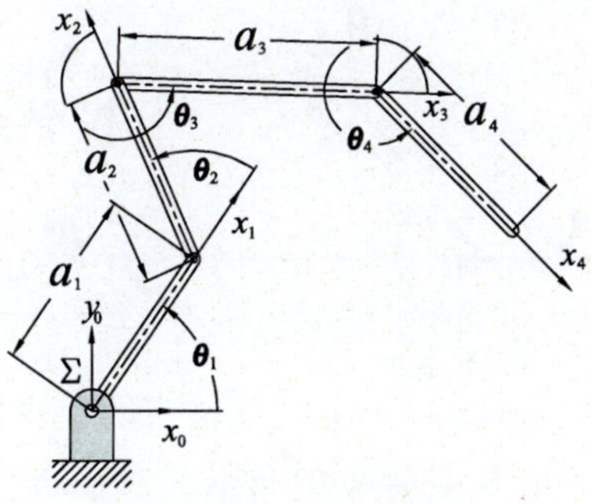

Fig. 2. Open 4R chain.

Table 1. DH parameters for open 4R chain.

Joint axis i	Link length a_i	Link angle θ_i	Link offset d_i	Link twist τ_i
1	a_1	θ_1	0	0
2	a_2	θ_2	0	0
3	a_3	θ_3	0	0
4	a_4	θ_4	0	0

$$x_0 = (2v_2v_3v_4 - 2v_2 - 2v_3 - 2v_4)v_1 + (-2v_3 - 2v_4)v_2 - 2v_3v_4 + 2, \tag{9}$$
$$x_3 = ((-2v_3 - 2v_4)v_2 - 2v_3v_4 + 2)v_1 - 2v_2v_3v_4 + 2v_2 + 2v_3 + 2v_4, \tag{10}$$
$$\begin{aligned} y_1 = {} & ((v_4(a_1 - a_2 + a_3 - a_4)v_3 - a_1 + a_2 + a_3 + a_4)v_2 + (-a_1 - a_2 + a_3 + a_4)v_3 \\ & -v_4(a_1 + a_2 + a_3 - a_4))v_1 + ((a_1 - a_2 + a_3 + a_4)v_3 + v_4(a_1 - a_2 - a_3 + a_4))v_2 \\ & +v_4(a_1 + a_2 - a_3 + a_4)v_3 - a_1 - a_2 - a_3 - a_4, \end{aligned} \tag{11}$$
$$\begin{aligned} y_2 = {} & (((a_1 - a_2 + a_3 + a_4)v_3 + v_4(a_1 - a_2 - a_3 + a_4))v_2 + v_4(a_1 + a_2 - a_3 + a_4)v_3 \\ & -a_1 - a_2 - a_3 - a_4)v_1 + (-v_4(a_1 - a_2 + a_3 - a_4)v_3 + a_1 - a_2 - a_3 - a_4)v_2 \\ & +(a_1 + a_2 - a_3 - a_4)v_3 + v_4(a_1 + a_2 + a_3 - a_4), \end{aligned} \tag{12}$$

where $v_i = \tan(\theta_i/2)$.

To close the four-bar mechanism, ${}^0_4\mathbf{T}$ is equated to the identity matrix which we also map using Eqs. (4, 5) onto Study's quadric, resulting in the following Study parameters after normalising:

$$I \mapsto [1:0:0:0:0:0:0:0]^\mathsf{T} \in P^7 \tag{13}$$

Equating the Study array of the overall transformation ${}^0_4\mathbf{T}$ to the Study array of the identity matrix, i.e. setting Eqs. (10–12) equal to zero, forces the coordinate frame of the end-effector to align with that of the base; but to satisfy the DH convention they are both rotated by π. Thus, the joint angles, θ_i, are measured as illustrated in Fig. 3. We select the three Equations that are equal to zero, (10–12), and manipulate them with Gröbner bases to eliminate the intermediate

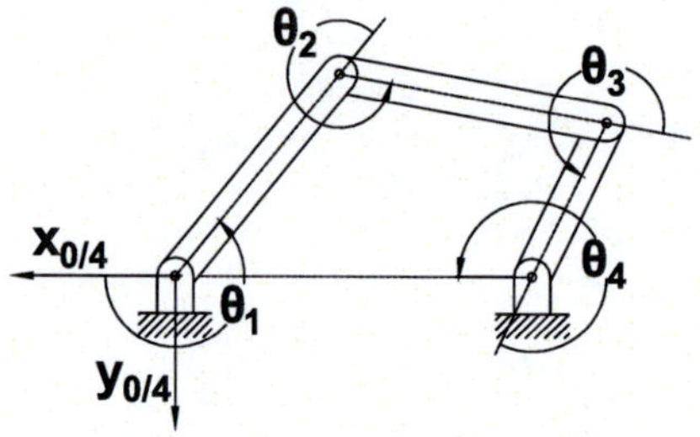

Fig. 3. Closed 4R kinematic chain.

joint angle parameters, v_2 and v_3. After collecting the input and output angle parameters v_1 and v_4, the following algebraic *IO* equation emerges

$$Av_1^2v_4^2 + Bv_1^2 + Cv_4^2 - 8a_1a_3v_1v_4 + D = 0, \tag{14}$$

where

$$\begin{aligned}
A = (a_1 - a_2 + a_3 - a_4)(a_1 + a_2 + a_3 - a_4) &= A_1A_2,\\
B = (a_1 + a_2 - a_3 - a_4)(a_1 - a_2 - a_3 - a_4) &= B_1B_2,\\
C = (a_1 - a_2 - a_3 + a_4)(a_1 + a_2 - a_3 + a_4) &= C_1C_2,\\
D = (a_1 + a_2 + a_3 + a_4)(a_1 - a_2 + a_3 + a_4) &= D_1D_2.
\end{aligned}$$

It can be shown that Eq. (14) is identical to Eq. (1) if the phase shift of the input and output angle as well as the different notation are considered.

4 Spherical Four-Bar Linkage

It will now be demonstrated that the same procedure can be applied to determine the *IO* equation for spherical linkages. The DH parameters for a spherical open 4R kinematic chain are listed in Table 2. Note that in the spherical case all link lengths, a_i, and offsets, d_i, are zero with strict adherence to the DH conventions for assigning parameters [3]. After evaluating the overall transformation matrix in terms of DH parameters by applying Eq. (3), the result can be mapped with Eqs. (4, 5) onto Study's quadric. Then setting $v_i = \tan(\theta_i/2)$ and $\alpha_i = \tan(\tau_i/2)$

Table 2. Open spherical 4R kinematic chain DH parameters.

Joint axis i	Link length a_i	Link angle θ_i	Link offset d_i	Link twist τ_i
1	0	θ_1	0	τ_1
2	0	θ_2	0	τ_2
3	0	θ_3	0	τ_3
4	0	θ_4	0	τ_4

into the result gives a Study array with non-zero entries for x_0, x_1, x_2 and x_3, while the y_i are all identically zero, as expected:

$$\begin{aligned}
x_0 = {} & ((2\alpha_4((v_2v_3v_4 + v_2 - v_3 + v_4)v_1 + (v_3 - v_4)v_2 + v_3v_4 + 1)\alpha_3 \\
& +(2v_2v_3v_4 - 2v_2 + 2v_3 + 2v_4)v_1 + (-2v_3 - 2v_4)v_2 + 2v_3v_4 - 2)\alpha_2 \\
& +((-2v_2v_3v_4 - 2v_2 - 2v_3 + 2v_4)v_1 + (2v_3 - 2v_4)v_2 - 2v_3v_4 - 2)\alpha_3 \\
& +(2((v_2v_3v_4 - v_2 - v_3 - v_4)v_1 + (v_3 + v_4)v_2 + v_3v_4 - 1))\alpha_4)\alpha_1 \\
& +(((2v_2v_3v_4 + 2v_2 - 2v_3 + 2v_4)v_1 + (-2v_3 + 2v_4)v_2 - 2v_3v_4 - 2)\alpha_3 \\
& -(2((v_2v_3v_4 - v_2 + v_3 + v_4)v_1 + (v_3 + v_4)v_2 - v_3v_4 + 1))\alpha_4)\alpha_2 \\
& +(2((v_2v_3v_4 + v_2 + v_3 - v_4)v_1 + (v_3 - v_4)v_2 - v_3v_4 - 1))\alpha_4\alpha_3 \\
& +(2v_2v_3v_4 - 2v_2 - 2v_3 - 2v_4)v_1 + (-2v_3 - 2v_4)v_2 - 2v_3v_4 + 2; \qquad (15)\\
x_1 = {} & ((((-2v_2v_3v_4 - 2v_2 + 2v_3 - 2v_4)v_1 + (-2v_3 + 2v_4)v_2 - 2v_3v_4 - 2)\alpha_3 \\
& +(2((v_2v_3v_4 - v_2 + v_3 + v_4)v_1 + (-v_3 - v_4)v_2 + v_3v_4 - 1))\alpha_4)\alpha_2 \\
& -2\alpha_4((v_2v_3v_4 + v_2 + v_3 - v_4)v_1 + (-v_3 + v_4)v_2 + v_3v_4 + 1)\alpha_3 \\
& +(-2v_2v_3v_4 + 2v_2 + 2v_3 + 2v_4)v_1 + (-2v_3 - 2v_4)v_2 - 2v_3v_4 + 2)\alpha_1 \\
& +((2((v_2v_3v_4 + v_2 - v_3 + v_4)v_1 + (-v_3 + v_4)v_2 - v_3v_4 - 1))\alpha_4\alpha_3 \\
& +(2v_2v_3v_4 - 2v_2 + 2v_3 + 2v_4)v_1 + (2v_3 + 2v_4)v_2 - 2v_3v_4 + 2)\alpha_2 \\
& +((-2v_2v_3v_4 - 2v_2 - 2v_3 + 2v_4)v_1 + (-2v_3 + 2v_4)v_2 + 2v_3v_4 + 2)\alpha_3 \\
& +(2((v_2v_3v_4 - v_2 - v_3 - v_4)v_1 + (-v_3 - v_4)v_2 - v_3v_4 + 1))\alpha_4; \qquad (16)\\
x_2 = {} & (((((-2v_3 + 2v_4)v_2 - 2v_3v_4 - 2)v_1 + 2v_2v_3v_4 + 2v_2 - 2v_3 + 2v_4)\alpha_3 \\
& -(2(((v_3 + v_4)v_2 - v_3v_4 + 1)v_1 + v_2v_3v_4 - v_2 + v_3 + v_4))\alpha_4)\alpha_2 \\
& +(2(((v_3 - v_4)v_2 - v_3v_4 - 1)v_1 + v_2v_3v_4 + v_2 + v_3 - v_4))\alpha_4\alpha_3 \\
& +((-2v_3 - 2v_4)v_2 - 2v_3v_4 + 2)v_1 + 2v_2v_3v_4 - 2v_2 - 2v_3 - 2v_4)\alpha_1 \\
& +(-(2(((v_3 - v_4)v_2 + v_3v_4 + 1)v_1 + v_2v_3v_4 + v_2 - v_3 + v_4))\alpha_4\alpha_3 \\
& +((2v_3 + 2v_4)v_2 - 2v_3v_4 + 2)v_1 - 2v_2v_3v_4 + 2v_2 - 2v_3 - 2v_4)\alpha_2 \\
& +(((-2v_3 + 2v_4)v_2 + 2v_3v_4 + 2)v_1 + 2v_2v_3v_4 + 2v_2 + 2v_3 - 2v_4)\alpha_3 \\
& -(2(((v_3 + v_4)v_2 + v_3v_4 - 1)v_1 + v_2v_3v_4 - v_2 - v_3 - v_4))\alpha_4; \qquad (17)\\
x_3 = {} & (((2(((v_3 - v_4)v_2 + v_3v_4 + 1)v_1 - v_2v_3v_4 - v_2 + v_3 - v_4))\alpha_4\alpha_3 \\
& +((-2v_3 - 2v_4)v_2 + 2v_3v_4 - 2)v_1 - 2v_2v_3v_4 + 2v_2 - 2v_3 - 2v_4)\alpha_2 \\
& +(((2v_3 - 2v_4)v_2 - 2v_3v_4 - 2)v_1 + 2v_2v_3v_4 + 2v_2 + 2v_3 - 2v_4)\alpha_3 \\
& +2\alpha_4(((v_3 + v_4)v_2 + v_3v_4 - 1)v_1 - v_2v_3v_4 + v_2 + v_3 + v_4))\alpha_1 \\
& +((((-2v_3 + 2v_4)v_2 - 2v_3v_4 - 2)v_1 - 2v_2v_3v_4 - 2v_2 + 2v_3 - 2v_4)\alpha_3 \\
& -(2(((v_3 + v_4)v_2 - v_3v_4 + 1)v_1 - v_2v_3v_4 + v_2 - v_3 - v_4))\alpha_4)\alpha_2 \\
& +2\alpha_4(((v_3 - v_4)v_2 - v_3v_4 - 1)v_1 - v_2v_3v_4 - v_2 - v_3 + v_4)\alpha_3 \\
& +((-2v_3 - 2v_4)v_2 - 2v_3v_4 + 2)v_1 - 2v_2v_3v_4 + 2v_2 + 2v_3 + 2v_4. \qquad (18)
\end{aligned}$$

Again, the open kinematic chain is closed by equating the Study array to the corresponding identity array in Study coordinates, i.e. setting Eqs. (16–18) equal to zero. Subsequently, we use Gröbner bases to eliminate the intermediate angle parameters v_2 and v_3 from Eqs. (16–18), and obtain the desired *IO* equation

$$Av_1^2v_4^2 + Bv_1^2 + Cv_4^2 + 8\alpha_1\alpha_3\left(\alpha_4^2+1\right)\left(\alpha_2^2+1\right)v_1v_4 + D = 0, \tag{19}$$

where

$$\begin{aligned}
A &= (\alpha_1\alpha_2\alpha_3 - \alpha_1\alpha_2\alpha_4 + \alpha_1\alpha_3\alpha_4 - \alpha_2\alpha_3\alpha_4 + \alpha_1 - \alpha_2 + \alpha_3 - \alpha_4)\\
&\quad (\alpha_1\alpha_2\alpha_3 - \alpha_1\alpha_2\alpha_4 - \alpha_1\alpha_3\alpha_4 - \alpha_2\alpha_3\alpha_4 - \alpha_1 - \alpha_2 - \alpha_3 + \alpha_4)\,,\\
B &= (\alpha_1\alpha_2\alpha_3 + \alpha_1\alpha_2\alpha_4 - \alpha_1\alpha_3\alpha_4 - \alpha_2\alpha_3\alpha_4 + \alpha_1 + \alpha_2 - \alpha_3 - \alpha_4)\\
&\quad (\alpha_1\alpha_2\alpha_3 + \alpha_1\alpha_2\alpha_4 + \alpha_1\alpha_3\alpha_4 - \alpha_2\alpha_3\alpha_4 - \alpha_1 + \alpha_2 + \alpha_3 + \alpha_4)\,,\\
C &= (\alpha_1\alpha_2\alpha_3 - \alpha_1\alpha_2\alpha_4 - \alpha_1\alpha_3\alpha_4 + \alpha_2\alpha_3\alpha_4 - \alpha_1 + \alpha_2 + \alpha_3 - \alpha_4)\\
&\quad (\alpha_1\alpha_2\alpha_3 - \alpha_1\alpha_2\alpha_4 + \alpha_1\alpha_3\alpha_4 + \alpha_2\alpha_3\alpha_4 + \alpha_1 + \alpha_2 - \alpha_3 + \alpha_4)\,,\\
D &= (\alpha_1\alpha_2\alpha_3 + \alpha_1\alpha_2\alpha_4 + \alpha_1\alpha_3\alpha_4 + \alpha_2\alpha_3\alpha_4 - \alpha_1 - \alpha_2 - \alpha_3 - \alpha_4)\\
&\quad (\alpha_1\alpha_2\alpha_3 + \alpha_1\alpha_2\alpha_4 - \alpha_1\alpha_3\alpha_4 + \alpha_2\alpha_3\alpha_4 + \alpha_1 - \alpha_2 + \alpha_3 + \alpha_4)\,.
\end{aligned}$$

It can be shown that Eq. (19) is identical to the corresponding trigonometric *IO* equation for spherical four-bar linkages found in [9].

5 Planar 4R Linkages as a Special Case of the Spherical 4R Linkage

The two algebraic *IO* equations for planar and spherical 4R linkages already suggest some similarities. As demonstrated in [8], the motion of the planar 4R linkage represents a special case of the spherical 4R linkage. To show that the same relationship is true for the *IO* equations, we consider the directions of the joint axes. While the joint axes of the spherical 4R linkage intersect in the centre of the sphere, the joint axes of the planar 4R linkage are all parallel. In Euclidean space E^3 parallel lines never intersect, however, they do meet in a point at infinity in any projective extension of E^3 [2,4]. This suggests that if the radius of a spherical linkage approaches infinity, the linkage becomes a planar mechanism in the limit [8]. As the link twist parameters α_i of the spherical *IO* equation are proportional to the ratios of the arc lengths to the sphere radius [12], we can make the following substitution in Eq. (19)

$$\alpha_i \propto \frac{a_i}{r}. \tag{20}$$

In the resulting equation the first two cubic factors simplify to

$$\begin{aligned}
\lim_{r\to\infty} -\frac{1}{r}&\left(\frac{a_1a_2a_3}{r^2} - \frac{a_1a_2a_4}{r^2} + \frac{a_1a_3a_4}{r^2} - \frac{a_2a_3a_4}{r^2} + a_1 - a_2 + a_3 - a_4\right)\\
&\left(-\frac{a_1a_2a_3}{r^2} + \frac{a_1a_2a_4}{r^2} + \frac{a_1a_3a_4}{r^2} + \frac{a_2a_3a_4}{r^2} + a_1 + a_2 + a_3 - a_4\right).
\end{aligned} \tag{21}$$

In the limit the only terms remaining inside the parentheses in Eq. (21) are

$$(a_1 - a_2 + a_3 - a_4)(a_1 + a_2 + a_3 - a_4) = A_1 A_2. \tag{22}$$

Proceeding with the other cubic factors in the same manner the algebraic *IO* equation for a spherical 4R mechanism leads directly to that of a planar 4R, Eq. (14), in the limit. As mentioned, this aligns with the results from [8], and further confirms the validity of the derived *IO* equations as well as the observation in [9] that there exists a connection between the planar and the spherical 4R *IO* equations via the RSSR linkage.

6 Conclusions

We have successfully demonstrated a general method to derive the algebraic *IO* equations for spherical and planar 4R linkages. It requires defining the DH parameters for an open 4R kinematic chain, mapping its coordinate transformation matrix onto Study's quadric, conceptually closing the 4R chain by equating the corresponding Study coordinates to the identity array and eliminating the intermediate joint angles using Gröbner bases. Moreover, we have shown that the planar 4R *IO* equation represents a special case of the spherical 4R by evaluating the limit at infinity of the equation.

References

1. Bottema, O., Roth, B.: Theoretical Kinematics. Dover Publications Inc., New York (1990)
2. Coxeter, H.S.M.: Non-Euclidean Geometry, 5th edn. University of Toronto Press, Toronto (1965)
3. Denavit, J., Hartenberg, R.S.: A kinematic notation for lower-pair mechanisms based on matrices. Trans ASME J. Appl. Mech. **23**, 215–221 (1955)
4. Fishback, W.T.: Projective and Euclidean Geometry, 2nd edn. Wiley, New York (1969)
5. Freudenstein, F.: Design of four-link mechanisms. Ph.D. thesis, Columbia University, New York (1954)
6. Hayes, M.J.D., Husty, M.L., Pfurner, M.: Input-output equation for planar four-bar linkages. In: International Symposium on Advances in Robot Kinematics, pp. 12–19. Springer (2018)
7. Husty, M.L., Karger, A., Sachs, H., Steinhilper, W.: Kinematik und Robotik. Springer, Heidelberg (1997)
8. McCarthy, J.M.: Planar and spatial rigid motion as special cases of spherical and 3-spherical motion. J. Mech. Transm. Autom. Des. **105**(3), 569–575 (1983)
9. McCarthy, J.M., Soh, G.S.: Geometric Design of Linkages. Interdiciplinaty Applied Mathematics, 2nd edn., pp. 155–161, pp. 267–269. Springer, New York (2011)
10. Rotzoll, M., Hayes, M.J.D., Husty, M.L.: An algebraic input-output equation for planar RRRP and PRRP linkages. Can. Comm. Theory Mach. Mech. (2019)
11. Study, E.: Geometrie der Dynamen. Teubner Verlag, Leipzig (1903)
12. Uicker, J.J., Pennock, G.R., Shigley, J.E.: Theory of Machines and Mechanisms, 5th edn. Oxford University Press, New York (2017)

Check for updates

The Forward Kinematics of the 4-1 Cable-Driven Parallel Robot with Non Elastic Sagging Cables

J-P. Merlet(✉)

INRIA Sophia-Antipolis, 2004 Route des Lucioles, Valbonne, France
Jean-Pierre.Merlet@inria.fr

Abstract. We consider a 3 dof translational cable-driven parallel robot (CDPR) with 4 cables connected at the same point on the load. We assume that the CDPR has a small or medium size and uses synthetic cables so that the elasticity of the cable material is negligible. We will investigate the forward kinematics (FK) of such a CDPR (i.e. determining the CDPR pose for given cable lengths) with the purpose of evaluating the influence of the sagging of the cables and showing that the FK has, in general, a single solution. We then consider other FK problems where only cable tensions or angles are measured and show their advantages and drawbacks. We then consider the FK problem where we fuse the measurements of the cable lengths with direct measurements of the pose of the platform provided at a low rate.

Keywords: Cable-driven parallel robot · Cable sagging · Forward kinematics

1 Introduction

Cable-driven parallel robots (CDPR) are a special class of parallel robot where the rigid legs/links of a classical parallel robot (CPR) are substituted by cables that can be wound or unwound. This type of actuation offers a lower mechanical complexity (passive joints may not be used) compared to CPR and, more importantly, a wider range of leg lengths, thereby allowing for a larger workspace. But this actuation is unilateral: a cable may pull a load but cannot push it so that kinematic analysis has to be mixed with static analysis to check if a cable is really acting on the platform. Furthermore assuming that cables have the same shape that a rigid leg and are inelastic(i.e. assuming that cables are *ideal*) is a rough approximation of the cable behavior especially if we have cables of large lengths. In this paper we will consider the forward kinematics (FK) (i.e. determining the pose of the CDPR for given cable lengths) of the so-called N–1 CDPR having N cables that are all attached to the same point B on the platform, whose center of mass is below this point. Provided that $N \geq 3$, such a CDPR is a 3-dof robot providing only translational motion of the load, which is appropriate for many

J. Lenarčič and B. Siciliano (Eds.): ARK 2020, SPAR 15, pp. 98–108, 2021.
https://doi.org/10.1007/978-3-030-50975-0_13

tasks [2,5], e.g. for 3D printing [4] or metrology [20]. One of our prototype of such a CDPR is shown in Fig. 1: it has been used during the whole 2019 summer for printing a wall of glass powder over an area of 20 × 9 m for an artistic exhibition. During this summer the robot has covered a distance of 5 km and has dispersed 1.5 ton of powder. Let us define a reference frame $O, \mathbf{x}, \mathbf{y}, \mathbf{z}$ where $\mathbf{z}$ is the vertical axis. If A_i denotes the winch output point of cable i, then the motion of the CDPR in the horizontal plane is restricted to lie in the convex hull of the projection of the A_i points in the plane (otherwise at least one of the cables will have to push the platform outside of the hull). One main use of a N–1 robot is to have a workspace that covers a rectangular room. In that case it makes sense to have the A_i located at the corner of the room and to use 4 cables: this is the 4-1 CDPR that we are interested in, with the additional assumption that all the A_is have the same altitude. In our examples we will use the following coordinates for the A_i: $A_1(-350, 200, 300), A_2(-350, -200, 300), A_3(350, -200, 300), A_4(350, 200, 300)$.

Solving the FK of CDPR is difficult even if ideal cables are assumed [1] and even more complex with more realistic model [18]. We assume that the cables have a small to moderate length and are made of synthetic material such as UHMWPE, which allow us to neglect the elasticity. Many complex cable models have been proposed [9,15,16,19] but given our assumption we may rely on the cable model presented in [10] that allows one to take into account the effect of the cable mass and has been experimentally validated on CDPR [17]. This model assumes that the cable lies in a vertical plane with a reference frame whose origin will be located at A, $\mathbf{x_r}$ being the horizontal and $\mathbf{z_r}$ being the vertical (Fig. 1). If x_b, z_b denote the coordinate of B in this frame we have

$$x_b = F_x\left(\frac{sinh^{-1}(F_z) - sinh^{-1}(F_z - \mu g \frac{L_0}{F_x})}{\mu g}\right) \qquad z_b = \frac{\sqrt{F_x^2 + F_z^2} - \sqrt{F_x^2 + (F_z - \mu g L_0)^2}}{\mu g} \tag{1}$$

where $F_x > 0, F_z$ are respectively the horizontal and vertical components of the force exerted on the cable at B, L_0 the cable length at rest and μ the linear density of the cable material (assumed to be the same for all cables). If F_z is negative, then the cable is partly supporting the weight of the platform while if F_z is positive, then the cable act as an additional weight.

We will denote by α the angle between the cable plane and the $\mathbf{x} - \mathbf{z}$ vertical plane. If x_a, y_a denote the x, y coordinate of A in the reference frame and x, y, z the coordinates of B in the same frame, then the angle α should satisfy the constraint:

$$(x - x_a)\sin\alpha + (y - y_a)\cos\alpha = 0 \tag{2}$$

which express that A, B lie in the same vertical plane. We will assume that the platform is only submitted to gravity an we consider the static equilibrium. If m denotes the platform mass we have

$$\sum_{j=1}^{j=4}\cos(\alpha_i)F_{xi} = 0 \quad \sum_{j=1}^{j=4}\sin(\alpha_i)F_{xi} = 0 \quad \sum_{j=1}^{j=4}F_{zi} + mg = 0 \tag{3}$$

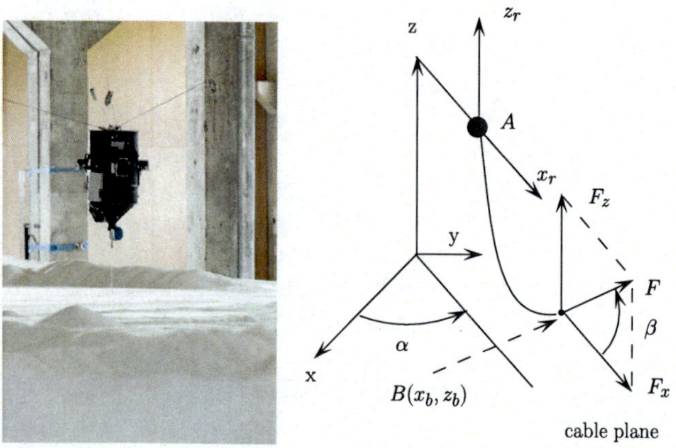

Fig. 1. An example of $4-1$ CDPR, the *Prince's tears: vitrifications*, that was used for printing the wall of glass powder for an artistic exhibition. The drum has an height of 82 cm and a radius of 35 cm was filled with 40 kg of glass powder. On the right notation for the sagging cable

Note that the 4-1 CDPR is redundantly actuated with 4 actuators for 3 dof and a companion paper will address the inverse kinematics of this CDPR while in this paper we will focus on its forward kinematics.

2 Forward Kinematics

The purpose of the forward kinematics (FK) is to determine the coordinates (x, y, z) of B in the reference frame being given the lengths L_0 of the cables. For a $N-1$ CDPR we have as unknowns the F_x, F_z, α ($3N$ unknowns) and the three x, y, z for a total of $3N + 3$ unknowns. In terms of constraints we have the $2N$ equations (1), the N constraints (2) and the 3 equilibrium conditions (3) for a total of $3N+3$ constraints. Hence solving the forward kinematics require to solve a square system, which in general will admit only a finite number of solutions. But before going into the solving of the FK we will investigate its maximum number of solution(s).

We will start our investigation by assuming that the cables are ideal. If we assume that all 4 cables under tension, then B is located at the apex of a pyramid with 4 facets. There are 2 possible locations for B but one is above the A plane and cannot be reached by a cable system. Let us assume now that one FK solution B_1 is such that a triplet of cables is under tension (leading to a single pose) while the remaining cable is slack. The question is now: is there a second FK solution $B_2 \neq B_1$ with the same cable lengths but with another triplet of cables under tension? Let us assume that the projection of B_1 in the $A_1A_2A_3A_4$ horizontal plane lies in the triangle 123 with cables 123 under tension and cable 4 is slack. Let B_{13} be the pose of the platform that will be obtained if only 1,3 were under tension and L_2 be the distance between B_{13} and A_2. As we have assumed that B_1 lies in the triangle 123, then the length of cable 2 must be shorter than the length L_2. If the second FK solution B_2 was lying in the triangle 134, then the length of cable 2 must be larger than L_2, which is contradictory and therefore B_2 cannot lie in the triangle 134.

The only possible place for B_2 is therefore either in the intersection of triangles 123 and 124, with cables 124 under tension and 3 slack or in the intersection of triangles 123 and 234, with cables 234 under tension and 1 being slack.

In the first case as 1 and 2 are under tension the point B moves on a vertical circle whose center lies between A_1 and A_2. The pose B_1 lies on this circle with 3 under tension and 4 slack. If 4 becomes under tension then B will move toward the vertical plane going through A_1, A_2, leading to an increase of the length of cable 3 compared to the one at B_1 which is impossible as this length is fixed: therefore there cannot a FK solution with 124 under tension. The same reasoning may be used for the second case, leading to an impossible increase of the length of cable 1. Using the same geometrical reasoning, that is not detailed because of the lack of space, it is possible to show that

- it is not possible to have a FK solution with 3 cables under tension and a second one with only 1 or 2 cables under tension
- it is not possible to have a FK solution with 2 cables under tension and another one with either a different pair of cable under tension or only one cable under tension. Finally if a FK solution has one cable under tension, then it is not possible to have another FK solution with another cable under tension.

In summary the FK of a 4-1 CDPR with ideal cables admits at most a single solution. If we move to sagging cable we notice that if $\mu \to 0$, then equations (1) are the same than for ideal cable. Consequently we may use a continuation process to determine the FK solution: starting from the single solution obtained for the ideal cables we may incrementally increase the value of μ and use the Newton scheme with as initial guess the one obtained at the previous step for determining the new FK solution. As we start from a single solution we will end up also at a single solution for the sagging cables unless we meet a singularity during the continuation process. Singularities of CDPR is a complex subject but it appears that they are constituted of the classical singularities of CPR to which should be added to singularities that are specific of the sagging cables but which occurs only at the CDPR workspace border [12,13]. We will assume here that a $N-1$ CDPR is never singular so that as **a CDPR with ideal cables has a single FK solution, then the CDPR with sagging cables will also have a single FK solution.**

2.1 Solving the FK

As we have a starting point for the continuation process we may use this method to find the FK solution. An alternate is based on the use of interval analysis (IA) as it is easy to find natural bounds for the unknowns apart of the upper bound for the F_x, that may be taken arbitrary large. But an important modification of the IA algorithm is that before bisecting a box in the IA algorithm we run a few iteration of the Newton scheme with as solution guess the mid-point of the box. If the Newton scheme converge to a solution with positive F_x, then we have found the FK solution. Our trials have shown that the computation of the continuation scheme (implemented in Maple) is between 5 and 7 s. Regarding the IA algorithm if we use as search space the whole area that is guaranteed to include the solution it runs in about [0.1–60] s with the advantage of being able to detect if the robot is close to a singular configuration, in which case the CDPR has to be stopped as we are unable to determine its pose. However in a controle scheme we may assume that the search space regarding the pose unknowns and the α's is much more restricted our trials have shown that the IA algorithms is real-time compliant with a computation ttime that is less than 5 ms. Hence both the IA and continuation schemes may be used for determining the FK solutions but during the real-time control we may use only the Newton scheme and the IA algorithm only if it does not converge.

2.2 Influence of the Sagging on the FK Solution

We may wonder if the sagging as an influence on the FK solution. To illustrate this difference we consider the pose T ($x = -200, y = -100, z = 150$), and a mass $M = 50$ kg for the platform and assume first that the cables are ideal. Reaching this pose with all 4 ideal cables under tension is impossible as this will require an infinitely accurate control of the cable lengths and therefore we must assume that at most 3 cables will be under tension. In this frame there are 2 inverse kinematic solutions, the first one with cable (1, 2, 3) under tension and the second one with (1, 2, 4) under tension with the following characteristics:

Cable	1	2	3	4	1	2	3	4
F_x	274.198	315.808	391.711	-	39.171	442.131	-	394.420
F_z	−122.625	−262.768	−105.107	-	−17.518	−367.875	–	−94.434
τ	300.369	410.830	405.568	-	42.910	575.162	-	405.568
α	65.905	50.238	74.980	-	65.905	50.238	-	76.535

We then assume sagging cables whose lengths are the one used for the ideal cables case and assume that $\mu = 0.000046$ kg/m (which corresponds to a Dyneema cable of diameter 3 mm). Solving the FK leads to a pose that is basically the same than for the previous case (the differences are less than 1.5e-6 cm), while for the other parameters we get: with significant differences in the cable tensions (τ is here the cable tension at B).

Cable	1	2	3	4		1	2	3	4
F_x	155.438	380.028	193.724	222.256	F_z	−69.431	−316.149	−51.851	−53.068
τ	170.240	494.340	200.543	228.503	α	65.931	50.243	75.016	76.571

2.3 Influence of Uncertainties

The FK is influenced by possible measurement errors on the L_0's and by uncertainties on the location of the A's. To evaluate the influence of the uncertainties on the L_0 we run a Monte-Carlo test, assigning a random error of ± 1 cm on the value of the L_0. For 31000 trials we get the FK solutions in the ranges $x \in [-201.128, -198.57]$, $y \in [-101.499, -98.504]$, $z \in [-148.01, 152.207]$, the maximal distance between the pose T and the FK solution being 2.8098 cm. Hence the sagging influence on the positioning is moderate. On the other hand the cable tensions exhibit a much larger range with $\tau_1 \in [37.216, 304.887]$, $\tau_2 \in [408.428, 583.208]$, $\tau_3 \in [0.698, 411.546]$ and $\tau_4 \in [0.864, 460.288]$.

To evaluate the influence of the uncertainties on the location of the A we assign a random noise in the range $[-1, 1]$ cm to each coordinate of the A's. After 31000 trial we get $x \in [-201.24, -198.2], y \in [-101.96, -98], z \in [-147.7, 153.1]$ with a maximal distance to T that is 3.5 cm. These positioning errors are larger by 20% than the one due to the influence of the sagging, which emphasize the need of a careful robot calibration. The ranges for the cable tensions are $\tau_1 \in [33.85, 308.63]$, $\tau_2 \in [406.647, 585.12]$, $\tau_3 \in [0.66, 413.24]$ and $\tau_4 \in [0.86, 463.22]$ which are roughly similar to the one obtained in the previous case. If we combine the uncertainties on the L_0's and A's we get $x \in [-201.67, -196.83]$, $y \in [-103, -96.89]$, $z \in [-146.51, 154.32]$ with a maximal distance between T and the solution of 4.927 cm that shows that the influence of both uncertainties do not exactly sum up.

3 FK with Other Inputs Than the L_0's

3.1 Measuring the Cable Tensions or Angles

Let us assume that instead of using the L_0 for the FK we intend to obtain the CDPR pose by measuring the 4 cable tensions at B. This FK variant leads to a system that is still square with 15 equations for 15 unknowns. We assume an accuracy of $\pm$2N on the tension measurements (which is well below to what is feasible). A Monte-Carlo simulation with 68000 trials has led to $x \in [-204.21, -195.56]$, $y \in [-102.08, -97.97]$, $z \in [147.84, 152.14]$ and a maximal distance to T of 5.069 cm. The evaluation of the L_0's are [365.74, 369.18], [229.94, 239.24], [575.35, 582.09], [640.33, 647.88]. Hence the positioning error is much larger than the one obtained by measuring the L_0's.

We may also consider solving the FK by measuring the α angles [8] for which we have shown that an accuracy of $\pm\,0.1°$ may be obtained [14]. This uncertainty led to $x \in [-202.3, -197.4]$, $y \in [-102.236, -97.85]$, $z \in [149.1, 150.86]$ with a maximal distance to T of 3.07 cm and an evaluation of the L_0's of [364.78, 370.35],

[233, 236.07], [576.26, 581.12], [642.81, 645.46]. These errors are much lower than the one obtained with the cable tensions but higher than the one obtained by measuring the L_0's. We note however an improvement on the evaluation of the z coordinates which suggests that it may be interesting to combine both measurements.

3.2 Using Low Rate Pose Measurements

As we have a 3 dof robot we may consider measuring directly the pose of the platform. Vision has been proposed for small-sized parallel robot [3,6] but is very sensitive to light condition and CDPR size [7]. Combining sensor data to improve the positioning accuracy has been proposed but only for elastic cables [11].

For the CDPR shown in Fig. 1 we have used 3 on-board RPLIDAR A1M8 lidars allowing 360° measurements that costs around 100 euros. A vertical one was used to measure the altitude of the CDPR that was evolving over a flat surface. The two other one were intended to measure the x, y coordinates by measuring points on the pillars that were surrounding the workspace. The accuracy on the z coordinate was very good (around 0.05 cm) as we were measuring around 300 points on the surface. The accuracy on the x, y was usually was very poor as only 1 or 2 pillars were in the range of the lidars, leading to only a few measurement points.

We will propose now a strategy to fuse lidars data and the sagging of the cables. Let us assume that 3 lidars allow us to obtain the x, y, z coordinates of B every Δt_l seconds and assume, for the time being, that these measurements are exact while the rotation angles of the drum is measured every Δt seconds with $\Delta t_l \gg \Delta t$. Our objective is to use the lidar data to improve the positioning accuracy of the CDPR along a trajectory. We assume a classical control law based on velocity control: at each sampling time we use the FK to determine the location of B and use the inverse jacobian matrix to compute the cable velocities that make B moves at the desired speed on the desired trajectory. For that purpose we need first to determine what are the cable lengths for the measured pose (as the 4–1 is redundant there are an infinite number of possible cable lengths for a given B) at each lidar sampling time. In between the lidar sampling time we will use as input of the FK the estimation of the L_0s based on the measurements of the drum rotation. We assume here that the cables are wound on the fly on the drum with possibly several layers. For this type of actuation the drum velocities required to obtain the desired cable velocities depend upon the drum radii that are therefore to be estimated. We propose a method that allows one both to estimate the cable lengths for a given measurement of the lidar and the drum radii.

At time $t = 0$ we assume that lidar measurements are available, while the current cable lengths L^0_{0i} and drum radii r^0_i are unknown. Consequently we have 16 unknowns $F^0_{xi}, F^0_{zi}, L^0_{0i}, r^0_i$, while Eqs. (1, 3) provide 11 constraints. At time $t = \Delta t_l$ we get new lidar measurements and we assume that Δt_l is small enough so that the drum radii have remained constant. Hence the cable lengths are $L^1_{0i} = L^0_{0i} + r^0_i \Delta\theta_i$ where $\Delta\theta_i$ is the measured drum rotation angle. Consequently we have 8 additional unknowns (F_{xi}^1, F_{zi}^1), while we get 11 additional constraints so that we have now 24 unknowns and 22 constraints. At time $t = 2\Delta t_l$ we have 8 additional unknowns and 11 additional constraints, so that we have now 32 unknowns and 33 constraints. Therefore we may determine the 8 unknowns L_{0i}^0, r_i^0 that minimize the cost function built as the sum of the square of the (1) equations under the 9 equality constraints (3). If the lidar measurements were exact the minimum of the cost function should be 0. However there will be uncertainties in the lidar measurements and to decrease their influence we will use as pose at time $0, \Delta t_l, 2\Delta t_l$ the FK solutions obtained for $L_{0i}^0, L_{0i}^1, L_{0i}^2$. At the following time $k\Delta t_l, k \geq 3$ we will repeat the process with the two latest lidar data and the new one.

Example: we consider the trajectory going from the pose (−200, −100, −150) to the pose (0, 0, 150) at a velocity of 3 cm/s and we assume a random uncertainty of ±0.2 cm on the lidar measurements with a sampling time Δt_l of 1 s while the CDPR controller sampling time is 5 ms. We illustrate in Fig. 2 the distance to the desired trajectory for a scheme were we assume a constant drum radius of 3.15 cm (case 1) and the proposed scheme using the lidars (case 2).

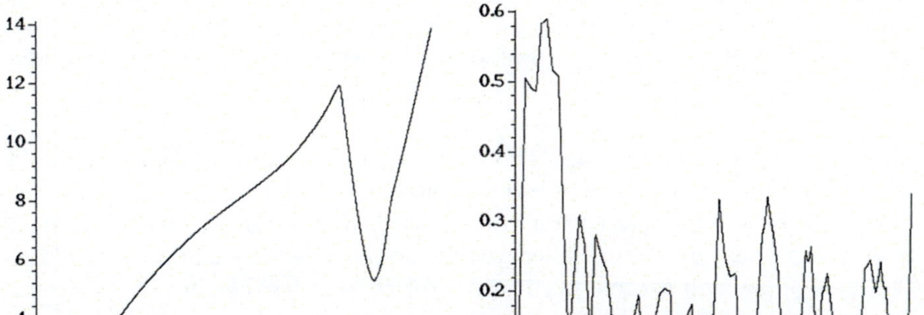

Fig. 2. Distance of the real trajectory to the desired one in cm as a function of time (s): on the left without using the lidars and on the right with the lidars

For case 1 the maximal distance is 13.833 cm, the mean distance is 6.851 cm with a variance of 10.2584. If we use the lidar the maximal error is 0.59 cm, the mean distance error is 0.2005 cm and the variance is 0.0161. Hence even with uncertainties on the lidar measurements we obtain a real improvement of the positioning. We then consider the estimation of the radius 4 and the estimation of τ_4 over time when using the lidars (Fig. 3). Regarding the tension we notice that we have a very significant difference between the estimation and the ground truth. Once again this shows that it is very difficult to simulate what will be the cable tensions of a CDPR. Regarding the estimation of the drum radius we notice an oscillatory effect but the mean value of the error is small (−0.0022 cm).

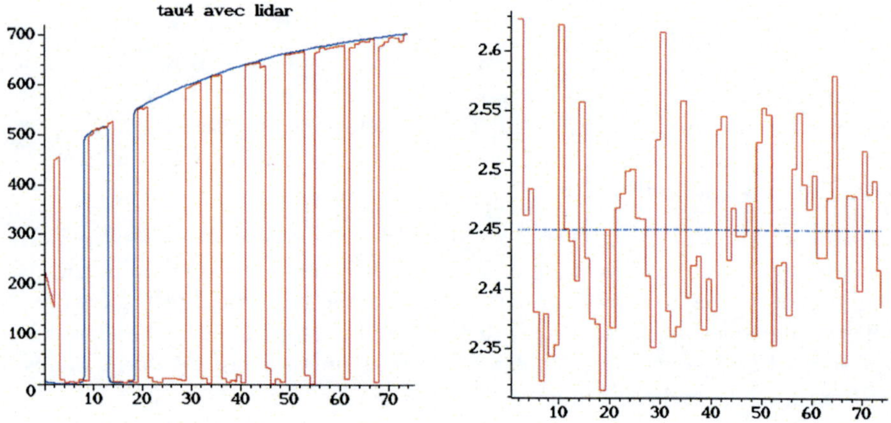

Fig. 3. On the left estimation of the cable 4 tension (N) as a function of time (s). In blue we have the real tension and in red the estimated one. On the right estimation of the radius of cable 4 drum (cm): in dashed blue the real value and in red the estimation

For small to medium-sized CDPR we may also use a drum with a spiral guide that allows for a much more accurate control of the cable lengths. In the next simulation we assume that such a drum is used. However we assume a randomly chosen initial bias on the cable lengths in the range [−0.5,−0.3] or [0.3, 0.5] cm. The control law will assume ideal cables for computing the current pose, while the real pose will be calculated by assuming sagging cables whose lengths will be affected by the initial bias. As shown in Fig. 4 the mean of the distance to the desired trajectory is 0.702 cm with a variance of 0.11e−3 and maximum value of 0.738 cm. Here again we observe significant differences in the cable tensions: for example tension of cable 1 is expected to lie in the range [0.22, 42.9]N while its real value is in the range [300.14, 702.63]N. In view of these results it appears that using the lidars may improve the accuracy. Hence we fuse the lidar data using a variant of the approach proposed for the drum with multiple layers. The simulation with the lidar uncertainties shows that the mean value of the distance to the desired trajectory becomes 0.158 cm with a variance of 0.0415 (Fig. 4). This figure shows also that in spite of the clear improvement of the positioning accuracy it remains very difficult to estimate the real cable tensions.

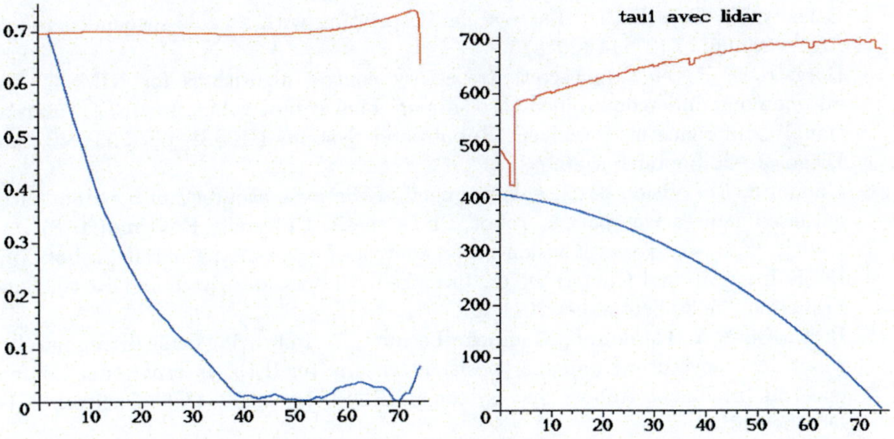

Fig. 4. On the left the positioning error (cm) for a drum with a spiral guide as a function of time (s): in red without the use of the lidars and in blue with the lidars. On the right in red the estimation of cable 2 tension (N) and in blue the real tension, as function of time (s)

4 Conclusion

To summarize this paper we have shown that taking into account the cable sagging in the FK based on the cable lengths for a N–1 CDPR does not significantly modify the platform pose but leads to drastic changes in the cable tensions. We have evaluated the influence of uncertainties for the FK's based on measuring the cable lengths (1), tension (2) and angles (3). By far the worst results have been obtained for (2), (1) being usually better than (3). The most promising approach is the proposed strategy to fuse the measurement of the L_0 together with using lidars data that provide an estimate of the platform pose, although at a relatively low rate.

Acknowledgements. This work was partly supported by the ANR CRAFT grant 18-CE10-0004.

References

1. Abbasnejad, G., Carricato, M.: Direct geometrico-static problem of underconstrained cable-driven parallel robots with n cables. IEEE Trans. Robot. **31**(2), 468–478 (2015)
2. Alikhani, A., et al.: Design of a large-scale cable-driven robot with translational motion. Robot. Comput. Integr. Manuf. **27**(2), 357–366 (2011)
3. Andreff, N., Dallej, T., Martinet, P.: Image-based visual servoing of a Gough-Stewart parallel manipulator using leg observations. Int. J. Robot. Res. **26**(7), 677–688 (2007)

4. Barnett, E., Gosselin, C.: Large-scale 3D printing with a cable-suspended robot. Addit. Manuf. **7**, 27–44 (2015)
5. Borgstrom, P., et al.: Discrete trajectory control algorithms for NIMS3D, an autonomous underconstrained three-dimensional cabled robot. In: IEEE International Conference on Intelligent Robots and Systems (IROS), pp. 253–240. San Diego, 22–26 September 2007
6. Chianura, M.: Vision-based enhancement of the pose accuracy of a redundantly actuated cable-driven parallel robot. Ph.D. thesis, University Bologna (2018)
7. Dallej, T., et al.: Towards vision-based control of cable-driven parallel robots. In: IEEE International Conference on Intelligent Robots and Systems (IROS), San-Francisco, 25–30 September 2011
8. Fortin-Côté, A., Cardou, P., Campeau-Lecours, A.: Improving cable driven parallel robot accuracy through angular position sensors. In: IEEE International Conference on Intelligent Robots and Systems (IROS), pp. 4350–4355, Daejeon, 9–14 October 2016
9. Hussein, H., Gouttefarde, M., Pierrot, F.: Static modeling of sagging cables with flexural rigidity and shear forces. In: ARK, Bologna, 1–5 July 2018
10. Irvine, H.M.: Cable Structures. MIT Press, Cambridge (1981)
11. Korayem, M., Yousefzadeh, M., Kian, S.: Precise end-effector pose estimation in spatial cable-driven parallel robots with elastic cables using data fusion methods. Measurement **130**, 177–190 (2018)
12. Merlet, J.P.: Computing cross-sections of the workspace of suspended cable-driven parallel robot with sagging cables having tension limitations. In: IEEE International Conference on Intelligent Robots and Systems (IROS), Madrid, 1–5 October 2018
13. Merlet, J.P.: Singularity of cable-driven parallel robot with sagging cables: preliminary investigation. In: IEEE International Conference on Robotics and Automation, Montréal, 20–24 May 2019
14. Merlet, J.P.: An experimental investigation of extra measurements for solving the direct kinematics of cable-driven parallel robots. In: IEEE International Conference on Robotics and Automation, Brisbane, 21–25 May 2018
15. Miermeister, P., et al.: An elastic cable model for cable-driven parallel robots including hysteresis effect. In: 2nd International Conference on cable-driven parallel robots (CableCon), pp. 17–28. Duisburg, 24–27 August 2014
16. Piao, J., et al.: Open-loop position control of a polymer cable-driven parallel robot via a viscoelastic cable model for high payload workspaces. Adv. Mech. Eng. **9**(12), 1687814017737199 (2017)
17. Riehl, N., et al.: On the determination of cable characteristics for large dimension cable-driven parallel mechanisms. In: IEEE International Conference on Robotics and Automation, pp. 4709–4714, Anchorage, 3–8 May 2010
18. Schmidt, V.: Modeling techniques and reliable real-time implementation of kinematics for cable-driven parallel robots using polymer fiber cables. Ph.D. thesis, Université Stuttgart, 20 June 2016
19. Tempel, P., Trautweing, F., Pott, A.: Experimental identification of stress-strain material models of UHMWPE fiber cables for improving cable tension control strategies. In: ARK, Bologna, 1–5 July 2018
20. Williams II, R., Albus, J., Bostelman, R.: 3D cable-based cartesian metrology system. J. Robot. Syst. **21**(5), 237–257 (2004)

The Geometrical Arrangement of Joint Constraints that Makes Natural Motion Possible: Experimental Verification on the Ankle

Michele Conconi(✉), Nicola Sancisi, and Vincenzo Parenti-Castelli

DIN - Department of Industrial Engineering,
Alma Mater Studiorum - University of Bologna,
Viale Risorgimento 2, 40136 Bologna, Italy
michele.conconi@unibo.it

Abstract. The passive motion of a joint, namely the articular motion under no external loads, provides insights into the joint physiology. It represents the baseline motion of an articulation before passive structures are loaded by external loads. Moreover, during natural motion, the strain energy density stored within ligaments and cartilage is minimized, reducing the risk of microdamage and the corresponding metabolic cost for tissue repairing. In a recent paper, we showed that the line of action of resultant forces of all the constraints provided by the passive structures in a joint must intersect the instantaneous helical axis to make the natural motion possible. In other words, the lines of action of all these constraints must cross the same line at each flexion angle to guarantee the natural motion of the joint. This geometrical property was proven theoretically and verified experimentally for the knee. To prove its generality, in this work we will verify the same property on nine ankle specimens.

1 Introduction

The relative motion of bones in diarthrodial joints is constrained by passive structures, namely the articular contacts, the ligaments, the joint capsule and in some cases additional fibrous structure such as menisci or labra. Despite this high number of constraints, it is still possible to continuously move human joints with virtually no external loads. In other words, an infinite number of configurations of stable equilibrium exists compatible with the constraints imposed by the passive structures. For the ankle, as well as for the knee, it was experimentally proven that these equilibrium configurations describe a well-defined one degree of freedom (dof) three-dimensional unresisted motion [15,19], also called the natural or passive motion. A very close approximation of this idealized motion can be measured experimentally by introducing the smallest amount of work in the system, i.e., the work necessary to overcome the small internal friction, gravitational effects, and possible residual soft tissue deformations [4,8].

J. Lenarčič and B. Siciliano (Eds.): ARK 2020, SPAR 15, pp. 109–116, 2021.
https://doi.org/10.1007/978-3-030-50975-0_14

Natural motion plays a fundamental role in joint physiology, as it shows several interesting properties. Experimental evidence showed that, along the natural motion of both knee and ankle some ligament fibers are almost isometric [15,20], while the shape of articular surfaces guarantee the best contact loads distribution along the natural motion [4,5]. Thus, during this motion the deformation energy stored in the passive structures is the smallest possible, providing an energetic optimum for the articulation. Moreover, since deformation energy is associated with microdamage occurrence and accumulation [2], the natural motion minimizes the metabolic cost for tissue repairing.

In order to have an unresisted motion, the articular constraints must satisfy some specific properties. A first requirement is ligament isometry. In fact, if ligament elongation occurs, a perfectly unresisted motion would necessarily imply a perfectly statically balanced system [11]: elastic energy would transfer between ligaments during the motion and no additional external work would be needed. It could be shown, however, that this condition could only be obtained by zero-free-length elastic elements [7] or more complex systems [7,11], both conditions not representative of the ankle human ligaments arrangement. It follows that no or negligible ligament elongation may occur during natural motion and elastic potential energy is almost constant. This confirms the isometric behavior observed experimentally and cited above.

In these conditions and considering friction as negligible [10], the principle of virtual work (PVW) provides a second property that must be fulfilled: constrained motion is possible only if each constraint makes no work against the infinitesimal allowed act of motion. In terms of screw theory, this is equivalent to requiring the reciprocity between the twist, representing the instantaneous helical axis (IHA) of motion of the joint, and all the resultant wrenches of passive constraint [12], independently of their number or type.

We hypothesized that the wrenches related to the resultants of each ligament and contact constraints are pure forces, or, equivalently, the pitches of the screws representing these wrenches are zero. This hypothesis is supported by the possibility to model the ankle natural motion with parallel mechanisms, defined by substituting ligament and contact constraints with rigid links connecting the tibia and the talus and exerting pure forces between the two bones [9,17]. We also hypothesized that the work produced by the passive constraints against the translation along the IHA is negligible. This second hypothesis also agrees with the representation of the ankle as a spherical mechanism [17], which implies small pitch and thus small translation along the IHA for the ankle natural motion. Finally, as previously stated, we consider the effect of ankle internal friction as negligible.

Under these hypotheses, it is possible to prove analytically that the application of the PVW implies that the line of action of all the constraint forces must intersect the ankle IHA over the whole natural motion range.

The same property has been previously derived analytically and proven experimentally, both in-vitro and in-vivo, for the knee joint [6]. This work aims at proving the generality of this property by verifying it experimentally on nine ankle specimens.

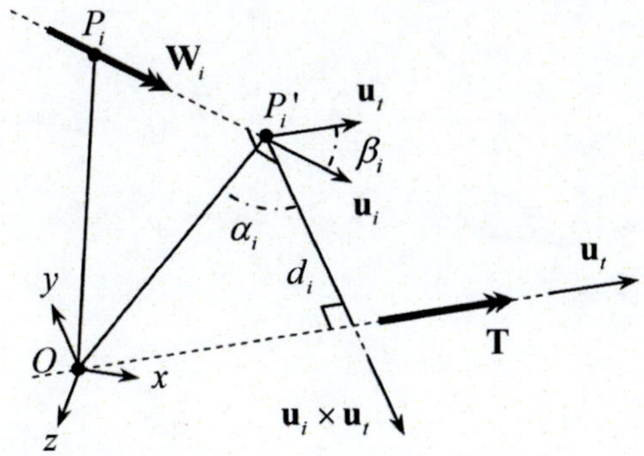

Fig. 1. The screws representing the IHA ($\boldsymbol{T}$) and the i-th wrench ($\boldsymbol{W}_i$) at the generic instant of time.

2 Materials and Methods

2.1 Theoretical Derivation

For the sake of clarity, the derivation proposed in [6] will be here recalled.

$\boldsymbol{T}$ and $\boldsymbol{W}_i$ are respectively the twist representing the ankle IHA during natural motion, and the wrench corresponding to the i-th passive constraint. In particular, if at each instant the origin of the reference system is chosen on the axis of $\boldsymbol{T}$ (Fig. 1), it is possible to write:

$$\boldsymbol{T} = t\begin{bmatrix} \boldsymbol{u}_t \\ h_t\boldsymbol{u}_t \end{bmatrix},\ \boldsymbol{W}_i = w_i\begin{bmatrix} \boldsymbol{u}_i \\ (P_i - O)\times \boldsymbol{u}_i + h_i\boldsymbol{u}_i \end{bmatrix}, \tag{1}$$

where $\boldsymbol{u}_t$ and $\boldsymbol{u}_i$ are the screw unit vectors of the IHA and the i-th wrench, respectively; h_t and h_i are the screw pitches of the IHA and the i-th wrench, respectively; P_i is a generic point on the line of the i-th wrench; t and w_i are the screw amplitudes of the IHA and the i-th wrench, respectively. The values of pitches and amplitudes in Eq. 1 determine the type of motion and constraint. In particular, null h_t and h_i correspond respectively to a pure rotation about the IHA and a resultant pure force exerted by the i-th constraint. By introducing the interchange operator $\boldsymbol{\Delta}$, the PVW can be expressed in terms of screws reciprocity as

$$(\boldsymbol{W}_i)^T\boldsymbol{\Delta}\boldsymbol{T} = 0,\ i = 1,\ldots,n,\ with\ \boldsymbol{\Delta} = \begin{bmatrix} \boldsymbol{0} & \boldsymbol{I}_3 \\ \boldsymbol{I}_3 & \boldsymbol{0} \end{bmatrix}, \tag{2}$$

Equation 2 is general and does not depend on the type (i.e., the wrench amplitude and pitch) and the number of constraints in the ankle, nor on the type (i.e., the twist amplitude and pitch) of the natural motion. Expansion of Eq. 2 leads to

$$(\boldsymbol{W}_i)^T\boldsymbol{\Delta}\boldsymbol{T} = \underbrace{(h_t + h_i)\boldsymbol{u}_i\cdot\boldsymbol{u}_t}_{I} + \underbrace{\boldsymbol{u}_t\cdot((P_i - O)\times\boldsymbol{u}_i))}_{II} = 0,\ i = 1,\ldots,n. \tag{3}$$

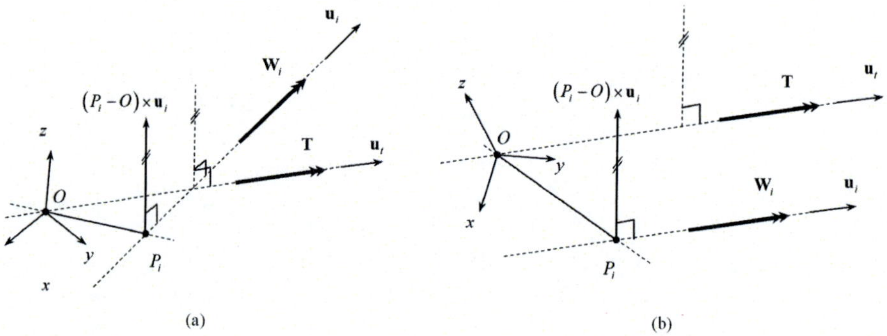

Fig. 2. Representation of the two geometrical conditions that satisfy Eq. 4: incident (a) and parallel (b) screws

Equation 3 is composed of two terms, representing the work produced by the wrench $\boldsymbol{W}_i$ through the components along (term *I*) and about (term *II*) the IHA, hereinafter denoted as longitudinal and transversal work, respectively. As previously noted, the IHA pitch for the ankle natural motion is small enough to allow the modeling of the ankle as a spherical mechanism [17]. Moreover, experimental evidence shows that over the ankle natural motion the IHA roughly passes through the tips of the malleoli [16], thus being almost orthogonal to the passive constraint lines and resulting in small values for $\boldsymbol{u}_i \cdot \boldsymbol{u}_t$. Finally, as discussed above, we hypothesized that passive constraints exert zero pitch wrenches. In these conditions, the longitudinal work can be reasonably assumed zero and Eq. 3 simplifies to

$$(\boldsymbol{W}_i)^T \boldsymbol{\Delta T} = \boldsymbol{u}_t \cdot ((P_i - O) \times \boldsymbol{u}_i)) = 0,\ i = 1, \ldots, n. \quad (4)$$

Equation 4 is satisfied if and only if $(P_i - O)$, $\boldsymbol{u}_i$ and $\boldsymbol{u}_t$ belong to the same plane, that is if $\boldsymbol{W}_i$ and $\boldsymbol{T}$ are incident (Fig. 2.a) or parallel (Fig. 2.b), this latter condition being in conflict with previous hypotheses. This means that the lines of action of each passive constraint that guides the natural motion at a given instant must all cross the IHA obtained at the same instant. This fundamental property guarantees the existence of the unresisted joint motion under the given assumptions.

As a confirmation and as a more general result, the distance d_i between reciprocal twist and i-th wrench lines can also be expressed as:

$$d_i = \frac{(h_t + h_i)}{tan(\beta_i)}\ i = 1, \ldots, n. \quad (5)$$

where β_i is the angle between $\boldsymbol{u}_i$ and $\boldsymbol{u}_t$. Equation 5 provides a general result and makes it possible to compute the exact distance between the line of action of each constraint and the IHA, when reciprocity holds.

2.2 In-Vitro Experimental Setup

A detailed description of the experimental procedure can be found in [4]. Nine fresh frozen amputated lower limb specimens comprising complete shank and foot, were declared free of anatomical defects by a surgeon, and were fixed through the tibia to a workbench for the measure of the tibio-talar natural motion. Starting from a rest position in maximum plantarflexion, the joint was extended to maximum dorsiflexion, thus producing the desired complete arc of joint motion. A standard stereophotogrammetric system (Stryker Navigation System; accuracy: ±0.5°, ±0.5 mm) was used for the acquisition of the tibio-talar relative motion. After opening the joint capsulae, the articular surfaces of the talus dome, the distal tibia, and distal fibula where digitized, together with origin and insertion of TiCaL and CaFiL.

2.3 Experimental Derivation of the IHAs and the Constraint Line of Action

The IHAs and the corresponding pitch were obtained from the measured natural motion, sampled at 1° of flexion [18].

At each flexion angle, the line of action of each ligament wrench was identified by the most isometric fiber over the measured natural motion, i.e. as the line connecting two points, one on the tibial or fibula and the other on the calcaneal insertion areas, that showed the smallest length variation during motion. The maximum percent length variation of the most isometric fiber in each ligament was computed as $\%\Delta L = 100 \cdot (L_{max} - L_{min})/L_{max}$. As for the line of action of the contact forces, the tibio-talar articulation were subdivided into four regions: the articulation of the talus with the medial malleolus, the medial and lateral articulation of the talar dome with the tibial plateau, and the talar articulation with the fibula, here considered as rigidly connected with the tibia. Every portion of these four regions whose distance from corresponding articulation was below the threshold of 5 mm was considered to be in contact. Contact pressure was assumed to vary linearly with this distance. This approach moves from the Winkler contact model [13], which proved its applicability to the study of the highly conforming articular contacts [1,3,14]. With this representation, the direction of the resultant contact force depends only on the relative geometry, while it is independent of the material properties. In particular, the directions of the contact forces were found as the average of the normals of the condylar contact surfaces, weighted proportionally to their distance from the tibial plateau, and applied to the geometrical centroid of the condylar contact surfaces.

In order to verify experimentally whether all wrench lines intersect the IHA along the natural motion, the distances between each constraint line of action and the IHA were determined geometrically from the experimental data as line-to-line distances.

Table 1. Maximal percent length variation of the most isometric fiber of each ankle ligament over the measured natural motion.

	Leg 1	Leg 2	Leg 3	Leg 4	Leg 5	Leg 6	Leg 7	Leg 8	Leg 9
CaFiL	0.2	1.1	0.6	0.3	5.0	3.9	2.6	0.8	5.1
TiCaL	0.2	0.3	0.4	1.8	1.2	0.7	0.5	1.1	0.8

Table 2. Mean value of the IHA-to-wrench axis distance over the measured natural motion.

	CaFiL to IHA	TiCaL to IHA	MM to IHA	MT to IHA	LT to IHA	LF to IHA
Leg 1	0.6 ± 0.7	0.5 ± 0.5	4.9 ± 1.5	1 ± 0.7	2.2 ± 1.1	4.8 ± 1.5
Leg 2	1.8 ± 1.2	0.6 ± 0.5	5 ± 1.5	1.3 ± 1.2	2.5 ± 1.9	3.7 ± 1.8
Leg 3	2.6 ± 5.5	1.4 ± 2.3	5.5 ± 3.9	1.1 ± 2.1	2.7 ± 3.4	4.5 ± 4
Leg 4	0.5 ± 1	1.8 ± 1.9	4.9 ± 1.6	0.8 ± 0.8	0.7 ± 0.8	3.2 ± 1.3
Leg 5	2.8 ± 3.7	0.6 ± 0.8	4.5 ± 1.2	1 ± 0.7	0.7 ± 0.8	2.9 ± 1.2
Leg 6	1.8 ± 1	1.3 ± 0.9	4.5 ± 1	1 ± 0.8	1.3 ± 0.8	3.1 ± 0.9
Leg 7	1.7 ± 1.6	3 ± 3.3	4.8 ± 3.6	2.7 ± 2	2.3 ± 1.4	3 ± 3.3
Leg 8	1.4 ± 0.9	0.8 ± 0.6	4 ± 1	0.8 ± 0.5	0.6 ± 0.4	2.1 ± 0.6
Leg 9	2.7 ± 1.4	0.4 ± 0.7	0.4 ± 1.4	3.3 ± 1.4	4.2 ± 1.6	1.9 ± 1

3 Results

Maximum percent length variations of isometric fibers are listed in Table 1. Values below or proximal to 5% were considered representative of isometric behavior [15]. All the ligaments stay below this threshold but two calcaneo-fibular ligaments, which however shows a length of 5 % and 5.1 %. The results are thus in agreement with the literature [15,20].

The mean distances between each constraint line of action and the IHA over the full flexion arc are reported in Table 2: all these mean distances stay below 5.5 mm. The small values of these distances can be considered as an experimental verification of the intersection property. A visualization of the intersection property is depicted also in Fig. 3 for a representative specimen (Leg 9).

4 Discussion

The goal of this paper was to present a geometrical characterization of the ankle natural motion, based on the principle of virtual work, according to which all the ligament and contact resultant forces must intersect the IHA at every flexion angle to allow a one dof unresisted motion at the ankle. The property was derived by exploiting the reciprocity between constraints and allowed motion, assuming that the screws representing the constraint wrenches have zero pitch, that their work done along the IHA is negligible, and that friction can be ignored. This characterization was previously developed for the knee joint. To prove its

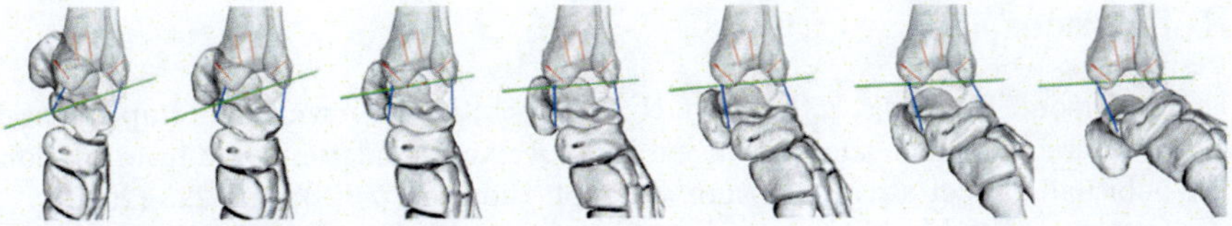

Fig. 3. Representation of the IHA (green line) and of the lines of action of the ligament (blue) and contact (red) wrenches every 10° of the analyzed range of ankle natural motion for a representative specimen (Leg 9). Spheres represent the closest point to the IHA on each wrench and their color goes from white ($d_i > 5$ mm) to the one of the corresponding line ($d_i = 0$ mm).

generality, the same is here experimentally verified on nine ankle specimens by measuring the distance between the constraint lines of action and the IHA.

The measured distances for all constraints and in all experiments were small enough to be considered as an experimental verification of the property of intersection between wrenches and IHA lines and of the hypotheses behind its derivation.

It is worth noticing that the properties here presented holds independently by the number of constraints. In particular, in the present analysis six constraints have been considered, all respecting the same properties. This implies that the ankle, as well as the knee, can be successfully represented as an overconstrained mechanism.

The presented results could be meaningfully applied in the multibody modelling of lower extremities, where it is often assumed that ligaments and contacts do not contribute to the energy balance of the joints, and thus to their equilibrium. This assumption is acceptable as long as the investigated motion is close to the joint natural one. For any other motion, ligament and contact constraints could contribute or not to the joint equilibrium, but this should be verified case by case, particularly when joint reaction analysis is performed.

5 Conclusion

This paper verified on nine ankle specimens that the lines of action of all the ligament and contact resultant forces must intersect the instantaneous helical axis at every flexion angle to allow a one dof unresisted motion at the ankle joint. The analysis and the results here obtained, together with the previous investigation on the knee, further confirm the generality of this property and its extendibility to the study of other joints. This fundamental property provides useful guidelines for the design of medical devices and treatments capable to provide the correct guidance of the joint motion and avoid unnatural tensioning of ligaments, thus providing a correct balancing of passive structures.

References

1. Anderson, D.D., Iyer, K.S., Segal, N.A., Lynch, J.A., Brown, T.D.: Implementation of discrete element analysis for subject-specific, population-wide investigations of habitual contact stress exposure. J. Appl. Biomech. **26**(2), 215–223 (2010)
2. Carter, D.R.: Mechanical loading history and skeletal biology. J. Biomech. **20**, 1095–1109 (1987)
3. Conconi, M., Castelli, V.P.: A sound and efficient measure of joint congruence. Proc. Inst. Mech. Eng. H **228**(9), 935–941 (2014)
4. Conconi, M., Leardini, A., Parenti-Castelli, V.: Joint kinematics from functional adaptation: a validation on the tibio-talar articulation. J. Biomech. **48**(12), 2960–2967 (2015)
5. Conconi, M., Sancisi, N., Parenti-Castelli, V.: Subject-Specific Model of Knee Natural Motion: A Non-invasive Approach, pp. 261–269. Springer (2018)
6. Conconi, M., Sancisi, N., Parenti-Castelli, V.: The geometrical arrangement of knee constraints that makes natural motion possible: theoretical and experimental analysis. J. Biomech. Eng. **141**(5), 051001 (2019)
7. Deepak, S., Ananthasuresh, G.: Perfect static balance of linkages by addition of springs but not auxiliary bodies. J. Mech. Robot. **4**(2), 021014 (2012)
8. Forlani, M., Sancisi, N., Parenti-Castelli, V.: A three-dimensional ankle kinetostatic model to simulate loaded and unloaded joint motion. J. Biomech. Eng. **137**(6), 061005 (2015)
9. Franci, R., Parenti-Castelli, V., Belvedere, C., Leardini, A.: A new one-dof fully parallel mechanism for modelling passive motion at the human tibiotalar joint. J. Biomech **42**(10), 1403–1408 (2009)
10. Fujie, H., Imade, K.: Effects of low tangential permeability in the superficial layer on the frictional property of articular cartilage. Biosurf Biotribol **1**(2), 124–129 (2015)
11. Herder, J.: Energy-free systems. theory, conception and design of statically balanced spring mechanisms. Ph.D. thesis, Delft University of Technology, Delft, The Netherlands (2001)
12. Hunt, K.: Kinematic Geometry of Mechanisms. Clarendon Press, Oxford (1978)
13. Johnson, K.: Contact Mechanics. Cambridge University Press, Cambridge (1985)
14. Kern, A.M., Anderson, D.D.: Expedited patient-specific assessment of contact stress exposure in the ankle joint following definitive articular fracture reduction. J. Biomech. **48**(12), 3427–3432 (2015)
15. Leardini, A., O'Connor, J., Catani, F., Giannini, S.: Kinematics of the human ankle complex in passive flexion; a single degree of freedom system. J. Biomech. **32**(2), 111–18 (1999)
16. Montefiori, E., Modenese, L., Di Marco, R., Magni-Manzoni, E.A.: An image-based kinematic model of the tibiotalar and subtalar joints and its application to gait analysis in children with juvenile idiopathic arthritis. J. Biomech. **85**, 27–36 (2019)
17. Sancisi, N., Baldisserri, B., Parenti-Castelli, V., Belvedere, C., Leardini, A.: One-degree-of-freedom spherical model for the passive motion of the human ankle joint. Med. Biol. Eng. Comput. **52**(4), 363–373 (2014)
18. Tsai, L.: Robot Analysis: The Mechanics of Serial and Parallel Manipulators. Wiley, Hoboken (1999)
19. Wilson, D., Feikes, J., Zavatsky, A., O'Connor, J.: The components of passive knee movement are coupled to flexion angle. J. Biomech. **33**(4), 465–473 (2000)
20. Wilson, D.R., Feikes, J.D., O'Connor, J.J.: Ligaments and articular contact guide passive knee flexion. J. Biomech. **31**(12), 1127–1136 (1998)

Development of a Vector Geometrical Model for PKM Identification

J.-B. Guyon(✉), B. Boudon, H. Chanal, and B. Blaysat

Université Clermont Auvergne, CNRS, SIGMA Clermont, Institut Pascal, 63000 Clermont-Ferrand, France
J-Baptiste.GUYON@uca.fr, helene.chanal@sigma-clermont.fr

Abstract. Geometric accuracy of parallel kinematic machine tool is an issue that limits their use in challenging context such as the aeronautic. One way to improve their accuracy consists at improving their geometrical identification process. For this purpose, a new geometric modeling method on the vectorial modeling of joint invariants is proposed in this article. This method introduces only independent geometric parameters and guarantees the robustness of the identification process. The studied machine-tool structure is modeled with links connected by joints. The geometrical end-effector pose is defined relying on geometrical defects and joint degrees of freedom. This new methodology is applied to an Exechon-like parallel kinematic machine tool, the Tripteor X7.

Keywords: Identification · PKM · Geometrical modeling · Sensitivity analysis

1 Introduction

The application of parallel kinematic machine tool (PKM) is still limited because of due to their lack of accuracy. Pritschow [12] presents six behaviors which have an influence on Parallel Kinematic Machines (PKM) accuracy: eigenmode excitation, elastic deformation from inertia and process load, thermal errors, gravity and geometrical transformation. Geometrical transformation error is the main concern of this article. In the review [7], Fu compares those errors between traditional serial kinematic machine-tool (SKM) and PKM. SKM geometrical errors are easily identified since they have directly an impact on tool pose accuracy. On the opposite the identification of PKM geometrical errors is complex due to the multiple closed-loop in the kinematic chain.

In order to reduce the errors of geometrical transformation, the review of Ramesh highlight the relevance of a proper geometrical identification when considering serial robots [13]. Moreover Wan shows the importance of geometrical identification for PKM in order to accurately reach the target point in the workspace [16].

A geometrical identification is performed in four steps: modeling, measurement, identification and correction [17]. We mainly focus here on the modeling

J. Lenarčič and B. Siciliano (Eds.): ARK 2020, SPAR 15, pp. 117–124, 2021.
https://doi.org/10.1007/978-3-030-50975-0_15

for improving the identification process. Geometrical modeling consists at defining a mathematical function that describes the geometrical behavior of a PKM.

In order to design the geometrical model, two strategies can be explored. The first strategy considers the PKM like a serial robot and models the global link between the actuator parameters and the position of the end-effector [3, 9]. The second strategy proposed by Song consists at splitting the PKM into several subsystems [15]. This strategy allows studying the influence of a given architecture to the geometric behavior. This is the strategy used in here with the use of only elementary joints with one degree of freedom. This leads to a model with only rotation and prismatic joints [6].

The definition of a geometric model introduces several parameters. The number of parameters should be reduced to improve the identification process. Reduction of the number can be achieved using independent modeling parameters for joint as in [4] or with sensitivity analysis [11]. Classical way for defining a geometrical models rely on the Denavit-Hartenberg method [1,8,14] . Nevertheless this model can be inadequate for modeling two adjacent, almost parallels joints for example. Moreover, it defines axes orientations with angle value [2]. Thus, introduced geometric parameters are angles and lengths and it decreases the identification process conditioning. For improving the conditioning of identification process, we choose to model joint axes orientations with the coordinates of their orientation vectors.

The PKM Tripteor X7 illustrates this work and is based on Exechon machine presented in [10]. The paper is organized as follows: the PKM Tripteor is firstly presented. The model methodology is introduced in the Sect. 3 with an application on the Tripteor X7. The Sect. 4 presents the sensibility analysis conducted on the developed model.

2 Tripteor X7

The Tripteor X7 is a three legs over-constrained PKM with a serial wrist. Two of those legs are identical. The identical legs,denoted here 1 and 3 are composed of a universal joint, a prismatic joint and then a rotational joint. The last leg ,numbered 2, features a spherical joint instead of the universal joint. Actuators of legs control the prismatic joints position. The wrist is composed of two controlled rotational joints. The Fig. 1 presents the design of the Tripteor X7 (left) and its joint parametrization (right).

3 Geometric Model

This section aims first at introducing the general concepts of the proposed modeling methodology. Its application to the Tripteor X7 is then discussed.

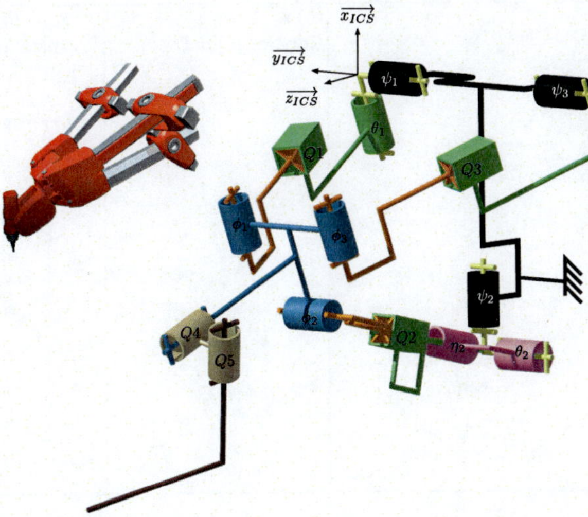

Fig. 1. Tripteor X7 (left), joint degrees of freedom (DoF) parameters of Tripteor X7 (right). This PKM is in our research institute workshop (Institut Pascal, France).

3.1 Principle

Let us consider two jointed links, denoted $i-1$ and i. Frame R_i (resp. R_{i-1}) is related to the link i (resp. $i-1$). An extra frame J_i is introduced here, it is aligned with joint axis and fixed to link $i-1$. Figure 2 shows the position and the orientation of the introduced reference frames. $(H_{R_{i-1}} H_{J_i})$ is the normal of straight lines $(A_{R_{i-1}}, \overrightarrow{z_{R_{i-1}}})$ and $(A_{J_i}, \overrightarrow{z_{J_i}})$.

Moreover, the positioning and the orientation of the axis of any joint are represented with homogeneous matrices. Matrix $\mathbf{D}_{J_i}^{R_{i-1}}$ (defined in Eq. (1)) denotes the change-of-basis matrix from the frame R_{i-1} to J_i. It traduces is used to describe the orientation and the position of the new frame J_i in the frame R_{i-1} (Fig. 2 (a)). $\mathbf{D}_{J_i}^{R_{i-1}}$ is the defect matrix with $\overrightarrow{z_{J_i}} = [I_{i-1} \quad J_{i-1} \quad K_{i-1}]_{R_{i-1}}$ and $\overrightarrow{A_{i-1}A_i} = a_{i-1} \cdot \overrightarrow{x_{i-1}} + b_{i-1} \cdot \overrightarrow{y_{i-1}} + c_{i-1} \cdot \overrightarrow{z_{i-1}}$. c_{i-1} is constraint as it always will be along a joint axis.

$$\mathbf{X}_{J_i} = \mathbf{D}_{J_i}^{R_{i-1}} \mathbf{X}_{R_{i-1}} = \begin{bmatrix} -\frac{J_{i-1}}{\sqrt{I_{i-1}^2+J_{i-1}^2}} & -\frac{I_{i-1}K_{i-1}}{\sqrt{I_{i-1}^2+J_{i-1}^2}} & I_{i-1} & a_{i-1} \\ \frac{I_{i-1}}{\sqrt{I_{i-1}^2+J_{i-1}^2}} & -\frac{J_{i-1}K_{i-1}}{\sqrt{I_{i-1}^2+J_{i-1}^2}} & J_{i-1} & b_{i-1} \\ 0 & \frac{I_{i-1}^2+J_{i-1}^2}{\sqrt{I_{i-1}^2+J_{i-1}^2}} & K_{i-1} & c_{i-1} \\ 0 & 0 & 0 & 1 \end{bmatrix}_{R_{i-1}} \mathbf{X}_{R_{i-1}} \quad (1)$$

Vector $\mathbf{X}_{J_i}$ represents the coordinate of the structural point in the J_i reference frame and $\mathbf{X}_{R_{I-1}}$ its coordinate in R_{i-1} reference frame. Matrix $\mathbf{D}_{J_i}^{R_{i-1}}$ remains undefined if $K_{R_{i-1}} = 1$. To circumvent this singular case, the rotation part of

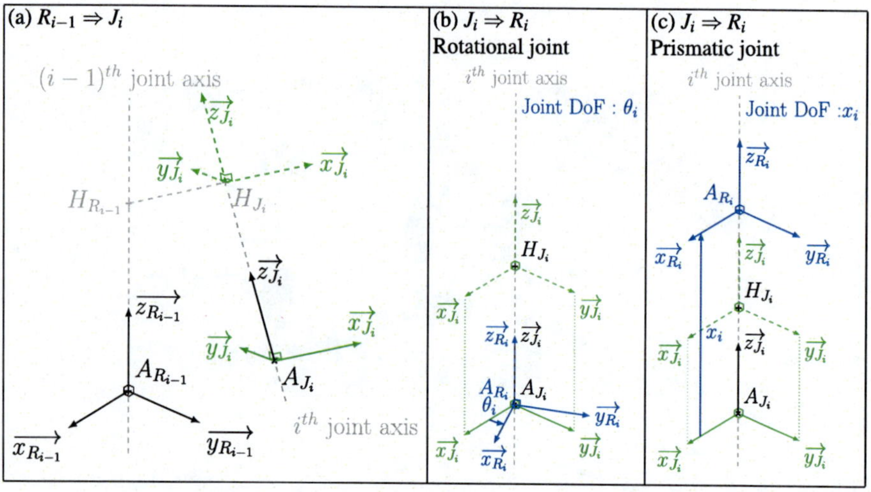

Fig. 2. Definition of the position and orientation of ith joint (a), application to rotational joint (b) and application to prismatic joint (c)

the homogeneous matrix is replaced by the 3×3 identity matrix. Since $\overrightarrow{z_{J_i}}$ is a unit vector, it satisfies:

$$\sqrt{I_{i-1}^2 + J_{i-1}^2 + K_{i-1}^2} = 1. \tag{2}$$

Axis $\overrightarrow{x_{J_i}}$ is a unit vector of direction the common normal of the straight line $(A_{R_{i-1}}, \overrightarrow{z_{R_{i-1}}})$ and the straight line $(A_{J_i}, \overrightarrow{z_{J_i}})$. Axis $\overrightarrow{y_{J_i}}$ is defined in order to obtain a direct orthonormal base.

The ideal i^{th} joint between frame J_i and R_i can be introduced now. It can be either a revolute joint (Fig. 2 (b)) or a prismatic joint (Fig. 2 (c)). Joints have invariant elements in their movement. Invariants for R joints are the rotation axis and a point of this axis. For a prismatic joint, the translation axis is the only invariant of the joint. An interested reader can refer to those invariant in [5]. We use invariants and the Eq. 2 to reduce parameters needed for the geometrical model:

- 4 parameters to identify geometric position and orientation of a revolute joint: $I_{i-1}, J_{i-1}, a_{i-1}, b_{i-1}$, plus θ_i as DoF;
- 2 parameters to identify geometric position and orientation of a prismatic joint: I_{i-1}, J_{i-1}, plus x_i as DoF.

Note that each joint can be split into a "one DoF joint": for example, a cylindrical joint is split into a R joint and a P joint. With the formalism of homogeneous matrices, we have the mobility matrices in Eq. (3) with rotation θ_i

on the left or translation x_i on the right for a link i.

$$\mathbf{X}_{R_i} = \begin{bmatrix} \cos(\theta_i) & -\sin(\theta_i) & 0 & 0 \\ \sin(\theta_i) & \cos(\theta_i) & 0 & 0 \\ 0 & 0 & 1 & 0 \\ 0 & 0 & 0 & 1 \end{bmatrix}_{R_{i-1}} \mathbf{X}_{J_i}, \qquad \mathbf{X}_{R_i} = \begin{bmatrix} 1 & 0 & 0 & 0 \\ 0 & 1 & 0 & 0 \\ 0 & 0 & 1 & x_i \\ 0 & 0 & 0 & 1 \end{bmatrix}_{R_{i-1}} \mathbf{X}_{J_i} \quad (3)$$

$\mathbf{X}_{R_i}$ represents the coordinate of the structural point in R_i reference frame. This principle is now applied on the Tripteor X7.

3.2 Application to Tripteor

The parameters θ_i and x_i will be replaced by the mobility parameters introduced in Fig. 2. In the case of the Tripteor X7, translation parameters are Q_1, Q_2 & Q_3. The rotation parameters are passive $(\psi_i, \theta_i, \phi_i)_{i \in \{1,2,3\}}$ & η_2 and active joint parameters Q_4 & Q_5. The geometric model is obtained by the product of the defect matrices and mobility matrices for each PKM leg. For the Tripteor X7, we got three matrices $\mathbf{T}_{ICS}^{MPS}$ due to each the three legs (MPS for *Mobile Platform System* and ICS for *Internal Coordinate System*) as well as a matrix $\mathbf{T}_T^{MPS}$ from MPS to T (for *Tool*) which describe the wrist geometric behavior.

$$\begin{cases} \mathbf{T}_{MPS}^{ICS} = \mathbf{T}_{MPS}^{\phi_1}\, \mathbf{M}_{\phi_1}^{15}\, \mathbf{D}_{15}^{Q1}\, \mathbf{M}_{Q1}^{14}\, \mathbf{D}_{14}^{\theta_1}\, \mathbf{M}_{\theta_1}^{12}\, \mathbf{D}_{12}^{\psi_1}\, \mathbf{M}_{\psi_1}^{11}\, \mathbf{D}_{11}^{ICS} & \text{Leg 1} \\ \mathbf{T}_{MPS}^{ICS} = \mathbf{T}_{MPS}^{\phi_2}\, \mathbf{M}_{\phi_2}^{25}\, \mathbf{D}_{25}^{Q2}\, \mathbf{M}_{Q2}^{24}\, \mathbf{D}_{24}^{\eta_2}\, \mathbf{M}_{\eta_2}^{23}\, \mathbf{D}_{23}^{\theta_2}\, \mathbf{M}_{\theta_2}^{22}\, \mathbf{D}_{22}^{\psi_2}\, \mathbf{M}_{\psi_2}^{21}\, \mathbf{D}_{21}^{ICS} & \text{Leg 2} \\ \mathbf{T}_{MPS}^{ICS} = \mathbf{T}_{MPS}^{\phi_3}\, \mathbf{M}_{\phi_3}^{35}\, \mathbf{D}_{35}^{Q3}\, \mathbf{M}_{Q3}^{34}\, \mathbf{D}_{34}^{\theta_3}\, \mathbf{M}_{\theta_3}^{32}\, \mathbf{D}_{32}^{\psi_3}\, \mathbf{M}_{\psi_3}^{31}\, \mathbf{D}_{31}^{ICS} & \text{Leg 3} \\ \mathbf{T}_T^{MPS} = \mathbf{T}_T^{Q5}\, \mathbf{M}_{Q5}^{7}\, \mathbf{D}_7^{Q4}\, \mathbf{M}_{Q4}^{6}\, \mathbf{D}_6^{MPS} & \text{Wrist} \end{cases} \quad (4)$$

With this model, we introduced 4 parameters for each R joint and 2 parameters for each P joint. Thus we respect respect the equation of [4] which defined the number of needed independent parameters for describing geometrical behavior of robot. To validate the independence of introduced parameters and the influence of over-constrained behavior of Tripteor X7 PKM, a sensitivity analysis is conducted

4 Sensibility Analysis on the Model of the Tripteor X7

To validate the independence of introduced parameters, a sensibility analysis is realized. Sensibility analysis allows to determine the relative influence of each parameter in the end-effector pose. To conduct this analysis, the direct model is differentiated with regard to each parameters. For this sensitivity analysis, we focus on the parallel mechanism. Thus, the sensitivity matrix S is defined from Eq. (5).

$$S = \frac{\partial \mathbf{T}_{end}^{start}}{\partial \xi} \quad (5)$$

Matrix $\mathbf{T}_{end}^{start}$ is the model developed in the previous section. ξ are parameters introduced in matrices $\mathbf{D}_{J_i}^{R_{i-1}}$: I_{i-1}, J_{i-1}, K_{i-1}, a_{i-1}, b_{i-1}, c_{i-1}. The sensitivity from each parameter is obtain from the relation $dX = Sd\xi$. We gathered the maximum of sensitivity across the workspace and obtained the Fig. 3 for the parallel part of the Tripteor X7.

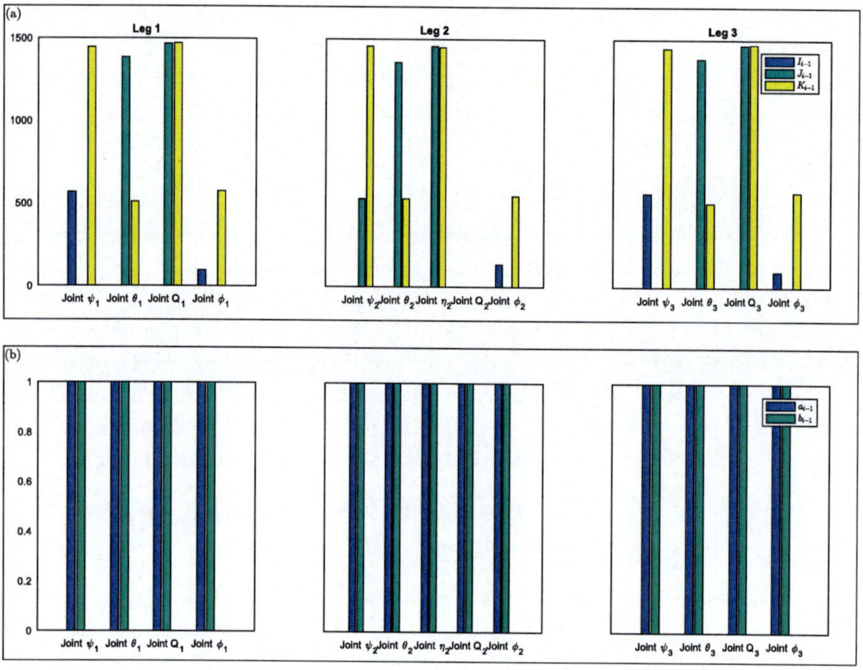

Fig. 3. Maximum sensitivity of orientation defects (a) and position defects (b) across the workspace

It appears orientation of joint frames (I_{i-1}, J_{i-1} and K_{i-1}) mostly got an important influence on the platform. It is to be noted that the values on the graph are with an unit defect, which shall never happened in the machine. However for every joint, one parameter of the triplet has no sensitivity so no influence. Also joint positioning parameters have an influence but lower as defect are small compared to nominal dimensions. As announced in 3.1, c_{i-1} is constrained so there is no need to study its sensitivity.

5 Conclusion

In this paper, we presented a new geometric model methodology which focus on the technological joint choice of a structure. This model also splits every joint into a defect parameters and a movement parameters. Defects are defined in order to avoid the use of angles for joint axes definitions.

Sensitivity analysis show that this model is relevant since the orientation and position of joints features influence to the whole machine accuracy and illustrate the need to weight unit vector parameters. A constant weight can be used as their sensitivity are similar. The work on the whole method is still in progress, further results on the numerical evaluation are expected.

Acknowledgements. This work is partially supported by the ECSASDPE H2020-MSCA-RISE-2016 n°734272 project.

References

1. Alici, G., Shirinzadeh, B.: A systematic technique to estimate positioning errors for robot accuracy improvement using laser interferometry based sensing. Mech. Mach. Theory **40**(8), 879–906 (2005). https://doi.org/10.1016/j.mechmachtheory.2004.12.012
2. Bai, Y., Zhuang, H., Roth, Z.: Experiment study of PUMA robot calibration using a laser tracking system. In: Proceedings of the 2003 IEEE International Workshop on Soft Computing in Industrial Applications, 2003. SMCia/03. IEEE (2003). https://doi.org/10.1109/smcia.2003.1231359
3. Chanal, H., Duc, E., Ray, P., Hascoët, J.: A new approach for the geometrical calibration of parallel kinematics machines tools based on the machining of a dedicated part. Int. J. Mach. Tools Manuf **47**(7–8), 1151–1163 (2007). https://doi.org/10.1016/j.ijmachtools.2006.09.006
4. Chen, G., Wang, H., Lin, Z.: Determination of the identifiable parameters in robot calibration based on the POE formula. IEEE Trans. Rob. **30**(5), 1066–1077 (2014). https://doi.org/10.1109/tro.2014.2319560
5. Dufailly, J., Poss, M.: Cotation fonctionnelle, chaîne de cotes, optimisation des tolérances: analyse fonctionnelle technique. Ellipses (2017)
6. Everett, L., Driels, M., Mooring, B.: Kinematic modelling for robot calibration. In: Proceedings of the IEEE International Conference on Robotics and Automation, Institute of Electrical and Electronics Engineers (1987). https://doi.org/10.1109/robot.1987.1087818
7. Fu, R., Jin, Y., Yang, L., Sun, D., Murphy, A., Higgins, C.: Review on structure-based errors of parallel kinematic machines in comparison with traditional NC machines. In: Communications in Computer and Information Science, pp. 249–256. Springer, Singapore (2018). https://doi.org/10.1007/978-981-13-2396-6_23
8. Hartenberg, R.S., Denavit, J.: A kinematic notation for lower pair mechanisms based on matrices (1955)
9. Huang, T., Chetwynd, D.G., Whitehouse, D.J., Wang, J.: A general and novel approach for parameter identification of 6-DOF parallel kinematic machines. Mech. Mach. Theory **40**(2), 219–239 (2005). https://doi.org/10.1016/j.mechmachtheory.2004.06.009
10. Jin, Y., et al.: Kinematic analysis and dimensional synthesis of Exechon parallel kinematic machine for large volume machining. J. Mech. Robot. **7**(4) (2015). https://doi.org/10.1115/1.4029499
11. Fan, K.C., Wang, H., Zhao, J.W., Chang, T.H.: Sensitivity analysis of the 3-PRS parallel kinematic spindle platform of a serial-parallel machine tool. Int. J. Mach. Tools Manuf **43**(15), 1561–1569 (2003). https://doi.org/10.1016/s0890-6955(03)00202-5

12. Pritschow, G., Eppler, C., Garber, T.: Influence of the dynamic stiffness on the accuracy of PKM. In: Chemnitz Parallel Kinematic Seminar, pp. 313–333 (2002)
13. Ramesh, R., Mannan, M., Poo, A.: Error compensation in machine tools — a review. Int. J. Mach. Tools Manuf **40**(9), 1235–1256 (2000). https://doi.org/10.1016/s0890-6955(00)00009-2
14. Renders, J.M., Rossignol, E., Becquet, M., Hanus, R.: Kinematic calibration and geometrical parameter identification for robots. IEEE Trans. Robot. Autom. **7**(6), 721–732 (1991). https://doi.org/10.1109/70.105381
15. Song, Y., Zhang, J., Lian, B., Sun, T.: Kinematic calibration of a 5-DoF parallel kinematic machine. Precis. Eng. **45**, 242–261 (2016). https://doi.org/10.1016/j.precisioneng.2016.03.002
16. Wan, A., Song, L., Xu, J., Liu, S., Chen, K.: Calibration and compensation of machine tool volumetric error using a laser tracker. Int. J. Mach. Tools Manuf **124**, 126–133 (2018). https://doi.org/10.1016/j.ijmachtools.2017.10.004
17. Yang, X., Wu, L., Li, J., Chen, K.: A minimal kinematic model for serial robot calibration using POE formula. Robot. Comput. Integr. Manuf. **30**(3), 326–334 (2014). https://doi.org/10.1016/j.rcim.2013.11.002

Wohlhart's Three-Loop Mechanism: An Overconstrained and Shaky Linkage

Andreas Müller(✉)

Johannes Kepler University, Linz, Austria
a.mueller@jku.at

Abstract. This paper revisits a three-loop spatial linkage that was proposed in an ARK 2004 paper by Karl Wohlhart (as extension of a two-loop linkage proposed by Eddie Baker in 1980) and later analyzed in an ARK 2006 paper by Diez-Martínez et al. A local analysis shows that this linkage has a finite degree of freedom (DOF) 3 (and is thus overconstrained) while in its reference configuration the differential DOF is 5. It is shown that its configuration space is locally a smooth manifold so that the reference configuration is not a c-space singularity. It is shown that the differential DOF is locally constant, which makes this linkage shaky (so that the reference configuration is not a singularity). The higher-order local analysis is facilitated by the computation of the kinematic tangent cone as well as a local approximation of the c-space.

Keywords: Multi-loop linkage · Configuration space · Kinematic singularities · Shaky · Higher-order analysis · Kinematic tangent cone

1 Introduction

In [1], Baker proposed a two-loop linkage as an example to demonstrate the application of his method for identification of the relative twist of any two bodies, and thus their instantaneous relative mobility, in a mechanism. This linkage was also used by Davies [2] to demonstrate his graph representation of the kinematic topology. Fayet [4] extended it to a three-loop linkage and used it as an example for application of his method to determine the 'relative screw space', i.e. the subspace of $se(3)$ defined by the possible relative twists. An efficient method for identification of the relative twists was proposed by Wohlhart [16], and later Diez-Martínez et al. [3] proposed a method for determination of the finite relative mobility. In both publications, a variant of Fayet's linkage shown in Fig. 1 was used as an example, and it was analyzed in the shown reference configuration. It was shown in [16] that in this reference configuration the differential DOF is 5. In [3], the finite DOF was found to be 3. The differential and local DOF being different may suggest that the reference configuration is a kinematic singularity. This question has not yet been addressed, and will be answered in this paper by means of a local mobility analysis. The latter determines the finite mobility at the reference configuration in Fig. 1 as well as the differential

J. Lenarčič and B. Siciliano (Eds.): ARK 2020, SPAR 15, pp. 125–132, 2021.
https://doi.org/10.1007/978-3-030-50975-0_16

mobility when the linkage performs finite motions through that configuration. It will be shown that this configuration is a regular point of the c-space, that it is not a kinematic singularity, and that the differential DOF is always 5, which renders the linkage shaky. A mechanism is shaky iff, in a regular configuration, its differential (instantaneous) DOF is higher than its local (finite) DOF. The analysis makes use of the kinematic tangent cone, which provides the mathematical framework for such a local analysis as introduced in [5,8,9,12], and a local approximation of the c-space. The computation is enabled by the closed form and recursive relations for the higher-order expressions in terms of joint screws [13]. A Mathematica© notebook with all computations for the linkage in Fig. 1 can be obtained from the author.

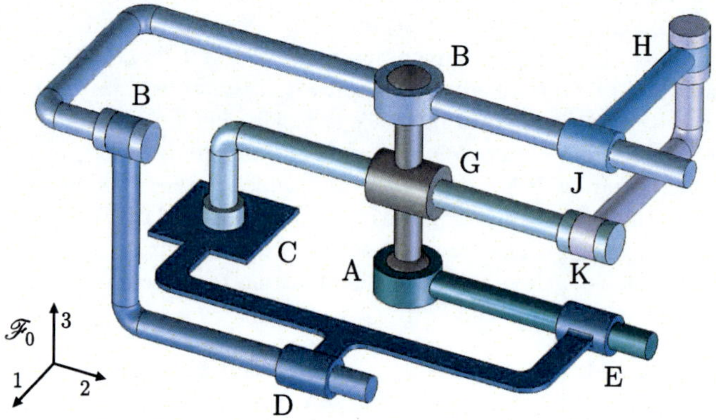

Fig. 1. Multiloop linkage presented by Fayet (CAD drawing curtesy of Jose Rico).

2 Kinematic Topology

The kinematic topology of a linkage is represented by a non-directed (multi)graph Γ [2,10]. Vertices represent bodies/links, and edges represent (1-DOF) joints. The linkage in Fig. 1 comprises 10 joints (2 spherical, 4 cylindrical, 1 planar, 3 revolute) and 8 links. The cylindrical, spherical, and planar joints are modeled as combination of helical joints [3,16]. The kinematic model thus consists of $n = 20$ helical joints (6 prismatic, 14 revolute) and $N = 17$ links. The topological graph possesses $\gamma = N - n + 1 = 3$ fundamental cycles (FC), also called fundamental loops, denoted with $\Lambda_l, l = 1, \ldots, \gamma$. A co-tree $\mathscr{H}$ can be introduced consisting of exactly one edge in each of the γ FCs. Eliminating the co-tree edges yields a spanning tree $\mathscr{G}$. For the kinematics modeling, an associated directed graph $\vec{\Gamma}$ is introduced. The edge directions indicate the polarity of joint variables. This defines the corresponding directed tree $\vec{\mathscr{G}}$ and co-tree $\vec{\mathscr{H}}$, as shown in Fig. 2 (vertex/body labels are omitted). FCs induce an ordering according to how edges appear when traversing a FC. All edges are aligned with the FCs except edges 15 and 16.

The correspondence of edges of $\vec{\Gamma}$ (i.e. revolute/prismatic joints in the model) to the multi-DOF joints of the linkage is as follows: *Joint A*: 1, 2, 3; *B*: 4, 5, 6; *C*: 7, 8, 9; *D*: 10, 11; *E*: 12, 13; *F*: 14; *G*: 15, 16; *H*: 17; *J*: 18, 19; *K*: 20.

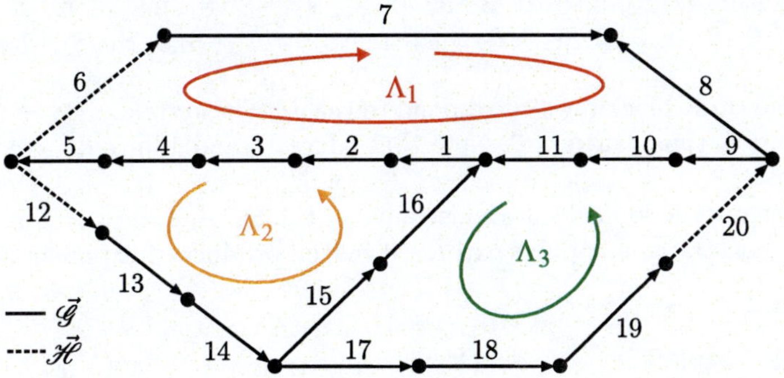

Fig. 2. Tree $\vec{\mathscr{G}}$, co-tree $\vec{\mathscr{H}}$, and FCs Λ_l of the topological graph $\vec{\Gamma}$ for the linkage in Fig. 1.

3 Kinematic Model

Edge i of the topological graph corresponds to one of the helical joints, modeled with a screw coordinate vector, denoted with $\mathbf{Y}_i, i = 1, \ldots, n$, and a joint variable, denoted with $q_i, i = 1, \ldots, n$. The latter are summarized in the vector $\mathbf{q} \in \mathbb{V}^n = \mathbb{R}^6 \times \mathbb{T}^{14}$. The configuration in Fig. 1 is used as zero reference configuration $\mathbf{q}_0 = \mathbf{0}$. The screw coordinate vectors are determined in this reference configuration, and represented in the spatial frame $\mathscr{F}_0$ when located at joint A (adopted from [3]):

$$
\begin{aligned}
&\mathbf{Y}_1 = (1,0,0,0,0,0)^T, \mathbf{Y}_2 = (0,1,0,0,0,0)^T, \mathbf{Y}_3 = \mathbf{Y}_6 = (0,0,1,0,0,0)^T \\
&\mathbf{Y}_4 = (1,0,0,0,1,0)^T, \mathbf{Y}_5 = (0,1,0,-1,0,0)^T, \mathbf{Y}_7 = (0,0,0,1,0,0)^T \\
&\mathbf{Y}_8 = \mathbf{Y}_{10} = \mathbf{Y}_{12} = \mathbf{Y}_{15} = \mathbf{Y}_{18} = (0,0,0,0,1,0)^T \\
&\mathbf{Y}_9 = (0,0,1,-1,0,0)^T, \mathbf{Y}_{11} = (0,1,0,0,0,1)^T \\
&\mathbf{Y}_{13} = (0,1,0,0,0,0)^T, \mathbf{Y}_{14} = (0,1,0,-1,0,1)^T, \mathbf{Y}_{16} = (0,1,0,-1/2,0,0)^T \\
&\mathbf{Y}_{17} = (0,0,1,1,1,0)^T, \mathbf{Y}_{19} = (0,1,0,-1,0,0)^T, \mathbf{Y}_{20} = (0,1,0,-1/2,1,0)^T .
\end{aligned}
$$

The instantaneous joint screw coordinates in an arbitrary configuration are denoted with $\mathbf{S}_i(\mathbf{q})$. They are found from the joint screw coordinates in the reference configuration by a frame transformation of $\mathbf{Y}_i$ to the current configuration [12,13]. For instance $\mathbf{S}_9(\mathbf{q}) = \mathbf{Ad}_{\exp(\mathbf{Y}_1 q_1)\cdot\ldots\cdot\exp(\mathbf{Y}_8 q_8)}\mathbf{Y}_9$, where $\mathbf{Ad}$ is the matrix transforming screw coordinates [15]. Indeed, $\mathbf{S}_i(\mathbf{0}) = \mathbf{Y}_i$ in the zero reference configuration.

Denote with $\mathbf{J}_l$ the constraint Jacobian in the velocity constraints $\mathbf{J}_l(\mathbf{q})\dot{\mathbf{q}} = \mathbf{0}$ for the FC $\Lambda_l, l = 1, \ldots, 3$. The overall constraint Jacobian for the $\gamma = 3$ FCs, according to the orientation of the topological graph, is the 18×20 matrix

$$\mathbf{J} = \begin{pmatrix} \mathbf{S}_1 & \mathbf{S}_2 & \mathbf{S}_3 & \mathbf{S}_4 & \mathbf{S}_5 & \mathbf{S}_6 & \mathbf{S}_7 & \mathbf{S}_8 & \mathbf{S}_9 & \mathbf{S}_{10} & \mathbf{S}_{11} & \mathbf{0} & \mathbf{0} & \mathbf{0} & \mathbf{0} & \mathbf{0} & \mathbf{0} & \mathbf{0} & \mathbf{0} & \mathbf{0} \\ \mathbf{S}_1 & \mathbf{S}_2 & \mathbf{S}_3 & \mathbf{S}_4 & \mathbf{S}_5 & \mathbf{0} & \mathbf{0} & \mathbf{0} & \mathbf{0} & \mathbf{0} & \mathbf{0} & \mathbf{S}_{12} & \mathbf{S}_{13} & \mathbf{S}_{14} & \mathbf{S}_{15} & \mathbf{S}_{16} & \mathbf{0} & \mathbf{0} & \mathbf{0} & \mathbf{0} \\ \mathbf{0} & \mathbf{0} & \mathbf{0} & \mathbf{0} & \mathbf{0} & \mathbf{0} & \mathbf{0} & \mathbf{0} & \mathbf{S}_9 & \mathbf{S}_{10} & \mathbf{S}_{11} & \mathbf{0} & \mathbf{0} & \mathbf{0} & -\mathbf{S}_{15} & -\mathbf{S}_{16} & \mathbf{S}_{17} & \mathbf{S}_{18} & \mathbf{S}_{19} & \mathbf{S}_{20} \end{pmatrix} \quad (1)$$

where the three block rows correspond to the $\mathbf{J}_l$. The system of $6\gamma = 18$ velocity constraints is thus $\mathbf{J}(\mathbf{q})\dot{\mathbf{q}} = \mathbf{0}$. Note that edges 15 and 16 are directed opposite to Λ_3.

The geometric loop closure constraints of the FC $\Lambda_l, l = 1, \ldots, \gamma$ is $f_l(\mathbf{q}) = \mathbf{I}$ with the constraint mappings expressed by the product of exponentials

$$f_1(\mathbf{q}) = \exp(\mathbf{Y}_1 q_1) \cdot \ldots \cdot \exp(\mathbf{Y}_{11} q_{11}) \quad (2)$$
$$f_2(\mathbf{q}) = \exp(\mathbf{Y}_1 q_1) \cdot \ldots \cdot \exp(\mathbf{Y}_5 q_5) \exp(\mathbf{Y}_{12} q_{12}) \cdot \ldots \cdot \exp(\mathbf{Y}_{16} q_{16}) \quad (3)$$
$$f_3(\mathbf{q}) = \exp(\mathbf{Y}_9 q_9) \exp(\mathbf{Y}_{10} q_{10}) \exp(\mathbf{Y}_{11} q_{11}) \exp(-\mathbf{Y}_{16} q_{16}) \exp(-\mathbf{Y}_{15} q_{15}) \cdot \exp(\mathbf{Y}_{17} q_{17}) \exp(\mathbf{Y}_{18} q_{18}) \exp(\mathbf{Y}_{19} q_{19}) \exp(\mathbf{Y}_{20} q_{20}) \quad (4)$$

where the ordering is defined by the FCs and the sign in the exponential by the direction of edges relative to the FCs. The c-space is then the real analytic variety

$$V := \{\mathbf{q} \in \mathbb{V}^{20} | f_l(\mathbf{q}) = \mathbf{I}, l = 1, \ldots, 3\}. \quad (5)$$

The principle aim of a mobility analysis consists in determining the dimension (i.e. the DOF) and the topology of V (i.e. the mobility). The differential DOF at $\mathbf{q} \in V$ is $\delta_{\mathrm{diff}}(\mathbf{q}) = \dim \ker \mathbf{J}(\mathbf{q}) = n - \mathrm{rank}\, \mathbf{J}(\mathbf{q})$. The differential mobility is described by vectors $\mathbf{x} \in \ker \mathbf{J}(\mathbf{q})$. The finite local DOF at $\mathbf{q} \in V$ is the local dimension of V, denoted $\delta_{\mathrm{loc}}(\mathbf{q}) = \dim_{\mathbf{q}} V$.

C-space singularities are non-smooth points of V (where V is not a smooth manifold). *Kinematic singularities* are points where the differential DOF is not constant in any neighborhood of $\mathbf{q}$ in V. C-space singularities are kinematic singularities. The opposite is not necessarily true. *Constraint singularities* are points where the constraint Jacobian is not full rank. C-space and kinematic singularities are constraint singularities. The opposite is not necessarily true (e.g. overconstrained mechanisms).

4 Smooth Finite Motions Through the Reference Configuration

A motion of the linkage corresponds to a curve in V. At a given configuration $\mathbf{q} \in V$, the tangents to smooth finite curves in V through $\mathbf{q}$, i.e. to smooth motions, form the kinematic tangent cone $C^{\mathrm{K}}_{\mathbf{q}} V := \{\dot{\sigma} | \sigma \in \mathscr{C}_{\mathbf{q}}\} \subset \mathbb{R}^n$, where $\mathscr{C}_{\mathbf{q}}$ is the class of smooth curves in V through $\mathbf{q} \in V$ [12]. With the parameterization in terms of joint variables it can be simply regarded as a formal definition of the set of all velocity vectors $\dot{\mathbf{q}}$ such that $\mathbf{q}(t)$ is a smooth curve in V and satisfies time derivative of the loop constraints of any order. It can thus be defined in terms of higher-order derivatives of the velocity constraints [5,8,9]. Denoting $H_l^{(1)}(\mathbf{q},\dot{\mathbf{q}}) := \mathbf{J}_l(\mathbf{q})\dot{\mathbf{q}}$ and $H_l^{(i)}(\mathbf{q},\dot{\mathbf{q}},\ldots,\mathbf{q}^{(i)}) := \frac{d^{i-1}}{dt^{i-1}}H_l^{(1)}(\mathbf{q},\dot{\mathbf{q}})$, the ith-order constraints for the FC Λ_l is

$$H_l^{(i)}(\mathbf{q},\dot{\mathbf{q}},\ldots,\mathbf{q}^{(i)}) = \mathbf{0}. \tag{6}$$

The kinematic tangent cone is then determined as

$$C_{\mathbf{q}}^{\mathrm{K}}V = K_{\mathbf{q}}^{\kappa} \subseteq \ldots \subseteq K_{\mathbf{q}}^{3} \subseteq K_{\mathbf{q}}^{2} \subseteq K_{\mathbf{q}}^{1} \tag{7}$$

where $K_{\mathbf{q}}^{i}$ is a cone (in $\mathbf{x}$) defined as

$$\begin{aligned} K_{\mathbf{q}}^{i} := \{\mathbf{x}|\exists \mathbf{y},\mathbf{z},\ldots \in \mathbb{R}^{20} : H_l^{(1)}(\mathbf{q},\mathbf{x}) = \mathbf{0}, H_l^{(2)}(\mathbf{q},\mathbf{x},\mathbf{y}) = \mathbf{0}, H_l^{(3)}(\mathbf{q},\mathbf{x},\mathbf{y},\mathbf{z}) = \mathbf{0},\ldots \\ \ldots, H_l^{(i)}(\mathbf{q},\mathbf{x},\mathbf{y},\mathbf{z},\ldots) = \mathbf{0}, l = 1,2,3\}. \end{aligned} \tag{8}$$

The sequence (7) terminates with finite κ, which is specific to the linkage and depends on the configuration. The crucial point of this formulation is that the higher-order mappings $H_l^{(i)}$ can be evaluated in closed form or recursively by means of simple vector operations [12,13]. A Mathematica© implementation is available at [11].

The kinematic tangent cone provides a local description of the manifolds intersecting at a c-space singularity $\mathbf{q}$. It thus allows to identify motion bifurcations, including non-transversal intersections [6]. The calculation shows that the first- and second-order cone is a 5-dimensional and 3-dimensional vector space, respectively:

$$\begin{aligned} K_{\mathbf{q}_0}^{1} = \{\mathbf{x} \in \mathbb{R}^{20}|\ & x_1 = 0, x_2 = u, x_3 = v, x_4 = 0, x_5 = w + 2v, x_6 = -v, x_7 = w/2, \\ & x_8 = r, x_9 = -v, x_{10} = s, x_{11} = w + 2v, x_{12} = -s, x_{13} = -w - u - 2v, \\ & x_{14} = -w - 2v, x_{15} = -r + s, x_{16} = w + 2v, x_{17} = v, \\ & x_{18} = -w - v - r + s, x_{19} = -w, x_{20} = w;\ r,s,u,v,w \in \mathbb{R}\}. \end{aligned}$$

$$\begin{aligned} K_{\mathbf{q}_0}^{2} = \{\mathbf{x} \in \mathbb{R}^{20}|\ & x_1 = 0, x_2 = -u, x_3 = 0, x_4 = 0, x_5 = 0, x_6 = 0, x_7 = 0, x_8 = r, \\ & x_9 = 0, x_{10} = s, x_{11} = 0, x_{12} = -x_{10}, x_{13} = u, x_{14} = 0, x_{15} = -r + s, \\ & x_{16} = 0, x_{17} = 0, x_{18} = -r + s, x_{19} = 0, x_{20} = 0;\ r,s,u \in \mathbb{R}\}. \end{aligned}$$

This shows that, at $\mathbf{q}_0$, the linkage can perform 5-dimensional first-order motions, so that the differential DOF is $\delta_{\mathrm{diff}}(\mathbf{q}_0) = \dim K_{\mathbf{q}_0}^{1}V = 5$. The higher-order cones are $K_{\mathbf{q}_0}^{i} = K_{\mathbf{q}_0}^{2}, i \geq 3$, and thus $C_{\mathbf{q}_0}^{\mathrm{K}}V = K_{\mathbf{q}_0}^{2}$. This shows that there is exactly one manifold passing through $\mathbf{q}_0$. Consequently, the linkage admits well-defined 3-dimensional smooth motions, but no bifurcation, and the local finite DOF has a lower bound $\delta_{\mathrm{loc}}(\mathbf{q}_0) \geq \dim C_{\mathbf{q}_0}^{\mathrm{K}}V = 3$. The DOF could be higher than 3 if $\mathbf{q}_0$ belongs to a subvariety of V (with higher dimension) which does not allow smooth motions, e.g. cusps. The latter is not captured by the kinematic tangent cone (or any analysis based on higher-order derivatives of constraints). This is addressed in Sect. 6.

5 Differential Mobility at Smooth Finite Motions

The analysis of the local finite mobility does not reveal the differential mobility when passing through $\mathbf{q}_0$. It remains to determine i) the differential DOF when the linkage performs finite motions through $\mathbf{q}_0$, and ii) whether the differential DOF is locally constant in a neighborhood of $\mathbf{q}_0$ in V. The first allows to conclude whether the linkage is shaky. The latter is crucial in order to assert whether $\mathbf{q}_0$ is a kinematic singularity. It was recently recognized (although this phenomenon has been know for a long time already) that a mechanism can exhibit kinematic singularities even when the c-space is a smooth manifold [14].

Denote with $\mathbf{J}_{\alpha\beta}$ the $k \times k$ submatrix of $\mathbf{J}$, containing elements J_{ij} according to the index sets $\boldsymbol{\alpha} = \{\alpha_1, \ldots, \alpha_k\}, \alpha_{i-1} < \alpha_i$ and $\boldsymbol{\beta} = \{\beta_1, \ldots, \beta_k\}, \beta_{j-1} < \beta_j$. The $\boldsymbol{\alpha\beta}$-minor of $\mathbf{J}$ of order k is then $m_{\alpha\beta}(\mathbf{q}) := \det \mathbf{J}_{\alpha\beta}(\mathbf{q})$. Its ith derivative is denoted with $M^{(i)}_{\alpha\beta}(\mathbf{q}, \dot{\mathbf{q}}, \ldots, \mathbf{q}^{(i)}) := \frac{d^i}{dt^i} m_{\alpha\beta}(\mathbf{q})$. The set of points where the constraint Jacobian has rank less than k can be defined in terms of the k-minors [9,12]

$$L_k := \{\mathbf{q} \in \mathbb{V}^{20} |\ f_l(\mathbf{q}) = \mathbf{I}, m_{\alpha\beta}(\mathbf{q}) = 0, l = 1, 2, 3 \\ \forall |\boldsymbol{\alpha}| = |\boldsymbol{\beta}| = k, \boldsymbol{\alpha} \subseteq \{1, \ldots, 6\}, \boldsymbol{\beta} \subseteq \{1, \ldots, n\}\}. \tag{9}$$

Of interest are the finite smooth motions with rank less than k. Tangents to such motions through $\mathbf{q} \in V$ form the kinematic tangent cone to L_k, determined by the sequence

$$C^{\mathrm{K}}_{\mathbf{q}} L_k = K^{k,\kappa}_{\mathbf{q}} \subset \ldots \subset K^{k,3}_{\mathbf{q}} \subset K^{k,2}_{\mathbf{q}} \subset K^{k,1}_{\mathbf{q}} \tag{10}$$

with the ith-order cone defined by the ith-order derivatives of constraints and minors

$$K^{k,i}_{\mathbf{q}} := \{\mathbf{x} | \exists \mathbf{y}, \mathbf{z}, \ldots \in \mathbb{R}^{20} : H^{(1)}_l(\mathbf{q}, \mathbf{x}) = H^{(2)}_l(\mathbf{q}, \mathbf{x}, \mathbf{y}) = \ldots = H^{(i)}_l(\mathbf{q}, \mathbf{x}, \mathbf{y}, \mathbf{z}, \ldots) = \mathbf{0}, \\ M^{(1)}_{\alpha\beta}(\mathbf{q}, \mathbf{x}) = M^{(2)}_{\alpha\beta}(\mathbf{q}, \mathbf{x}, \mathbf{y}) = \ldots = M^{(i)}_{\alpha\beta}(\mathbf{q}, \mathbf{x}, \mathbf{y}, \mathbf{z}, \ldots) = 0, \\ \forall \boldsymbol{\alpha} \subseteq \{1, \ldots, 6\}, \boldsymbol{\beta} \subseteq \{1, \ldots, n\}, |\boldsymbol{\alpha}| = |\boldsymbol{\beta}| = k, l = 1, 2, 3\}. \tag{11}$$

Calculation of the derivatives of the minors is again possible efficiently by simple vector operations [13], also available as Mathematica© implementation [11].

In the reference configuration $\mathbf{q}_0$, the Jacobian has rank $\mathbf{J}(\mathbf{q}_0) = 15$, while it can not have rank higher than $3 \cdot 6 = 18$. Thus $L_k, k = 16, 17, 18$ must be investigated. The computation yields that all derivatives of the minors of order $k = 16, 17, 18$, $M^{(i)}_{\alpha\beta} = 0, |\boldsymbol{\alpha}| = |\boldsymbol{\beta}| = 16, 17, 18$ are zero for solutions on (6). It is concluded that the Jacobian has locally constant rank 15 for any smooth motion through $\mathbf{q}_0$. Thus the linkage has a locally constant differential DOF $\delta_{\text{diff}}(\mathbf{q}) = n - \text{rank}\ \mathbf{J}(\mathbf{q}) = 20 - 15 = 5, \mathbf{q} \in U(\mathbf{q}_0) \cap V$. Since its differential and local DOF are permanently different, the linkage is shaky of degree $\delta_{\text{diff}} - \delta_{\text{loc}} = 2$.

6 Local Approximation of the Configuration Space

The above analysis does not capture singularities where no smooth motions are possible, and the reference configuration $\mathbf{q}_0$ may still be a singularity where V is not locally the intersection of manifolds. This happens at a dead-point/motion-reversal point (also called stationary singularity). At such configurations the c-space is not the intersection of manifolds so that no tangent can be defined at that point, e.g. at cusps [7,12]. This can be checked by means of a local approximation of the c-space V using the kth-order Taylor-series expansion of the constraint mappings at $\mathbf{q}$

$$f_l(\mathbf{q}+\mathbf{x}) = f_l(\mathbf{q}) + \sum_{k\geq 1} \frac{1}{k!} \mathrm{d}^k f_{l,\mathbf{q}}(\mathbf{x}). \tag{12}$$

Since $f_l(\mathbf{q}) = \mathbf{I}$, for $\mathbf{q} \in V$, the kth-order approximation of V at $\mathbf{q}$ is given by

$$V_{\mathbf{q}}^k := \{\mathbf{x} \in \mathbb{R}^n | \mathrm{d}f_{l,\mathbf{q}}(\mathbf{x}) + \frac{1}{2}\mathrm{d}^2 f_{l,\mathbf{q}}(\mathbf{x}) + \ldots + \frac{1}{k!}\mathrm{d}^k f_{l,\mathbf{q}}(\mathbf{x}) = \mathbf{0}, l = 1, \ldots, 3\}. \tag{13}$$

An efficient recursive and explicit expression of the differentials $\mathrm{d}^k f_{l,\mathbf{q}}$ was reported in [12,13], and a Mathematica© implementation can be found in [11]. The computation yields $V_{\mathbf{q}_0}^1 = K_{\mathbf{q}_0}^1$ and $V_{\mathbf{q}_0}^2 = K_{\mathbf{q}_0}^2$, which confirms that the linkage can only perform the smooth 3-dimensional motions.

7 Conclusion

It was shown that, in the reference configuration, the three-loop linkage presented by Wohlhart [16] has finite local DOF 3, a locally constant differential DOF 5, and that the reference configuration is regular. Consequently, the linkage can perform finite 3-dimensional smooth motions. Since the differential DOF exceeds the local DOF at all regular points, it is shaky of degree 2.

The linkage is also overconstrained. The generic (topological) DOF (of any spatial linkage with the given topology) is given by the Chebychev-Kutzbach-Grübler formula, which can be expressed as $\delta_{\mathrm{top}} = \sum_{i \in J} \delta_i - 6\gamma = n - 6\gamma$. Applied to this linkage, the generic DOF is $\delta_{\mathrm{top}} = 20 - 6 \cdot 3 = 2$, so that it is overconstrained of degree 1.

In conclusion, the reference configuration is a constraint singularity but not a kinematic singularity or a c-space singularity. It should be remarked that the three-loop linkage shown in Fig. 4 of [4] (repeated as Fig. 3 in [16]) shows similar interesting properties. The presented analysis only reveals the mobility at the reference configuration (which is arbitrary though). A global analysis will have to resort to an algebraic formulation and use tools from geometric algebra.

Acknowledgements. This work has been support by the LCM K2 Center for Symbiotic Mechatronics within the framework of the Austrian COMET-K2 program.

References

1. Baker, J.E.: On relative freedom between links in kinematic chains with cross-jointing. Mech. Mach. Theory **15**(5), 397–413 (1980)

2. Davies, T.: Kirchhoff's circulation law applied to multi-loop kinematic chains. Mech. Mach. Theory **16**(3), 171–183 (1981)
3. Diez-Martínez, C.R., Rico, J.M., Cervantes-Sánchez, J.J., Gallardo, J.: Mobility and connectivity in multiloop linkages. In: Lennarčič, J., Roth, B. (eds.) Advances in Robot Kinematics, pp. 455–464. Springer, Netherlands (2006)
4. Fayet, M.: Mécanismes multi-boucles détermination des espaces de torseurs cinématiques dans un mécanisme multi-boucles quelconque. Mech. Mach. Theory **30**(2), 201–217 (1995)
5. Lerbet, J.: Analytic geometry and singularities of mechanisms. ZAMM Z. Angew. Math. Mech. **78**(10b), 687–694 (1999)
6. Lopez-Custodio, P., Müller, A., Kang, X., Dai, J.: Tangential intersection of branches of motion. Mech. Mach. Theory **147**, 103730 (2020)
7. Lopez-Custodio, P., Müller, A., Rico, J., Dai, J.: A synthesis method for 1-DOF mechanisms with a cusp in the configuration space. Mech. Mach. Theory **132**, 154–175 (2019)
8. Müller, A.: Local kinematic analysis of closed-loop linkages -mobility, singularities, and shakiness. ASME J. Mech. Rob. **8**(4), 041013, 11 p. (2016)
9. Müller, A.: Higher-order analysis of kinematic singularities of lower pair linkages and serial manipulators. ASME J. Mech. Rob. **10**(1), 011008, 13 p. (2018)
10. Müller, A.: Topology, kinematics, and constraints of multi-loop linkages. Robotica **36**(11), 1641–1663 (2018)
11. Müller, A.: Data for: an overview of formulae for the higher-order kinematics of lower-pair chains with applications in robotics and mechanism theory. Mendeley Data, v1 (2019)
12. Müller, A.: Local investigation of mobility and singularities of linkages. In: Müller, A., Zlatanov, D. (eds.) Singular Configurations of Mechanisms and Manipulators, CISM 589. Springer, Cham (2019)
13. Müller, A.: An overview of formulae for the higher-order kinematics of lower-pair chains with applications in robotics and mechanism theory. Mech. Mach. Theory **142**, 103594 (2019)
14. Müller, A., Li, Z.: Mechanism singularities revisited from an algebraic viewpoint. In: 43rd Mechanisms and Robotics Conference (MR)/ASME International Design Engineering Technical Conferences (IDETC), Anaheim, 18–21 August 2019
15. Selig, J.: Geometric Fundamentals of Robotics. Springer, New York (2005)
16. Wohlhart, K.: Screw spaces and connectivities in multiloop linkages. In: Lenarčič, J., Galletti, C. (eds.) On Advances in Robot Kinematics, pp. 97–104. Springer, Netherlands (2004)

Invariants for Multi-twists, Screw Systems and Serial Manipulators

Peter Donelan(✉)

School of Mathematics and Statistics, Victoria University of Wellington, Wellington, New Zealand
peter.donelan@vuw.ac.nz

Abstract. The Euclidean group of proper isometries $SE(3)$ acts on its Lie algebra, the vector space of twists by the adjoint action. This extends to multi-twists and screw systems. Invariants of these actions encode geometric information about the objects and are fundamental in applications to robot kinematics. This paper explores relations between known invariants and applies them to serial manipulators.

Keywords: Euclidean group · Adjoint action · Multi-twists · Screw systems · Invariant polynomials · Serial manipulators

1 Introduction

Twists, vectors of twists and screw systems all play an important role in mathematical models of robot manipulator kinematics. They describe the infinitesimal capabilities of joints, links, and end-effector or platform for a given configuration of a manipulator. Properties of these objects, especially those that are invariant under the Euclidean group, are therefore fundamental. By way of examples:

1. a 1-degree of freedom (dof) holonomic, lower-pair joint may be represented by a twist or screw, which remains invariant under motions of the joint and simultaneous motion of the set of links (and their joints) in space;
2. the Denavit–Hartenberg parameters for serial manipulators are design parameters that are assumed to be unchanged under arbitrary movements of the manipulator;
3. the infinitesimal motion of the platform or end-effector of a manipulator with $k < 6$-dof is, in a given configuration, described by a k-system, which should be independent of the choice of coordinates.

The aim of this paper is to bring together information about polynomial invariants in each of these situations. A central role is given to the multi-twist invariants [5,6] and their connection to the invariants of Selig [12].

J. Lenarčič and B. Siciliano (Eds.): ARK 2020, SPAR 15, pp. 133–141, 2021.
https://doi.org/10.1007/978-3-030-50975-0_17

2 The Euclidean Group: Twists, Multi-twists and Screw Systems

The displacements of a spatial rigid body are described by the (special) Euclidean group $SE(3)$, a 6-dimensional Lie group, identified via choice of coordinate frames as the semi-direct product of rotations about the origin, $SO(3)$, and translations $\mathbb{R}^3$. The action of $(A, \mathbf{a}) \in SO(3) \ltimes \mathbb{R}^3 \cong SE(3)$ transforms body coordinates $\mathbf{x} \in \mathbb{R}^3$ to ambient coordinates $\mathbf{X}$ via $\mathbf{x} \mapsto \mathbf{X} = A\mathbf{x} + \mathbf{a}$. A simultaneous change of ambient and body coordinate frames, $\rho = (R, \mathbf{r})$, transforms $\alpha \in SE(3)$ by *conjugation*:

$$\alpha \mapsto \rho \circ \alpha \circ \rho^{-1} = (RAR^{-1}, -RAR^{-1}\mathbf{r} + R\mathbf{a} + \mathbf{r}). \tag{1}$$

Instantaneous displacement at $\alpha \in SE(3)$ is described by the tangent space to the group, in particular that at the identity, its *Lie algebra* $\mathfrak{se}(3)$. An element $\mathbf{s} \in \mathfrak{se}(3)$ is a *twist* and can be represented by a pair of 3-vectors $(\boldsymbol{\omega}, \mathbf{v})$, its *twist coordinates.* The vector $\boldsymbol{\omega} = (\omega_1, \omega_2, \omega_3)^T$ can be identified in a standard way with a skew-symmetric 3×3 matrix Ω, such that for $\mathbf{x} \in \mathbb{R}^3$, $\Omega\mathbf{x} = \boldsymbol{\omega} \times \mathbf{x}$. It can be convenient to represent $\mathbf{s}$ by $(\Omega, \mathbf{v})$. This defines a vector field $\dot{\mathbf{x}} = \Omega\mathbf{x} + \mathbf{v}$ on $\mathbb{R}^3$, whose solution, with initial condition $\mathbf{x}(0) = \mathbf{x}_0$, is $\exp(t\mathbf{s})\mathbf{x}_0$ in $SE(3)$. These are the displacements generated by 1-dof joints, giving the correspondence between joints and twists. The motion is the same for any non-zero multiple of $\mathbf{s}$ so joints are often identified with *screws*, elements of the projective Lie algebra $\mathbb{P}\mathfrak{se}(3)$.

Differentiating (1) gives the *adjoint action* Ad of $SE(3)$ on $\mathfrak{se}(3)$—how infinitesimal displacements transform under change of coordinates $\rho \in SE(3)$:

$$\operatorname{Ad}\rho(\Omega, \mathbf{v}) = (R\Omega R^{-1}, -R\Omega R^{-1}\mathbf{r} + R\mathbf{v}). \tag{2}$$

Differentiating again, with respect to ρ, and evaluating at the identity gives the *adjoint representation* of the Lie algebra. For $\mathbf{s}_i = (\Omega_i, \mathbf{v}_i)$, $i = 1, 2$, $\operatorname{ad}\mathbf{s}_1(\mathbf{s}_2) = (\Omega_1\Omega_2 - \Omega_2\Omega_1, \Omega_1\mathbf{v}_2 - \Omega_2\mathbf{v}_1)$. We also write $\operatorname{ad}\mathbf{s}_1(\mathbf{s}_2) = [\mathbf{s}_1, \mathbf{s}_2]$, the *Lie bracket* of $\mathfrak{se}(3)$, given in twist coordinates $\mathbf{s}_i = (\boldsymbol{\omega}_i, \mathbf{v}_i)$ by $[\mathbf{s}_1, \mathbf{s}_2] = (\boldsymbol{\omega}_1 \times \boldsymbol{\omega}_2, \boldsymbol{\omega}_1 \times \mathbf{v}_2 + \mathbf{v}_1 \times \boldsymbol{\omega}_2)$.

Euclidean displacements can also be represented by dual quaternions [3]. Briefly, a dual quaternion $\check{q} = q_0 + \varepsilon q_1$ has conjugate $\check{q}^* = q_0^* + \varepsilon q_1^*$ (where q^* is the quaternion conjugate of q) and is *unit* if $\check{q}\check{q}^* = 1$. The unit dual quaternions, $\mathbf{D}S^3$, form a 6-dimensional Lie group. The Lie algebra of $\mathbf{D}S^3$ is the set of elements $\check{\omega} = \boldsymbol{\omega} + \varepsilon\mathbf{v}$, where $\boldsymbol{\omega}, \mathbf{v}$ are pure imaginary quaternions and the Lie bracket in dual quaternion form is the dualised vector product, $[\check{\omega}_1, \check{\omega}_2] = (\boldsymbol{\omega}_1 + \varepsilon\mathbf{v}_1) \times (\boldsymbol{\omega}_2 + \varepsilon\mathbf{v}_2) = \boldsymbol{\omega}_1 \times \boldsymbol{\omega}_2 + \varepsilon(\boldsymbol{\omega}_1 \times \mathbf{v}_2 + \mathbf{v}_1 \times \boldsymbol{\omega}_2)$. There is a 2:1 homomorphism $\theta : \mathbf{D}S^3 \to SE(3)$, $\theta(\check{q}).\mathbf{v} = q_0\mathbf{v}q_0^* + (q_1q_0^* - q_0q_1^*)$, which determines a Lie algebra isomorphism and the adjoint actions of the two groups are equivalent.

The adjoint action of $SE(3)$ induces several other actions of interest. A *multi-twist* is a k-tuple of twists, $(\mathbf{s}_1, \ldots, \mathbf{s}_k) \in \mathfrak{se}(3)^k$ and there is a simultaneous action by $\rho \in SE(3)$:

$$\rho \cdot (\mathbf{s}_1, \ldots, \mathbf{s}_k) = (\operatorname{Ad}\rho(\mathbf{s}_1), \ldots, \operatorname{Ad}\rho(\mathbf{s}_k)). \tag{3}$$

A manipulator or robot arm with several degrees of freedom, as we shall see in Sect. 5, is associated with multi-twists. The manipulator's instantaneous capability is a subspace of the Lie algebra $\mathfrak{se}(3)$, i.e. a *screw system of order* k, or k-system [9,10]. These form a *Grassmannian manifold* $G(k, \mathfrak{se}(3))$ for fixed k and since the adjoint action is linear, there is an induced action of $SE(3)$ on it. Given a k-system $S \in G(k, \mathfrak{se}(3))$ $(1 \leq k \leq 6)$ and $\mathbf{s}_1, \ldots, \mathbf{s}_k$ a basis for S, there is a well-defined one-to-one function into the projective exterior product (see, for example, [12]):

$$\pi : G(k, \mathfrak{se}(3)) \to \mathbb{P}\left(\wedge^k \mathfrak{se}(3)\right), \quad S \mapsto [\mathbf{s}_1 \wedge \cdots \wedge \mathbf{s}_k], \tag{4}$$

and $\rho \cdot [\mathbf{s}_1 \wedge \cdots \wedge \mathbf{s}_k] = [\mathrm{Ad}\,\rho(\mathbf{s}_1) \wedge \cdots \wedge \mathrm{Ad}\,\rho(\mathbf{s}_k)]$. Let $\mathbf{e}_1, \ldots, \mathbf{e}_6$ denote the standard basis for $\mathfrak{se}(3)$, then $\mathbf{s}_1 \wedge \cdots \wedge \mathbf{s}_k = \sum p_{i_1 \cdots i_k} \mathbf{e}_{i_1} \wedge \cdots \wedge \mathbf{e}_{i_k}$, where the sum is over sequences $i_1, \ldots, i_k$, $1 \leq i_1 < \cdots < i_k \leq 6$. The coefficients $p_{i_1 \cdots i_k}$ may be written in terms of twist coordinates for $\mathbf{s}_i$ and are called *Plücker coordinates* for the exterior product. They satisfy algebraic relations that define the Grassmannian $G(k, \mathfrak{se}(3))$ as a real algebraic variety and a manifold of dimension $k(6-k)$.

3 Invariants of Multi-twists

The orbits of group actions are important, both mathematically and practically, and invariant theory is a key tool in identifying them. Suppose V is a finite-dimensional real vector (or projective) space and G a group acting on V by linear transformations. If $x_1, \ldots, x_n$ are coordinates for V, the polynomial functions on V form an algebra $\mathbb{R}[x_1, \ldots, x_n]$, abbreviated as $\mathbb{R}[\mathbf{x}]$. There is an induced action of G on $\mathbb{R}[\mathbf{x}]$: for $g \in G$, $f \in \mathbb{R}[\mathbf{x}]$ and $\mathbf{x} \in V$, $(g \cdot f)(\mathbf{x}) = f(g \cdot \mathbf{x})$. An *invariant polynomial* is an element $f \in \mathbb{R}[\mathbf{x}]$ such that for all $g \in G$, $g \cdot f = f$, and they form a subalgebra $\mathbb{R}[\mathbf{x}]^G$. For many groups G, e.g. finite, semisimple or reductive, $\mathbb{R}[\mathbf{x}]^G$ is *finitely generated* [14]. This is called the First Fundamental Theorem of Invariant Theory, in respect of G. There exist non-reductive groups for which finite generation does not hold and the Euclidean group is not reductive, though we know invariant rings are finitely generated for some representations. For the adjoint action of $SE(3)$, the polynomial ring is $\mathbb{R}[\omega_1, \omega_2, \omega_3, v_1, v_2, v_3]$, (or $\mathbb{R}[\omega, \mathbf{v}]$). In the case $n = 3$ of a theorem [8] for $SE(n)$ we have:

Theorem 1. *The invariant ring* $\mathbb{R}[\boldsymbol{\omega}, \mathbf{v}]^{SE(3)}$ *is the polynomial ring* $\mathbb{R}[\boldsymbol{\omega}.\boldsymbol{\omega}, \boldsymbol{\omega}.\mathbf{v}]$.

The generating invariants are multiples of the Killing and Klein forms, $\boldsymbol{\omega}.\boldsymbol{\omega} = -\frac{1}{2}\mathbf{s}^T A \mathbf{s}, \boldsymbol{\omega}.\mathbf{v} = \frac{1}{2}\mathbf{s}^T B \mathbf{s}$, where $A = \begin{pmatrix} -2I_3 & O_3 \\ O_3 & O_3 \end{pmatrix}$, $B = \begin{pmatrix} O_3 & I_3 \\ I_3 & O_3 \end{pmatrix}$.

The forms $Q_{\alpha,\beta} = \alpha A + \beta B$, $\alpha, \beta \in \mathbb{R}$ are non-degenerate if and only if $\beta \neq 0$ and then are indefinite of index 3. The equations $q_{\alpha,\beta}(\mathbf{s}) = \mathbf{s}^T Q_{\alpha,\beta} \mathbf{s} = 0$ define a family of hypersurfaces, the *pitch quadrics*, in the screw space $\mathbb{P}\mathfrak{se}(3)$. For $\beta \neq 0$, these are parametrised by the invariant *pitch of the screw* $h = \alpha/\beta = \boldsymbol{\omega}.\mathbf{v}/\boldsymbol{\omega}.\boldsymbol{\omega}$, and we write $q_h = 0$. For $\beta = 0$, set $h = \infty$ and the variety $q_\infty = 0$ is the projective plane $\boldsymbol{\omega} = \mathbf{0}$, which lies in the intersection of all the pitch quadrics.

The Principle of Transference [11,13] may be used to find invariants of multi-twists by dualising the vector invariants of the rotation group $SO(3)$ [15]. Specifically, given $f \in \mathbb{R}[\boldsymbol{\omega}_1, \dots, \boldsymbol{\omega}_k]$, its dual form $\check{f}$ is given by replacing its variables by dual quantitites $\boldsymbol{\omega}_i + \varepsilon \mathbf{v}_i$, expanding its Maclaurin series and using $\varepsilon^2 = 0$:

$$\check{f}(\boldsymbol{\omega}_1 + \varepsilon \mathbf{v}_1, \cdots, \boldsymbol{\omega}_k + \varepsilon \mathbf{v}_k) = f(\boldsymbol{\omega}_1, \cdots, \boldsymbol{\omega}_k) + \varepsilon \left(\sum_{r=1}^{k} \mathbf{v}_r \frac{\partial f}{\partial \boldsymbol{\omega}_r} (\boldsymbol{\omega}_1, \dots, \boldsymbol{\omega}_k) \right). \tag{5}$$

The dual part, $\tilde{f}$, is the partial polarisation of f. For example, the dual of the invariant form $f(\boldsymbol{\omega}) = \boldsymbol{\omega}.\boldsymbol{\omega}$ is $\check{f}(\boldsymbol{\omega}, \mathbf{v}) = \boldsymbol{\omega}.\boldsymbol{\omega} + \varepsilon\, \boldsymbol{\omega}.\mathbf{v}$, giving the Killing and Klein forms, respectively. The following theorem [6] realises multi-twist invariants:

Theorem 2. *If f is a vector invariant of the adjoint action of $SO(3)$, then the primal and dual parts of $\check{f}$ are vector invariants of the adjoint action of $SE(3)$. The primal and dual parts of the dualisation of any $SO(3)$ syzygy are syzygies for $SE(3)$.*

Dualising the generators for vector invariants [15] of $SO(3)$ gives:

1. for $1 \leq i \leq j \leq m$,

$$\check{I}_{ij} := I_{ij} + \varepsilon \tilde{I}_{ij} := \boldsymbol{\omega}_i.\boldsymbol{\omega}_j + \varepsilon(\boldsymbol{\omega}_i.\mathbf{v}_j + \boldsymbol{\omega}_j.\mathbf{v}_i) = -\tfrac{1}{2}\left(\mathbf{s}_i^T Q_\infty \mathbf{s}_j + \varepsilon \mathbf{s}_i^T Q_0 \mathbf{s}_j\right); \tag{6}$$

2. for $m \geq 3$ and any $1 \leq i < j < k \leq m$, setting $[\mathbf{u}\ \mathbf{v}\ \mathbf{w}] = \mathbf{u}.(\mathbf{v} \times \mathbf{w})$,

$$\begin{aligned} \check{I}_{ijk} := I_{ijk} + \varepsilon \tilde{I}_{ijk} &:= [\boldsymbol{\omega}_i\ \boldsymbol{\omega}_j\ \boldsymbol{\omega}_k] + \varepsilon\left([\boldsymbol{\omega}_i\ \boldsymbol{\omega}_j\ \mathbf{v}_k] + [\boldsymbol{\omega}_i\ \mathbf{v}_j\ \boldsymbol{\omega}_k] + [\mathbf{v}_i\ \boldsymbol{\omega}_j\ \boldsymbol{\omega}_k]\right) \\ &= -\tfrac{1}{2}\left(\mathbf{s}_i^T Q_\infty [\mathbf{s}_j, \mathbf{s}_k] + \varepsilon \mathbf{s}_i^T Q_0 [\mathbf{s}_j, \mathbf{s}_k]\right), \end{aligned} \tag{7}$$

For $m = 2$, there are 6 (quadratic) invariants and they generate the ring of invariants. For $m = 3$, there are 14 invariants, but is not known whether the First Fundamental Theorem holds for $m \geq 3$. However, the theorem in [6] that, for $m = 3$, every polynomial invariant can be written as a rational expression in $I_{ij}, \tilde{I}_{ij}$ and I_{123}, extends to the general case $m > 3$, still with only the one cubic invariant. The invariants are connected by three types of dualised syzygy [15], which we do not specify here.

4 Invariants of Screw Systems

Hunt's classification [10] of Ball's screw systems is given a mathematical basis by Gibson and Hunt (GH) [9], where a normal form of basis twists for each screw system is given. Classes are unions of orbits and the normal forms include invariant parameters (moduli). For a generic 2-system (GH-type IA) the moduli h_1, h_2 are the principal pitches, which can be found by solving the discriminant $\det(\mathbf{S}^T Q_h \mathbf{S}) = 0$, where $\mathbf{S}$ is the 6×2 matrix whose columns are spanning twists.

Type IB systems have a unique infinite pitch screw and contain a unique screw of every pitch whose axes are coplanar. The normal form has a modulus q, which is the tangent of the angle this plane makes with the infinitesimal translation. The generic 3-system, type IA, has normal form with moduli the three principal pitches $h_1 < h_2 < h_3$ corresponding to the degenerate or singular intersections with the pitch quadrics, which can be found analogously with 2-systems.

Selig [12] explores the invariant theory for screw systems along these lines. For a 2-system $S = [\mathbf{s}_1 \wedge \mathbf{s}_2]$, the coefficients $\iota_1, \iota_2, \iota_3$ of $\det \mathbf{S}^T Q_{\alpha,\beta} \mathbf{S} = \iota_1 \beta^2 + \iota_2 \alpha\beta + \iota_3 \alpha^2$ are invariants of the twist-pair $\mathbf{s}_1, \mathbf{s}_2$. Further, if M is a change of basis for the 2-system then $\iota'_r = (\det M)^2 \, \iota_r$ for $r = 1, 2, 3$, so they are screw system invariants. They are sufficient to distinguish most, but not all, of the 2-system orbits. For type IA, the principal pitches h_1, h_2 are the solutions of the quadratic $\iota_3 h^2 + \iota_2 h + \iota_1 = 0$, giving $h_1 + h_2 = -\iota_2/\iota_3$, $h_1 h_2 = \iota_1/\iota_3$. The modulus q for IB systems, which separates orbits, cannot be found from $\iota_1, \iota_2, \iota_3$ since these invariants vanish for all q. We show below that these are the only invariants. (The expressions i_4, i_5 in [12], Sect. 8.4 are not true invariants.) Invariants for 3-systems are also identified in [12].

We adapt this approach using the invariant ring for k twists. Form the quadratic vector invariants into $k \times k$ symmetric matrices $\mathbf{I} = (I_{ij})$, $\tilde{\mathbf{I}} = (\tilde{I}_{ij})$. A change of basis $M = (m_{ij}) \in GL(k)$ transforms $\mathbf{S} = (\mathbf{s}_1, \ldots, \mathbf{s}_k) \mapsto \mathbf{S}'_i = \mathbf{S}M$ and, by (6), the invariants transform in the following way:

$$I'_{ij} = -\tfrac{1}{2}\mathbf{s}'_i Q_\infty \mathbf{s}'_j = -\tfrac{1}{2}\sum_r\sum_s m_{ri} m_{sj} (\mathbf{s}_r^T Q_\infty \mathbf{s}_j) = -\tfrac{1}{2}\sum_r\sum_s m_{ri} m_{sj} I_{rs} = (M^T \mathbf{I} M)_{ij} \tag{8}$$

and similarly for $\tilde{I}'_{ij}$, replacing Q_∞ by Q_0. Thus, the action of $GL(k)$ on the quadratic k-twist invariants is the same as its action on a pair of k-ary quadratic forms. In the case $k = 2$, the invariant ring is:

$$\mathbb{R}[\boldsymbol{\omega}_1, \mathbf{v}_1, \boldsymbol{\omega}_2, \mathbf{v}_2]^{SE(3)} = \mathbb{R}[I_{11}, I_{12}, I_{22}, \tilde{I}_{11}, \tilde{I}_{12}, \tilde{I}_{22}], \tag{9}$$

and the 6×6 representation of $M \in GL(2)$ is the double symmetric square of M:

$$\begin{pmatrix} \tilde{M} & O \\ O & \tilde{M} \end{pmatrix} \quad \text{where} \quad \tilde{M} = \begin{pmatrix} m_{11}^2 & 2m_{11}m_{21} & m_{21}^2 \\ m_{11}m_{12} & m_{11}m_{22} + m_{12}m_{21} & m_{21}m_{22} \\ m_{12}^2 & 2m_{12}m_{22} & m_{22}^2 \end{pmatrix}. \tag{10}$$

From classical invariant theory [14], we have the following invariants:

$$j_1 = \tilde{I}_{11}\tilde{I}_{22} - \tilde{I}_{12}^2, \quad j_2 = I_{11}\tilde{I}_{22} + I_{22}\tilde{I}_{11} - 2I_{12}\tilde{I}_{12}, \quad j_3 = I_{11}I_{22} - I_{12}^2. \tag{11}$$

A simple way to view these is as the coefficients of $\det(\alpha\mathbf{I} + \beta\tilde{\mathbf{I}})$. Substituting the Plücker coordinate expressions for the 2-twist invariants and for the coordinates p_{ij} confirms that $j_r = i_r$, $r = 1, 2, 3$ (up to non-zero multiples). By (9), this argument shows that j_1, j_2, j_3 form a generating set for the invariants of 2-systems.

For $k = 3$, in addition to 12 quadratic 3-twist invariants there are two cubic invariants. Since $I_{123} = \omega_1.(\omega_2 \times \omega_3)$ is the volume of the parallelepiped spanned by these vectors, it satisfies for $M \in GL(3)$: $I'_{123} = (\det M) I_{123}$. Polarising this equation shows that $\tilde{I}_{123}$ is also an invariant. Now consider the action of $GL(3)$ on the polynomial ring generated by I_{ij}, $\tilde{I}_{ij}$ that corresponds to the action on a pair of ternary quadratics. A set of generating invariants is the coefficients of $\det(\alpha \mathbf{I} + \beta \tilde{\mathbf{I}})$:

$$
\begin{aligned}
k_1 &= \tilde{I}_{11}\tilde{I}_{22}\tilde{I}_{33} - \tilde{I}_{11}\tilde{I}_{23}^2 - \tilde{I}_{22}\tilde{I}_{13}^2 - \tilde{I}_{33}\tilde{I}_{12}^2 + 2\tilde{I}_{12}\tilde{I}_{13}\tilde{I}_{23} \\
k_2 &= I_{11}\tilde{I}_{22}\tilde{I}_{33} + I_{22}\tilde{I}_{11}\tilde{I}_{33} + I_{33}\tilde{I}_{11}\tilde{I}_{22} - I_{11}\tilde{I}_{23}^2 - I_{22}\tilde{I}_{13}^2 - I_{33}\tilde{I}_{12}^2 \\
&\quad + 2I_{12}(\tilde{I}_{13}\tilde{I}_{23} - \tilde{I}_{12}\tilde{I}_{33}) + 2I_{13}(\tilde{I}_{12}\tilde{I}_{23} - \tilde{I}_{13}\tilde{I}_{22}) + 2I_{23}(\tilde{I}_{12}\tilde{I}_{13} - \tilde{I}_{11}\tilde{I}_{23}) \\
k_3 &= I_{11}I_{22}\tilde{I}_{33} + I_{11}I_{33}\tilde{I}_{22} + I_{22}I_{33}\tilde{I}_{11} - I_{12}^2\tilde{I}_{33} - I_{13}^2\tilde{I}_{22} - I_{23}^2\tilde{I}_{11} \\
&\quad + 2(I_{13}I_{23} - I_{12}I_{33})\tilde{I}_{12} + 2(I_{12}I_{23} - I_{13}I_{22})\tilde{I}_{13} + 2(I_{12}I_{13} - I_{11}I_{23})\tilde{I}_{23} \\
k_4 &= I_{11}I_{22}I_{33} - I_{11}I_{23}^2 - I_{22}I_{13}^2 - I_{33}I_{12}^2 + 2I_{12}I_{13}I_{23}.
\end{aligned}
\tag{12}
$$

These satisfy the identities $k_4 = I_{123}^2, k_3 = 2I_{123}\tilde{I}_{123}$, being the two syzygies for 3-twist invariants [6]. If, in fact, (6,7) generate 3-twist invariants, then $I_{123}, \tilde{I}_{123}, k_1, k_2$ generate the invariant polynomials for 3-systems, however this remains a conjecture. The invariants correspond to Selig's: $\sqrt{-\frac{1}{2}\det \Upsilon_\infty}$, i_0, $\det \Upsilon_0$, Φ, respectively. As for 2-systems, these invariants determine the principal pitches.

5 Invariants and Serial Manipulators

Brockett's product of exponentials formulation for a m-dof serial manipulator [1] requires the choice of m twists, $\mathbf{s}_1, \ldots, \mathbf{s}_m$, representing the joints in the home configuration. In principle, they may have any pitch, but in practice are usually pitch zero or infinity, i.e. revolute or prismatic. The forward kinematics for the end-effector is $f : \mathbb{R}^m \to SE(3)$, $f(u_1, \ldots, u_m) = \exp(u_1\mathbf{s}_1)\exp(u_2\mathbf{s}_2)\cdots\exp(u_m\mathbf{s}_m)$, where $u_1, \ldots, u_m$ are joint variables. As the arm moves, so do the joints (other than $\mathbf{s}_1$) and the multi-twist changes. At configuration $\mathbf{u} = (u_1, \ldots, u_m)$:

$$\mathbf{s}_2(\mathbf{u}) = \mathrm{Ad}\,(\exp(u_1\mathbf{s}_1))\mathbf{s}_2, \ldots, \mathbf{s}_m(\mathbf{u}) = \mathrm{Ad}\,(\exp(u_1\mathbf{s}_1)\cdots\exp(u_{m-1}\mathbf{s}_{m-1}))\mathbf{s}_m. \tag{13}$$

From the point of view of invariants, in addition to the joint adjoint action on multi-twists, we must also consider the internal action of earlier joints on later joints. Notice that motion around the first joint alone gives an equivalent multi-twist, since we may apply $\mathrm{Ad}\,(\exp(-u_1\mathbf{s}_1))$ to the updated twists in (13) and $\mathbf{s}_1$ is fixed by this. Also, clearly motion about the final joint has no effect.

An alternative formalism for serial kinematics is Denavit–Hartenberg (DH) parameters [7]. A coordinate frame is chosen for each link in such a way that the joint twists have a standard form. The exponential motions generated by each joint must be connected by corresponding changes of coordinates, expressed in terms of the DH parameters: *link length*, *link angle* and *offset*. It is implicitly recognised that these parameters are invariant under global coordinate change (the adjoint action on the multi-twist), which we refer to as *static invariance*, and the internal motion of the manipulator or *kinematic invariance*. The connection between the static, multi-twist, invariants and the DH parameters is established in [4,5]. Explicit formulae are obtained for each in terms of the invariants (6, 7). For a multi-twist $(\mathbf{s}_1, \ldots, \mathbf{s}_m)$,

- invariants $I_{ii}, \tilde{I}_{ii}$, $i = 1, \ldots, m$ tell us about the individual joints and their pitch h_i.
- for each pair $\mathbf{s}_i, \mathbf{s}_j$, $1 \leq i < j \leq m$, the invariants $I_{ij}, \tilde{I}_{ij}$ determine the link length l_{ij}—the perpendicular distance between the axes of the twists—and link angle θ_{ij}—the angle between these axes, where these are defined.
- for each triple $\mathbf{s}_i, \mathbf{s}_j, \mathbf{s}_k$, $1 \leq i < j < k \leq m$, the quadratic plus cubic invariants $I_{ijk}, \tilde{I}_{ijk}$, determine the offsets d_{ijk}, the distance along the axis of $\mathbf{s}_j$ between the feet of the perpendiculars from $\mathbf{s}_i, \mathbf{s}_k$, again so long as this is well defined.

Explicit formulae for the screw and DH parameters are:

$$h_i = \frac{\tilde{I}_{ii}}{I_{ii}},\ l_{ij} = \frac{I_{ij}(I_{ii}\tilde{I}_{jj} + \tilde{I}_{ii}I_{jj}) - I_{ii}I_{jj}\tilde{I}_{ij}}{I_{ii}I_{jj}\sqrt{I_{ii}I_{jj} - I_{ij}^2}},\ \cos\theta_{ij} = \frac{I_{ij}}{\sqrt{I_{ii}I_{jj}}},\ l_{ij}\sin\theta_{ij} = \frac{\tilde{I}_{ij}}{\sqrt{I_{ii}I_{jj}}}$$

$$d_{ijk} = \frac{\tilde{I}_{ijk}I_{jj}(I_{ij}I_{jk} - I_{ik}I_{jj}) + I_{ijk}\left(\frac{1}{2}\tilde{I}_{jj}(I_{ij}I_{jk} + I_{ik}I_{jj}) + I_{jj}(\tilde{I}_{ik}I_{jj} - I_{ij}\tilde{I}_{jk} - \tilde{I}_{ij}I_{jk})\right)}{\sqrt{I_{jj}}(I_{ii}I_{jj} - I_{ij}^2)(I_{jj}I_{kk} - I_{jk}^2)}. \tag{14}$$

We have $I_{ii} = 0$ if and only if $\boldsymbol{\omega}_i = 0$, so that $\mathbf{s}_i$ is prismatic. The invariant expression $I_{ii}I_{jj} - I_{ij}^2 = \|\boldsymbol{\omega}_i \times \boldsymbol{\omega}_j\|^2$, so vanishes if $\boldsymbol{\omega}_i = 0$, $\boldsymbol{\omega}_j = 0$ (i.e. $\mathbf{s}_i$ or $\mathbf{s}_j$ prismatic) or $\boldsymbol{\omega}_i \| \boldsymbol{\omega}_j$—so no unique common perpendicular between the joint axes.

Verifying that an expression is invariant with respect to the action of a (connected) Lie group is equivalent to showing that it is invariant under the infinitesimal action of the Lie algebra, found by differentiating at the identity. Specifically, since the derivative of the Euclidean group's adjoint action is the Lie bracket, if $f(\mathbf{s}_i, \ldots, \mathbf{s}_m)$ is a multi-twist invariant, then it is an invariant of the motion about joint r, $2 \leq r \leq m-1$, of a serial manipulator if and only if:

$$\nabla f(\mathbf{s}_1, \ldots, \mathbf{s}_m) \cdot (\mathbf{0}, \ldots, \mathbf{0}, [\mathbf{s_r}, \mathbf{s_{r+1}}], \ldots, [\mathbf{s_r}, \mathbf{s_m}]) = \mathbf{0}. \tag{15}$$

This is the $6m$-dimensional gradient vector multiplied by the column consisting of r zero 6-vectors, followed by $m - r$ brackets in the Lie algebra $\mathfrak{se}(3)$.

Theorem 3. *For a serial manipulator with joint twists* $\mathbf{s}_1, \ldots, \mathbf{s}_m$, *the only static invariants that are kinematically invariant are* $I_{ii}, \tilde{I}_{ii}$, $1 \leq i \leq m$ *and* $I_{i,i+1}, \tilde{I}_{i,i+1}$, $1 \leq i \leq m-1$.

In particular, the invariants $I_{ik}, \tilde{I}_{ik}$, $k \geq i+2$ are not kinematically invariant under the action of joint j for $i < j < k$. The cubic invariants $I_{ijk}, \tilde{I}_{ijk}$ are not generally kinematically invariant for any i, j, k. It follows from (14) that pitches of joints and link lengths and angles for successive joints are kinematically invariant, while offsets are not. However, we have the following special case.

Theorem 4. *For a serial manipulator as in Theorem 3, the invariants* $I_{i,i+1,i+2}, \tilde{I}_{i,i+1,i+2}$ *for* $1 \leq i \leq m-2$ *are kinematically invariant so long as* $\tilde{I}_{i+1} = 0$. *Thus, the offset for 3 successive joints is invariant if the middle joint is revolute.*

6 Conclusion

We have shown that the invariant polynomials of multi-twists play a valuable role in understanding the screw systems and the kinematics of serial manipulators. They may throw light on other aspects of manipulator kinematics, such as *persistence* [2] in which the screw system remains constant under motion of the manipulator's own motion. Given their algebraic simplicity, they provide an alternative to DH parameters that are well-adapted to product-of-exponentials formalism.

References

1. Brockett, R.: Robotic manipulators and the product of exponentials formula. In: Fuhrmann, P.A. (ed.) Mathematical Theory of Networks and Systems, vol. 58, pp. 120–129. Springer, Heidelberg (1984)
2. Carricato, M., Rico Martínez, J.M.: Persistent screw systems. Advances in Robot Kinematics: Motion in Man and Machine, pp. 185–194. Springer, Dordrecht (2010)
3. Clifford, W.K.: Preliminary sketch of biquaternions. Proc. London Math. Soc. **4**, 381–395 (1873)
4. Crook, D., Donelan, P.: Polynomial invariants and SAGBI bases for multi-screws, 10 p. arXiv 2001.05417 (2012)
5. Daher, M., Donelan, P.: Invariant properties of the Denavit-Hartenberg parameters. Interdisciplinary Applications of Kinematics, Mechanisms and Machine Science, vol. 26, pp. 43–51. Springer, Cham (2015)
6. Daher, M., Donelan, P.: Invariants of the k-fold adjoint action of the Euclidean isometry group. J. Geom. **107**, 169–185 (2015)
7. Denavit, J., Hartenberg, R.S.: Kinematic notation for lower pair mechanisms based on matrices. J. Appl. Mech. **22**, 215–221 (1955)
8. Donelan, P., Gibson, C.G.: First-order invariants of Euclidean motions. Acta Appl. Math. **24**, 233–251 (1991)
9. Gibson, C.G., Hunt, K.H.: Geometry of screw systems–1: screws: genesis and geometry/2: classification of screw systems. Mech. Mach. Theory **25**, 1–27 (1990)
10. Hunt, K.H.: Kinematic Geometry of Mechanisms. Oxford University Press, Oxford (1978)
11. Rico Martínez, J.M., Duffy, J.: The principle of transference: history, statement and proof. Mech. Mach. Theory **28**, 165–177 (1993)

12. Selig, J.M.: Geometric Fundamentals of Robotics. Springer, Cham (2004)
13. Study, E: Geometrie der Dynamen. Teubner, Leipzig (1903)
14. Sturmfels, B.: Algorithms in Invariant Theory. Springer, Cham (2008)
15. Weyl, H.: The Classical Groups: Their Invariants and Representations. Princeton University Press, Princeton (1997)

Singularity-Free Extraction of a Dual Quaternion from Orthogonal Dual Tensor

Daniel Condurache(✉)

Technical University of Iasi, D. Mangeron Street no. 59, 700050 Iasi, Romania
daniel.condurache@tuiasi.ro

Abstract. The parameterization of a rigid-body motion can be done using multiple algebraic entities. A very important criterion when choosing a parameterization method is the number of algebraic equations and variables. Recently, orthogonal dual tensors and dual quaternion proved to be a complete tool for computing rigid body displacement and motion parameters. The present research is focused on developing new methods for recovering kinematic data when the state of features attached to a body during a rigid displacement is available. The proof of concept is sustained by computational solutions both for the singularity-free extraction of a dual quaternion from feature-based representation of motion and for the recovery algorithms of the dual quaternion.

Keywords: Dual algebra · Dual tensor · Dual quaternion · Motion parameters

1 Introduction

The present research is focused on developing a new method to generate orthogonal dual tensor using dual rigid basis, for rigid body pose computation when dual quaternions are used for rigid body motion parameterization. The main advantage of working with dual rigid basis is the possibility for rigid body motion parameterization without any attached orthonormal reference frames [1, 2]. The proposed technique offers a new algebraic solution for the following problem given measurements of a rigid body find the correspondent pose expressed by dual quaternions through direct computation. Also, the use of dual rigid basis increases the flexibility of the chosen method for parameterization of the rigid body motion. The proof of concept is sustained by computational solutions both for the singularity-free extraction of a dual quaternion from feature-based representation of motion and for the recovery algorithms of the dual quaternion.

2 Rigid Body Motion Parameterization Through Orthogonal Dual Tensors and Dual Quaternions

Let the orthogonal dual tensor set be denoted by

J. Lenarčič and B. Siciliano (Eds.): ARK 2020, SPAR 15, pp. 142–149, 2021.
https://doi.org/10.1007/978-3-030-50975-0_18

$$\underline{\mathbb{SO}}_3 = \{\underline{\boldsymbol{R}} \in \mathbf{L}(\underline{V}_3, \underline{V}_3) | \underline{\boldsymbol{R}}\underline{\boldsymbol{R}}^T = \underline{\boldsymbol{I}}, \det \underline{\boldsymbol{R}} = 1\} \tag{1}$$

where $\underline{\mathbb{SO}}_3$ is the set of real special orthogonal dual tensors and $\underline{\boldsymbol{I}}$ is the unit orthogonal dual tensor. The internal structure of any orthogonal dual tensor $\underline{\boldsymbol{R}} \in \underline{\mathbb{SO}}_3$ is illustrated in a series of results which were detailed in our previous work [1]:

Theorem 1 *(Structure Theorem). For any $\underline{\mathbf{R}} \in \underline{\mathbb{SO}}_3$, a unique decomposition is viable* [1]

$$\underline{\mathbf{R}} = (I + \varepsilon\tilde{\rho})Q \tag{2}$$

where $Q \in \mathbb{SO}_3$ and $\boldsymbol{\rho} \in V_3$ are called ***structural invariants****. For any orthogonal dual tensor $\underline{\mathbf{R}}$ defined as in Eq. (1), a dual number $\underline{\alpha}$ and a dual unit vector $\underline{\mathbf{u}}$ can be computed in order to have* [1, 2]:

$$\underline{\boldsymbol{R}}(\underline{\alpha}, \underline{\boldsymbol{u}}) = \boldsymbol{I} + \sin \underline{\alpha}\underline{\tilde{\boldsymbol{u}}} + (1 - \cos \underline{\alpha})\underline{\tilde{\boldsymbol{u}}}^2 = \exp(\underline{\alpha}\underline{\tilde{\boldsymbol{u}}}). \tag{3}$$

$\underline{\alpha}$ and $\underline{\mathbf{u}}$ are called ***natural invariants*** *of dual orthogonal tensor $\underline{\mathbf{R}}$.*

Next, we introduce an isomorphism between the Lie group SE_3 and the Lie group $\underline{\mathbb{SO}}_3$:

Theorem 2 *(Isomorphism theorem). The special Euclidean group $(SE_3, \cdot)$ and $(\underline{\mathbb{SO}}_3, \cdot)$ are connected via the isomorphism*

$$\phi : SE_3 \to \underline{\mathbb{SO}}_3, \phi(g) = (I + \varepsilon\tilde{\rho})Q \tag{4}$$

where $g = \begin{bmatrix} Q & \rho \\ 0 & 1 \end{bmatrix}$.

Proof. For any $g_1, g_2 \in SE_3$, the map defined in (4) yields:

$$\phi(g_1 \cdot g_2) = \phi(g_1) \cdot \phi(g_2) \tag{5}$$

Let $\underline{\boldsymbol{R}} \in \underline{\mathbb{SO}}_3$. Based on **Theorem** 1, which ensures an unique decomposition, we can conclude that the only choice for g, such that $\phi(g) = \underline{\mathbf{R}}$, is $g = \begin{bmatrix} Q & \rho \\ 0 & 1 \end{bmatrix}$. This underlines that ϕ is a bijection and keeps all the internal operations.

Remark 1. *The inverse of ϕ is*

$$\phi^{-1} : \underline{\mathbb{SO}}_3 \leftrightarrow SE_3; \phi^{-1}(\underline{\mathbf{R}}) = \begin{bmatrix} Q & \rho \\ 0 & 1 \end{bmatrix} \tag{6}$$

where $Q = Re(\underline{\mathbf{R}}), \boldsymbol{\rho} = vect(Du(\underline{\mathbf{R}}) \cdot Q^{\mathrm{T}})$.

Taking into account the Lie group structure of $\underline{\mathbb{SO}}_3$ [2] and the result presented in the previous theorems, it can be concluded that any orthogonal dual tensor $\underline{\boldsymbol{R}} \in \underline{\mathbb{SO}}_3$ can be used to globally parameterize displacements of rigid bodies.

A dual quaternion can be defined as an associated pair of a dual scalar quantity and a free dual vector [3–7]:

$$\hat{\underline{\mathbf{q}}} = \left(\underline{q}, \underline{\mathbf{q}}\right), \underline{q} \in \underline{\mathbb{R}}, \underline{\mathbf{q}} \in \underline{V}_3 \tag{7}$$

The set of dual quaternions will be denoted $\underline{\mathbf{Q}}$ and is a $\underline{\mathbb{R}}$-module of rank 4, if dual quaternion addition and multiplication with dual numbers are considered [2, 8].

The product of two dual quaternions $\hat{\underline{\mathbf{q}}}_1 = \left(\underline{q}_1, \underline{\mathbf{q}}_1\right)$ and $\hat{\underline{\mathbf{q}}}_2 = \left(\underline{q}_2, \underline{\mathbf{q}}_2\right)$ is defined by

$$\hat{\underline{\mathbf{q}}}_1\hat{\underline{\mathbf{q}}}_2 = \left(\underline{q}_1 \cdot \underline{q}_2 - \underline{\mathbf{q}}_1 \cdot \underline{\mathbf{q}}_2, \underline{q}_1\underline{\mathbf{q}}_2 + \underline{q}_2\underline{\mathbf{q}}_1 + \underline{\mathbf{q}}_1 \times \underline{\mathbf{q}}_2\right) \tag{8}$$

Taking into account the above properties results that the $\underline{\mathbb{R}}$-module $\underline{\mathbf{Q}}$ becomes an associative, non-commutative linear dual algebra of order 4 over the ring of dual numbers. For any dual quaternion defined by Eq. (7), the followings can be computed: the conjugate denoted by $\hat{\underline{\mathbf{q}}}^* = \left(\underline{q}, -\underline{\mathbf{q}}\right)$ and the norm denoted by $\left|\hat{\underline{\mathbf{q}}}\right| = \sqrt{\hat{\underline{\mathbf{q}}}\hat{\underline{\mathbf{q}}}^*}$.

Let $\mathbb{U}$ denote the set of real unit quaternions ($|\hat{\mathbf{q}}| = 1$) and $\underline{\mathbb{U}}$ denote the set of dual unit quaternions $\left(\left|\hat{\underline{\mathbf{q}}}\right| = 1\right)$. The scalar part and the vector part of a unit dual quaternion are also called **dual Euler parameters.**

Using the internal structure of any element from $\underline{\mathbb{SO}}_3$ the following theorem is valid [2]:

Theorem 3. *The Lie groups* $\underline{\mathbb{U}}$ *and* $\underline{\mathbb{SO}}_3$ *are linked by a surjective homomorphism*

$$\Delta : \underline{\mathbb{U}} \rightarrow \underline{\mathbb{SO}}_3, \Delta\left(\underline{q} + \underline{\mathbf{q}}\right) = \underline{\boldsymbol{I}} + 2\underline{q}\tilde{\underline{\mathbf{q}}} + 2\tilde{\underline{\mathbf{q}}}^2 \tag{9}$$

An important property of the previous homomorphism is that for $\hat{\underline{\mathbf{q}}}$ and $-\hat{\underline{\mathbf{q}}}$ we can associate the same orthogonal dual tensor, which shows that Eq. (9) is not injective and $\underline{\mathbb{U}}$ is a double cover of $\underline{\mathbb{SO}}_3$. Depending on the natural invariants, the dual quaternion corresponding of the orthogonal dual tensor from Eq. (3) is:

$$\hat{\underline{\mathbf{q}}} = \cos\frac{\underline{\alpha}}{2} + \underline{\mathbf{u}}\sin\frac{\underline{\alpha}}{2}, \forall \hat{\underline{\mathbf{q}}} \in \underline{\mathbb{U}} \tag{10}$$

3 Singularity-Free Extraction of a Dual Quaternion from Orthogonal Dual Tensor

In this section we will propose a computational solution both the singularity-free extraction of a dual quaternion from orthogonal dual tensor.

Let be $\underline{\boldsymbol{R}} \in \underline{S\mathbb{O}}_3$, the orthogonal dual tensor that models a rigid body motion and $\pm\hat{\underline{\mathbf{q}}}$ corresponding dual quaternion. We build the symmetric tensor:

$$\underline{\boldsymbol{S}} = \underline{\boldsymbol{R}} + \underline{\boldsymbol{R}}^{\mathrm{T}} + (1 - \mathrm{trace}\underline{\boldsymbol{R}})\underline{\boldsymbol{I}} \quad (11)$$

the following theorem takes place:

Theorem 4: *For any $\underline{\boldsymbol{R}} \in \underline{S\mathbb{O}}_3$, the following identities take place:*

$$\underline{\boldsymbol{R}}\underline{\boldsymbol{S}} = \underline{\boldsymbol{S}}\underline{\boldsymbol{R}} = \underline{\boldsymbol{S}} \quad (12)$$

$$\underline{\boldsymbol{S}} = 4\underline{\mathbf{q}} \otimes \underline{\mathbf{q}} \quad (13)$$

Proof. From **Theorem** 2 it follows: $\underline{\boldsymbol{S}} = \underline{\boldsymbol{I}} + \sin\underline{\alpha}\tilde{\underline{\mathbf{u}}} + (1 - \cos\underline{\alpha})\tilde{\underline{\mathbf{u}}}^2 + \underline{\boldsymbol{I}} - \sin\underline{\alpha}\tilde{\underline{\mathbf{u}}} + (1 - \cos\underline{\alpha})\tilde{\underline{\mathbf{u}}}^2 = 2\underline{\boldsymbol{I}} + 2(1 - \cos\underline{\alpha})\tilde{\underline{\mathbf{u}}}^2$. Taking into account Eq. (9), it follows:

$$\underline{\boldsymbol{S}} = (3 - \mathrm{trace}\underline{\boldsymbol{R}})\underline{\mathbf{u}} \otimes \underline{\mathbf{u}} \quad (14)$$

Because $\underline{\boldsymbol{R}}\underline{\mathbf{u}} = \underline{\mathbf{u}}\underline{\boldsymbol{R}} = \underline{\mathbf{u}}, \forall \underline{\boldsymbol{R}} \in \underline{S\mathbb{O}}_3$, from Eq. (14) it follows Eq. (12).

Knowing $3 - \mathrm{trace}\underline{\boldsymbol{R}} = 4\cos^2\frac{\underline{\alpha}}{2}$ and $\underline{\mathbf{q}} = \cos\frac{\underline{\alpha}}{2}\underline{\mathbf{u}}$, and taking into account Eq. (14), it follows Eq. (13).

Remark 2. *From Eq.* (14) *it follows* $\underline{\mathbf{S}} \neq \underline{\boldsymbol{O}}$ *if and only if* $Re[\mathrm{trace}\underline{\boldsymbol{R}} - 3] \neq 0$, *namely* $\underline{\mathbf{S}} = \underline{\boldsymbol{O}} \Leftrightarrow \underline{\mathbf{R}} = \boldsymbol{I} + \varepsilon\tilde{\rho}$, *so* $\underline{\mathbf{R}}$ *models a pure translation. Except this case, it follows from Eq.* (14)*, the screw unit dual vector* $\underline{\mathbf{u}}$ *corresponding to the orthogonal dual tensor* $\underline{\boldsymbol{R}} = \underline{\boldsymbol{R}}(\underline{\alpha}, \underline{\mathbf{u}}) \in \underline{S\mathbb{O}}_3$:

$$\underline{\mathbf{u}} = \pm\frac{\underline{\boldsymbol{S}}\underline{\mathbf{v}}}{|\underline{\boldsymbol{S}}\underline{\mathbf{v}}|}, \forall \underline{\mathbf{v}} \in \underline{V}_3, Re(\underline{\boldsymbol{S}}\underline{\mathbf{v}}) \neq 0 \quad (15)$$

In the following, we will note by $\hat{\underline{\mathbf{q}}}_1 \otimes \hat{\underline{\mathbf{q}}}_2$ the dyadic product of two dual quaternions, defined by:

$$\left(\hat{\underline{\mathbf{q}}}_1 \otimes \hat{\underline{\mathbf{q}}}_2\right)\hat{\underline{\mathbf{q}}} = \left(\hat{\underline{\mathbf{q}}}_2 \cdot \hat{\underline{\mathbf{q}}}\right)\hat{\underline{\mathbf{q}}}_1, \forall \hat{\underline{\mathbf{q}}} \in \underline{\boldsymbol{Q}} \quad (16)$$

The following theorem take place:

Theorem 5. *For* $\forall \underline{\boldsymbol{R}} \in \underline{S\mathbb{O}}_3$, *the identity takes place:*

$$\frac{1}{4}\begin{bmatrix} 1 + \mathrm{trace}\underline{\boldsymbol{R}} & 2(\mathrm{vect}\underline{\boldsymbol{R}})^{\mathrm{T}} \\ 2\mathrm{vect}\underline{\boldsymbol{R}} & \underline{\boldsymbol{S}} \end{bmatrix} = \hat{\underline{\mathbf{q}}} \otimes \hat{\underline{\mathbf{q}}} \quad (17)$$

Proof. Let be a unit dual quaternion $\hat{\underline{\mathbf{q}}} = \underline{\mathrm{q}} + \underline{\mathbf{q}} = \cos\frac{\underline{\alpha}}{2} + \underline{\mathbf{u}}\sin\frac{\underline{\alpha}}{2}$, then $1 + \mathrm{trace}\underline{\boldsymbol{R}} = 2(1 + \cos\underline{\alpha}) = 4\cos^2\frac{\underline{\alpha}}{2} = 4\underline{\mathrm{q}}^2$. From Eq. (9) it follows $\mathrm{vect}\underline{\boldsymbol{R}} = 2\underline{\mathrm{q}}\underline{\mathbf{q}}$. The left member of the Eq. (17) will be denoted by $\hat{\underline{\mathbf{S}}}$ and taking into account the previous equations and Eq. (13) it follows:

$$\hat{\underline{\mathbf{S}}} = \frac{1}{4}\begin{bmatrix} 4\underline{\mathrm{q}}^2 & 4\underline{\mathrm{q}}\underline{\mathbf{q}}^T \\ 4\underline{\mathrm{q}}\underline{\mathbf{q}} & 4\underline{\mathbf{q}} \otimes \underline{\mathbf{q}} \end{bmatrix} = \hat{\underline{\mathbf{q}}} \otimes \hat{\underline{\mathbf{q}}} \tag{18}$$

The previous **Theorem** allows singularity-free extraction of a dual quaternion from a orthogonal dual tensor. Thus, let be $\hat{\underline{\mathbf{v}}}$ a dual quaternion such that $Re\hat{\underline{\mathbf{v}}} \neq \hat{\underline{0}}$. From Eq. (16) and Eq. (17) it follows:

$$\underline{\mathbf{q}} = \frac{\hat{\underline{\mathbf{S}}}\hat{\underline{\mathbf{v}}}}{|\hat{\underline{\mathbf{S}}}\hat{\underline{\mathbf{v}}}|}; \forall \underline{\mathbf{v}} \in \underline{Q}, Re\hat{\underline{\mathbf{S}}}\hat{\underline{\mathbf{v}}} \neq \hat{\underline{0}} \tag{19}$$

Remark 3. *Let be* $\{\hat{\underline{\mathbf{e}}}_0, \hat{\underline{\mathbf{e}}}_1, \hat{\underline{\mathbf{e}}}_2, \hat{\underline{\mathbf{e}}}_3\}$ *an orthonormal basis in the dual quaternions set:*

$$\begin{aligned} \hat{\underline{\mathbf{e}}}_0 &= [1,0,0,0]^{\mathrm{T}} \\ \hat{\underline{\mathbf{e}}}_1 &= [0,1,0,0]^{\mathrm{T}} \\ \hat{\underline{\mathbf{e}}}_2 &= [0,0,1,0]^{\mathrm{T}} \\ \hat{\underline{\mathbf{e}}}_3 &= [0,0,0,1]^{\mathrm{T}} \end{aligned} \tag{20}$$

In this case, the dual matrix corresponding with the dual orthogonal tensor $\underline{\boldsymbol{R}}$ *is:*

$$\underline{\boldsymbol{R}} = \left(\underline{\mathrm{r}}_{ij}\right)_{i,j=\overline{1,3}}, \underline{\mathrm{r}}_{ij} \in \underline{\mathbb{R}} \tag{21}$$

The matrix $\hat{\underline{\mathbf{S}}} = \frac{1}{4}\begin{bmatrix} 1+\mathrm{trace}\underline{\boldsymbol{R}} & 2(\mathrm{vect}\underline{\boldsymbol{R}})^{\mathrm{T}} \\ 2\mathrm{vect}\underline{\boldsymbol{R}} & \underline{\boldsymbol{S}} \end{bmatrix} = [\underline{\mathbf{s}}_0, \underline{\mathbf{s}}_1, \underline{\mathbf{s}}_2, \underline{\mathbf{s}}_3]$, *where the column matrix* $\underline{\mathbf{s}}_k, k = \overline{0,3}$, *has the expressions:*

$$\begin{aligned} \underline{\mathbf{s}}_0 &= \frac{1}{4}[1+\underline{\mathrm{r}}_{11}+\underline{\mathrm{r}}_{22}+\underline{\mathrm{r}}_{33}, \underline{\mathrm{r}}_{32}-\underline{\mathrm{r}}_{23}, \underline{\mathrm{r}}_{13}-\underline{\mathrm{r}}_{31}, \underline{\mathrm{r}}_{21}-\underline{\mathrm{r}}_{12}]^{\mathrm{T}} \\ \underline{\mathbf{s}}_1 &= \frac{1}{4}[\underline{\mathrm{r}}_{32}-\underline{\mathrm{r}}_{23}, 1+\underline{\mathrm{r}}_{11}-\underline{\mathrm{r}}_{22}-\underline{\mathrm{r}}_{33}, \underline{\mathrm{r}}_{21}+\underline{\mathrm{r}}_{12}, \underline{\mathrm{r}}_{31}+\underline{\mathrm{r}}_{13}]^{\mathrm{T}} \\ \underline{\mathbf{s}}_2 &= \frac{1}{4}[\underline{\mathrm{r}}_{13}-\underline{\mathrm{r}}_{31}, \underline{\mathrm{r}}_{21}+\underline{\mathrm{r}}_{12}, 1-\underline{\mathrm{r}}_{11}+\underline{\mathrm{r}}_{22}-\underline{\mathrm{r}}_{33}, \underline{\mathrm{r}}_{31}+\underline{\mathrm{r}}_{23}]^{\mathrm{T}} \\ \underline{\mathbf{s}}_3 &= \frac{1}{4}[\underline{\mathrm{r}}_{21}-\underline{\mathrm{r}}_{12}, \underline{\mathrm{r}}_{31}+\underline{\mathrm{r}}_{13}, \underline{\mathrm{r}}_{32}+\underline{\mathrm{r}}_{23}, 1-\underline{\mathrm{r}}_{11}-\underline{\mathrm{r}}_{22}+\underline{\mathrm{r}}_{33}]^{\mathrm{T}} \end{aligned} \tag{22}$$

Knowing $\hat{\underline{\mathbf{S}}}\hat{\underline{\mathbf{e}}}_k = \hat{\underline{\mathbf{s}}}_k, k = \overline{0,3}$, and using Eq. (19) it follows:

$$\hat{\underline{\mathbf{q}}} = \frac{\hat{\underline{\mathbf{s}}}_k}{\|\underline{\mathbf{s}}_k\|}, k = \overline{0,3} \; if \; Re\|\hat{\underline{\mathbf{s}}}_k\|^2 \neq \underline{0} \tag{23}$$

Noting $\|\underline{\mathbf{s}}\| = \max_{k=\overline{1,3}} Re\|\underline{\hat{\mathbf{s}}}_k\|$, it follows:

$$\underline{\hat{\mathbf{q}}} = \frac{\underline{\hat{\mathbf{s}}}}{\|\underline{\hat{\mathbf{s}}}\|} \tag{24}$$

The Eq. (24) is the final solution of the singularity-free extraction of dual quaternion from orthogonal dual tensor problem.

Remark 4. *The dual matrix corresponding to dual quaternion* $\underline{\hat{\mathbf{q}}}$ *will be noted with* $\underline{\hat{\mathbf{q}}} = \left[\underline{q}_0, \underline{q}_1, \underline{q}_2, \underline{q}_3\right]^T; \underline{q}_k \in \underline{\mathbb{R}}, k = \overline{0,3}$ *and taking into account the Eqs.* (18) *and* (22) *it follows:*

$$\underline{q}_k^2 = \|\underline{\hat{\mathbf{s}}}_k\|^2, k = \overline{0,3} \tag{25}$$

The Eq. (25) generalizes to the case of dual quaternions, the result was obtained in the case of the real quaternions [9]. In [9] the results were obtained using Cayley's factorization in Lie group $\mathbb{S}O_4$.

4 Dual Quaternions and Dual Vector Recovery Solutions

A basis $\mathbf{B} = \{\underline{\mathbf{e}}_1, \underline{\mathbf{e}}_2, \underline{\mathbf{e}}_3\}$ from $\underline{V}_3^{\mathbb{R}}$, which has the values $\underline{\mathbf{e}}_{0i} = \underline{\mathbf{e}}_i(t_0)$ at initial time t_0, is named dual rigid basis if:

$$\begin{cases} \underline{\mathbf{e}}_i \cdot \underline{\mathbf{e}}_j = \underline{\mathbf{e}}_{0i} \cdot \underline{\mathbf{e}}_{0j}, i,j = \overline{1,3} \\ \langle \underline{\mathbf{e}}_1, \underline{\mathbf{e}}_2, \underline{\mathbf{e}}_3 \rangle = \langle \underline{\mathbf{e}}_{01}, \underline{\mathbf{e}}_{02}, \underline{\mathbf{e}}_{03} \rangle \end{cases} \tag{26}$$

Remark 5. *For a dual basis* $\mathbf{B} = \{\underline{\mathbf{e}}_1, \underline{\mathbf{e}}_2, \underline{\mathbf{e}}_3\}$, *the set* $\mathbf{B}^* = \{\underline{\mathbf{e}}^1, \underline{\mathbf{e}}^2, \underline{\mathbf{e}}^3\}$ *represents its reciprocal*

$$\underline{\mathbf{e}}^1 = \frac{\underline{\mathbf{e}}_2 \times \underline{\mathbf{e}}_3}{\langle \underline{\mathbf{e}}_1, \underline{\mathbf{e}}_2, \underline{\mathbf{e}}_3 \rangle}, \underline{\mathbf{e}}^2 = \frac{\underline{\mathbf{e}}_3 \times \underline{\mathbf{e}}_1}{\langle \underline{\mathbf{e}}_1, \underline{\mathbf{e}}_2, \underline{\mathbf{e}}_3 \rangle}, \underline{\mathbf{e}}^3 = \frac{\underline{\mathbf{e}}_1 \times \underline{\mathbf{e}}_2}{\langle \underline{\mathbf{e}}_1, \underline{\mathbf{e}}_2, \underline{\mathbf{e}}_3 \rangle} \tag{27}$$

If $\mathbf{B} = \{\underline{\mathbf{e}}_1, \underline{\mathbf{e}}_2, \underline{\mathbf{e}}_3\}$ is a dual rigid basis then $\mathbf{B}^* = \{\underline{\mathbf{e}}^1, \underline{\mathbf{e}}^2, \underline{\mathbf{e}}^3\}$, its reciprocal dual basis, is also rigid [1].

Theorem 6. *If* $\mathbf{B} = \{\underline{\mathbf{e}}_1, \underline{\mathbf{e}}_2, \underline{\mathbf{e}}_3\}$ *is a dual rigid basis in* $\underline{V}_3^{\mathbb{R}}$ *and* $\underline{\mathbf{e}}_i$ *are continuous functions, then the dual tensor*

$$\underline{\boldsymbol{R}} = \underline{\mathbf{e}}_i \otimes \underline{\mathbf{e}}_0^i \tag{28}$$

is proper orthogonal and uniquely defined by $\underline{\boldsymbol{R}}(\underline{\mathbf{e}}_{0i}) = \underline{\mathbf{e}}_i, i = \overline{1,3}$ [1, 10].

Remark 6. *Based on the linear invariants of Eq.* (28) *results*:

$$\text{vect}\underline{\boldsymbol{R}} = \frac{1}{2}\underline{\mathbf{e}}_0^i \times \underline{\mathbf{e}}_i, \text{trace}\underline{\boldsymbol{R}} = \underline{\mathbf{e}}_0^i \cdot \underline{\mathbf{e}}_i \tag{29}$$

Previous, the Einstein's rule for mute indexes summation was considered. The proofs for **Theorem 6** and **Remark 6** are presented in [1].

Remark 7. *Consider as inputs information about the features-based representation of a rigid body finite motion given by the dual rigid basis* [1], *taking into account the Eqs.* (28) *and* (29) *the tensor* $\hat{\underline{\mathbf{S}}}$ *it will be written*:

$$\hat{\underline{\mathbf{S}}} = \frac{1}{4}\begin{bmatrix} 1 + \underline{\mathbf{e}}_i \cdot \underline{\mathbf{e}}_0^i & \left(\underline{\mathbf{e}}_0^i \times \underline{\mathbf{e}}_i\right)^{\mathrm{T}} \\ \underline{\mathbf{e}}_0^i \times \underline{\mathbf{e}}_i & \underline{\mathbf{e}}_i \otimes \underline{\mathbf{e}}_0^i + \underline{\mathbf{e}}_0^i \otimes \underline{\mathbf{e}}_i + \left(1 - \underline{\mathbf{e}}_i \cdot \underline{\mathbf{e}}_0^i\right)\boldsymbol{I} \end{bmatrix} \tag{30}$$

As we have shown $\hat{\underline{\mathbf{S}}}$ allow the singularity-free determination of dual unit quaternion.

For singularity-free extraction of a dual quaternion from featured-based representation of rigid body motion we can have the following algorithm:

Algorithm:	**Singularity-free extraction of a dual quaternion from feature-based representation of motion**
Input data:	Featured-based representation of a rigid body motion (points and lines)
	1. The rigid bases $\mathrm{B} = \{\underline{\mathbf{e}}_1, \underline{\mathbf{e}}_2, \underline{\mathbf{e}}_3\}$ and $\mathrm{B}^0 = \{\underline{\mathbf{e}}_{01}, \underline{\mathbf{e}}_{02}, \underline{\mathbf{e}}_{03}\}$ are determined 2. The rigidity conditions from Eq. (26) are checked 3. Taking into account Eq. (30), the dual matrix $\hat{\underline{\mathbf{S}}}$ is calculated 4. Taking into account Eq. (24), the dual quaternion is calculated
Output data:	$\hat{\underline{\mathrm{q}}}$ – unit dual quaternion

5 Conclusion

The research presented in this paper is focused on developing new solutions to recover unit dual quaternion from feature-based representation of rigid motion. Dual rigid bases can be constructed using different combinations of points and lines which represent rigid body features. The combination between a dual rigid basis and its reciprocal provides a natural computational instrument that can be used to solve many problems in the kinematics of rigid bodies. Our approach does not necessitate the computation of the dual tensor when the unit dual quaternion needs to be recovered. This may imply an advantage over existing solutions (which usually iterate the computation of the dual matrix associated with the dual tensor) when point and lines features of the rigid body are available. The theoretical solutions have a form suitable for a direct implementation into numerical codes and should provide useful tools for the development of future applications.

References

1. Condurache, D., Burlacu, A.: Dual tensors based solutions for rigid body motion parameterization. Mech. Mach. Theory **74**, 390–412 (2014)
2. Condurache, D., Burlacu, A.: Dual lie algebra representations of the rigid body motion. In: AIAA/AAS Astrodynamics Specialist Conference, AIAA Paper, San Diego (2014)
3. Pennestri, E., Valentini, P.: Dual quaternions as a tool for rigid body motion analysis: a tutorial with an application to biomechanics. Arch. Mech. Eng. **LVII**, 187–205 (2010)
4. Angeles, J.: The application of dual algebra to kinematic analysis. Comput. Methods Mech. Syst. **161**, 3–31 (1998)
5. Dong, H., Hu, Q., Akella, M.R.: Dual-quaternion-based spacecraft autonomous rendezvous and docking under six-degree-of-freedom motion constraints. J. Guidance Control Dyn. **41** (5), 1150–1162 (2018)
6. Leclercq, G., Lefevre, P., Blohm, G.: 3D kinematics using dual quaternions: theory and applications in neuroscience. Front. Behav. Neurosci. **1–25**, 7 (2013)
7. Müller, A.: Coordinate mappings for rigid body motions. J. Comput. Nonlinear Dyn. **12**(2), 021010 (2017)
8. Condurache, D.: A Davenport dual angles approach for minimal parameterization of the rigid body displacement and motion. Mech. Mach. Theory **140**, 104–122 (2019)
9. Sarabandi, S., Perez-Garcia, A., Thomas, F.: On Cayley's factorization with an application to the orthonormalization of noisy rotation matrices. Adv. Appl. Clifford Algebras **29**, 49 (2019)
10. Condurache, D.: Orthogonal dual tensor method for solving the AX = XB sensor calibration problem. Mech. Mach. Theory **104**, 382–404 (2016)

Singularities in the Image-Based Visual Servoing of Five Points

Abhilash Nayak(✉) and Sébastien Briot

CNRS, Laboratoire des Sciences du Numérique de Nantes (LS2N), Nantes, France
{Abhilash.Nayak,Sebastien.Briot}@ls2n.fr

Abstract. Image-based Visual Servoing (IBVS) hinges on the observation of geometric entities such as 3D points on a target object and their projected image on the camera to control a robot. It includes a mapping between two vector spaces: the time variation of the projected 2D points on the image plane of the camera and the instantaneous spatial velocity of the camera induced by the motion of the robot. The configuration of the camera and the points is said to be singular if the matrix that maps the two vector spaces, the so-called interaction matrix, loses rank, leading to problems in control. Utilizing the results on singularities in visual servoing of three and four 3D points, the goal of this article is to determine the configuration of five points leading to singularities. The problem boils down to examining the vanishing conditions of all the minors of the interaction matrix. It turns out that it is not necessary to consider all the minors. Thus, the simplest set of equations is considered and tools from computational algebraic geometry are used to derive a necessary condition leading to singularities in IBVS of five points.

Keywords: IBVS · Interaction matrix · Singularities · Algebraic geometry

1 Introduction

Visual servoing is the closed-loop position control of a robot using vision. A particular technique relevant to this article is when a camera is attached to the end-effector (EE) of a robot, called as the *eye-in-hand configuration* [5]. Thus, the motion of the robot induces the motion of the camera which obtains the vision data from the object to control the robot. One could look at the features drawn on the object such as points, lines or curves to relatively position the EE w.r.t their projection in the camera space [2,3,8]. This kind of control which depends on the use of these geometric entities as perceived by the camera is called *Image-Based Visual Servoing (IBVS)*. A concise account of basic and advanced approaches in IBVS can be found in Chaumette and Hutchinson [4,5]. Essentially, the control of the robot is based on the rate of change of the image observed by the vision system induced by the motion of the manipulator EE. There could be instances when the image observed by the vision system is

J. Lenarčič and B. Siciliano (Eds.): ARK 2020, SPAR 15, pp. 150–157, 2021.
https://doi.org/10.1007/978-3-030-50975-0_19

stationary although the EE is moving, which leads to controllability issues. They are called singularities in the IBVS [4,5].

Although it is crucial to determine these singularities (in order to avoid them), there exists hardly any research in this regard due to the complexity of the mathematical problem. The singularities in the observation of three 3D points were first identified geometrically in Michel and Rives [12]. It was perceived that more than three points could be used to avoid the occurrence of singularities. We proved recently that this assertion was false: we showed in [1] that there are always a finite set of singular positions of the camera given four non-degenerate points. Therefore, the goal of this article is to take it a step further and to check the occurrence of singularities (if any) when five non-degenerate 3D points are observed.

2 Interaction Matrix Corresponding to IBVS of Points

In the context of this article, the set of features observed on an object used to control the robot are points. The corresponding interaction matrix is recalled in this section.

Following the standard pin-hole camera model with a focal length of 1 unit shown in Fig. 1, let the camera center C be at the origin of the EE coordinate frame. Let the camera axis be along the z-axis and the image plane be parallel to the xy-plane. For a point $M(x, y, z)$ expressed in the EE coordinate frame, the mapping between EE twist (i.e. $\mathbf{v}_c = [v_c, \boldsymbol{\omega}_c]^T$, where $\mathbf{v}$ is the instantaneous linear velocity of the optical center C and $\boldsymbol{\omega}$ is the instantaneous angular velocity of the camera) and the rate of change of image features on the image plane (i.e. $\dot{\mathbf{s}} = [\dot{u}, \dot{v}]^T$, where (u, v) are the coordinates of the projection of M onto the image plane) in Fig. 1 is expressed as follows [6,10]:

$$\dot{\mathbf{s}} = \mathbf{L}(\mathbf{s})\mathbf{v}_c, \quad \begin{bmatrix} \dot{u} \\ \dot{v} \end{bmatrix} = \begin{bmatrix} -z & 0 & x & xy & -x^2 - z^2 & yz \\ 0 & -z & y & y^2 + z^2 & -xy & -zx \end{bmatrix} \begin{bmatrix} v_c \\ \boldsymbol{\omega}_c \end{bmatrix}.$$

The interaction matrix here is expressed in terms of the 3D coordinates of the observed point whereas the standard way to express it is in terms of u, v and the depth z. Although the final results should be identical, the results with the current parametrization are easier to interpret geometrically.

For n-points, $\mathbf{L}$ is a $(2n \times 6)$ matrix formed by the concatenation of rows of the matrix in Eq. (1). Henceforth, the dimension of a matrix is written as its subscript. The singular configurations in IBVS of n-points correspond to the cases in which $\mathbf{L}_{(2n\times6)}$ has a non-zero kernel [10]. To control 6 degrees of freedom of a robot, it is necessary to observe at least three 3D points. However, in this case, the pose of the camera in the control scheme is not unique [4]. As a result, generally, more than three points are observed. It has been established that for three non-collinear points, $\mathbf{L}_{(6\times6)}$ is rank deficient when the camera center lies on an affine cylinder that passes through the three points with its axis normal to the plane containing the points [12]. For four non-degenerate (of which no three points are collinear nor all four coplanar) points, it was shown in [1] that

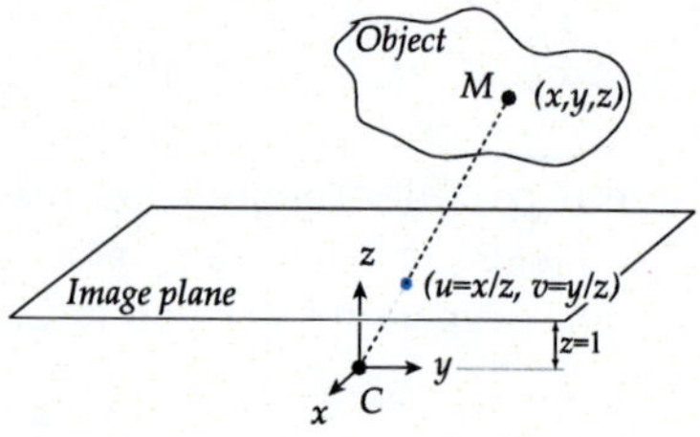

Fig. 1. The pin-hole camera model

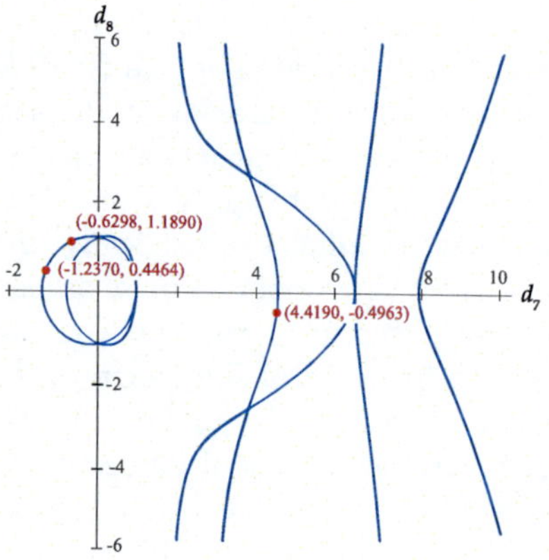

Fig. 2. $V(\mathscr{K}) \in V(res)$, only three elements of $V(\mathscr{K})$ in Eq. (16) are shown for clarity.

any configuration of them leads to a finite number of singular positions of the camera.

The goal is to find the configurations of five non-degenerate points and the camera center such that $\text{rank}(\mathbf{L}_{(10\times6)}) < 6$.

3 Algebraic Description of the Problem

Without any loss of generality, the affine coordinates of the five points in the object coordinate frame can be considered as follows:

$$\begin{aligned} M_1 &= [0,0,0]^T, \quad M_2 = [1,0,0]^T, \quad M_3 = [d_1, d_2, 0]^T, \\ M_4 &= [d_3, d_4, d_5]^T, \quad M_5 = [d_6, d_7, d_8]^T. \end{aligned} \tag{1}$$

Let the coordinates of the camera expressed in the object frame be $C = [x, y, z]^T$. The orientation between the object and the camera coordinate frames is assumed constant. This is due to the fact that the relative configuration of the points and the camera center are invariant to rotations. For any other object orientation given by a rotation matrix $\mathbf{R}$, the singularities can be determined by rotating the singular position coordinates of the camera obtained for the corresponding fixed orientation by $\mathbf{R}$ as proven in [1,2].

If only the first three points are considered (with $d_2 \neq 0$), the singular positions of the camera correspond to the vanishing determinant of $\mathbf{L}_{(6\times6)}$ [6]:

$$C_{123} := x^2 d_2 - x d_2 + y^2 d_2 + (-{d_1}^2 - {d_2}^2 + d_1)y = 0. \tag{2}$$

It is the equation of a cylinder in $\mathbb{R}[x, y, z]$ with its cross-section being the circumcircle of $\Delta M_1 M_2 M_3$ and axis normal to the plane containing M_1, M_2 and M_3 [12].

In case of four points, the interaction matrix is no longer square. Hence, the singular positions of the camera correspond to the vanishing of all of the (8×6) minors of $\mathbf{L}_{(8\times6)}$ [1]. There are 28 minors and they are polynomials in variables x, y, z and parameters d_1, d_2, d_3, d_4, d_5. This set of polynomials generates a *polynomial ideal* $\mathscr{L}_4 \subset \mathbb{R}[x, y, z, d_1, d_2, d_3, d_4, d_5]$. It is an infinite set that consists of all combinations of those minors with coefficients chosen from $\mathbb{R}[x, y, z, d_1, d_2, d_3, d_4, d_5]$ and is denoted by angular brackets[1]. As shown in [1], four of the minors must be singular cylinders corresponding to the four sets of three points. Let them generate an ideal

$$\mathscr{C}_4 = \langle C_{123}, C_{124}, C_{134}, C_{234}\rangle \subset \mathbb{R}[x, y, z, d_1, d_2, d_3, d_4, d_5], \tag{3}$$

where C_{ijk} is the cylinder of singularities in $\mathbb{R}[x, y, z]$ corresponding to points M_i, M_j and M_k. The sets of common zeros to the minors correspond to the so called *varieties* of the ideals, denoted as $V(\mathscr{L}_4)$ and $V(\mathscr{C}_4)$, respectively. Geometrically, $V(\mathscr{C}_4)$ is the intersection of the four cylinders $C_{123}, C_{124}, C_{134}, C_{234}$ and $V(\mathscr{C}_4) \supseteq V(\mathscr{L}_4)$. It was shown in [1] that when $d_2 \neq 0$ and $d_5 \neq 0$, $V(\mathscr{L}_4) = V(\mathscr{C}_4)$. Furthermore, it was also proved that there are at least 2 and at most 6 singular positions of the camera for any given non-degenerate configuration of four points.

The interest of the current article is to explore the outcome of adding a fifth point on the singular positions of the camera. In this vein, the (10×6) interaction matrix $\mathbf{L}_{(10\times6)}$ obtained when observing five points has 210 principal minors that generate the ideal $\mathscr{L}_5 \subset \mathbb{R}[x, y, z, d_i, i = 1, ..., 8]$. Among these minors, 10 cylinders corresponding to the singularities of 10 sets of three points can be picked out to generate the ideal

$$\mathscr{C}_5 = \langle C_{123}, C_{124}, C_{125}, C_{134}, C_{135}, C_{145}, C_{234}, C_{235}, C_{245}, C_{345}\rangle \subset \mathbb{R}[x, y, z, d_i]. \tag{4}$$

Unlike for four points, the computational software such as `Maple` and `Singular` fail to prove that $V(\mathscr{L}_5) = V(\mathscr{C}_5)$ for any $d_i, i = 1, ..., 8$, due to the large number of variables and equations. However, since the intersection of 10 cylinders is the largest variety (i.e. $V(\mathscr{C}_5) \supseteq V(\mathscr{L}_5)$), it will be considered for further analysis in Sect. 4.2 (not in 4.1).

4 Relative Configuration of Five Points for the Occurrence of Singularities

It is clear by now that starting from three points, addition of a point to be observed decreases the dimension of the singular manifold (cylinder in case of three points, a finite set of points in case of four points) for the placement of the

[1] The terms from algebraic geometry are defined briefly in the context of this article as and when they are used. For formal definitions and examples, the reader is directed to the excellent book by Cox, Little and O'Shea [7].

camera [9]. So, it is expected that adding a fifth point should in general not lead to any singularities. However, it is possible to choose a fifth point in such a way that the position of the camera coincides with the singularities corresponding to the first four points. This section aims at determining these relations between $d_i, i = 1, ..., 8$ such that the 10 cylinders in Eq. (4) have a non-zero intersection. This is shown initially by fixing some of the d_i values. Subsequently, a necessary condition with arbitrary d_i values is derived. All the calculations done in this section can be found in the `Maple` sheet [11].

4.1 First Four Points Are Fixed

In this sub-section, the coordinates of the five points are chosen arbitrarily as follows:

$$\{d_1 = 2,\ d_2 = 7,\ d_3 = 5,\ d_4 = -6,\ d_5 = 4,\ d_6 = 3,\ d_7 = -11,\ d_8 = 13\} \quad (5)$$

When all five points are fixed according to (5), the *Gröbner basis* (set of generators of an ideal such that it has a nicer structure and one of its applications is to solve a set of polynomial equations [7]) of ideal $\mathscr{L}_5$ yields $\langle 1 \rangle$ implying:

$$V(\mathscr{L}_5) = \emptyset \quad \text{for} \quad \mathscr{L}_5 \subset \mathbb{R}[x, y, z]. \quad (6)$$

In other words, for a set of non-degenerate ($d_2, d_5, d_8 \neq 0$ and so on) five points, there are no singular positions of the camera. Therefore, the conditions for $V(\mathscr{L}_5) \neq \emptyset$ is explored in the following by letting the parameters d_i arbitrary one at a time. If d_8 is not substituted, the solution set is still empty. However, if both d_7 and d_8 are left arbitrary (i.e. $\mathscr{L}_5 \subset \mathbb{R}[d_7, d_8, x, y, z]$), the solution set is finite:

$$\begin{aligned} V(\mathscr{L}_5) = \{&d_7 = 4.4190, d_8 = -.4963, x = 3.4390, y = 5.8530, z = 25.1100,\\ &d_7 = -1.2730, d_8 = .4464, x = 3.4390, y = 5.8530, z = 25.1100,\\ &d_7 = 7.4800, d_8 = 24.7200, x = 3.4390, y = 5.8530, z = 25.1100,\\ &d_7 = -.6298, d_8 = 1.1890, x = -2.4390, y = 1.4330, z = 3.8180\}. \end{aligned} \quad (7)$$

The above set gives the placement of the fifth point and the camera such that there are singularities in the IBVS of five points. Furthermore, if M_5 is left arbitrary, the dimension of the following variety is 1:

$$\dim(V(\mathscr{L}_5)) = 1 \quad \text{for} \quad \mathscr{L}_5 \subset \mathbb{R}[d_6, d_7, d_8, x, y, z]. \quad (8)$$

Therefore, if the first four points are fixed, there exists a curve on which the fifth point lies such that the singularities appear in the IBVS of five points. To obtain the equations describing this curve, x, y, z must be eliminated. This amounts to computing the *elimination ideal* (Geometrically, it amounts to projecting the corresponding variety onto a lower dimensional subspace):

$$\mathscr{K}_s = \mathscr{L}_5 \cap \mathbb{R}[d_6, d_7, d_8]. \quad (9)$$

$\mathscr{K}_s$ is generated by 9 polynomials that are functions of only d_6, d_7 and d_8. Recall that this ideal is calculated when the first four points are fixed according to (5). On the other hand, due to the complexity of the equations, `Maple` fails to calculate this ideal symbolically (when all the points are arbitrary):

$$\mathscr{K} = \mathscr{L}_5 \cap \mathbb{R}[d_i],\ i = 1, ..., 8, \tag{10}$$

where $\mathscr{K}$ is the ideal obtained after eliminating x, y, z from polynomials generating $\mathscr{L}_5$. However, a workaround is proposed in the next sub-section to obtain a necessary relation between the parameters $d_i, i = 1, ..., 8$ that leads to singularities. Nonetheless, obtaining necessary and sufficient relation(s) between the parameters d_i is analogous to finding the elimination ideal in Eq. (10) and is still an open question. Some ideas to obtain it are discussed in the section on conclusions.

4.2 All Five Points Are Chosen Arbitrarily

The goal of this sub-section is to come up with a necessary condition such that $V(\mathscr{L}_5) \neq \emptyset$. Since $\mathscr{L}_5 \supseteq \mathscr{C}_5$ (dually $V(\mathscr{L}_5) \subseteq V(\mathscr{C}_5)$), it is enough to determine the necessary relations between the coordinates of five points so that the cylinders $C_{ijk} = 0$ in $\mathscr{C}_5$ have a non-zero intersection.

It is already known from [1] that the Gröbner basis of $\mathscr{C}_4$, a sub-ideal of $\mathscr{C}_5$ in Eq. (3) yields a univariate sextic polynomial in variable x and parameters $d_i, i = 1, .., 5$ that can be transformed to a cubic G_{C4} with a different variable q:

$$G_{C4} = a_3q^3 + a_2q^2 + a_1q + a_0,\ a_i \in \mathbb{R}[d_1, d_2, d_3, d_4, d_5], \tag{11}$$

where q is an affine function of x. Another similar sub-ideal of $\mathscr{C}_5$ can be considered: $\mathscr{C}_{4b} = \langle C_{123}, C_{125}, C_{135}, C_{235}\rangle \subset \mathbb{R}[x, y, z, d_1, d_2, d_6, d_7, d_8]$. The varieties $V(\mathscr{C}_4)$ and $V(\mathscr{C}_{4b})$ are intersections of cylinders C_{ijk} corresponding to points M_1, M_2, M_3, M_4 and M_1, M_2, M_3, M_5, respectively. As a consequence, simply replacing the coordinates of point M_4 by that of M_5 (i.e. $d_3 \to d_6,\ d_4 \to d_7,\ d_5 \to d_8$) in G_{C4}, gives another cubic polynomial G_{C4b} in variable q:

$$G_{C4b} = b_3q^3 + b_2q^2 + b_1q + b_0,\ b_i \in \mathbb{R}[d_1, d_2, d_6, d_7, d_8]. \tag{12}$$

Moreover, if the 10 cylinders $C_{ijk} = 0$ that generate $\mathscr{C}_5$ in Eq. (4) share a non-zero solution set, it is necessary that the solutions lie on the 7 cylinders that belong to both the ideals $\mathscr{C}_4$ and $\mathscr{C}_{4b}$ as follows [11]:

$$\begin{aligned} \langle C_{123}, C_{124}, C_{125}, C_{134}, C_{135}, C_{234}, C_{235}\rangle = \mathscr{C}_4 \cup \mathscr{C}_{4b} \subseteq \mathscr{C}_5 \subseteq \mathscr{L}_5 \\ \implies V(\mathscr{C}_4) \cap V(\mathscr{C}_{4b}) \supseteq V(\mathscr{C}_5) \supseteq V(\mathscr{L}_5) \end{aligned} \tag{13}$$

The condition for $G_{C4} = 0$ and $G_{C4b} = 0$ to have the same solutions (so that $V(\mathscr{C}_5), V(\mathscr{L}_5) \neq \emptyset$) is that their resultant w.r.t q must be zero. The resultant is a 58 degree polynomial in $d_i, i = 1, ..., 8$.

$$res = -d_1^{32}d_4^{12}d_6^2d_7^4d_8^8 - 2d_1^{32}d_4^{12}d_6^2d_7^2d_8^{10} - d_1^{32}d_4^{12}d_6^2d_8^{12} + \cdots \tag{14}$$

It implies that, $res = 0$ is a necessary condition for $V(\mathscr{L}_5)$, $V(\mathscr{C}_5) \neq 0$. As a result, the d_i values satisfying the generators of $\mathscr{K}$ in Eq. (10) must also satisfy $res = 0$. Dually, res must belong to the ideal $\mathscr{K}$:

$$res \in \mathscr{K} \iff V(\mathscr{K}) \in V(res) \tag{15}$$

The above relation can be verified by substituting parameters $d_i, i = 1, ..., 6$ from (5). The reason for substituting only first six parameters is to be able to visualize the results since in this case, $V(\mathscr{K})$ is a finite set as follows:

$$\begin{aligned} V(\mathscr{K}) = \{&(-1.2370, 0.4464),\ (4.4190, -0.4963), \\ &(7.4800, 24.7200),\ (-0.6298, 1.1890)\}. \end{aligned} \tag{16}$$

It corresponds exactly to coordinates of the fifth point in (7) that lead to singularities. The inverse condition number of the matrix $\mathbf{L}^T_{(6\times10)}\mathbf{L}_{(10\times6)}$ for the aforementioned configuration of five points is of the range 10^{-19} implying they are indeed singularities in the IBVS. $V(res)$ or simply $res = 0$ is a curve in the d_7d_8-plane and $V(\mathscr{K}) \in V(res)$ as shown in Fig. 2 on page 3. Moreover, it can be verified in `Maple` that $res \in \mathscr{K}$.

For practical purposes, it is enough to have a necessary condition. The five points must be chosen such that they do not satisfy the relation $res = 0$, which assures that there are no singularities. However, the converse is not true, i.e. any $d_i, i = 1, ..., 8$ satisfying $res = 0$ is not sufficient for the occurrence of singularities. Therefore, to obtain necessary and sufficient conditions (It is evident from Eq. (9) that there is more than one relation) to avoid the singularities is still an open question and perhaps the symmetry of the cylinder equations holds the key.

5 Conclusions

Image-based visual servoing of three and four points is always known to have singularities while their occurrence in the observation of five points was still an open question. We showed that the observation of five points is mostly free of singularities when their relative configuration is non-degenerate. However, there are some special configurations which might lead to problems in the velocity control. This article focused on determining those configurations as relations between the affine coordinates of the points. The problem was formulated as determining the set of common zeros to the principal minors of the interaction matrix. It turned out that they do not have common zeros when the five points are chosen arbitrarily. Therefore, a scenario was considered in which the first four points are fixed and conditions on the fifth point such that the minors have common zeros were derived. Additionally, using Gröbner basis computations, a necessary relation between the coordinates of all five points for the occurrence of singularities was derived.

Acknowledgements. This work was supported by the French ANR project SESAME (funding ID: ANR-18-CE33-0011).

References

1. Pascual-Eduardo, B., Nayak, A., Briot, S., Kermorgant, O., Martinet, P., Din, M.S.E., Chaumette, F.: Complete singularity analysis when observing four image points. Int. J. Comput. Visi. (2020, submitted). https://uncloud.univ-nantes.fr/index.php/s/qi2tHqMc4sm89w3
2. Briot, S., Chaumette, F., Martinet, P.: Revisiting the determination of the singularity cases in the visual servoing of image points through the concept of hidden robot. IEEE Trans. Robot. **33**(2), 536–546 (2017)
3. Briot, S., Martinet, P., Chaumette, F.: Singularity cases in the visual servoing of three image lines **2**(2), 412–419 (2017)
4. Chaumette, F., Hutchinson, S.: Visual servo control Part I: basic approaches. IEEE Robot. Autom. Mag. **13**(4), 82–90 (2006)
5. Chaumette, F., Hutchinson, S.: Visual servo control, Part II: advanced approaches. IEEE Robot. Autom. Mag. **14**(1), 109–118 (2007)
6. Chaumette, F., Hutchinson, S.: Visual servoing and visual tracking. In: Siciliano, B., Khatib, O. (eds.) Handbook of Robotics. Springer, Cham (2008)
7. Cox, D.A., Little, J., O'Shea, D.: Ideals, Varieties, and Algorithms: An Introduction to Computational Algebraic Geometry and Commutative Algebra, 3/e (Undergraduate Texts in Mathematics). Springer, Berlin (2007)
8. Espiau, B., Chaumette, F., Rives, P.: A new approach to visual servoing in robotics. IEEE Trans. Robot. Autom. **8**(3), 313–326 (1992)
9. Feddema, J., Lee, C., Mitchell, O.: Automatic selection of image features for visual servoing of a robot manipulator. In: Proceedings of the 1989 IEEE International Conference on Robotics and Automation (ICRA 1989), pp. 832–837. Scottsdale (1989)
10. Hutchinson, S., Hager, G., Corke, P.: A tutorial on visual servo control. IEEE Trans. Robot. Autom. **12**, 651–670 (1996)
11. Maple-file: Maple file with calculations (2020). https://uncloud.univ-nantes.fr/index.php/s/QcHAtkzQimJW278
12. Michel, H., Rives, P.: Singularities in the determination of the situation of a robot effector from the perspective view of 3 points. Technical report, INRIA (1993)

Real-Time Motion-Planning in Dynamic Environments via Enhanced Velocity Obstacle

Amir Shahidi(✉), Katrin Peitsch, Stefan Klimmek, Mathias Hüsing, and Burkhard Corves

Institute of Mechanism Theory, Machine Dynamics and Robotics, RWTH Aachen University, Aachen, Germany
shahidi@igmr.rwth-aachen.de

Abstract. Freely networked production lines are the essence of the factories of the future. In this context, the mobility of the robotic agents (i.e. mobile manipulators) and logistic units for transferring the goods and the products in a highly dynamic environment shall be considered as a challenging problem. The challenge lies mainly on the recognition of the changes in the environment and fast proper reaction to these changes. The available solutions to the problem are mostly based on the information from embedded local sensoric systems of the mobile units. In this paper, we present a solution based on a global networked sensoric system. We use the depth cameras to monitor the environment of a population of the robotic agents. We apply enhanced method of the Velocity Obstacle to perform a collision free path planning for mobile agents. To save space and computation time, we enhance the collision cones of the obstacle velocities by defining oblique ellipsoidal cones instead of straight circular cones. The aim is to make the process for real-time scenarios as simple, quick and practical as possible. The described methods will be implemented in a dynamic laboratory and the results will be discussed.

Keywords: Motion planning · Dynamic environment · Velocity obstacle

1 Introduction

Motion planning has been one of the most important parts of control in robotics. Besides the kinematic issues, the knowledge about the shape of the environment is of great importance to address the problem. There are many available and widely used motion planning algorithms in robotics of which *Exact and Approximate Cell Decomposition*, *Control-based Methods*, *Potential Fields*, *Randomized Planning* and *Sampling-based Motion Planning* can be named. For a preview for motion planning in robotics, we refer to [4, 14]. In this study, we use the method of the *Velocity Obstacle* for performing the motion planning. The Velocity Obstacle

J. Lenarčič and B. Siciliano (Eds.): ARK 2020, SPAR 15, pp. 158–165, 2021.
https://doi.org/10.1007/978-3-030-50975-0_20

is a planning method especially for dynamic environments with varying numbers of stationary and moving obstacles, to predict and avoid possible collisions. The behavior of the agents depends on the velocities of both agents and the obstacles. Velocity Obstacle was introduced first in 1998 for a 2D path-planning [8]. The idea was the definition of cones (so called collision cones) that put up the hulls bounding an obstacle. The obstacles were considered to be expanded by the agent's footprint. The collision cones shall be transformed according to the obstacle's velocity vector (Velocity Obstacle). A collision shall be avoided, when the agent's velocity vector lies outside these Velocity Obstacles. A three-dimensional version of the Velocity Obstacles were introduced in [12]. The straight circular cones are constructed in such a way that they cover the biggest bounding sphere around the obstacles. Similar application were used in [3,5]. We generalize the method of 3D Velocity Obstacle to the oblique ellipsoidal cones. This way we gain the advantage of saving the space to construct narrow passages that may have been lost in Velocity Obstacle Cones constructed by spheres (*cf.* Sect. 3).

To perform path-planning in dynamic environments, we need the information of the free space, through which the agents can plan their motion. With the help of depth cameras it is possible to record and analyze the environment in the form of point clouds. The point clouds must first be filtered and then divided into separate objects in the space. The filtering of the point cloud is referred to the reduction of the noise and the amount of data to the relevant part, such as simplifying the surfaces that belong to an identical cluster (*e.g.* walls, floors and ceilings). The RANSAC (RANdom SAmple Consensus) algorithm is often used to remove such flat surfaces [9]. The RANSAC algorithm iteratively forms a 3D surface from randomly selected data and assigns all corresponding points to the surface [2,9,16]. Another common (but computationally more expensive) method of filtering the data is the usage of the normal vectors of the points [11].

In a dynamic environment, it is important to consider that the objects can enter and exit the environment, therefore the number of objects in the environment should be flexible. Cluster analysis techniques can generally be divided into four groups: hierarchical, partitioning, grid-based and density-based cluster analysis [10]. Here the density-based cluster analysis method DBSCAN (Density-Based Spatial Clustering of Applications with Noise) algorithm is used and explained in more detail [6]. For further information on other methods, interested readers are referred to [1,7,10,13].

In a density-based cluster analysis, the data are analyzed according to their relative density to each other. The advantage of density-based cluster analysis methods is that they are independent of the number of clusters [10]. Moreover, clusters of arbitrary shapes are found and noise is filtered. However, density-based algorithms have a weakness for data sets with different densities. We reduce the point cloud to a more uniform density, so that the weaknesses of density-based methods do not come into play. The DBSCAN algorithm requires merely the two input parameters *eps* and *MinPts*. The parameter *eps* denotes the maximum distance between two points within a cluster and *MinPts* determines the minimum number of points in a cluster. Should the parameter *MinPts* be chosen greater than 1, the DBSCAN algorithm can be used to filter noise from a data set [6]. In Sect. 2 the algorithm will be explained in more detail for the description of a dynamic environment.

2 Analysis of the Environment

In this section, we start with the principles of the description of the static environment via DBSCAN algorithm and develop the idea to the case were the moving objects can be recognized and categorized in dynamic environments.
Static environment

First, the environment of the robotic agents is assumed to be static with rigid objects (holding footprint). The processing of the point clouds is done with the help of the free program library "Point Cloud Library PCL" [15]. After filtering the point cloud to the relevant parts of the environment, the point cloud is reduced to a more uniform density. By reducing the point cloud, computational time and memory shall be saved, as the clustering of the point cloud is highly dependent on the density of the points.

The calculation of the parameters *eps* and *MinPts* of the DBSCAN algorithm can be done by calculating the k-Nearest-Neighbor classification [6]. In the k-Nearest-Neighbor calculation, all neighbors are determined for each point and sorted according to their distance from each other. Then the kth neighbor is selected for each point and the distance is saved as *k-dist*. The points are sorted and plotted according to the size of *k-dist*. The k-Nearest-Neighbor and *k-dists* are calculated for different values of k. The value of *k-dist*, where the plotted curve bends upwards, represents the parameter *eps*. The corresponding k of this curve can be assumed as *MinPts*. In the exemplary graph shown in Fig. 1, the possible choices for the parameters *eps* and *k-dist* would be 0.4 to 0.6 and 4 to 6 respectively.

The clustered points are described by their bounding hulls (tightest enclosing convex hull of all points) and the position of the centroid. The centroid of a cluster is determined to be the center of the hull.
Dynamic environment

A dynamic analysis of the environment carried out by iterative performance of the static analysis and comparison between the recognized clusters of successive records. The comparison of the clusters is done by the size of the bounding hull and the position of the centroids relative to each other. We divide the clusters into five subgroups as follows:

- Rigid/Changed: Centroid not moved - Dimension changed (*e.g.*, industrial robot),
- Rigid/Unaltered: Centroid not moved - Dimension not changed (*e.g.* table),
- Moved/Changed: Centroid moved - Dimension changed (*e.g.* mob. manipulator),
- Moved/Unaltered: Centroid moved - Dimension not changed (*e.g.* mob. units),
- New: When an object entering to the environment.

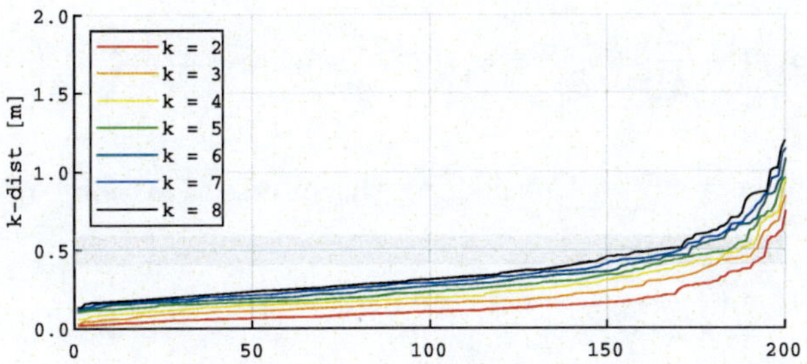

Fig. 1. Determining the parameters *eps* and *MinPts* for the DBSCAN algorithm (the horizontal axis shows the amount of the points sorted by *k-dist*)

The velocities of the moving objects can be calculated knowing the amount of the displacement and the frequency of the record. The free space required for motion planning shall be constructed subtracting the recognized occupied space from the whole space.

3 Enhanced Velocity Obstacle

The method of Velocity Obstacles for 3D environments, as introduced in [12], demands the description of the tightest straight circular cone with a base covering the spherical bounding hull of the obstacles. The loss of space in this manner of description is relatively high, as generally smaller bounding hulls around the obstacles can be defined. For instance, one can use (oblique) pyramids in place of cones, with rectangular bases defined by rectangular cuboid bounding hulls around the obstacles (see Fig. 3). Even though the computational cost of this method is less than defining spheres and spheroids, the loss of space is still relatively high for slender objects, for which spheroids shall be considered as a suitable description. To define a spheroid given the spatial coordinates of the clustered points, the major axis shall be defined as the longest distance between the clustered points or via regression methods (which are more computationally demanding) and the minor axis shall be considered as the maximum orthogonal distance of the points from this axis. In the dynamic environment this process should be carried out for each iteration of record though, which is a computational burden. These expenses can be reduced using rectangular cuboid bounding hulls and extracting its smallest boundary spheroidal hull (see Fig. 2). Due to symmetry of the spheroids, we can solve the problem in 2D (ellipse $\frac{x^2}{p^2} + \frac{z^2}{q^2} = 1$) and embed it to the 3D case (spheroid $\frac{x^2}{p^2} + \frac{y^2}{p^2} + \frac{z^2}{q^2} = 1$). It is to find the smallest ellipse (p, q) that encloses the rectangle (l, w) (see Fig. 2 for notations). In other words the rectangle's area inside the ellipsoid shall be maximized. Denote $x = \frac{w}{2}$. Considering the rectangle's area and its derivation with respect to the variable x, we have

$$A = 2x \cdot 2q\sqrt{1-\frac{x^2}{p^2}} \Rightarrow \frac{\partial A}{\partial x} = 4q\sqrt{1-\frac{x^2}{p^2}} + 4xq\frac{1}{\sqrt{1-\frac{x^2}{p^2}}} \cdot \left(-\frac{x}{p^2}\right) \stackrel{!}{=} 0, \quad (1)$$

that results in $x = \pm\frac{p}{\sqrt{2}}$. Considering the intersection point of rectangle and ellipse

$$(\frac{p}{\sqrt{2}}, \frac{q}{\sqrt{2}}) \stackrel{!}{=} (\frac{w}{2}, \frac{l}{2}) \Rightarrow p = \frac{w\sqrt{2}}{2}, \quad q = \frac{l\sqrt{2}}{2}. \quad (2)$$

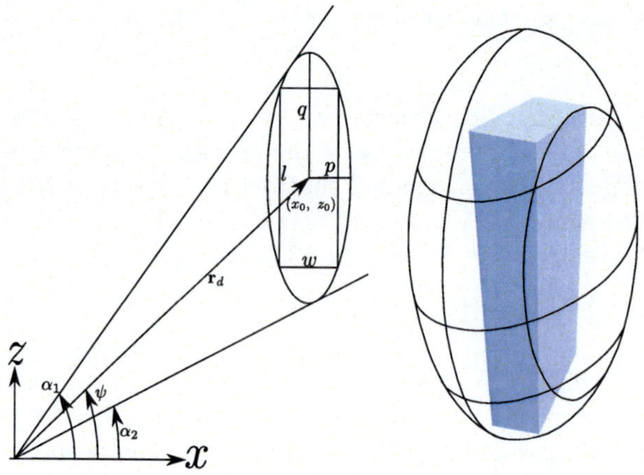

Fig. 2. Left: Description of the ellipses; Right: The boundary spheroidal hull

To describe the oblique ellipsoidal collision cones, we need to define the angles α_1 and α_2 to construct the tangent lines $z = m_i x, m_i = \tan(\alpha_i), i \in \{1, 2\}$ (see Fig. 2 for notations). For an ellipse with center point at (x_0, z_0), we have

$$\frac{(x-x_0)^2}{p^2} + \frac{(z-z_0)^2}{q^2} = 1 \Rightarrow z = q\sqrt{1-\frac{(x-x_0)^2}{p^2}} + z_0 \stackrel{!}{=} m_i x \quad (3)$$

$$\Rightarrow x_{1,2} = \frac{m_i z_0 p^2 + x_0 q^2}{m_i^2 p^2 + q^2} \pm \underbrace{\sqrt{\left(\frac{m_i z_0 p^2 + x_0 q^2}{m_i^2 p^2 + q^2}\right)^2 - \frac{z_0^2 p^2 + x_0^2 q^2 - p^2 q^2}{m_i^2 p^2 + q^2}}}_{\stackrel{!}{=}\, 0}. \quad (4)$$

In order for the line to be a tangent line, there can only be one point of intersection, hence the square-root-term ought to be zero. Therefore

$$\sqrt{\left(\frac{m_i z_0 p^2 + x_0 q^2}{m_i^2 p^2 + q^2}\right)^2 - \frac{z_0^2 p^2 + x_0^2 q^2 - p^2 q^2}{m_i^2 p^2 + q^2}} = 0, \tag{5}$$

$$\Rightarrow \left(\frac{m_i z_0 p^2 + x_0 q^2}{m_i^2 p^2 + q^2}\right)^2 = \frac{z_0^2 p^2 + x_0^2 q^2 - p^2 q^2}{m_i^2 p^2 + q^2}, \tag{6}$$

$$\Rightarrow m_i = \frac{-x_0 z_0}{p^2 - x_0^2} \pm \sqrt{\frac{x_0^2 z_0^2}{(p^2 - x_0^2)^2} - \frac{q^2 - z_0^2}{p^2 - x_0^2}}, \tag{7}$$

$$S(x_s, z_s) = \left(\frac{m_i z_0 p^2 + x_0 q^2}{m_i^2 p^2 + q^2},\ q\sqrt{1 - \frac{\left(\frac{m_i z_0 p^2 + x_0 q^2}{m_i^2 p^2 + q^2} - x_0\right)^2}{p^2}} + z_0 \right), \tag{8}$$

with $S(x_s, z_s)$ denoting the intersection point.

Knowing the angle $\psi = \sin^{-1} \frac{(\boldsymbol{r}_d \cdot \boldsymbol{n}_z)}{||\boldsymbol{r}_d||}$ ($\boldsymbol{n}_z$ denoting the unit vector in z-direction), the following parameters can be obtained to describe the cone $f(x, y, z) = \frac{x^2}{a^2} + \frac{y^2}{b^2} - \frac{z^2}{c^2}$ and identifying the points inside the cone (*i.e.* $f(x, y, z) < 0$)

$$a = \frac{p\,\|\boldsymbol{r}_d\|}{\sqrt{\|\boldsymbol{r}_d\|^2 - p^2}}, \quad c = \|\boldsymbol{r}_d\| \cdot \cos\left(\psi - \frac{\alpha_1 + \alpha_2}{2}\right), \quad b = c \cdot \tan\left(\frac{\alpha_1 - \alpha_2}{2}\right). \tag{9}$$

The formation of Velocity Obstacles would be concluded by transferring the cones into the coordinate system of the agent.

4 Implementation and Discussion

The method has been implemented in the laboratory of Internet of Production[1] at IGMR of RWTH Aachen University[2], where we perform experimental studies on the theme. We record the environment via four depth cameras Visionary-T from the company SICK that provide a 3D point cloud, which serves as a basis for the description of the environment. An instance of the processed data, according to the procedure described in Sect. 2 is presented in Fig. 3. The colored points (except for black ones) represent the detected objects in the environment. A color-Scala scheme is utilized to establish a distinguishment between the velocities of the obstacles (Scala from blue (rigid objects) to red (moving with 5 m/s)). The black points are uniformly distributed in the recognized free space of the environment and determine the available space for the motion planning. Furthermore, the same strategy can be used to integrate the available information

1 www.iop.rwth-aachen.de.

2 www.igmr.rwth-aachen.de.

of the free space in the context of other motion planning algorithms like *Search-based* or *Sampling-based Motion Planning*. The points in free space can also be weighted depending on theirs relative position to a moving object. The closer a point is to a moving cluster and the higher the velocity of the object in its vicinity, the smaller is its weighting factor wf $(0 < wf \leq 1)$[3].

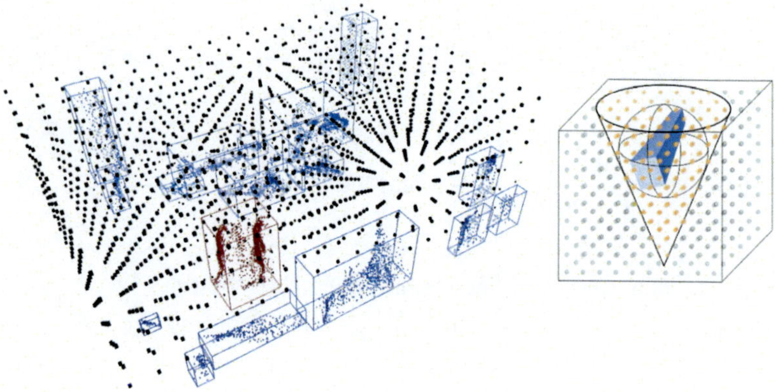

Fig. 3. Left: Depiction of the moving object in the room and representation of the free space with weighting factors; Right: The cones of the Velocity Obstacle

After the detection of a moving obstacle, the Velocity Obstacles of the objects, for each individual agent, will also be constructed using the methods introduced in Sect. 3. The space lying inside the collision cones (shown with orange points in Fig. 3) should also be removed from the free space to perform a smooth and collision-free motion planning.

5 Conclusion

We have presented a simple method for recognition of the obstacles in 3D dynamic environments. The speed of the analysis may vary based on the amount of the recorded data from the sensoric systems. In addition, we have introduced an enhanced model for the Velocity Obstacles, by defining oblique ellipsoidal collision cones using spheroidal hulls for slender obstacles. A simple mathematical procedure is introduced to reduce the computational effort and make the process real-time capable for motion-planning scenarios. For a case study of the gathered data from 4 depth cameras (750 points) we have achieved a processing rate of 1 Hz.

[3] A video of the process shall be found under www.igmr.rwth-aachen.de/index.php/en/rob-en/rob-iop-en.

Acknowledgements. The Authors would like to thank for the kind support of German Research Foundation DFG (Deutsche Forschungsgemeinschaft) under Germany's Excellence Strategy – EXC-2023 Internet of Production – 390621612.

References

1. Arslan, O., Guralnik, D.P., Koditschek, D.E.: Coordinated robot navigation via hierarchical clustering. IEEE Trans. Robot. **32**(2), 352–371 (2016)
2. Cantzler, H., Fisher, R.B., Devy, M.: Quality enhancement of reconstructed 3D models using coplanarity and constraints. In: Joint Pattern Recognition Symposium, pp. 34–41. Springer (2002)
3. Choi, M., Rubenecia, A., Shon, T., Choi, H.H.: Velocity obstacle based 3D collision avoidance scheme for low-cost micro UAVs. Sustainability **9**(7), 1174 (2017)
4. Choset, H.M., Hutchinson, S., Lynch, K.M., Kantor, G., Burgard, W., Kavraki, L.E., Thrun, S.: Principles of Robot Motion: Theory, Algorithms, and Implementation. MIT Press, Cambridge (2005)
5. Conroy, P., Bareiss, D., Beall, M., van den Berg, J.: 3-D reciprocal collision avoidance on physical quadrotor helicopters with on-board sensing for relative positioning. arXiv preprint arXiv:1411.3794 (2014)
6. Ester, M., Kriegel, H.P., Sander, J., Xu, X., et al.: A density-based algorithm for discovering clusters in large spatial databases with noise. KDD **96**, 226–231 (1996)
7. Ester, M., Sander, J.: Knowledge Discovery in Databases: Techniken und Anwendungen. Springer, Heidelberg (2013)
8. Fiorini, P., Shiller, Z.: Motion planning in dynamic environments using velocity obstacles. Int. J. Robot. Res. **17**(7), 760–772 (1998)
9. Fischler, M.A., Bolles, R.C.: Random sample consensus: a paradigm for model fitting with applications to image analysis and automated cartography. In: Readings in Computer Vision, pp. 726–740. Elsevier (1987)
10. Han, J., Kamber, M., Tung, A.K.: Spatial clustering methods in data mining. In: Geographic Data Mining and Knowledge Discovery, pp. 188–217 (2001)
11. Hoppe, H., DeRose, T., Duchamp, T., McDonald, J., Stuetzle, W.: Surface reconstruction from unorganized points, vol. 26. ACM (1992)
12. Jenie, Y., Kempen, E.J.V., de Visser, C., Chu, Q.P.: Selective velocity obstacle method for cooperative autonomous collision avoidance system for unmanned aerial vehicles. In: AIAA Guidance, Navigation, and Control Conference 2013 (2013)
13. Klasing, K., Wollherr, D., Buss, M.: A clustering method for efficient segmentation of 3D laser data. In: 2008 IEEE International Conference on Robotics and Automation, pp. 4043–4048. IEEE (2008)
14. LaValle, S.M.: Planning Algorithms. Cambridge University Press, Cambridge (2006)
15. Rusu, R.B., Cousins, S.: 3D is here: point cloud library (PCL). In: 2011 IEEE International Conference on Robotics and Automation, pp. 1–4. IEEE (2011)
16. Yang, Q., Fan, S., Wang, L., Wang, Y.: Road detection by RANSAC on randomly sampled patches with slanted plane prior. In: 2016 IEEE 13th International Conference on Signal Processing (ICSP), pp. 929–933. IEEE (2016)

Algebraic Analysis of 3-RRC Parallel Manipulators

Abhilash Nayak[1], Martin Pfurner[1], Huiping Shen[2], and Manfred Husty[1(✉)]

[1] University of Innsbruck, Innsbruck, Austria
abhilash.un@outlook.com,
{Martin.Pfurner,Manfred.Husty}@uibk.ac.at
[2] Changzhou University, Changzhou, China
shp65@126.com

Abstract. The 3-RRC parallel manipulator consist of three identical serial revolute-revolute-cylindrical legs and had been introduced by L.W. Tsai in 1997. Several papers concerning the kinematic and dynamic behaviour of the device have been published. Contrary to previous results it is shown in this paper that an algebraic approach, using kinematic mapping and algebraic methods allows to give a complete answer to the kinematic features of the manipulator. This results in a simple and general solution of the direct and inverse kinematics, a description of the workspace, the input and the output singularities in the Cartesian and the joint space.

Keywords: 3-RRC parallel manipulator · Direct kinematics · Inverse kinematics · Workspace · Singularities

1 Introduction

In [8] Tsai proposed a 3-RRC (R stands for rotational joint and C for cylinder joint) parallel mechanism (PM) that can be used for three-translation motion operations. The structure of the PM is simple and symmetrical with a small number of motion pairs. The rotation axes of the three kinematic pairs in one limb are parallel to each other, and all of them are parallel to the base frame plane. Zhao et al. [13] performed the velocity analysis of this PM, which shows that this manipulator has no input-output motion decoupling characteristics. Jin [3] created a special 3-RRC PM, as shown in Fig. 1, via keeping the structure of the chain unchanged but optimizing the layout of three chains between the frame platform and moving platform, that is, all axes of motion pairs of the two chains of are parallel, and the third chain is arranged orthogonally to the previous two chains. Furthermore the forward kinematics of the PM was also solved. Yin [12] established the kinematic equations of the PM using vector loop methods. The direct position solution of the PM is obtained in symbolic form. It is proven furthermore that the PM has partial motion decoupling characteristics and the influences of structural parameters on the workspace of the PM were

J. Lenarčič and B. Siciliano (Eds.): ARK 2020, SPAR 15, pp. 166–173, 2021.
https://doi.org/10.1007/978-3-030-50975-0_21

discussed. But no singularity analysis was included in the paper. Liu [4,5] et al. established closed-form dynamic equations of the PM based on the Lagrange formulation and discussed dynamic properties of the manipulator, including it's driving force/torque and energy consumption.

From the point view of topology [10], and based on the "union" operation rules of the Position and Orientation Characteristics (POC) equation of the topological design method of a PM [10,11], the POC elements, i.e., motion output elements of each RRC chain are three-translation and one-rotation. When two such 3-RRC chains are arranged in non-parallel way no common rotation motion of the two RRC chains will be produced by the "intersection" operation rules of the POC equation of the topological design method of a PM [10,11]. Therefore, it is enough that as long as two RRC chains are arranged non parallel, pure translation output motion of the moving platform of PM can be guaranteed. But for the moving platform to produce three-translation output, three degrees-of-freedom are required, thus a third chain is needed. But the third chain can be replaced by un-constrained RSS or SPS chain instead of the constrained chain RRC. A complete analysis of this design variant is presented in [6].

In this paper a complete kinematic analysis of this well known PM using an algebraic method is performed. This includes the direct and inverse kinematics, a description of the workspace, as well as the input and output singularities. In Sect. 2 the design of the PM is introduced, Sect. 3 provides the kinematic constraint equations which describe the overall motion behaviour of the PM. In Sect. 4 the direct and inverse kinematics is performed without assigning values to the design parameters and in Sect. 5 a complete singularity analysis is given, yielding equations for the singularity sets in kinematic image space as well as in joint space. The paper finishes with some conclusions.

2 Architecture of 3-RRC PMs

R, C and S denote Revolute, Cylindrical and Spherical joints, respectively. A 3-RRC PM consists of three RRC limbs as shown in Fig. 1. The R-joint axes

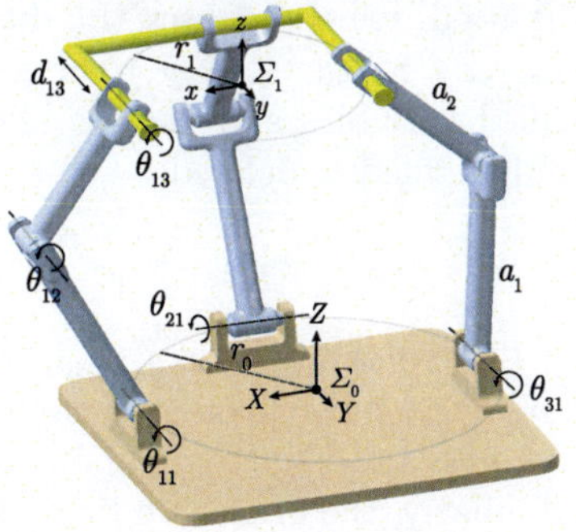

Fig. 1. A 3-RRC PM

attached to the base are tangential to the incircle of a square with side length $2r_0$, while those attached to the moving platform are tangential to the incircle of a square with side length $2r_1$. Two of the axes are parallel to each other while the third axis is orthogonal to the other two.

Fixed and moving coordinate frames Σ_0 and Σ_1 are attached to corresponding base and moving platform square centers. The Z (resp. z) axis points vertically upwards and the X (resp. x) axis to the contact point of the incircle and the first (resp. last) R-joint axis of the first limb. The identical link lengths in each limb are named a_1 and a_2. Therefore, the four design parameters r_0, r_1, a_1 and a_2 describe the complete architecture of this type of parallel manipulator. The R-joints attached to the base are considered to be active with inputs $v_{i1} = \tan\left(\frac{\theta_{i1}}{2}\right),\ i \in \{1, 2, 3\}$.

3 Constraint Equations

In this section, constraint equations are obtained using the Linear Implicitization Algorithm (LIA) [9].

3.1 3-RRC PM

A transformation matrix between Σ_0 and Σ_1 can be obtained by traversing via each RRC limb:

$$\mathbf{T} = \mathbf{F_0 M_1 G_1 M_2 G_2 M_3 G_3 F_1}, \tag{1}$$

where $\mathbf{F}_0$ and $\mathbf{F}_1$ represent the fixed transformation matrices in the base and the moving platform, respectively. $\mathbf{M}_i, i = 1, .., 3$ represent rotations about the R-joint axes and $\mathbf{G}_i, i = 1, .., 3$ consist of Denavit-Hartenberg parameters necessary to relate the local coordinate frames assigned to each joint. Further details on the introduction of these transformation matrices can be found in [2,7]. Therefore, $\mathbf{T}$ for the first RRC limb turns out be a function of $r_0, r_1, a_1, a_2, d_{13}, v_{11} = \tan\left(\frac{\theta_{11}}{2}\right), v_{12} = \tan\left(\frac{\theta_{12}}{2}\right)$ and $v_{13} = \tan\left(\frac{\theta_{13}}{2}\right)$, where v_{12}, v_{13} and d_{13} are passive joint variables as shown in Fig. 1. To obtain the constraint equations of the RRC limb only in terms of active joint variables and design parameters, LIA is used, which equates the dual quaternion form of $\mathbf{T}$ to Study parameters $(x_0, x_1, x_2, x_3, y_0, y_1, y_2, y_3)$ and gives a set of polynomials as functions of user-specified variables and degrees. The obtained polynomials describe manifolds on Study's quadric defined by

$$x_0 y_0 + x_1 y_1 + x_2 y_2 + x_3 y_3 = 0. \tag{2}$$

For the first limb, asking LIA to eliminate v_{12}, v_{13} and d_{13} yields polynomials that constitute an ideal $\mathscr{I}_1$. The only linear polynomials in $\mathscr{I}_1$ are x_1 and x_3 implying that this limb is constrained to rotate about the X or Z axis, which is evident from its design. Similarly, the other two limbs lead to $\mathscr{I}_2$ and $\mathscr{I}_3$.

Due to parallel arrangement of the limbs, polynomials describing the constrained motions of the entire 3-RRC PM are nothing but the union of ideals

$\mathscr{I} = \mathscr{I}_1 \cup \mathscr{I}_2 \cup \mathscr{I}_3$. $\mathscr{I}$ consists of x_1, x_2, x_3 which affirms the fact that this the PM only exhibits pure translational motions. Assuming the normalization condition $x_0 = 1$ to avoid simultaneous vanishing of $x_i, i = \{0, 1, 2, 3\}$ also leads to $y_0 = 0$ due to Eq. (2). Making these substitutions in $\mathscr{I}$ leads to the following final constraint equations:

$$f_1 := \left(y_1 - \frac{a_1(v_{11}^2 - 1) + (-r_0 + r_1)(v_{11}^2 + 1)}{2({v_{11}}^2 + 1)}\right)^2 + \left(y_3 - \frac{a_1 v_{11}}{{v_{11}}^2 + 1}\right)^2 - a_2^2 = 0, \tag{3a}$$

$$f_2 := \left(y_2 + \frac{2a_1 v_{21} + (-r_0 + r_1)(v_{21}^2 + 1)}{2({v_{21}}^2 + 1)}\right)^2 + \left(y_3 + \frac{a_1(v_{21}^2 - 1)}{2({v_{11}}^2 + 1)}\right)^2 - a_2^2 = 0, \tag{3b}$$

$$f_3 := \left(y_1 - \frac{a_1(v_{31}^2 - 1) + (r_0 - r_1)(v_{31}^2 + 1)}{2({v_{31}}^2 + 1)}\right)^2 + \left(y_3 - \frac{a_1 v_{31}}{{v_{31}}^2 + 1}\right)^2 - a_2^2 = 0. \tag{3c}$$

In fact, they represent three cylinders as shown in Fig. 2. $f_1 = 0$ and $f_3 = 0$ have their axes parallel to the Y-axis and the axis of $f_2 = 0$ is parallel to the X-axis.

4 Direct and Inverse Kinematics

When, the inputs are given, the direct kinematics problem boils down to finding the intersection points between three cylinders in case of the 3-RRC PM.

4.1 3-RRC PM

A Gröbner basis of $\mathscr{I}$ over the ring $k[y_1, y_2, y_3]$ with a lexicographic order $y_1 < y_2 < y_3$ consists of three generators. The first generator is a univariate quadratic polynomial in y_3, second one is quadratic in y_2 but linear in y_3 and the third one is linear in y_1 and y_3. Thus, there could be at most 4 assembly modes. Moreover, the first generator only has v_{11} and v_{31} as inputs. It shows that the vertical translation can be controlled solely by motors in the first and third limb.

For an arbitrary set of design parameters $\{a_1 = 2, a_2 = 3, r_0 = 5, r_1 = 4\}$ and inputs $v_{11} = -2+\sqrt{2}$, $v_{21} = -1+2\sqrt{2}$, $v_{31} = -1+\frac{1}{\sqrt{2}}$, Fig. 2 shows the four real assembly modes as intersections of three cylinders. $f_1 = f_3 = 0$ have parallel axes and intersect in two lines shown in red. Similarly, $f_1 = f_2 = 0$ intersect in a quartic intersection curve also shown in red. The intersections of these manifolds are four points that are solutions to the direct kinematics problem. The four assembly modes of the PM are shown. For better visualization, the square moving platform is rendered as a triangle shown in yellow. The proximal and distal links are colored black and green, respectively. The three cylinder equations can be homogenized and it turns out that they also share 4 absolute points $(y_1, y_2, y_3) = (\pm i, \pm i, 1)$ at infinity but is not relevant in this context.

Note that the Gröbner base can be computed completely general without assigning values to the design variables.

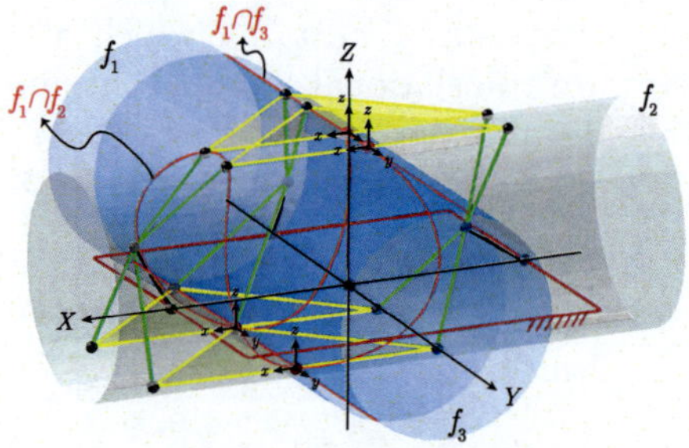

Fig. 2. Four real assembly modes of a 3-RRC PM

4.2 Working Modes

Inverse kinematics of the 3-RCC PM is fairly simple and corresponds to the elbow-up and elbow-down postures of each limb. Since there are three limbs in each PM, there could be a maximum of 8 working modes. Furthermore, closed form inverse kinematics solutions with general design parameters can be obtained from Eqs. (3a)–(3c). It is evident that they are quadratic in $v_{i1}, i = \{1, 2, 3\}$ verifying the maximum number of working modes.

5 Singularities

When the inputs are fixed and if the moving platform has a motion (albeit infinitesimal), the configuration is said to be in an output singularity. On the other hand, if the moving platform is fixed and at least one limb has a motion, the PM is in an input singular configuration. They can be algebraically determined as conditions on rank deficiency of input and output Jacobian matrices $\mathbf{J}_I = \dfrac{\partial f_i}{\partial y_j}$ and $\mathbf{J}_O = \dfrac{\partial f_i}{\partial v_{j1}}, i, j \in \{1, 2, 3\}$ related as follows [1]:

$$\mathbf{J}_O \begin{bmatrix} \dot{y}_1 \\ \dot{y}_2 \\ \dot{y}_3 \end{bmatrix} + \mathbf{J}_I \begin{bmatrix} \dot{v}_{11} \\ \dot{v}_{21} \\ \dot{v}_{31} \end{bmatrix} = 0. \tag{4}$$

For the PMs under study, both $\mathbf{J}_I$ and $\mathbf{J}_O$ are square matrices. Hence, the solution set of their vanishing determinants along with constraint equations must yield singular configurations.

5.1 Output Singularities of the 3-RRC PM

Determinant of $\mathbf{J}_O$ is given by:

$$\det(\mathbf{J}_O) := 128\,(-r_0{v_{21}}^2 + r_1{v_{21}}^2 + 2\,{v_{21}}^2y_2 + 2\,a_1v_{21} - r_0 + r_1 + 2\,y_2)(2\,r_0{v_{11}}^2{v_{31}}^2y_3$$
$$- 2\,r_1{v_{11}}^2{v_{31}}^2y_3 + {a_1}^2{v_{11}}^2v_{31} - {a_1}^2v_{11}{v_{31}}^2 - a_1r_0{v_{11}}^2v_{31} - a_1r_0v_{11}{v_{31}}^2 + a_1r_1{v_{11}}^2v_{31}$$
$$+ a_1r_1v_{11}{v_{31}}^2 - 2\,a_1{v_{11}}^2v_{31}y_1 + 2\,a_1v_{11}{v_{31}}^2y_1 - 2\,a_1{v_{11}}^2y_3 + 2\,a_1{v_{31}}^2y_3 + 2\,r_0{v_{11}}^2y_3$$
$$+ 2\,r_0{v_{31}}^2y_3 - 2\,r_1{v_{11}}^2y_3 - 2\,r_1{v_{31}}^2y_3 + {a_1}^2v_{11} - {a_1}^2v_{31} - a_1r_0v_{11} - a_1r_0v_{31} + a_1r_1v_{11}$$
$$+ a_1r_1v_{31} + 2\,a_1v_{11}y_1 - 2\,a_1v_{31}y_1 + 2\,r_0y_3 - 2\,r_1y_3) = 0. \tag{5}$$

Eliminating actuated joint variables v_{11}, v_{21} and v_{31} from Eqs. (3a)–(3c) and (5) leads to:

$$a_1a_2(r_0 - r_1) \cdot h_1(y_2, y_3) \cdot h_2(y_2, y_3) \cdot h_3(y_1, y_3) \cdot h_4(y_1, y_3) = 0. \tag{6}$$

The four factors $h_i, i = 1, 2, 3, 4$ correspond to four cylinders in the kinematic image space. The axes of $h_1 = h_2 = 0$ are parallel to the X-axis while that of $h_3 = h_4 = 0$ parallel to the Y-axis. It is noteworthy that not all working modes corresponding to the triplet $\{y_1, y_2, y_3\}$ satisfying the above equation are necessarily singular configurations.

The output singularities can also be expressed as functions of inputs. The second factor of Eq. (5) is solved for y_2 and substituted in Eqs. (3a)–(3c). Eliminating y_1 and y_3 from these equations results in an equation only in terms of $v_{j1}, j \in \{1, 2, 3\}$ and design parameters, too long to display here. This equation corresponds to singular configurations in which the moving platform center is on surfaces of the cylinders $h_3 = h_4 = 0$ in Eq. (6) (Fig. 4).

Similarly, the third factor of Eq. (5) is solved for y_1 and substituted in Eqs. (3a)–(3c). This results in the first and the last equations to be functions of only y_3, v_{31} and v_{11}. Eliminating y_3 from them yields an equation only in terms of the first and third input variables. It suggests that once v_{11} and v_{31} are chosen in such a way that they satisfy the relation and the PM is in an output singularity, changing v_{21} can only move the moving platform parallel to the Y-axis (along surfaces of the cylinders $h_1 = h_2 = 0$ in Eq. (6)) and the PM remains in an output singular configuration.

5.2 Input Singularities of the 3-RRC PM

Determinant of $\mathbf{J}_I$ is given by:

$$\det(\mathbf{J}_I) := \;\; 8\,({a_1}^2v_{11} - 2\,a_1r_0v_{11} + 2\,a_1r_1v_{11} - 4\,a_1v_{11}y_1 - {a_2}^2v_{11} + {r_0}^2v_{11} - 2\,r_0r_1v_{11}$$
$$+ 4\,r_0v_{11}y_1 + {r_1}^2v_{11} - 4\,r_1v_{11}y_1 + 4\,v_{11}{y_1}^2 + 4\,v_{11}{y_3}^2 - 4\,a_1y_3)({a_1}^2v_{21} + 4\,a_1v_{21}y_3$$
$$- {a_2}^2v_{21} + {r_0}^2v_{21} - 2\,r_0r_1v_{21} - 4\,r_0v_{21}y_2 + {r_1}^2v_{21} + 4\,r_1v_{21}y_2 + 4\,v_{21}{y_2}^2 + 4\,v_{21}{y_3}^2$$
$$- 2\,a_1r_0 + 2\,a_1r_1 + 4\,a_1y_2)({a_1}^2v_{31} + 2\,a_1r_0v_{31} - 2\,a_1r_1v_{31} - 4\,a_1v_{31}y_1 - {a_2}^2v_{31} + {r_0}^2v_{31}$$
$$- 2\,r_0r_1v_{31} - 4\,r_0v_{31}y_1 + {r_1}^2v_{31} + 4\,r_1v_{31}y_1 + 4\,v_{31}{y_1}^2 + 4\,v_{31}{y_3}^2 - 4\,a_1y_3) = 0. \tag{7}$$

The three factors can also be obtained by partial differentiation of constraint Eqs. (3a)–(3c) w.r.t v_{11}, v_{21} and v_{31}, respectively. By eliminating v_{11} from the first factor and Eq. (3a) yields

$$(a_1{}^2 + 2\,a_1 a_2 + a_2{}^2 - r_0{}^2 + 2\,r_0 r_1 - 4\,r_0 y_1 - r_1{}^2 + 4\,r_1 y_1 - 4\,y_1{}^2 - 4\,y_3{}^2)$$
$$(a_1{}^2 - 2\,a_1 a_2 + a_2{}^2 - r_0{}^2 + 2\,r_0 r_1 - 4\,r_0 y_1 - r_1{}^2 + 4\,r_1 y_1 - 4\,y_1{}^2 - 4\,y_3{}^2) = 0, \tag{8}$$

where the two factors are again cylinders corresponding to fully extended and folded configurations of the first limb. Similarly, the limb singularities for the other two limbs also correspond to cylinders and their enclosure gives the translational workspace boundary of the PM as shown in Fig. 3 when $\{a_1 = 2,\, a_2 = 3, r_0 = 5, r_1 = 4\}$.

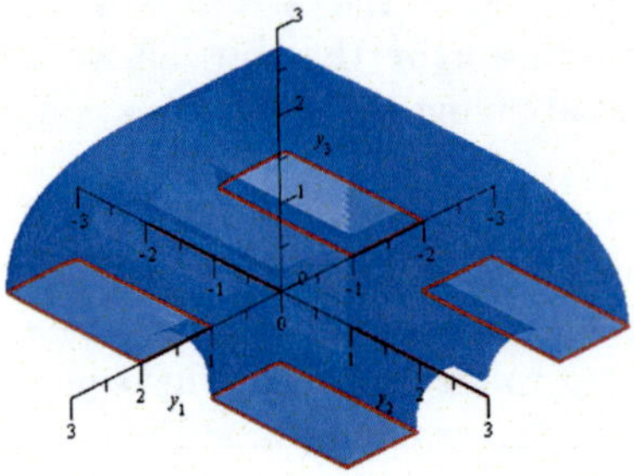

Fig. 3. Upper half translational workspace boundary of the 3-RRC PM.

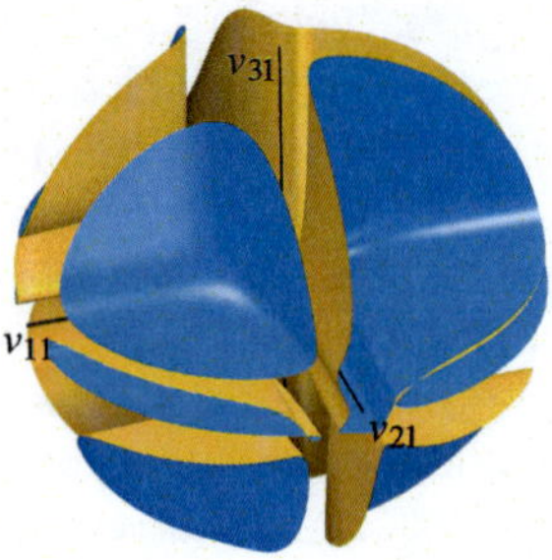

Fig. 4. Output singularity surface in joint space, corresponding to the second factor in Eq. 5 ($a_1 = 2, a_2 = 3, r_0 = 5, r_1 = 4$)

If necessary, also the input singularity conditions of the 3-RRC PM can be derived solely in terms of its inputs as follows for the first limb:

$$(v_{11}{}^2 v_{31}{}^2 - 2\,v_{11}{}^2 + 10\,v_{11} v_{31} - 9\,v_{31}{}^2 - 2)(2\,v_{11}{}^2 v_{31}{}^2 - v_{11}{}^2 + 2\,v_{11} v_{31} - 1) = 0. \tag{9}$$

6 Conclusion

In this paper it was shown that an algebraic approach using kinematic mapping, algebraic constraint equations and some simple techniques from algebraic geometry yield an overall picture of the motion behaviour of a well known parallel manipulator. It is believed that this approach can be successfully applied to any kind of spatial mechanism or manipulator. This investigation has also shown that this translational manipulator, contrary to other manipulators that have been designed for translational motion, has no other operation mode.

Acknowledgements. Abhilash Nayak acknowledges the support by the Austrian Science Fund (FWF): Project P28349-N32.

References

1. Gosselin, C., Angeles, J.: Singularity analysis of closed-loop kinematic chains. IEEE Trans. Robot. Autom. **6**(3), 281–290 (1990)
2. Husty, M., Karger, A., Sachs, H., Steinhilper, W.: Kinematik und Robotik, 1st edn. Springer, Heidelberg (1997)
3. Jin, Q., Yang, T.L.: Position analysis for a class of novel 3-D of translational parallel robot mechanisms. J. Southeast Univ. **31**(5), 33–38 (2001). (in Chinese)
4. Liu, S., Zhu, Z., Li, Z., Liu, C., Yu, Y.: Kinematics and dynamics of 3-RRC flexible parallel manipulator. In: 2010 International Conference on E-Product E-Service and E-Entertainment, pp. 1–4 (2010)
5. Liu, S., Zhu, Z., Sun, Z., Cao, G.: Kinematics and dynamics analysis of a three-degree-of-freedom parallel manipulator. J. Center South Univ. **21**, 2660–2666 (2014)
6. Nayak, A., Pfurner, M., Shen, H., Husty, M.: Complete kinematic analysis of a 2-RRC-1-RSS parallel manipulator, Belfast. Submitted to Parallel (2020)
7. Pfurner, M.: Analysis of spatial serial manipulators using kinematic mapping. Doctoral thesis, University Innsbruck (2006)
8. Tsai, L.W.: Multi-degree-freedom mechanism for machine tools and the like. U.S. Patent No. 5656905 (1997)
9. Walter, D.R., Husty, M.L.: On implicitization of kinematic constraint equations. Mach. Des. Res. **26**, 218–226 (2010)
10. Yang, T.L., Liu, A., Luo, Y., Shen, H., Hang, L., Qiong, J.: Theory and Application of Robot Mechanism Topology. Science Press, Beijing (2012). (in Chinese)
11. Yang, T.L., Liu, A., Luo, Y., Shen, H., Hang, L., Qiong, J.: Topology Design of Robot Mechanisms. Springer, Singapore (2018)
12. Yin, X.Q.: Workspace analysis of a 3-DOF translational parallel mechanism 3-RRC. China Mech. Eng. **14**(18), 1531–1533 (2003). (in Chinese)
13. Zhao, T.S., Dai, J.S., Huang, Z.: Geometric analysis of overconstrained parallel manipulators with three and four degrees of freedom. JSME Int. J. Ser. C **45**(3), 730–740 (2002)

Magneto-Inertial Data Sensory Fusion Based on Jacobian Weighted-Left-Pseudoinverse

Janez Podobnik(✉), Marko Munih, and Matjaž Mihelj

Faculty of Electrical Engineering, University of Ljubljana, Tržaška c. 25, 1000 Ljubljana, Slovenia
janez.podobnik@fe.uni-lj.si

Abstract. This paper presents a method for sensory fusion of magneto-inertial data for estimation of joint angles of serial kinematic chain with rotational degrees-of-freedom. Method is named Magneto-Inertial tracking based on JAcobian PseudoInverse (*MIJAPI*). Method incorporates a kinematic model of the mechanism and the estimation relies on the inverse kinematics solution based on the Jacobian inverse utilizing the Moore-Penrose weighted left pseudoinverse of the mechanism. Jacobian matrix is used to solve an overdetermined system in a least squares approach due to available redundant measurements resulting from constraints related to attachments of magneto-inertial sensors and kinematic model.

Keywords: Sensory fusion · Motion analysis · Motion estimation

1 Introduction

This paper presents a sensory fusion method for estimating joint angles of serial kinematic chains with rotational degrees-of-freedom (DoF) of a human body or a serial robot using measurements from magneto-inertial measurement units (IMU) attached to segments of the mechanism. The aim is to estimate joint angles using IMU measurements as an inverse kinematic problem. IMUs typically consist of a triad of gyroscopes, a triad of accelerometers and a triad of magnetometers. They allow measurement of 3-DoF segment orientation. Majority of methods for sensory fusion of magneto-inertial data estimate orientation based on a single IMU and do not account for kinematic constraints imposed by the joints of the mechanism [8].

Segments are connected by rotational DoFs, which can be joints with one, two or three degrees-of-freedom. When considering body-segment orientation measurements with IMUs attached to each segment and segments are linked with joints that have less than three DoF, the resulting system becomes overdetermined, since number of measurements is larger than number of the unknown

J. Lenarčič and B. Siciliano (Eds.): ARK 2020, SPAR 15, pp. 174–181, 2021.
https://doi.org/10.1007/978-3-030-50975-0_22

quantities. The problem can be solved using least squares to estimate segment orientations based on IMU data [1]. The solution of the presented inverse kinematic problem considering the weighted least squares leads to the implementation of the algorithm based on the Moore-Penrose weighted left pseudoinverse [7] of the mechanism Jacobian matrix. The Jacobian pseudoinverse can then be used in a closed-loop algorithm for calculating mechanism joint angles from the orientation and angular velocity of IMU sensors with the same method as for calculating the inverse kinematics of the robot manipulator with Jacobian inverse. The novelty and central idea of the method is that when the number of available measurements is larger than the number of degrees-of-freedom of the mechanism redundant measurements and knowledge of the kinematics of the mechanism can be used to improve the estimation of joint angles in presence of measurement noise, artefacts and disturbances.

2 Methods

2.1 Forward Kinematic and Differential Kinematic Model

Consider a general serial mechanism with N revolute 1-DoF joints (see Fig. 1). Mechanism links are selectively equipped with sensors S_i (IMUs consisting of a 3-axis accelerometer, gyroscopes and magnetometer). All sensors form a set $S = \{S_i : i = 1 \ldots M \leq N\}$. Sensor S_0 is assumed to measure the base orientation. The goal is to compute the joint angles of the mechanism. Considering the Denavit-Hartenberg notation [3], the orientation of the link j relative to the link $j-1$ is defined as

$$\mathbf{R}_j^{j-1}(\vartheta_j) = Rot(\mathbf{z}_{j-1}, \vartheta_j) \cdot Rot(\mathbf{x}_j, \alpha_j), \tag{1}$$

where $Rot()$ represents a rotation matrix, ϑ_j represents the angle of rotation between the links j and $j-1$, which is measured from $\mathbf{x}_{j-1}$ to $\mathbf{x}_j$, about the $\mathbf{z}_{j-1}$ vector and α_j represents the angle measured about the $\mathbf{x}_j$ axis between the vectors $\mathbf{z}_{j-1}$ and $\mathbf{z}_j$. The orientation of the link j relative to the base frame O_0 can be computed as

$$\mathbf{R}_j = \prod_{k=1}^{j} \mathbf{R}_k^{k-1}(\vartheta_k). \tag{2}$$

Angular velocity of link j relative to the base and expressed in the base coordinate frame O_0 can be computed as a function of joint velocities

$$\omega_j = \sum_{k=1}^{j} \mathbf{z}_{k-1}\dot{\vartheta}_k = \mathbf{J}_j\dot{\boldsymbol{\vartheta}}, \tag{3}$$

where $\mathbf{z}_{j-1}$ represents the joint axis vector that is given by the third column of the rotation matrix $\mathbf{R}_{j-1}$. Matrix $\mathbf{J}_j$ represents the Jacobian matrix for link j. Consider a set of measurable link velocities $\Omega_{\mathscr{M}}$ (i.e. links that are equipped with sensors S_i) as a subset of all link velocities. Cardinality $M = |\Omega_{\mathscr{M}}|$ defines the

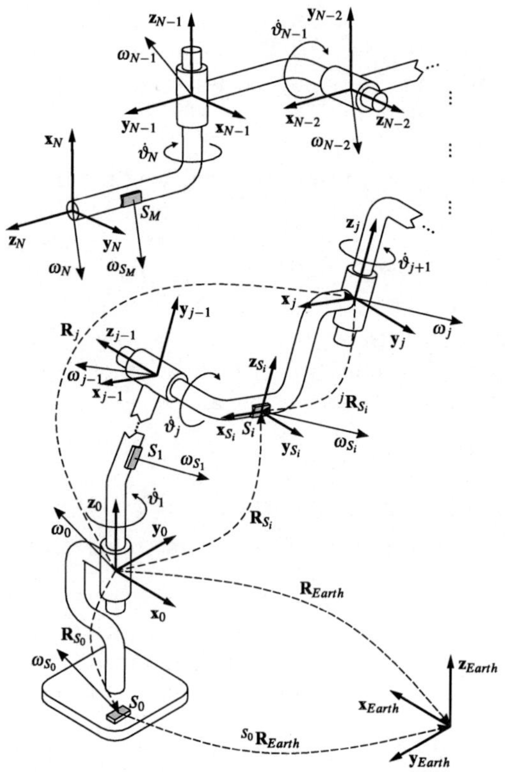

Fig. 1. General serial mechanism with revolute joints. Mechanism links are selectively equipped with sensors S_i being magneto-inertial measurement units.

number of measurable link velocities (without taking into account the base link velocity). A totally ordered set of indices defines the measurable link velocities

$$\mathscr{M} = \{m_i : i = 1 \dots M\}, \tag{4}$$

meaning that sensor S_i measures rotational velocity of link m_i, that is ω_{m_i}. Considering (3) and the set of measurable velocities, the following matrix equation can be defined

$$\left[\omega_{m_1} \cdots \omega_{m_M}\right]^T = \left[\mathbf{J}_{m_1} \cdots \mathbf{J}_{m_M}\right]^T \dot{\boldsymbol{\vartheta}} \tag{5}$$

Sensor S_i measures rotational velocity of link m_i, that is ω_{m_i}. Above equation can be rewritten in a compact form as

$$\Omega_{\mathscr{M}} = \mathbb{J}_{\mathscr{M}} \dot{\boldsymbol{\vartheta}}, \tag{6}$$

where $\Omega_{\mathscr{M}}$ represents a vector of all measurable link velocities and $\mathbb{J}_{\mathscr{M}}$ is a block-triangular Jacobian matrix related to measurable link velocities. The rectangular submatrix $\mathbb{J}_{\mathscr{M}_{ii}}$ of size $3 \times (m_i - m_{i-1})$ defines the following relation

$$\omega_{m_i} = \omega_{m_{i-1}} + \mathbb{J}_{\mathscr{M}_{ii}} \left[\dot{\vartheta}_{m_{i-1}+1} \cdots \dot{\vartheta}_{m_i}\right]^T, \tag{7}$$

where $m_0 = 0$ ($\omega_0 = \mathbf{0}$, if stationary base is assumed).

2.2 Jacobian Pseudo-Inverse Based Inverse Kinematics

The inverse kinematics aims to compute the joint velocities and angles that correspond to the sensors angular velocities and orientations. Equation (6) must be inverted [5], leading to

$$\dot{\boldsymbol{\vartheta}} = \mathbf{J}_{\mathcal{M}}^{\dagger} \Omega_{\mathcal{M}}, \tag{8}$$

where $\dot{\boldsymbol{\vartheta}}$ is the optimal estimation of joint velocities in the least-square-error sense and $\mathbf{J}_{\mathcal{M}}^{\dagger}$ is the Moore-Penrose weighted left pseudo-inverse of the Jacobian matrix $\mathbf{J}_{\mathcal{M}}$ defined as

$$\mathbf{J}_{\mathcal{M}}^{\dagger} = (\mathbf{J}_{\mathcal{M}}^{T} \mathbf{W}_{\mathcal{M}}^{-1} \mathbf{J}_{\mathcal{M}})^{-1} \mathbf{J}_{\mathcal{M}}^{T} \mathbf{W}_{\mathcal{M}}^{-1}, \tag{9}$$

where $\mathbf{W}_{\mathcal{M}}$ is a weighting matrix (typically diagonal). The matrix $\mathbf{J}_{\mathcal{M}}$ must have a full rank, which equals the number of joints N. If all measurements are equally reliable, meaning are equally weighted, weighting matrix $\mathbf{W}_{\mathcal{M}}$ becomes identity matrix. However, in practice some measurements are less reliable than others, which can be represented with different weights. In (8) we assumed that measured link velocities are expressed in the base coordinate frame O_0, which is not the case, since measurements are expressed in the body coordinate frame of the sensor. The orientation of the sensor S_i, which is attached to the link m_i, in relation to the base frame O_0 is $\mathbf{R}_{S_i} = \mathbf{R}_{m_i} {}^{m_i}\mathbf{R}_{S_i}$, where rotation matrix ${}^{m_i}\mathbf{R}_{S_i}$ represents orientation of the sensor S_i relative to the link m_i. Relation between the velocity expressed in the sensor body coordinate frame and the velocity expressed in the base frame is then

$$\omega_{m_i} = \mathbf{R}_{S_i} \omega_{S_i}, \tag{10}$$

where ω_{S_i} is the rotational velocity as measured by the sensor S_i and expressed in the sensor body coordinate frame. Equation (10) in compact form is

$$\Omega_{\mathcal{M}} = \mathbb{R}_S \Omega_S, \tag{11}$$

where $\mathbb{R}_S$ is a block-diagonal matrix having main diagonal blocks matrices $\mathbf{R}_{S_i}$.

By combining (8) and (11) the following relation can be written

$$\Omega_S = \mathbb{R}_S^T \mathbf{J}_{\mathcal{M}} \dot{\boldsymbol{\vartheta}} = \mathbf{J}_S \dot{\boldsymbol{\vartheta}}, \tag{12}$$

where $\mathbf{J}_S = \mathbb{R}_S^T \mathbf{J}_{\mathcal{M}}$ is the Jacobian matrix expressed in the sensor body coordinate frames. Matrix $\mathbb{R}_S^T$ is invertible, thus $\text{rank}(\mathbf{J}_S) = \text{rank}(\mathbf{J}_{\mathcal{M}})$. The weighting matrix $\mathbf{W}_{\mathcal{M}}$ defines velocity weights relative to the base coordinate frame and can be expressed as

$$\mathbf{W}_{\mathcal{M}} = \mathbb{R}_S \mathbf{W}_S \mathbb{R}_S^T, \tag{13}$$

where $\mathbf{W}_S$ is a weighting matrix related to velocities expressed in the sensor body coordinate frames. By defining the weighted left pseudo-inverse Jacobian matrix related to the sensor body frame velocities as

$$\mathbf{J}_S^{\dagger} = (\mathbf{J}_S^T \mathbf{W}_S^{-1} \mathbf{J}_S)^{-1} \mathbf{J}_S^T \mathbf{W}_S^{-1}, \tag{14}$$

Equation (8) leads to final expression

$$\dot{\boldsymbol{\vartheta}} = (\mathbf{J}_S^T \mathbf{W}_S^{-1} \mathbf{J}_S)^{-1} \mathbf{J}_S^T \mathbf{W}_S^{-1} \mathbf{R}_S^T \Omega_{\mathcal{M}} = \mathbf{J}_S^{\dagger} \Omega_S. \tag{15}$$

2.3 Prediction and Correction of the Estimated Joint Angles

The block diagram of the *MIJAPI* inverse kinematics approach is shown in Fig. 2.

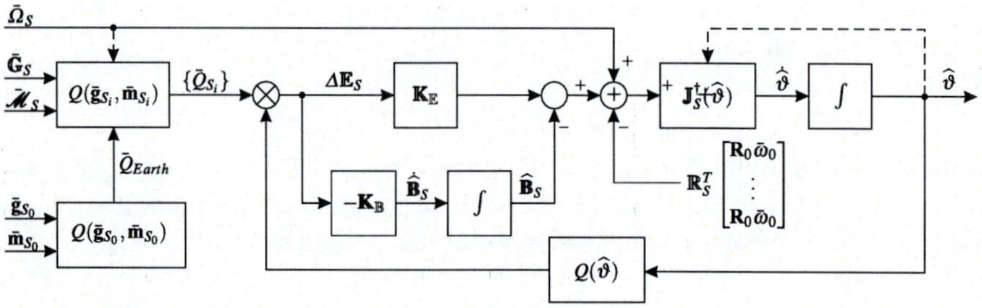

Fig. 2. Jacobian pseudo-inverse kinematics algorithm. Input into algorithm are all measured velocities $\bar{\omega}_{S_i}$ combined into vector $\bar{\Omega}_S$, all measured accelerations $\bar{\mathbf{g}}_{S_i}$ combined into vector $\bar{\mathbf{G}}_S$, all measured magnetic field vectors $\bar{\mathbf{m}}_{S_i}$ combined into vector $\bar{\mathcal{M}}_S$.

The angular velocity measured by the sensor S_i (bar over the variable will be used throughout the paper to indicate measured quantities) equals

$$\bar{\omega}_{S_i} = \omega_{S_i} + \mathbf{b}_{S_i} + \mathbf{w}_{\omega S_i}, \tag{16}$$

where ω_{S_i} indicates the real sensor angular velocity, $\mathbf{b}_{S_i}$ and $\mathbf{w}_{\omega S_i}$ represent the gyroscope bias and noise. With all measured velocities $\bar{\omega}_{S_i}$ combined into vector $\bar{\Omega}_S$, joint angular velocities and joint angle predictions can be computed as

$$\hat{\dot{\boldsymbol{\vartheta}}} = \mathbf{J}_S^{\dagger} \bar{\Omega}_S \quad \text{and} \quad \hat{\boldsymbol{\vartheta}} = \hat{\boldsymbol{\vartheta}}_0 + \int_0^t \hat{\dot{\boldsymbol{\vartheta}}} dt. \tag{17}$$

However, due to the measurement noise, in particular gyroscope bias $\mathbf{b}_{S_i}$, in practical applications this would only work for short time periods and exactly known initial joint positions $\hat{\boldsymbol{\vartheta}}_0$.

The estimated sensor orientation $\widehat{\mathbf{R}}_{S_i} = \mathbf{R}_{S_i}(\hat{\boldsymbol{\vartheta}})$ in relation to the base coordinate frame O_0 is

$$\widehat{\mathbf{R}}_{S_i} = \left(\prod_{j=1}^{m_i} Rot(\mathbf{z}_{j-1}, \hat{\vartheta}_j) \cdot Rot(\mathbf{x}_j, \alpha_j) \right) {}^{m_i}\mathbf{R}_{S_i}. \tag{18}$$

The orientation matrix $\widehat{\mathbf{R}}_{S_i}$ can be expressed in terms of a quaternion $\widehat{Q}_{S_i} = Q(\widehat{\mathbf{R}}_{S_i})$.

Accelerometers of the sensor S_i measures a three-axial linear acceleration vector $\bar{\mathbf{a}}_{S_i}$ and magnetometers of the sensor S_i measure a three-axial magnetic field vector $\bar{\mathbf{m}}_{S_i}$. The measured vectors, $\bar{\mathbf{a}}_{S_i}$ and $\bar{\mathbf{m}}_{S_i}$, expressed in the sensor body coordinate frame, provide an estimate of the absolute sensor orientation in relation to the Earth coordinate frame O_{Earth}. The q-method [2,4] can be applied to compute the relative sensor orientation in terms of a quaternion $Q(\bar{\mathbf{a}}_{S_i}, \bar{\mathbf{m}}_{S_i})$ representing the orientation of the Earth coordinate frame in relation to the sensor body coordinate frame. Sensor orientation in relation to the base coordinate frame O_0 is

$$\bar{Q}_{S_i} = \bar{Q}_{Earth}(Q(\bar{\mathbf{a}}_{S_i}, \bar{\mathbf{m}}_{S_i}))^{-1}, \tag{19}$$

where the quaternion $\bar{Q}_{Earth} = Q(\mathbf{R}_{S_0})Q(\bar{\mathbf{a}}_{S_0}, \bar{\mathbf{m}}_{S_0})$, represents the orientation of the Earth coordinate frame (defined by the gravity and magnetic field vectors) in relation to the base frame O_0. The quaternion $\bar{Q}_{S_i}$ is a measure of the absolute sensor S_i orientation and can be applied to correct the predicted signal $\widehat{\boldsymbol{\vartheta}}$ values.

Prediction of sensor orientation $\widehat{Q}_{S_i}$ can be obtained from predicted joint angles $\widehat{\boldsymbol{\vartheta}}$. The estimated absolute sensor orientation $\bar{Q}_{S_i}$ is obtained from (19) by replacing $\bar{\mathbf{a}}_{S_i}$ with $\bar{\mathbf{g}}_{S_i}$. The error quaternion $\Delta\bar{Q}_{S_i}^{S_i} = \{\Delta\bar{\eta}_{S_i}, \Delta\bar{\epsilon}_{S_i}\}$ can be computed as

$$\Delta\bar{Q}_{S_i}^{S_i} = \widehat{Q}_{S_i}^{-1}\bar{Q}_{S_i}. \tag{20}$$

The error quaternion $\Delta\bar{Q}_{S_i}^{S_i}$ is expressed in the body coordinate frame of the estimated sensor orientation, $\Delta\bar{\eta}_{S_i}$ represents the quaternion scalar part and $\Delta\bar{\epsilon}_{S_i}$ represents its vector part. By taking into account the orientation estimation error $\Delta\bar{\epsilon}_{S_i}$ and the estimated gyroscope bias $\widehat{\mathbf{b}}_{S_i}$ the (17) can be rewritten as

$$\dot{\widehat{\boldsymbol{\vartheta}}} = \mathbf{J}_S^{\dagger}\left(\bar{\Omega}_S + \mathbf{K}_{\mathbb{E}}\Delta\bar{\mathbf{E}}_S - \widehat{\mathbf{B}}_S\right), \tag{21}$$

where $\mathbf{K}_{\mathbb{E}}$ is a positive definite gain matrix,

$$\Delta\bar{\mathbf{E}}_S = \left[\Delta\bar{\epsilon}_{S_1} \cdots \Delta\bar{\epsilon}_{S_M}\right]^T \quad \text{and} \quad \widehat{\mathbf{B}}_S = \left[\widehat{\mathbf{b}}_{S_1} \cdots \widehat{\mathbf{b}}_{S_M}\right]^T. \tag{22}$$

The gyroscope bias can be estimated from the error vector $\Delta\mathbf{E}_S$ with the following dynamics

$$\dot{\widehat{\mathbf{B}}}_S = -\mathbf{K}_{\mathbb{B}}\Delta\bar{\mathbf{E}}_S. \tag{23}$$

3 Results and Discussion

The *MIJAPI* concept was validated on a 7 DoF kinematic model of a human arm: shoulder (3-DoF), elbow (1-DoF) and wrist (3-DoF) joints connected by upper arm (IMU 1), forearm (IMU 2) and hand (IMU 3) segments. The performance of the *MIJAPI* method (see Fig. 3) was validated in a simulation study by comparing it to the Unscented Kalman filter (UKF) incorporating the arm model with 7 DoF. Simulation signals were obtained by superimposing noise extracted from real IMUs on ideal values computed from the kinematic model of

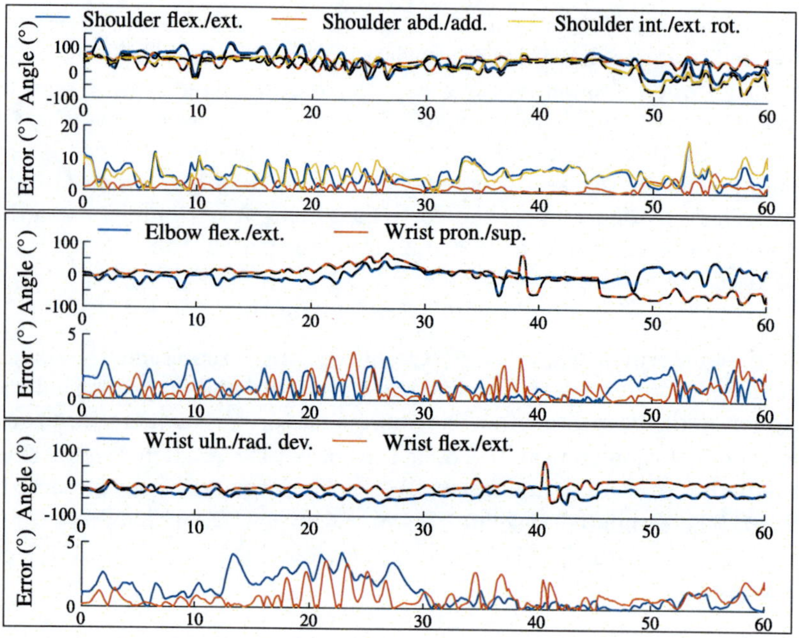

Fig. 3. Figure shows estimated trajectories from IMUs and trajectories from optical tracking system (superimposed dashed black lines) and errors between estimated and measured trajectories for 7 joint arm angles.

the human arm: a) high-frequency noise was added to ideal signals, b) gyroscope signals were compromised with real-sensor bias values and scaling factors on all axes, c) accelerometer signals were disturbed with high amplitude noise resulting from interactions of real IMUs with hard surface, d) magnetometer signals were manipulated with noise resulting from external magnetic fields. Analysis results are presented in terms of signal-to-noise ratios (SNR) for input and output signals. SNR is defined as $\mathrm{SNR} = 10\log_{10}(A^2_{signal}/A^2_{noise})$, where A is root mean square amplitude. Noise is computed as difference between ideal and corrupted input signals. SNR is calculated for each of the seven joint angles. From this set of 7 SNR values mean and minimal SNR values for joint angles are determined. When evaluating the performance of the method both mean as well as minimal SNR need to be taken into account. While for one method mean SNR can be high, which means that in average estimation of the angles is very good for this method, one of the estimated angles can be estimated poorly. Mean SNR for *MIJAPI* was 19 dB, minimal SNR was 15 dB. For UKF mean SNR was 16 dB, minimal SNR was 8 dB. Note that SNR of 20 dB corresponds to 1% of error and SNR of 15 dB dB corresponds to 3% of error. Results show that the method *MIJAPI* performs better than the methods based on UKF. *MIJAPI* method performs better in average by about 3 dB compared to UKF method. Additionally, the *MIJAPI* method is computationally approximately 50 times more efficient than UKF method. Validation of the *MIJAPI* concept is in more details covered in [6].

4 Conclusion

The paper presents a novel algorithm for estimation of joint angles based on magneto-inertial measurements. The estimation requires the complete kinematic model of the mechanism including constraints introduced by various types of rotational joints. The computation is based on the Moore-Penrose left pseudoinverse of the mechanism Jacobian matrix that relates segment angular velocities to joint velocities and presents the weighted least squares solution to the problem. The results of the experimental validation demonstrate that the *MIJAPI* algorithm is capable of estimating joint angles in the presence of various disturbances. The *MIJAPI* algorithm is computationally more efficient than the corresponding solutions based on the Kalman filter.

Acknowledgements. The authors acknowledge the financial support from the Slovenian Research Agency (research core funding No. P2-0228).

References

1. Ben-Israel, A., Greville, T.: Generalized Inverses - Theory and Applications. Springer, Cham (2003)
2. Davenport, P.: A vector approach to the algebra of rotations with applications. Technical report, X-546-65-437, NASA (1965)
3. Denavit, J., Hartenberg, R.S.: A kinematic notation for lower-pair mechanisms based on matrices. ASME J. Appl. Mech. **23**, 215–221 (1955)
4. Markley, F.L., Mortari, D.: Quaternion attitude estimation using vector observations. J. Astronaut. Sci. **48**(2/3), 359–380 (2000)
5. Meyer Jr., C.D.: Generalized inverses of block triangular matrices. SIAM J. Appl. Math. **19**(4), 741–750 (1970)
6. Mihelj, M., Podobnik, J., Munih, M.: Sensory fusion of magnetoinertial data based on kinematic model with Jacobian weighted-left-pseudoinverse and kalman-adaptive gains. IEEE Trans. Instrum. Meas. **68**(7), 2610–2620 (2019). https://doi.org/10.1109/TIM.2018.2867891
7. Siciliano, B., Sciavicco, L., Villani, L., Oriolo, G.: Robotics - Modelling, Planning and Control, p. 57. Springer (2009). Appendix A.7
8. Zhou, H., Hu, H.: Reducing drifts in the inertial measurements of wrist and elbow positions. IEEE Trans. Instrum. Meas. **59**(3), 575–585 (2010)

Evaluating the Snappability of Bar-Joint Frameworks

G. Nawratil(✉)

Institute of Discrete Mathematics and Geometry,
Center for Geometry and Computational Design,
TU Wien, Vienna, Austria
nawratil@geometrie.tuwien.ac.at

Abstract. It is well-known that there exist bar-joint frameworks (without continuous flexions) whose physical models can snap between different realizations due to non-destructive elastic deformations of material. We present a method to measure these snapping capability – shortly called *snappability* – based on the total elastic strain energy of the framework by computing the deformation of all bars using Hooke's law. The presented theoretical results give further connections between shakiness and snapping beside the well-known technique of averaging and deaveraging.

Keywords: Snapping framework · Multistability · Model flexor · Elastic deformation

1 Introduction

We consider a framework in the Euclidean n-dimensional space $\mathbb{E}^n$ which consists of a knot set $\mathscr{K} = \{K_1, K_2, \ldots, K_s\}$ and an abstract graph G on $\mathscr{K}$ fixing the combinatorial structure. We denote the edge connecting K_i to K_j by e_{ij} with $i < j$ and collect all indices of knots edge-connected to K_i in the knot neighborhood N_i. Moreover we denote the number of edges in the graph by b and assign[1] a length $L_{ij} \in \mathbb{R}_{>0}$ to each edge e_{ij}. In general this assignment does not determine the shape of the framework uniquely thus a framework has different incongruent realizations. For example, a triangular framework has in general two realizations in $\mathbb{E}^2$, which are not congruent with respect to the group of direct isometries. If we consider the isometry group then this number halves.

By materializing all edges by straight bars and linking them by S_n-joints[2] in the knots, we end up with a so-called bar-joint framework. We assume that (1) all bars are uniform made of a homogeneous isotropic material deforming

[1] This assignment corresponds to the definition of the intrinsic metric of the framework.

[2] S_n denotes the spherical joint, which enables the group of spherical motions SO(n) of $\mathbb{E}^n$. Note that a S_2-joint equals a rotational joint (R-joint).

J. Lenarčič and B. Siciliano (Eds.): ARK 2020, SPAR 15, pp. 182–189, 2021.
https://doi.org/10.1007/978-3-030-50975-0_23

at constant volume (i.e. Poisson's ratio $\nu = 1/2$) and that (2) all S_n-joints are without clearance.

In the following we consider the configuration of knots $\mathbf{k} := (\mathbf{k}_1, \ldots, \mathbf{k}_s)$, where $\mathbf{k}_i$ denotes the n-dimensional coordinate vector of the knot K_i, which together with the graph G implies the framework's realization $G(\mathbf{k})$. In the rigidity community (e.g. [2]) each edge e_{ij} is assigned with a *stress (coefficient)* $\omega_{ij} \in \mathbb{R}$. If in every knot $i \in \{1, \ldots, s\}$ the so-called *equilibrium condition*

$$\sum_{i<j\in N_i} \omega_{ij}(\mathbf{k}_i - \mathbf{k}_j) + \sum_{i>j\in N_i} \omega_{ji}(\mathbf{k}_i - \mathbf{k}_j) = \mathbf{o} \tag{1}$$

is fulfilled, where $\mathbf{o}$ denotes the n-dimensional zero-vector, then the b-dimensional vector $\omega = (\ldots, \omega_{ij}, \ldots)^T$ is refereed as *self-stress* (or *equilibrium stress*). According to Gluck [4] and Roth [16] the existence of a non-zero self-stress[3] corresponds to the infinitesimal flexibility (shakiness) of the framework's realization $G(\mathbf{k})$.

Shakiness (of order one[4]) can also be seen as the limiting case where two realizations of a framework coincide [19,21,23]. In contrast a realization $G(\mathbf{k}')$ is called a snapping realization if it is *close enough* to another incongruent realization such that the physical model can snap (flip/jump) into this realization due to non-destructive elastic deformations of material. The open problem in this context is the meaning of *closeness*, which is tackled in this article. In more detail, we present a method to measure the snapping capability (shortly called *snappability*) of a realization $G(\mathbf{k}')$, based on the total elastic strain energy of the framework.

Before we plunge in medias res we provide a short review on snapping (also called *multistable*; cf. [5]) structures. During the last years the interest in these structures has increased due to practical applications (e.g. [7,15,18]).

It is pointed out in [21] that there is a direct connection between shakiness and snapping through the technique of *averaging* and *deaveraging*, respectively (cf. [17, page 1604] and [8]). The latter allows to construct snapping frameworks in any dimension. Moreover for snapping bipartite frameworks in $\mathbb{E}^n$ an explicit result in terms of confocal hyperquadrics is known (cf. [21, page 112] under consideration of [20]). Most results are known for the dimension $n = 3$, which are summarized next.

There is a series of papers of Walter Wunderlich on snapping spatial structures (octaeder, Bennett mechanisms, antiprisms, icosaeder, dodecaeder), which are reviewed in [19]. In this context also the paper [5] should be cited, where *buckling polyhedral surfaces* and *Siamese dipyramids* are introduced. Snapping structures are also related to so-called *model flexors*[5] (cf. [11]) as in some cases the model flexibility can be reasoned by the snapping through different realizations. Examples for this phenomenon are the so-called *four-horn* [24] or the

[3] ω differs from the b-dimensional zero vector.

[4] Each additional coinciding realization raises the order of the infinitesimal flexibility by one [23].

[5] Mathematically these structures do not posses a continuous flexibility but due to free bendings without visible distortions of materials their physical models flex.

already mentioned *Siamese dipyramids.* The latter are studied in more detail in [6], especially how relative variations on the edge lengths produce significant relative variations in the spatial shape. The authors of [6] also suggested estimates to quantify these intrinsic and extrinsic variations.

2 Physical Model of Deformation

First we consider a single bar e_{ij} and apply equal but opposite directed forces F_{ij} to it's ends pointing outwards/inwards the bar, which imply a tensile/compression stress leading to an expansion/decrease of the bar. According to Hooke's law, which can be applied due to the elastic deformation during the process of snapping, the tensile/compression stress δ_{ij} in a uniform bar equals the product of the modulus of elasticity[6] $E_{ij} > 0$ and the Cauchy/engineering strain[7] ε_{ij}; i.e.

$$\delta_{ij} = E_{ij}\varepsilon_{ij} \quad \text{with} \quad \varepsilon_{ij} = \frac{l_{ij} - L_{ij}}{L_{ij}} \quad \text{for} \quad i < j \tag{2}$$

where $l_{ij} \in \mathbb{R}_{>0}$ is the deformed length (stressed length) of the bar e_{ij}, while L_{ij} is the original length (unstressed length) of the bar e_{ij}. Note that $\delta_{ij} > 0$ corresponds to a stretching and $\delta_{ij} < 0$ to a compression. Taking Eq. (2) into account the well-known relation $F_{ij} = \delta_{ij}A_{ij}$, where A_{ij} is the cross-sectional area of the bar e_{ij}, can be rewritten as

$$F_{ij} = E_{ij}A_{ij}\frac{l_{ij} - L_{ij}}{L_{ij}} \quad \text{for} \quad i < j. \tag{3}$$

This force acts on the end points K_i and K_j of the involved bar e_{ij} by the force vectors

$$\mathbf{f}_{ij} = F_{ij}\frac{\mathbf{k}_i - \mathbf{k}_j}{\|\mathbf{k}_i - \mathbf{k}_j\|} \quad \text{and} \quad \mathbf{f}_{ji} = -\mathbf{f}_{ij}, \tag{4}$$

where $\|.\|$ denotes the *standard norm.* Moreover, the elastic strain energy[8] stored in the bar e_{ij} can be computed by

$$U_{ij} = \tfrac{1}{2}F_{ij}(l_{ij} - L_{ij}) \quad \overset{(3)}{\Longrightarrow} \quad U_{ij} = \frac{1}{2}\frac{E_{ij}A_{ij}}{L_{ij}}(l_{ij} - L_{ij})^2 \tag{5}$$

according to [12, page 512], where $E_{ij}A_{ij}/L_{ij}$ is the axial stiffness of the bar e_{ij}. As a consequence the total deformation energy U of the framework reads as

$$U := \sum_{i<j} U_{ij} = \frac{1}{2}\sum_{i<j}\frac{E_{ij}A_{ij}}{L_{ij}}(l_{ij} - L_{ij})^2. \tag{6}$$

[6] In this paper we assume $E_{ij} > 0$ as for conventional structural material E_{ij} is positive.
[7] Equals in this case the engineering normal strain ε_{ij} of a material line segment e_{ij}.
[8] Note that elastic strain energy is a form of potential energy.

Remark 1. In the study of rigidity and stability of frameworks similar functions appear; e.g. the function

$$\Phi = \frac{1}{2}\sum_{i<j} \frac{F_{ij}}{L_{ij}} \left(l_{ij}^2 - L_{ij}^2\right) \tag{7}$$

was used by Kötter [9]. Connelly and Whiteley [3] based their results on a potential energy function $\sum_{ij} H_{ij}\|\mathbf{k}_i - \mathbf{k}_j\|^2$ where H_{ij} is any real-valued function of one variable, which has a local minimum at $\|\mathbf{k}_i - \mathbf{k}_j\|^2$. ⋄

Expressing l_{ij} in dependence of the knots K_i and K_j the system of partial derivatives

$$\frac{\partial U}{\partial k_{i,1}} = 0, \quad \frac{\partial U}{\partial k_{i,2}} = 0, \quad \ldots, \quad \frac{\partial U}{\partial k_{i,n}} = 0, \tag{8}$$

where $(k_{i,1}, \ldots, k_{i,n})$ is the coordinate vector of $\mathbf{k}_i$, equals the condition $\sum_{j\in N_i} \mathbf{f}_{ij} = \mathbf{o}$ for $i = 1, \ldots, s$. With respect to our physical model the meaning of the real values ω_{ij} in Eq. (1) is F_{ij}/l_{ij}, which is the so-called *force density* with respect to the stressed length [22]. Moreover it should be pointed out that each *critical point* of the total elastic strain-energy $U(\mathbf{k})$ corresponds to a self-stressed framework realization $G(\mathbf{k})$, which is called *deformed* for $U(\mathbf{k}) > 0$ and *undeformed* for $U(\mathbf{k}) = 0$.

Remark 2. The system of equations $\sum_{j\in N_i} \mathbf{f}_{ij} = \mathbf{o}$ was also obtained by Linkwitz and Schek [10, page 149ff.], who studied the form finding problem for cable networks. Within their proposed *force density method* [22] values are assigned to the mentioned *force densities* rendering these equations linear in the knot coordinates. ⋄

2.1 Metric Interpretation of U

This physical model implies in the b-dimensional space $\mathbb{R}^b$ of bar lengths the following scalar product $\langle .,. \rangle_P$: $\mathbb{R}^b \times \mathbb{R}^b \to \mathbb{R}$ with $(\mathbf{x}, \mathbf{y}) \mapsto \langle \mathbf{x}, \mathbf{y}\rangle_P := \mathbf{x}^T P \mathbf{y}$ and

$$\mathbf{x} := \begin{pmatrix} \vdots \\ x_{ij} \\ \vdots \end{pmatrix}, \quad \mathbf{y} := \begin{pmatrix} \vdots \\ y_{ij} \\ \vdots \end{pmatrix}, \quad P := \begin{pmatrix} \ddots & & \\ & \frac{E_{ij}A_{ij}}{2L_{ij}} & \\ & & \ddots \end{pmatrix}, \tag{9}$$

as the involved $b \times b$ diagonal matrix P is positive definite. Consequently this scalar product induces a norm $\|.\|_P$: $\mathbb{R}^b \to \mathbb{R}_{\geq 0}$ with $\mathbf{x} \mapsto \|\mathbf{x}\|_P := \sqrt{\langle \mathbf{x}, \mathbf{x}\rangle_P}$ and a metric (distance function) $d_P(.,.)$: $\mathbb{R}^b \times \mathbb{R}^b \to \mathbb{R}_{\geq 0}$ with $(\mathbf{x}, \mathbf{y}) \mapsto d_P(\mathbf{x}, \mathbf{y}) := \|\mathbf{x} - \mathbf{y}\|_P$. Thus U of Eq. (6) can be seen as squared distance between the deformed framework $\mathbf{l} = (\ldots, l_{ij}, \ldots)^T$ and the initial one $\mathbf{L} = (\ldots, L_{ij}, \ldots)^T$; i.e. $U = d_P(\mathbf{l}, \mathbf{L})^2$. It should be pointed out that this distance only depends on the intrinsic metric of the framework and is therefore independent of the actual realization.

In order to reduce the distance measure d_P to its geometric core we set $E_{ij} = 1$ and $A_{ij} = A$ for all bars in the remainder of the article.

3 Snappability of Realizations

If we have a realization $G(\mathbf{k})$ with non-zero self-stress and the corresponding total elastic energy $U(\mathbf{k})$ is not at a local minimum, then small perturbations would deform the framework according to the *minimum total potential energy principle*. Therefore we are interested in the set $\mathscr{S}$ of *stable* realizations, i.e. realizations which are at a local minimum of the total elastic strain-energy. Note that the set $\mathscr{S}$ is not empty as it contains at least the undeformed framework realizations.

3.1 Computation of the Set $\mathscr{S}$

For the computation of the set $\mathscr{S}$ for a given framework with a given intrinsic metric the following approach is used. We introduce new variables q_{ij} fulfilling the side condition $\Lambda_{ij} = 0$ with $\Lambda_{ij} := q_{ij}^2 - \|\mathbf{k}_i - \mathbf{k}_j\|^2$ and make the Lagrange ansatz

$$F(\mathbf{k}, \mathbf{q}, \lambda) := \frac{A}{2} \sum_{i<j} \frac{1}{L_{ij}} (q_{ij} - L_{ij})^2 + \sum_{i<j} \lambda_{ij} \Lambda_{ij} \qquad (10)$$

with the b-dimensional vectors $\mathbf{q} := (\ldots, q_{ij}, \ldots)^T$ and $\lambda := (\ldots, \lambda_{ij}, \ldots)^T$, where the latter is composed of the Lagrange multipliers λ_{ij}. By taking the partial derivatives of F with respect to the $(sn + 2b)$ variables $(\mathbf{k}, \mathbf{q}, \lambda)$ we obtain a system of equations, which is of algebraic nature. Therefore we can use homotopy continuation method (e.g. Bertini; cf. [1]), as other approaches (e.g. Gröbner base, resultant based elimination) are not promising due to the number of unknowns and degree of equations. First of all we can restrict to the obtained real critical points of F with all $q_{ij} > 0$ as only these correspond to realizations. This resulting set $\mathscr{R}$ of realizations is split into a set $\mathscr{M}$ and its absolute complement $\mathscr{M}^c = \mathscr{R} \backslash \mathscr{M}$, where the elements of $\mathscr{M}$ correspond to local minima of $U(\mathbf{k})$. They can be identified by the so-called *second derivative test*; i.e. all eigenvalues of the *Hessian matrix* of the function $U(\mathbf{k})$ are positive. Finally the desired set $\mathscr{S}$ can be obtained as the quotient $\mathscr{M}/\mathrm{SE}(n)$, where $\mathrm{SE}(n)$ denotes the group of direct isometries of $\mathbb{E}^n$. In the same way we define the set $\mathscr{S}^c := \mathscr{M}^c/\mathrm{SE}(n)$ of *unstable* realizations, which is needed later on.

3.2 Measuring the Snappability

The evaluation of the *snappability* has to be based on the intrinsic metric of the framework, as minor changes of this metric can heavily effect its spatial shape (cf. examples of the *four-horn* [24] and *Siamese dipyramids* [6]). Our intrinsic metric approach towards the determination of the *snappability* of an undeformed realization is based on the following theorem:

Theorem 1. *If a framework snaps out of a stable realization $G(\mathbf{k}')$ by applying the minimum energy needed to it, then the corresponding deformation of the realization has to pass a shaky realization $G(\mathbf{k}'')$ at the maximum state of deformation.*

Proof. We think of U as a graph function over the space $\mathbb{R}^{sn}$ of knots $\mathbf{k}$; i.e. the ordered pair $(\mathbf{k}, U(\mathbf{k}))$. In order to get out of the valley of the local minimum $(\mathbf{k}', U(\mathbf{k}'))$, which corresponds to the given stable realization $G(\mathbf{k}') \in \mathscr{S}$, with a minimum of energy needed, one has to pass a *saddle point* $(\mathbf{k}'', U(\mathbf{k}''))$ of the graph, which corresponds to a realization $G(\mathbf{k}'') \in \mathscr{S}^c$. As all realizations of $\mathscr{S}^c$ are self-stressed and deformed ($\Rightarrow$ no zero self-stress) they are infinitesimally flexible. □

Based on the pseudometric $d_S\colon \mathbb{R}^b \times \mathbb{R}^b \to \mathbb{R}_{\geq 0}$ with $(\mathbf{l}', \mathbf{l}'') \mapsto |d_p(\mathbf{l}'', \mathbf{L})^2 - d_p(\mathbf{l}', \mathbf{L})^2|$ the snappability index $s(\mathbf{k}')$ of $G(\mathbf{k}') \in \mathscr{S}$ can be quantified as follows

$$s(\mathbf{k}') = \frac{|U(\mathbf{k}'') - U(\mathbf{k}')|}{AL} = \frac{|U(\mathbf{l}'') - U(\mathbf{l}')|}{AL} = \frac{d_S(\mathbf{l}', \mathbf{l}'')}{AL} \tag{11}$$

where $\mathbf{l}' = (\ldots, l'_{ij}, \ldots)^T \in \mathbb{R}^b$ and $\mathbf{l}'' = (\ldots, l''_{ij}, \ldots)^T \in \mathbb{R}^b$ are the bar lengths of the $G(\mathbf{k}') \in \mathscr{S}$ and $G(\mathbf{k}'') \in \mathscr{S}^c$, respectively, using the notation of the above proof and $L = \sum_{i<j} L_{ij}$ denotes the framework's total length. Note that due to the division by the framework's volume AL, which is constant according to the assumed Poisson's ratio $\nu = 1/2$, the snappability index $s(\mathbf{k}')$ can be interpreted as the change of the *elastic strain energy density* $U/(AL)$. Therefore $s(\mathbf{k}')$ is invariant with respect to scaling (taking into account $E_{ij} = 1$; cf. Sect. 2.1) and enables the comparison of frameworks, which differ in the number of knots, the combinatorial structure and intrinsic metric. Note that the minimum of the obtained snappability indices over all undeformed realizations can be seen as the snappability index of the framework.

Before we give the algorithm for computing the snappability index $s(\mathbf{k})$ of an undeformed realization $G(\mathbf{k})$, it should be noted that the concrete curve $(\mathbf{k}_t, U(\mathbf{k}_t))$ on the graph connecting $(\mathbf{k}, 0)$ and the saddle point $(\mathbf{k}'', U(\mathbf{k}''))$ does not play a role as long as the deformation energy U_{ij} of each bar e_{ij} is monotonic increasing with respect to the curve parameter t. This is due to the fact that along each curve of this possible set $\mathscr{C}$ of curves the same amount of mechanical work (namely the minimum work needed) is performed on the framework to reach the saddle point.

Algorithm. Given is an undeformed realization $G(\mathbf{k})$. Let us assume that the unstable realization $G(\mathbf{k}'') \in \mathscr{S}^c$ yields the minimal function value $U(\mathbf{k}'')$. We consider the simplest possible path in $\mathbb{R}^b$, namely the straight line segment from $\mathbf{L}$ to $\mathbf{l}''$ and parametrize it with respect to the time $t \in [0, 1]$ yielding $\mathbf{l}_t := \mathbf{L} + t(\mathbf{l}'' - \mathbf{L})$. This path corresponds to different 1-parametric deformations of realizations in $\mathbb{E}^n$. If among these a deformation $G(\mathbf{k}_t)$ with the property

$$G(\mathbf{k}_t)\big|_{t=0} = G(\mathbf{k}), \quad G(\mathbf{k}_t)\big|_{t=1} = G(\mathbf{k}'') \tag{12}$$

exists ($\Rightarrow$ $(\mathbf{k}_t, U(\mathbf{k}_t)) \in \mathscr{C}$), then the undeformed realization $G(\mathbf{k})$ can be left over the unstable realization $G(\mathbf{k}'')$ and we get a value for $s(\mathbf{k})$. Computationally the property (12) can be checked e.g. by a parameter homotopy approach (e.g. Bertini; cf. [1, Sec. 6]). If such a deformation does not exist then we redefine $\mathscr{S}^c$ as $\mathscr{S}^c \setminus \{G(\mathbf{k}'')\}$ and run again the procedure explained in this paragraph until we end up with a value for $s(\mathbf{k})$. In the case of $\mathscr{S}^c = \varnothing$ we set $s(\mathbf{k}) = \infty$.

Remark 3. Note that our computational approach using Bertini does not recognize if the tracked path between the real starting point and real endpoint of the homotopy is entirely real. More generally it is an open problem to check if there exists at least one real curve $\in \mathscr{C}$, which corresponds to such a continuous real deformation. ◇

After applying the mechanical minimum work needed to deform $G(\mathbf{k}')$ into $G(\mathbf{k}'')$ as described in Theorem 1, the framework will relax according to the *minimum total potential energy principle.* Therefore this self-acting deformation will end up in a stable realization or it get stuck on the way to such a local minimum by reasons of reality. Note that a realization at the boarder of reality has to be infinitesimal flexible, as a real solution of an algebraic set of equations can only change over into a complex one through a double root. This results in the following theorem:

Theorem 2. *A snap of a framework described in Theorem 1 ends up in a realization $G(\mathbf{k}''')$ which is either undeformed or a deformed one with a shakiness.*

Remark 4. Note that $s(\mathbf{k}')$ is independent of the final snapping realization $G(\mathbf{k}''')$. ◇

3.3 Pinned Frameworks

From the mechanical point of view it makes sense to fix a system to the ground by pinning a subset $\mathscr{P}$ of the knot set $\mathscr{K}$. The results of this paper also hold for this scenario of so-called *pinned graphs* (or *grounded graphs*) due to their fundamental properties summarized in [14, Sect. 2.1]. One only has to keep in mind that bars between pinned knots cannot be deformed and that the equilibrium condition (1) only has to hold for knots $\in \mathscr{K}\setminus\mathscr{P}$ as pinned knots can counterbalance any force.

4 Conclusion and Open Problems

The total elastic strain energy of the framework (based on a physical model for the deformation of bars using Hooke's law) serves as base for the presented snappability index and the theoretical results of Theorem 1 and 2, which give further connections between shakiness and snapping beside the technique of averaging and deaveraging.

Due to limitation of pages we refer for a trivial example (comparison of a pinned and unpinned triangular framework) as well as a more sophisticated one (pinned 3-legged planar parallel manipulator) on the computation of the snappability index to the corresponding arXiv publication [13, Sects. 5.1 and 5.2].

Note that our approach neglects the possibility of collision of bars during the framework's deformation. A further open problem is mentioned in Remark 3.

Acknowledgements. The research is supported by Grant No. P 30855-N32 of the Austrian Science Fund FWF. Thanks to Hellmuth Stachel for constructive feedback on the final draft.

References

1. Bates, D.J., Hauenstein, J.D., Sommese, A.J., Wampler, C.W.: Numerically Solving Polynomial Systems with Bertini. SIAM, Philadelphia (2013)
2. Connelly, R.: Rigidity. In: Gruber, P.M., Wills, J.M. (eds.) Handbook of Convex Geometry, pp. 223–271. Elsevier (1993)
3. Connelly, R., Whiteley, W.: The stability of tensegrity frameworks. Int. J. Space Struct. **7**(2), 153–163 (1992)
4. Gluck, H.: Almost all simply connected closed surfaces are rigid. In: Glaser, L.C., Rushing, T.B. (eds.) Geometric Topology, pp. 225–239. Springer (1975)
5. Goldberg, M.: Unstable polyhedral structures. Math. Mag. **51**(3), 165–170 (1978)
6. Gorkavyy, V., Fesenko, I.: On the model flexibility of Siamese dipyramids. J. Geom. **110**, 7 (2019)
7. Haghpanah, B., Salari-Sharif, L., Pourrajab, P., Hopkins, J., Valdevit, L.: Multistable shape-reconfigurable architected materials. Adv. Mater. **28**(36), 7915–7920 (2016)
8. Izmestiev, I.: Statics and kinematics of frameworks in Euclidean and non-Euclidean geometry. In: Alberge, V., Papadopoulos, A. (eds.) Eighteen Essays in Non-Euclidean Geometry, pp. 191–233. EMS Publishing House (2019)
9. Kötter, E.: Über die Möglichkeit, n Punkte in der Ebene oder im Raume durch weniger als $2n-3$ oder $3n-6$ Stäbe von ganz unveränderlicher Länge unverschieblich miteinander zu verbinden. In: Boost, H., et al. (eds.) Festschrift Heinrich Müller-Breslau, pp. 61–80, Alfred Kröner Verlag Leipzig (1912)
10. Linkwitz, K., Schek, H.-J.: Einige Bemerkungen zur Berechnung von vorgespannten Seilnetzkonstruktionen. Ingenieur-Archiev **40**(3), 145–158 (1971)
11. Milka, A.D.: Linear bendings of right convex polyhedra. Matematicheskaya fizika, anliz, geometriya **1**(1), 116–130 (1994)
12. Mittemeijer, E.J.: Fundamentals of Materials Science. Springer, Heidelberg (2011)
13. Nawratil, G.: Evaluating the snappability of bar-joint frameworks. arXiv:2001.04430 (2020)
14. Nixon, A., Schulze, B., Sljoka, A., Whiteley, W.: Symmetry adapted assur decomposition. Symmetry **6**(3), 516–550 (2014)
15. Rafsanjani, A., Akbarzadeh, A., Pasini, D.: Snapping mechanical metamaterials under tension. Adv. Mater. **27**(39), 5931–5935 (2015)
16. Roth, B.: Rigid and flexible frameworks. Am. Math. Mon. **88**(1), 6–21 (1981)
17. Schulze, B., Whiteley, W.: Rigidity and scene analysis. In: Goodman, J.E., et al. (eds.) Handbook of Discrete and Computational Geometry, 3rd edn, pp. 1593–1632. CRC Press (2017)
18. Shang, X., Liu, L., Rafsanjani, A., Pasini, D.: Durable bistable auxetics made of rigid solids. J. Mater. Res. **33**(3), 300–308 (2018)
19. Stachel, H.: W. Wunderlichs Beiträge zur Wackeligkeit. Technical report no. 22, Institute of Geometry, TU Wien (1995)
20. Stachel, H.: Configuration theorems on bipartite frameworks. Rendiconti del Circolo Matematico di Palermo (Ser. 2) **70**(II), 335–351 (2002)
21. Stachel, H.: What lies between rigidity and flexibility of structures. Serb. Archit. J. **3**(2), 102–115 (2011)
22. Tibert, A.G., Pellegrino, S.: Review of form-finding methods for tensegrity structures. Int. J. Space Struct. **26**(3), 241–255 (2011)
23. Wohlhart, K.: Degrees of shakiness. Mech. Mach. Theory **34**(7), 1103–1126 (1999)
24. Wunderlich, W., Schwabe, C.: Eine Familie von geschlossenen gleichflächigen Polyedern, die fast beweglich sind. Elem. Math. **41**(4), 88–93 (1986)

Method for Selecting Self-aligning Mechanisms Enumerated by Matroid Contractions

Rodrigo Luis Pereira Barreto[1(✉)], Andrea Piga Carboni[2], Roberto Simoni[2], and Daniel Martins[1]

[1] Federal University of Santa Catarina, Florianópolis, Brazil
r.barreto@posgrad.ufsc.br, daniel.martins@ufsc.br
[2] Federal University of Santa Catarina, Campus Universitario, Joinville, Brazil
{andrea.piga,roberto.simoni}@ufsc.br

Abstract. This paper presents a novel method to select self-aligning mechanisms enumerated using Davies' method and matroid theory. Davies' method and screw theory are used to model mechanisms in terms of freedoms and constraints. Matroids are used to enumerate self-aligning mechanisms using the model created with Davies' method as input. Matroid contraction is then applied in order to select a set of feasible self-aligning mechanisms. The herein proposed method is applied to a well known mechanism, the Quadrupteron. Using matroid contraction, design requirements are employed to select a set of feasible self-alingning mechanisms. A novel familiy of self-aligning mechanisms, called Quadriflex family, is finally proposed.

Keywords: Self-aligning mechanisms · Matroid theory · Contraction

1 Introduction

Mechanisms and robots generally need strict manufacturing and assembly tolerances to operate properly with no additional freedoms nor extra constraints. Managing tolerances is important in the assembly of the mechanical systems, as poor manufactured links may prevent the assembly of the mechanism. However, when a self-aligning mechanism is manufactured, these issues decrease due to the lack of redundant constraints. Generally, self-aligning mechanisms require lower tolerances and simpler assembling of components. Also, due to the lack of redundant constraints, internal stresses caused by misalignment are avoided [4]. A self-aligning mechanism does not present redundant constraints, while redundant constraints are those constraints whose elimination does not change the mobility of the mechanism [17].

Reshetov [17] introduced a method to create self-aligning mechanisms using a mechanism with redundant constraints as a seed mechanism. The method consists in analyzing the mechanism, identifying the redundant constraints and

J. Lenarčič and B. Siciliano (Eds.): ARK 2020, SPAR 15, pp. 190–198, 2021.
https://doi.org/10.1007/978-3-030-50975-0_24

then removing these redundant constraints by increasing the freedom of a set of proper selected joints. Furthermore, the identification of redundant constraints is not correct for some classes of mechanisms [5]. In [4] and [10] design of mechanisms without redundant constraints is discussed, but both approaches require considerable knowledge of the types of joints requested to design new mechanisms. Moreover, a proper method for enumeration of self-align mechanisms is not presented, and the results achieved mostly rely on designer intuition.

Davies' formulation [1–3] has been used first to model a selected mechanism in terms of freedoms and constraints, and then to derive the corresponding self-aligning mechanisms. Screw theory is used in Davies' method for modelling the freedoms and constraints. Different tools are then employed in order to select the enumerated mechanism, such as greedy algorithm [5] or binary cobases matrix [1,2]. These methods require some knowledge on screw theory and Davies' method, but analyzing and generating new mechanisms is straightforward by using matrices or existing libraries for open-source software. The above cited methods for selecting self-aligning mechanisms are generally limited due to the huge number of solutions obtained for complex mechanisms. The number of self-aligning mechanisms derived from a redundant constrained mechanism increases exponentially with the number of links and joins.

In this paper, a new method is proposed in order to select self-aligning mechanisms enumerated using Davies' method and matroid theory. The Quadrupteron robot, a member of the well known multipteron family, is used as a seed mechanism for the development of new self-aligning mechanisms, thus creating the Quadriflex mechanisms family.

2 Brief Review of Davies' Method for Self-aligning Mechanisms Enumeration

In this section the Davies' method is briefly introduced. Davies [6–8] proposed an adaptation of Kirchhoff's circulation and cutset laws to multibody systems. Screw theory and graph theory are employed to model the freedoms and constraints of the joints. Several studies and applications based on the Davies' method can be found in the literature [12–14]. A multibody systems is represented by a coupling network, *i.e.* a coupling graph, in which links are represented as a node while joints are represented by edges of the graph. The fundamental cutset law for a coupling network can thus be written as:

$$\left[\hat{\boldsymbol{A}}_N\right]_{\lambda k,C}[\boldsymbol{\Psi}]_C=[\boldsymbol{0}]_{\lambda k} \tag{1}$$

where k is the number of the cutsets in the coupling network, C the sum of the degree of constraint c_i imposed by the couplings while $\boldsymbol{\Psi}$ is formed by the action screws with unknown magnitudes imposed by the couplings. The network unit action matrix $[\hat{\boldsymbol{A}}_N]$ can be written as:

$$\left[\hat{A}_N\right]_{\lambda k,C} = \begin{bmatrix} \left[\hat{A}_D\right]_{\lambda,C} & [Q_1]_{C,C} \\ \left[\hat{A}_D\right]_{\lambda,C} & [Q_2]_{C,C} \\ \vdots & \vdots \\ \left[\hat{A}_D\right]_{\lambda,C} & [Q_K]_{C,C} \end{bmatrix} \tag{2}$$

$$\left[\hat{A_D}\right] = \left[\overbrace{\$^a_{a1}\ \$^a_{a2} \cdots}^{\text{Coupling a}} \Big| \overbrace{\$^a_{b1} \cdots\cdots}^{\text{Coupling b}} \Big| \overbrace{\cdots}^{\cdots} \Big| \overbrace{\cdots\ \$^a_C}^{\cdots} \right] \tag{3}$$

where $[Q_i]_{C,C}$, $i = 1, 2, \ldots, k$ are diagonal matrices formed by the rows of the cutset matrix $[Q_A]_{k,C}$, created using the action graph G_A. Each column of the unit action matrix $[\hat{A}_D]$ is a constraint as in Eq. (3).

As each column is determined by a different constraint imposed by each coupling, the dependence and independence of the mechanism's constraints can be studied though the network action matrix $[\hat{A}_N]$. A set of independent columns in matrix $[\hat{A}_N]$ identifies a set of leading variables of a linear homogeneous system of Eq. 1, which can be described as a set of independent constraints of the mechanism. A set of independent constraints with a number of elements equal to the rank of the matrix forms a basis for the constraint space of the mechanism. Using a basis for the constraint space of the mechanism, we can derive a mechanism without redundant constraints, *i.e.* a self-aligning mechanism. Therefore, using the network action matrix $[\hat{A}_N]$ of a mechanism with C_N redundant constraints, new self-aligning mechanisms kinematically equivalent to the original mechanism can be created by selecting and removing the redundant constraints not present in the basis of the constraint space of the mechanism. Matroids, which are presented in the next section, are a powerful tool for listing sets of independent columns of matrices.

3 Method Based on Matroid Contraction to Select Self-aligning Mechanisms

In this section the properties and characteristics of matroids relevant to this work are briefly presented. Matroids can be defined by multiple structures, such as independent sets, bases, cicles, rank function, among others. Considering a finite set E and a collection of subsets $\mathcal{I}$ of E, matroids own the following properties [16]:

(1) $\emptyset \in \mathcal{I}$ - the empty set belongs to $\mathcal{I}$.
(2) $J \in \mathcal{I}$, and $I \subseteq J \Rightarrow I \in \mathcal{I}$ - a subset I, contained in the subset J, is contained in the collection of subsets $\mathcal{I}$ if the subset J is contained in $\mathcal{I}$.
(3) $I, J \in \mathcal{I}, |I| < |J| \Rightarrow (I \cup \{v\}) \in \mathcal{I}$ for some $\{v\} \in J \backslash I$ - if a subset I is smaller than the subset J, we can find an element from J that does not belong to I so that the union of this element to I is contained in $\mathcal{I}$.

Recalling the structure of the network action matrix $[\widehat{\mathbf{A}}_\mathbf{N}]$, the bases of a matroid defined over $[\widehat{\mathbf{A}}_\mathbf{N}]$ can be regarded as the sets of self-aligning mechanisms kinematically equivalent to the original mechanism represented by $[\widehat{\mathbf{A}}_\mathbf{N}]$ [5].

Applying matroids to enumerate self-aligning mechanisms may result in a great number of results, many of which are not feasible. For instance, a total of 1 510 bases were found when enumerating self-aligning clamping devices [2], while 15 704 self-aligning mechanisms were enumerated using a leg-rest mechanism of a hospital bed as the seed mechanism [1]. In the literature, different methods are proposed for the selection of basws. The greedy algorithm was first used to select self-aligning mechanisms in [5] while a new method called the cobases binary matrix was proposed in [1]. The main disadvantage of these methods is that the enumeration of all matroid bases can be time consuming and not necessary for the selection of an optimal mechanism. When increasing the number of elements of a matroid, the number of bases increases exponentially. Thus, complex seed mechanisms require huge computational time for enumerating bases, and even more time to select the bases.

In this work, we propose using matroid contraction as a method to select self-aligning mechanisms without the need of enumerating every basis. Contraction can be defined using the rank function of matroids while the rank of a matroid created using a matrix is the equal to the rank of the matrix in linear algebra [16]. Consider two subsets Y and X of the ground set E. The rank function R of every subset Y of the matroid $\mathcal{M}$ is defined by:

$$R(Y) = r(X \cup Y) - r(X) \tag{4}$$

Equation 4 is valid for any subset Y of the set $E - X$. The new matroid with rank function $R(Y)$ on the set $E - X$ is denoted by $\mathcal{M}/X$ and is known as the contraction of $\mathcal{M}$ to $E - X$. The contraction of elements can be regarded as the removal of a row of the matrix used to create the matroid [15]. In other words, contraction of an element can be thought as first selecting only bases that have that element and removing the bases that do not present that element. In this way, when the element is removed the number of bases and elements of the matroid decrease. The concept of matroid contraction can be usefully employed in the synthesis of mechanisms. Applying Davies' method, the network action matrix $[\hat{\boldsymbol{A}}_N]$ of a mechanism with redundant constraints is first created. Then by means of matroid theory, we can enumerate every possible self-aligning mechanisms kinematically equivalent to the seed mechanism. The elements of a basis of the matroid $\mathcal{M}_{A_N}$ are the constraints (a_R, a_S, ...) of the joints. Furthermore, constraints can be selected based on some design requirements and can be thus be contracted from the matroid. Recall that contraction can be regarded as the selection of the bases with a desired element. Thus, applying matroid contraction to an element of the matroid, *i.e.* a constraint, we are in fact selecting all bases that present that constraint. Therefore, a set of feasible bases can be found through a certain number of successive contractions. In this way, the contracted constraints are maintained into the bases, which can be regarded as a set of new self-aligning mechanisms based on the original seed mechanism.

4 Quadrupteron Robot

The Quadrupteron robot was first introduced by Richard et al. [18] as a partial decoupled 4-DOF 3T1R parallel manipulator, with translations on the three coordinated axes and a rotation along the z-axis. Several studies have been conducted on this parallel manipulator, such as forward displacement analysis [11] and acceleration benchmarking [9]. A structural representation of the Quadrupteron is shown in Fig. 1a while the corresponding coupling graph is presented in Fig. 1c.

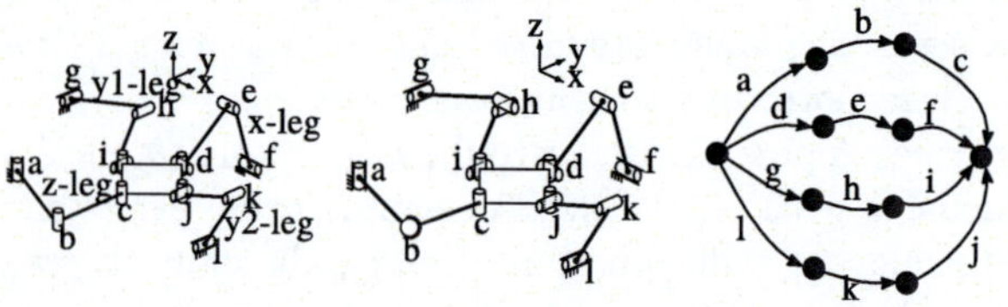

Fig. 1. Quadrupteron robot, Quadriflex robot and their graph representation.

This parallel manipulator has four legs, whose name comes from the axis of the cylindrical joints of each leg: the first leg is called the z-leg and is constituted by joints a, b and c; the second leg is called the x-leg and is constituted by joints d, e and f; the third leg is called y1-leg and is formed by joints g, h and i; the last leg is called y2-leg and joints j, k and l are present in this leg. Joints a, f, g and l are cylindrical, joints d, i and j are universal while joints b, c, e h and k are revolute joints. With the types of the joints, the unit action matrix $[\hat{\boldsymbol{A}}_{\boldsymbol{D}}]$ can be created, Eq. 5:

$$
\begin{aligned}
[\hat{\mathbf{A}}_{\mathbf{D}}]_{6,52} = [&\overbrace{a_R\ a_S\ a_U\ a_V}^{\text{Coupling a}}\ \overbrace{b_R\ b_S\ b_U\ b_V\ b_W}^{\text{Coupling b}}\ \overbrace{c_R\ c_S\ c_U\ c_V\ c_W}^{\text{Coupling c}}\ \overbrace{d_S\ d_U\ d_V\ d_W}^{\text{Coupling d}} \dots \\
&\dots \overbrace{e_S\ e_T\ e_U\ e_V\ e_W}^{\text{Coupling e}}\ \overbrace{f_S\ f_T\ f_V\ f_W}^{\text{Coupling f}}\ \overbrace{g_R\ g_T\ g_U\ g_W}^{\text{Coupling g}}\ \overbrace{h_R\ h_T\ h_U\ h_V\ h_W}^{\text{Coupling h}} \dots \\
&\dots \overbrace{i_R\ i_U\ i_V\ i_W}^{\text{Coupling i}}\ \overbrace{j_R\ j_U\ j_V\ j_W}^{\text{Coupling j}}\ \overbrace{k_R\ k_T\ k_U\ k_V\ k_W}^{\text{Coupling k}}\ \overbrace{l_R\ l_T\ l_U\ l_W}^{\text{Coupling l}}\ ]
\end{aligned} \quad (5)
$$

where the constraints are described by the joint and the mode of constraints, *e.g.* a_R is a moment constraint in the x-axis of the joint a. In this paper, the action screws are written in axis-coordinates in the form $[R\,S\,T\,U\,V\,W]$, in which the first three elements are moment constraints in the x, y and z axes and the last three elements are force constraints in the same axes. Table 1 presents three examples of the constraints present in joints of the Quadrupteron mechanism used in Eq. 5. Joint a is a revolute joint in the z-axis, thus it presents the constraints R (M_x), S (M_y), U (F_x), V (F_y) and W (F_z). In Table 1 we can also see the constraints present in the universal joint d and cylindrical joint g. The constraints not present in each of the joints are represented by –. Using the unit

Table 1. Constraints of the joints a, d and g.

Joint	Type	M_x	M_y	M_z	F_x	F_y	F_z
a	Revolute	a_R	a_S	–	a_U	a_V	a_W
d	Universal	–	d_S	–	d_U	d_V	d_W
g	Cylindrical	g_R	–	g_T	g_U	–	g_W

action matrix $[\hat{\boldsymbol{A}}_{\boldsymbol{D}}]$ and the graph from Fig. 1c, the network action matrix $[\hat{\boldsymbol{A}}_{\boldsymbol{N}}]$ is created in order to analyze the Quadrupteron and derive the self-aligning mechanisms.

5 Selection of Self-aligning Mechanisms Based on the Quadrupteron

In this section, the enumeration of self-aligning mechanisms derived from the original manipulator, regarded as a seed mechanism, is presented. Every matroid operation is done using the open-source software SageMath, because it already presents a library for matroid operations. Using the network action matrix $[\hat{\boldsymbol{A}}_{\boldsymbol{N}}]$, a matroid $\mathcal{M}_{A_N}$ is created, which has 162 different bases. In other words, 162 different self-aligning mechanisms can be derived from the Quadrupteron. However, not all of these 162 mechanisms are relevant, so we will use matroid contraction to select a set of feasible mechanisms, which best fit a set of design requirements. The first requirement is that the actuated joints must keep the original constraints, *i.e.* must not receive additional freedoms. This requirement is defined based on the choice of using commercial off-the-shelf actuators which cannot receive addition freedoms. Thus, the constraints of the actuated joints a, f, g, and l must be kept in the self-aligning mechanisms. In other words, the elements of the matroid corresponding to these constraints must be contracted. First, the constraints from joint a are contracted, resulting in a new matroid with 105 bases ($\mathcal{M}_{A_{N1}} = \mathcal{M}_{A_N}/\{a_R, a_S, a_U, a_V\}$). Then, the same operation is performed on the constraints of joint f ($\mathcal{M}_{A_{N2}} = \mathcal{M}_{A_{N1}}/\{f_S, f_T, f_V, f_W\}$), creating a matroid with 84 bases. Repeating the same operation for joints g and l ($\mathcal{M}_{A_{N3}} = \mathcal{M}_{A_{N2}}/\{g_R, g_T, g_U, g_W\}$ and $\mathcal{M}_{A_{N4}} = \mathcal{M}_{A_{N3}}/\{l_R, l_T, l_U, l_W\}$) results in matroids with 64 and 48 bases, respectively.

Regard that the 48 bases of the contracted matroids represent self-aligning mechanisms derived form the constrained Quadrupteron, which all kept the original cylindrical joints a, f, g and l for actuation. These remaining 48 bases represent 29.6% of the total of original bases. However, it is still a considerable number of bases to select. A further reduction of the number of feasible mechanism can be interesting. Considering that legs y1 and y2 are symmetrical, a subset of the 48 self-aligning mechanism are duplicated. In order to avoid the enumeration of the duplicated mechanism, only one of the two symmetrical legs is selected for the elimination of constraints. Thus, we can contract the constraints of the other leg. In this paper, the y1-leg is selected to receive freedoms, while y2-leg does not receive any freedoms, *i.e.* the corresponding constraints are contracted. Therefore, the constraints of joints j and k from y2-leg are contracted. We start by contracting the constraints from joint j ($\mathcal{M}_{A_{N_5}} = \mathcal{M}_{A_{N_4}}/\{j_R, j_U, j_V, j_W\}$) achieving 32 different bases, and then we create a matroid with 16 bases by repeating the operation with the constraints of the joint k ($\mathcal{M}_{A_{N_6}} = \mathcal{M}_{A_{N_5}}/\{k_R, k_T, k_U, k_V, k_W\}$).

Finally, applying the described operations of matroid contraction, we obtained 16 feasible self-aligning mechanisms based on the Quadrupteron out of the original 162 bases, reducing the amount of results to 9.9% of the original number. Considering the remaining 16 mechanisms, there are 2 pairs of mechanisms which present the same joints aligned in different axes. Thus, these mechanism can be discarded without loosing any different mechanism. The 14 enumerated self-aligning mechanisms are presented in Table 2. In Fig. 1b the Mechanism 1 from Table 2 is presented, in which joint b is now a spherical joint S while joint h is an universal joint U.

Table 2. Quadriflex Mechanisms.

	joint a	joint b	joint c	joint d	joint e	joint f	joint g	joint h	joint i	joint j	joint k	joint l
Mechanism 1	C	S	R	U	R	C	C	U	U	U	R	C
Mechanism 2	C	S	R	U	R	C	C	R	S	U	R	C
Mechanism 3	C	U	U	U	R	C	C	U	U	U	R	C
Mechanism 4	C	U	U	U	R	C	C	R	S	U	R	C
Mechanism 5	C	U	R	S	R	C	C	U	U	U	R	C
Mechanism 6	C	U	R	S	R	C	C	R	S	U	R	C
Mechanism 7	C	U	R	U	U	C	C	U	U	U	R	C
Mechanism 8	C	U	R	U	U	C	C	R	S	U	R	C
Mechanism 9	C	R	S	U	R	C	C	U	U	U	R	C
Mechanism 10	C	R	S	U	R	C	C	R	S	U	R	C
Mechanism 11	C	R	U	S	R	C	C	U	U	U	R	C
Mechanism 12	C	R	U	S	R	C	C	R	S	U	R	C
Mechanism 13	C	R	U	U	U	C	C	U	S	U	R	C
Mechanism 14	C	R	U	U	U	C	C	R	S	U	R	C

6 Conclusion

This paper presented a novel method to select self-aligning mechanisms using matroid contraction. First, the modelling of mechanisms according to Davies' method was introduced. Then matroid theory was discussed and finally matroid contraction was presented. The Quadrupteron parallel manipulator was introduced and modelled using Davies' method. The proposed method was applied to the Quadrupteron, regarded as the seed mechanism for the enumerating method. A set of novel self-aligning mechanism based on the Quadrupteron was presented, called the Quadriflex mechanisms family.

The proposed method was able to reduce the set of self-aligning mechanisms from 162 different possible mechanisms to a set of 16, by decreasing the number of elements in the ground set of the matroid. In this way, the computational cost related to enumerating matroid bases was effectively decreased. Thus, using this method, complex mechanisms can be analyzed and the corresponding bases enumerated, without the need of enumerating huge number of irrelevant self-aligning mechanisms. Using the built-in matroid library of the SageMath open-source software, the proposed method is easily applied and using it does not require programming skills nor a deep knowledge of matroids.

Acknowledgements. This work was funded by CAPES under project PGPTA n° 59/2014 and CNPq.

References

1. Artmann, V.N., Barreto, R.L., Carboni, A.P., Simoni, R., Martins, D.: Type synthesis of self-aligning mechanisms applied to the leg rest section of an hospital bed. In: IFToMM World Congress on Mechanism and Machine Science, pp. 1413–1422. Springer (2019)
2. Artmann, V.N., Barreto, R.L., Martins, D., Simoni, R.: Analysis and elimination of redundant constraints in clamping devices. In: Proceedings of the 25th International Congress of Mechanical Engineering - COBEM, 20–25 October 2019, pp. 147–161. ABCM (2019)
3. Barreto, R.L., Martins, D., Simoni, R., Dai, J.S., et al.: Self-aligning analysis of the metamorphic palm of the KCL/TJU metamorphic hand. In: 2018 International Conference on Reconfigurable Mechanisms and Robots (ReMAR), pp. 1–8. IEEE (2018)
4. Blanding, D.L.: Exact Constraint: Machine Design Using Kinematic Principles. ASME, New York (1999)
5. Carboni, A.P., Simas, H., Martins, D.: Analysis of self-aligning mechanisms by means of matroid theory. In: International Symposium on Multibody Systems and Mechatronics, pp. 61–73. Springer (2017)
6. Davies, T.: Kirchhoff's circulation law applied to multi-loop kinematic chains. Mech. Mach. Theory **16**(3), 171–183 (1981)
7. Davies, T.: Couplings, coupling networks and their graphs. Mech. Mach. Theory **30**(7), 991–1000 (1995)
8. Davies, T.: Dual coupling networks. Proc. Inst. Mech. Eng. Part C J. Mech. Eng. Sci. **220**(8), 1237–1247 (2006)
9. Gosselin, C.M., Masouleh, M.T., Duchaine, V., Richard, P.L., Foucault, S., Kong, X.: Parallel mechanisms of the multipteron family: kinematic architectures and benchmarking. In: Proceedings of the 2007 IEEE International Conference on Robotics and Automation, pp. 555–560. IEEE (2007)
10. Kamm, L.: Designing Cost-Efficient Mechanisms: Minimum Constraint Design, Designing with Commercial Components, and Topics in Design Engineering. McGraw-Hill, New York (1990)
11. Kong, X., Gosselin, C.: Forward displacement analysis of a quadratic 4-DoF 3T1R parallel manipulator. Meccanica **46**(1), 147–154 (2011)
12. Laus, L., Simas, H., Martins, D.: Efficiency of gear trains determined using graph and screw theories. Mech. Mach. Theory **52**, 296–325 (2012)

13. Mejia, L., Simas, H., Martins, D.: Force capability in general 3 DoF planar mechanisms. Mech. Mach. Theory **91**, 120–134 (2015)
14. Moreno, G., Nicolazzi, L.C., Vieira, R.D.S., Martins, D.: Stability of long combination vehicles. Int. J. Heavy Veh. Syst. **25**(1), 113–131 (2018)
15. Oxley, J.G.: Matroid Theory, vol. 3. Oxford University Press, Oxford (2006)
16. Recski, A.: Matroid Theory and Its Applications in Electric Network Theory and in Statics, vol. 6. Springer, Heidelberg (2013)
17. Reshetov, L.: Self-aligning mechanism, 2nd revised edition edn. MIR, Moscow (1979). Translated from Russian by Leo M. Sachs
18. Richard, P.L., Gosselin, C.M., Kong, X.: Kinematic analysis and prototyping of a partially decoupled 4-DoF 3T1R parallel manipulator (2006)

A Real-Time Capable Forward Kinematics Algorithm for Cable-Driven Parallel Robots Considering Pulley Kinematics

João Cavalcanti Santos(✉) and Marc Gouttefarde

LIRMM, University of Montpellier, CNRS, Montpellier, France
{joao.cavalcanti-santos,marc.gouttefarde}@lirmm.fr

Abstract. A real-time capable Forward Kinematics (FK) algorithm for Cable-Driven Parallel Robots (CDPRs) considering the pulley kinematics is proposed. The algorithm applies iteratively QR decomposition to solve a linearized version of the least squares problem representing the FK. Differential kinematics delivers an analytical expression for the Jacobian matrix of CDPRs considering the pulley kinematics. This Jacobian matrix is used to construct the linearization of the FK problem. Experimental and numerical results address the convergence capabilities of the proposed algorithm.

Keywords: Cable-Driven Parallel Robots · Forward Kinematics · Pulley kinematics

1 Introduction

Cable Driven Parallel Robots (CDPRs) may present several advantages compared to rigid links parallel robots such as large workspace, high payload capabilities, reduced moving mass and high dynamic capabilities. Therefore, CDPRs have been used as flight simulators [12], robotic cranes [2], large radio telescopes [13], rehabilitation devices [19], among others.

In spite of these advantages, positioning precision may be one prominent drawback of CDPRs. In comparison to rigid-link robots, the computation of the end-effector pose relying solely on the joint (winch) positions may be troublesome and inaccurate [10]. A solution for this issue would be the application of an external measurement system capturing the pose of the platform. This can be done, for instance, using a motion-capture system Bonita developed by Vicon [4] or a multi-cameras setup [5]. Nevertheless, the implementation of such measurement systems within uncontrolled environments may lead to issues related to occlusions and lighting variations. Furthermore, these solutions also depend on additional sensors, which may be costly.

Therefore, several studies sought a reliable and accurate algorithm able to compute the CDPR platform pose for a given set of winch positions, i.e., Forward

J. Lenarčič and B. Siciliano (Eds.): ARK 2020, SPAR 15, pp. 199–208, 2021.
https://doi.org/10.1007/978-3-030-50975-0_25

Kinematic (FK) Algorithms. The FK problem may be solved with interval analysis [1,8]. This approach allows one to find all the possible FK problem solutions, taking into account that some cables may be slack. Even though this method is considered mathematically complex, real-time capability has been presented [9]. Iterative numerical schemes are another typical approach [3,11,14]. In general, the FK problem is formulated as a nonlinear optimization problem. The iterative algorithm (usually, Levenberg-Marquadt) delivers a platform pose that locally minimize modeling errors, taking the winch positions as input. To this end, a differential kinematic model is necessary.

The previous works discussed above do not address the influence of the pulley kinematics. Indeed, Pott showed in [15] that relevant errors may be obtained assuming fixed proximal anchor points. In [17], the FK problem considering the pulley kinematics is solved but the equations are not derived explicitly. In this paper, these equations are obtained (in Sect. 3) and QR decomposition is used to iteratively solve the linearized version of the nonlinear FK problem (Sect. 4). This linearization is based on the differential kinematic model obtained with the differentiation of the Inverse Kinematics (IK) model (Sect. 2). Therefore, the contributions of this paper may be summarized as follows: (i) an analytical expression for the differential kinematics of CDPRs taking into account the pulley kinematics and (ii) the solution and implementation of the resultant least squares problem using QR decomposition on a CDPR prototype operated with industry-grade control software. Section 5 presents experimental results that demonstrate its applicability in a industrial real-time environment and numerical studies that address its convergence capability.

2 Inverse Kinematic Model

A six degree-of-freedom (DoF) CDPR is considered but the proposed algorithm can be applied to CDPRs with fewer DoFs. The platform is driven by m cables. The main variables and parameters of the kinematic model are introduced in Fig. 1. The position of each cable attachment point B_i on the platform is given by the vector $\mathbf{b}_i$, with $i = 1, ..., m$. Each point A_i is a fixed point which lies on the circumference of pulley i. Each pulley i is able to rotate around the line passing through A_i tangent to the pulley circumference. The constant position of A_i is given by vector $\mathbf{a}_i$ in the fixed reference frame.

The platform pose is given by the vector $\mathbf{x} = \left[\mathbf{p}^T \ \varphi^T\right]^T$. The position of the platform is $\mathbf{p} = \left[p_x \ p_y \ p_z\right]^T$. Its orientation is represented by the vector $\varphi = \left[\alpha \ \beta \ \gamma\right]^T$. The elements of this vector are Euler angles so that $\mathbf{b}_i = \mathbf{p} + \mathbf{R}_z(\gamma)\,\mathbf{R}_y(\beta)\,\mathbf{R}_x(\alpha)\,\mathbf{b}_{pi}$, where $\mathbf{b}_{pi}$ is the vector of the platform attachment point i written in the coordinate system $\mathcal{O}_p$, attached to the platform, and $\mathbf{R}_x$, $\mathbf{R}_y$ and $\mathbf{R}_z$ are the rotation matrices around x, y and z axis, respectively. The vector of cable lengths is $\boldsymbol{l} = \left[l_i \ ... \ l_m\right]^T$, where l_i is the length of cable i and the cable elongations are neglected. The IK model gives the motor positions for a given pose of the platform, i.e. compute $\boldsymbol{l}$ for a given $\mathbf{x}$. The remainder of this section recalls the IK considering the influence of the pulley geometry. Similar models were notably presented in [3,15].

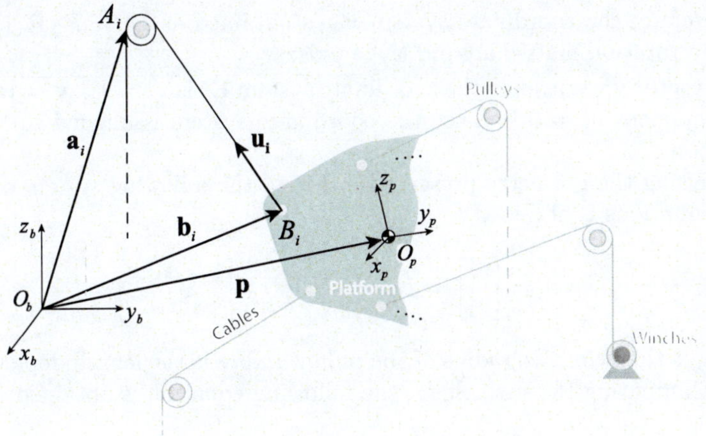

Fig. 1. Main geometric notation

The studied pulley geometry is illustrated in Fig. 2. One drawing pulley i among the m CDPR pulleys is considered and subscript i is dropped in the remainder of this section. Consider the definition of coordinate systems $\mathbf{E}_0 = [\mathbf{e}_1^0\ \mathbf{e}_2^0\ \mathbf{e}_3^0]$ and $\mathbf{E}_1 = [\mathbf{e}_1^1\ \mathbf{e}_2^1\ \mathbf{e}_3^1]$. The pulley may rotate over the fixed axis $\mathbf{e}_3^0$ (with angle ρ). $\mathbf{E}_0$ is any fixed coordinate system with $\mathbf{e}_3^0$ aligned with the rotation axis of the pulley. $\mathbf{E}_1$ rotates with the pulley and $\mathbf{e}_3^1 = \mathbf{e}_3^0$. $\mathbf{E}_1$ is thus pose dependent and its orientation is computed in the inverse kinematics.

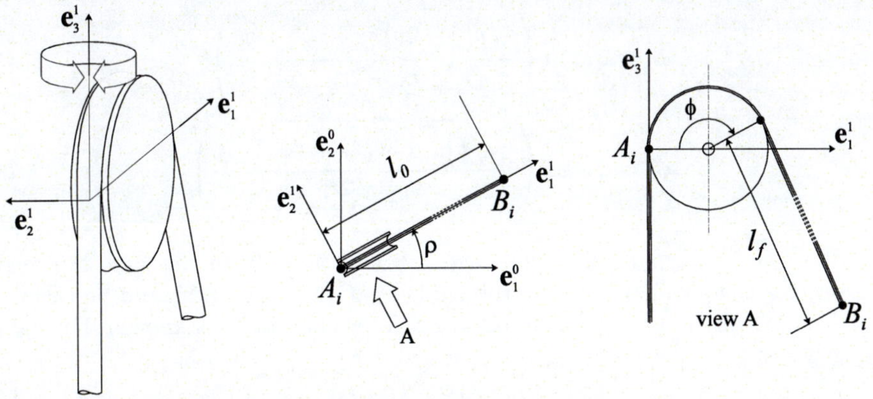

Fig. 2. Pulley geometry and coordinate systems

Define $\mathbf{v} = \mathbf{b}_i - \mathbf{a}_i$, written in the CDPR fixed reference frame. Vector $\mathbf{a}_i$ is constant for a given CDPR and $\mathbf{b}_i$ depends on the platform pose. Vector $\mathbf{v}$ can be written in the coordinate system $\mathbf{E}_0$ as $\mathbf{v}^0 = \mathbf{E}_0^T\,\mathbf{v}$. The angle ρ shown in Fig. 2 is obtained as $\text{atan2}(v_2^0, v_1^0)$, with v_1^0 and v_2^0 the first and second components of $\mathbf{v}^0$. Therefore, the coordinate system $\mathbf{E}_1$ is obtained as $\mathbf{E}_1 = \mathbf{E}_0\,\mathbf{R}_z(\rho)$, where $\mathbf{R}_z$ is the rotation matrix around the z axis.

The vector $\mathbf{v}^1$ written in the coordinate system $\mathbf{E}_1$ is $\mathbf{v}^1 = \mathbf{E}_1^T\,\mathbf{v} = \left[v_1^1\; 0\; v_3^1\right]^T$. The component $v_2^1 = 0$ because the coordinate system is aligned to the pulley orientation.

Analysing the geometry presented in Fig. 2, the following system of trigonometric equations is obtained

$$\mathbf{v}^{1r} = \begin{bmatrix} v_1^1 \\ v_3^1 \end{bmatrix} = l_f \begin{bmatrix} \sin\phi \\ \cos\phi \end{bmatrix} + r_p \begin{bmatrix} 1 - \cos\phi \\ \sin\phi \end{bmatrix} \tag{1}$$

where r_p is the primitive radius of the pulley and l_f is the length depicted in the Fig. 1. Eliminating the variable l_f, the following equation is obtained:

$$v_3^1 \sin\phi + (r_p - v_1^1)\cos\phi - r_p = 0 \tag{2}$$

The following transformations can be used to simplify (2):

$$\sin\phi = \frac{2\,\tan(\phi/2)}{1 + \tan^2(\phi/2)}, \quad \cos\phi = \frac{1 - \tan^2(\phi/2)}{1 + \tan^2(\phi/2)} \tag{3}$$

which leads to

$$\tan^2(\phi/2)\,(v_1^1 - 2\,r_p) + \tan(\phi/2)\,(2\,v_3^1) - v_1^1 = 0 \tag{4}$$

Equation (4) can be solved as a quadratic equation in $\tan(\phi/2)$. Then, using the inverse tangent function, two solutions are obtained in the interval $[-\pi, \pi]$:

$$\phi_1 = 2\,\text{atan}\left(\frac{\sqrt{{v_1^1}^2 - 2v_1^1 r_p + {v_3^1}^2} - v_3^1}{v_1^1 - 2r_p}\right) \tag{5}$$

$$\phi_2 = 2\,\text{atan}\left(\frac{-\sqrt{{v_1^1}^2 - 2v_1^1 r_p + {v_3^1}^2} - v_3^1}{v_1^1 - 2r_p}\right) \tag{6}$$

If $v_1^1 > 2r_p$ (typical situation), $\phi_1 > 0$ and $\phi_2 < 0$. In this case the angle respecting the geometry of the problem is ϕ_1. Otherwise, ϕ_2 should be taken.

The length l_f may be easily obtained from (1) once ϕ is known. The cable length l_i is given by

$$l_i = r_p\,\phi + l_f + l_{ci} \tag{7}$$

where l_{ci} is a constant additional length. This constant additional length may take into account the total distance between fixed pulleys for instance.

3 Differential Kinematics

The Jacobian matrix $\mathbf{J}$ is used in the solution of the FK problem since it relates $\delta \boldsymbol{l}$ to $\delta \mathbf{x}$ as

$$\delta \boldsymbol{l} = \mathbf{J}\, \delta \mathbf{x} \tag{8}$$

where $\delta(\cdot)$ denotes the differential variation of $(\cdot)$. This section presents the analytical expression of the matrix $\mathbf{J}$ obtained from the equations in Sect. 2. As in Sect. 2, subscript i is dropped in the remainder of this section. The vector $\mathbf{v} = \mathbf{b}_i - \mathbf{a}_i$ is computed as

$$\mathbf{v} = \mathbf{p} + \mathbf{R}(\varphi)\, \mathbf{b}_{pi} - \mathbf{a}_i \tag{9}$$

Differentiating this expression with respect to the platform pose, the following equation is obtained

$$\delta \mathbf{v} = \underbrace{\left[\mathbf{I}_3 \ \tfrac{\partial}{\partial \varphi}\left(\mathbf{R}(\varphi)\, \mathbf{b}_{pi}\right)\right]}_{\frac{\partial \mathbf{v}}{\partial \mathbf{x}}} \delta \mathbf{x} \tag{10}$$

where

$$\frac{\partial}{\partial \varphi}\left(\mathbf{R}(\varphi)\, \mathbf{b}_{pi}\right) = \begin{bmatrix} b_{piy}\, \sigma_1 + b_{piz}\, \sigma_2 & b_{piz}\, c_\alpha\, c_\beta\, c_\gamma - b_{pix}\, c_\gamma\, s_\beta + b_{piy}\, c_\beta\, c_\gamma\, s_\alpha & b_{piz}\, \sigma_3 - b_{piy}\, \sigma_4 - b_{pix}\, c_\beta\, s_\gamma \\ -b_{piy}\, \sigma_3 - b_{piz}\, \sigma_4 & b_{piz}\, c_\alpha\, c_\beta\, s_\gamma - b_{pix}\, s_\beta\, s_\gamma + b_{piy}\, c_\beta\, s_\alpha\, s_\gamma & b_{piz}\, \sigma_3 - b_{piy}\, \sigma_2 + b_{pix}\, c_\beta\, c_\gamma \\ b_{piy}\, c_\alpha\, c_\beta - b_{piz}\, c_\beta\, s_\alpha & -b_{pix}\, c_\beta - b_{piz}\, c_\alpha\, s_\beta - b_{piy}\, s_\alpha\, s_\beta & 0 \end{bmatrix}$$

with

$$\sigma_1 = (s_\alpha\, s_\gamma + c_\alpha\, c_\gamma\, s_\beta), \quad \sigma_2 = (c_\alpha\, s_\gamma - c_\gamma\, s_\alpha\, s_\beta) \tag{11}$$

$$\sigma_3 = (c_\gamma\, s_\alpha - c_\alpha\, s_\beta\, s_\gamma), \quad \sigma_4 = (c_\alpha\, c_\gamma + s_\alpha\, s_\beta\, s_\gamma) \tag{12}$$

and $c_\theta = \cos\theta$ and $s_\theta = \sin\theta$. Defining $\mathbf{v}^{0r} = \left[v_1^0\ v_2^0\right]^T$ and $\mathbf{E}_{0r} = \left[\mathbf{e}_1^0\ \mathbf{e}_2^0\right]$, one may write

$$\mathbf{v}^{0r} = \mathbf{E}_{0r}^T\, \mathbf{v} = l_0 \begin{bmatrix} \cos\rho \\ \sin\rho \end{bmatrix} \tag{13}$$

with l_0 the length depicted in Fig. 2. Differentiating (13) and isolating $\delta\rho$, this equation leads to the derivative of ρ with respect to $\mathbf{x}$

$$\delta\rho = \underbrace{\left(\frac{1}{l_0}\left[-\sin\rho\ \cos\rho\right] \mathbf{E}_{0r}^T\, \frac{\partial \mathbf{v}}{\partial \mathbf{x}}\right)}_{\frac{\partial \rho}{\partial \mathbf{x}}} \delta \mathbf{x} \tag{14}$$

Defining

$$\mathbf{R}_{zr}^T(\rho) = \begin{bmatrix} \cos\rho & \sin\rho & 0 \\ 0 & 0 & 1 \end{bmatrix}$$

Equation (1) may be rewritten as follows

$$\mathbf{v}^{1r} = \mathbf{R}_{zr}^{T}(\rho)\,\mathbf{E}_0^T\,\mathbf{v} = l_f \begin{bmatrix}\sin\phi \\ \cos\phi\end{bmatrix} + r_p \begin{bmatrix}1-\cos\phi \\ \sin\phi\end{bmatrix} \tag{15}$$

Differentiating this equation, the following relation is obtained

$$\overbrace{\left(\mathbf{D}_{Rz}(\rho)\,\mathbf{E}_0^T\mathbf{v}\frac{\partial\rho}{\partial\mathbf{x}} + \mathbf{R}_{zr}^{T}(\rho)\,\mathbf{E}_0^T\,\frac{\partial\mathbf{v}}{\partial\mathbf{x}}\right)}^{\frac{\partial\mathbf{v}^{1r}}{\partial\mathbf{x}}}\delta\mathbf{x}$$
$$= \left(r_p\begin{bmatrix}\sin\phi \\ \cos\phi\end{bmatrix} + l_f\begin{bmatrix}\cos\phi \\ -\sin\phi\end{bmatrix}\right)\delta\phi + \begin{bmatrix}\sin\phi \\ \cos\phi\end{bmatrix}\delta l_f \tag{16}$$

with

$$\mathbf{D}_{Rz}(\rho) = \begin{bmatrix}-\sin\rho & \cos\rho & 0 \\ 0 & 0 & 0\end{bmatrix} \tag{17}$$

The differential $\delta\phi$ can be isolated in (16) leading to the derivative of ϕ

$$\delta\phi = \underbrace{\left(\frac{1}{l_f}\begin{bmatrix}\cos\phi & -\sin\phi\end{bmatrix}\frac{\partial\mathbf{v}^{1r}}{\partial\mathbf{x}}\right)}_{\frac{\partial\phi}{\partial\mathbf{x}}}\delta\mathbf{x} \tag{18}$$

Moreover, this expression can be substituted for $\delta\phi$ in (16) and δl_f is thereby obtained as

$$\delta l_f = \underbrace{\left(\begin{bmatrix}\sin\phi & \cos\phi\end{bmatrix}\frac{\partial\mathbf{v}^{1r}}{\partial\mathbf{x}} - r_p\frac{\partial\phi}{\partial\mathbf{x}}\right)}_{\frac{\partial l_f}{\partial\mathbf{x}}}\delta\mathbf{x} \tag{19}$$

Differentiation of (7) leads to

$$\delta l_i = \underbrace{\left(r_p\frac{\partial\phi}{\partial\mathbf{x}} + \frac{\partial l_f}{\partial\mathbf{x}}\right)}_{\frac{\partial l_i}{\partial\mathbf{x}}}\delta\mathbf{x} \tag{20}$$

Finally, the computations detailed above can be applied to each kinematic chain $i = 1, ..., m$, and the derivatives are concatenated to form the expression of the Jacobian matrix

$$\delta\boldsymbol{l} = \begin{bmatrix}\delta l_1 \\ \vdots \\ \delta l_m\end{bmatrix} = \underbrace{\begin{bmatrix}\frac{\partial l_1}{\partial\mathbf{x}} \\ \vdots \\ \frac{\partial l_m}{\partial\mathbf{x}}\end{bmatrix}}_{\mathbf{J}}\delta\mathbf{x} \tag{21}$$

4 Forward Kinematic Model

The FK consists in computing the platform pose for a given set of winch motor positions. In the case of redundant CDPRs with $m > n$, n being the number of degrees of freedom of the mobile platform, the kinematic model is overdetermined. In general, due to modeling and measurement inaccuracies, it is impossible to find a platform pose that satisfies exactly the kinematic constraints set by the m kinematic chains. Therefore, the FK may be formulated as the problem of minimizing the error between the measured cable lengths and the cable lengths computed by the IK (Eq. (7)). Let $\boldsymbol{l}_m$ be the set of measured cable lengths and $\hat{\boldsymbol{l}}(\mathbf{x})$ the cable lengths obtained with the IK model for a given pose $\mathbf{x}$. The error to be minimized is defined as $e(\mathbf{x}) = \|\hat{\boldsymbol{l}}(\mathbf{x}) - \boldsymbol{l}_m\|_2$, where the 2-norm is denoted by $\|\cdot\|_2$. Therefore, for a given $\boldsymbol{l}_m$, the FK Algorithm should produce the solution $\mathbf{x}^* = \arg\min_{\mathbf{x}} \|\hat{\boldsymbol{l}}(\mathbf{x}) - \boldsymbol{l}_m\|_2$.

In general, the function $e(\mathbf{x}) = \|\hat{\boldsymbol{l}}(\mathbf{x}) - \boldsymbol{l}_m\|_2$ possesses several local minima. The proposed FK algorithm takes an initial guess $\mathbf{x}_g$ and find a $\mathbf{x}^*$ that locally minimizes the function $e(\mathbf{x})$. The proposed algorithm assumes that $\mathbf{x}_g$ is sufficiently close to the current platform pose and consists of an iterative scheme. For an iteration k with platform pose $\mathbf{x}_k$, the next iteration takes $\mathbf{x}_{k+1} = \mathbf{x}_k + \Delta\mathbf{x}_k$. Considering the approximation $\Delta\hat{\boldsymbol{l}} \approx \mathbf{J}(\mathbf{x}_k)\,\Delta\mathbf{x}$, a reasonable choice for $\Delta\mathbf{x}_k$ is

$$\Delta\mathbf{x}_k = \arg\min_{\Delta\mathbf{x}} \left\|\mathbf{J}(\mathbf{x}_k)\,\Delta\mathbf{x} - \left(\boldsymbol{l}_m - \hat{\boldsymbol{l}}(\mathbf{x}_k)\right)\right\|_2 \tag{22}$$

The minimization problem (22) is a Linear Least Squares problem. The solution of (22) might be computed solving $\mathbf{J}(\mathbf{x}_k)^T\mathbf{J}(\mathbf{x}_k)\,\Delta\mathbf{x} = \mathbf{J}(\mathbf{x}_k)^T\left(\boldsymbol{l}_m - \hat{\boldsymbol{l}}(\mathbf{x}_k)\right)$ for $\Delta\mathbf{x}$. However, this solution may lead to numerical instability [16]. Accordingly, this problem is typically solved using SVD or QR factorizations [6,16]. The SVD may be preferred because of its greater diagnostic capability in pathological cases. Nevertheless, QR factorization presents faster computing time. The latter being critical in real-time applications, the QR factorization is preferred. Therefore, the Jacobian matrix $\mathbf{J}$ is decomposed using the QR factorisation such that $\mathbf{J} = \mathbf{Q}\,\mathbf{R}$, with $\mathbf{Q}$ orthogonal and $\mathbf{R}$ upper triangular. The solution of (22) is then obtained from the back substitution of the system $\mathbf{R}\,\Delta\mathbf{x} = \mathbf{Q}^T\left(\hat{\boldsymbol{l}}(\mathbf{x}_k) - \boldsymbol{l}_m\right)$. Note that the proposed algorithm does not present numerical damping. For the studied scenarios, this was not necessary.

Once $\Delta\mathbf{x}_k$ is obtained, the next pose $\mathbf{x}_{k+1} = \mathbf{x}_k + \Delta\mathbf{x}_k$ is computed. This procedure is repeated until $\|\Delta\mathbf{x}\|_2 < \epsilon$, for ϵ the desired tolerance.

5 Experimental Results

The proposed FK algorithm has been implemented in the HRPCable prototype shown in Fig. 3. This prototype has a 6-DoF platform fully constrained by 8 cables. The control was programmed in C++ and runs in an industrial PC Beckhoff C6920 equipped with 2.4 GHz i7 core processor. The platform pose control loop runs at 125 Hz. An inner feedback loop running at 2 kHz control the cable tensions, the latter being measured by means of force sensors (load pins) placed in the drawing pulleys. Figure 3 depicts the performed trajectory. The threshold ϵ was defined as $\epsilon = 1 \times 10^{-6}$.

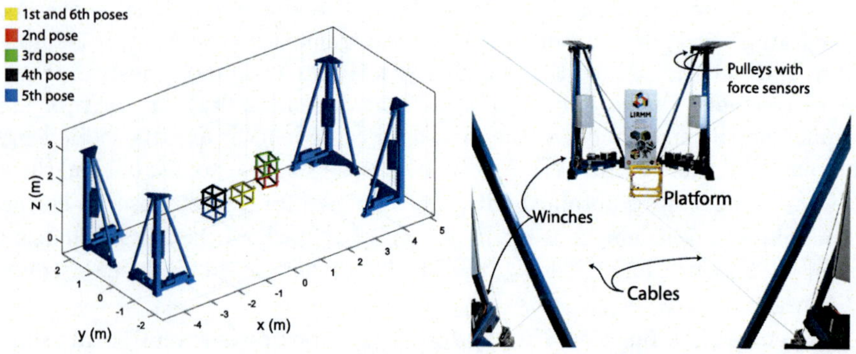

Fig. 3. HRPCable prototype and its CAD view presenting the performed trajectory

In addition to the components shown in Fig. 3, a Metris K600 camera system is used to measure the pose (6-DoF) of the platform in real time with a precision of 70 µm. The measurements obtained with the K600 camera system were used as reference to compare with the estimations obtained with the proposed FK algorithm. These results are shown in Fig. 4. This figure also shows the necessary number of iterations during the trajectory.

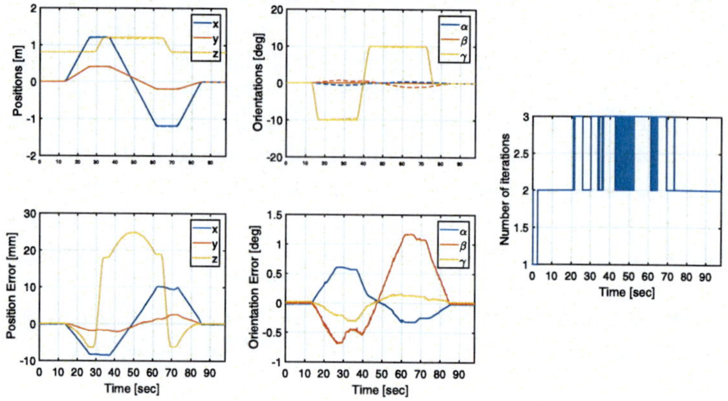

Fig. 4. Experimental results: solid lines represent the poses estimated with the proposed FK algorithm and dashed lines represent the poses measured by the Metris K600 camera system

For a given pair of vectors $\mathbf{x}^*$ and $\boldsymbol{l}^*$ consistent with the constraints presented in Sect. 2, strictly evaluating the capability (denoted here as convergence capability) of an FK iterative algorithm to quickly and reliably find $\mathbf{x}^*$ for given $\boldsymbol{l}^*$ is not a trivial task and is out of the scope of this paper. The FK problem of a Stewart-Gough platform (topologically very similar to CDPRs) may have up to 40 solutions [18] and redundant configurations do not necessarily present a reduced number of possible solutions compared to the non-redundant case [7]. Therefore, it is necessary to prove that the algorithm is able to converge to $\mathbf{x}^*$, and not just to one of the (potentially) many $\mathbf{x}$ that are consistent with $\boldsymbol{l}^*$. Nevertheless, it is still interesting to address this capability as follows.

The FK algorithm has been applied to IK solutions corresponding to more than 30×10^3 poses equally spaced across the workspace keeping the initial guess constant equal to $\mathbf{x}_g = \left[0\ 0\ .8\ 0\ 0\ 0\right]^T$ for all the poses. The workspace of the prototype is a $4 \times 2 \times 1.4\ \text{m}^3$ $(x,\ y,\ z)$ cuboid, considering $-10° \leqslant \gamma \leqslant 10°$, $\alpha = \beta = 0$. The algorithm obtained a pose with errors smaller than the proposed tolerance within 7 iterations for all these poses.

6 Conclusions

The present work introduced an explicit expression for the Jacobian matrix of CDPRs considering the pulley kinematics. Furthermore, a Forward Kinematics (FK) algorithm has been implemented based on this explicit expression. The Jacobian matrix is used to linearize the FK problem formulating it as a least squares problem, the latter being solved iteratively by means of QR decomposition. The algorithm was implemented in a prototype with industrial real-time environment. The proposed algorithm was successfully tested in numerous poses distributed in the workspace of the robot. Nevertheless, the measured position and orientation accuracy indicates that there is room for improvements.

Acknowledgements. This work was supported by the European Union's H2020 Program (H2020/2014–2020) under the grant agreement No. 732513 (HEPHAESTUS project).

This work also has been sponsored by the French government research program Investissements d'avenir through the Robotex Equipment of Excellence (ANR-10-EQPX-44).

References

1. Berti, A., Merlet, J.P., Carricato, M.: Solving the direct geometrico-static problem of underconstrained cable-driven parallel robots by interval analysis. Int. J. Robot. Res. **35**(6), 723–739 (2016). https://doi.org/10.1177/0278364915595277
2. Bostelman, R., Albus, J., Dagalakis, N., Jacoff, A., Gross, J.: Applications of the NIST RoboCrane. In: Proceedings of the 5th International Symposium on Robotics and Manufacturing, pp. 14–18 (1994)
3. Bruckmann, T., Mikelsons, L., Brandt, T., Hiller, M., Schramm, D.: Wire robots part I: kinematics, analysis & design. I-Tech Education and Publishing, Vienna (2008)

4. Chellal, R., Cuvillon, L., Laroche, E.: A kinematic vision-based position control of a 6-DoF cable-driven parallel robot. In: Cable-Driven Parallel Robots, pp. 213–225. Springer (2015)
5. Dallej, T., Gouttefarde, M., Andreff, N., Hervé, P.E., Martinet, P.: Modeling and vision-based control of large-dimension cable-driven parallel robots using a multiple-camera setup. Mechatronics **61**, 20–36 (2019)
6. Golub, G.H., Loan, C.F.V.: Matrix Computations, 4th edn (2013)
7. Husty, M., Mielczarek, S., Hiller, M.: A redundant spatial Stewart-Gough platform with a maximal forward kinematics solution set. In: Advances in Robot Kinematics, pp. 147–154. Springer (2002)
8. Merlet, J.P.: Solving the forward kinematics of a Gough-type parallel manipulator with interval analysis. Int. J. Robot. Res. **23**(3), 221–235 (2004)
9. Merlet, J.P.: On the real-time calculation of the forward kinematics of suspended cable-driven parallel robots. In: 4th IFToMM World Congress on the Theory of Machines and Mechanisms, Taipei, Taiwan (2015)
10. Merlet, J.P.: Simulation of discrete-time controlled cable-driven parallel robots on a trajectory. IEEE Trans. Robot. **33**(3), 675–688 (2017). https://doi.org/10.1109/TRO.2017.2664888
11. Miermeister, P., Kraus, W., Pott, A.: Differential kinematics for calibration, system investigation, and force based forward kinematics of cable-driven parallel robots. In: Cable-Driven Parallel Robots, pp. 319–333. Springer (2013)
12. Miermeister, P., Lächele, M., Boss, R., Masone, C., Schenk, C., Tesch, J., Kerger, M., Teufel, H., Pott, A., Bülthoff, H.H.: The CableRobot simulator large scale motion platform based on cable robot technology. In: 2016 IEEE/RSJ International Conference on Intelligent Robots and Systems (IROS), pp. 3024–3029. IEEE (2016)
13. Nan, R., Li, D., Jin, C., Wang, Q., Zhu, L., Zhu, W., Zhang, H., Yue, Y., Qian, L.: The five-hundred-meter aperture spherical radio telescope (FAST) project. Int. J. Mod. Phys. D **20**(06), 989–1024 (2011)
14. Pott, A.: An algorithm for real-time forward kinematics of cable-driven parallel robots. In: Advances in Robot Kinematics: Motion in Man and Machine, pp. 529–538. Springer (2010)
15. Pott, A.: Influence of pulley kinematics on cable-driven parallel robots. In: Latest Advances in Robot Kinematics, pp. 197–204. Springer (2012)
16. Press, W.H., Teukolsky, S.A., Vetterling, W.T., Flannery, B.P.: Numerical recipes in C++. The art of scientific computing **2**, 1002 (1992)
17. Schmidt, V.L.: Modeling Techniques and Reliable Real-Time Implementation of Kinematics for Cable-Driven Parallel Robots Using Polymer Fiber Cables. Fraunhofer Verlag, Stuttgart (2017)
18. Wampler, C.W.: Forward displacement analysis of general six-in-parallel SPS (Stewart) platform manipulators using soma coordinates. Mech. Mach. Theory **31**(3), 331–337 (1996)
19. Zanotto, D., Rosati, G., Minto, S., Rossi, A.: Sophia-3: a semiadaptive cable-driven rehabilitation device with a tilting working plane. IEEE Trans. Robot. **30**(4), 974–979 (2014). https://doi.org/10.1109/TRO.2014.2301532

Stiffness Oriented Tension Distribution Algorithm for Cable-Driven Parallel Robots

Etienne Picard[1], Stéphane Caro[2(✉)], Franck Plestan[3], and Fabien Claveau[4]

[1] IRT Jules Verne, 44340 Bouguenais, France
etienne.picard@irt-jules-verne.fr
[2] Centre National de la Recherche Scientifique (CNRS), 44321 Nantes, France
stephane.caro@ls2n.fr
[3] École Centrale de Nantes - LS2N, UMR CNRS 6004, 44321 Nantes, France
franck.plestan@ec-nantes.fr
[4] IMT-Atlantique - LS2N, UMR CNRS 6004, 44300 Nantes, France
fabien.claveau@imt-atlantique.fr

Abstract. A novel criterion is introduced in this paper to determine the set of cable tensions for Cable-Driven Parallel Robots (CDPRs) with the aim of maximizing the robot stiffness along a specific direction. Based on the feasible polygon of the CDPR and its stiffness matrix, an algorithm selects the set of admissible cable tensions leading to the smallest moving-platform displacement, the moving-platform being subject to an external wrench. The proposed tension distribution is implemented in a control scheme and experimented on a fully-constrained CDPR for a window cleaning application.

Keywords: Cable-Driven Parallel Robot · Tension Distribution Algorithm · Stiffness · Control · Experiments

1 Introduction

Cable-Driven Parallel Robots (CDPRs) are a particular class of parallel robots whose moving-platform (MP) is connected to a fixed base frame by cables, as illustrated in Fig. 1. Both the orientation and the position of the MP are controlled by cables coiled on motorized winches. Passive pulleys may guide the cables from the winches to the cable exit points. CDPRs have several advantages such as a relatively low mass of moving parts, high dynamics and a potential large workspace. As a consequence, they can be used in several applications such as heavy load handling [1], painting and sandblasting of large structures [8], fast pick-and-place operations [11], haptic devices [6], support structures for giant telescopes [18], and search and rescue deployable platforms [14].

However, the use of cables introduces specific control challenges. In particular, the position control of fully-constrained CDPRs, whose cables coming from both the top and the bottom of the frame, requires more attention than

J. Lenarčič and B. Siciliano (Eds.): ARK 2020, SPAR 15, pp. 209–217, 2021.
https://doi.org/10.1007/978-3-030-50975-0_26

purely suspended ones. Small errors in the computation of the necessary cable lengths to achieve a desired MP pose can lead to slack cables or high tensions in opposing cables. A tension distribution algorithm (TDA) can then be implemented to balance a wrench between all cables while maintaining the static or dynamic equilibrium of the MP, and to enforce boundaries on the cable tensions. The most commonly used TDAs for calculating the force distribution among the cables of classical CDPRs are the following: (*i*) Gradient-based optimization approach [17]; (*ii*) Quadratic programming [5]; (*iii*) Linear programming Methods [4]; (*iv*) Minimizing norm solution with Dykstra method [9]; (*v*) Centroid/Barycentric approach [13]. A comparison between existing TDAs is given in [16]. It is noteworthy that the cable tension distribution is influenced by the cable elasticity and the existing TDAs do not consider the CDPR stiffness. Accordingly, the main contribution of this paper lies in the cable tension distribution based on a CDPR stiffness performance index, namely, a novel *Stiffness Oriented Tension Distribution Algorithm* (SOTDA).

A CDPR prototype was assembled in the framework of the IRT Jules Verne ROCKET project for a window cleaning application, denoted as ROWC (ROCKET Window Cleaning) and shown in Fig. 1. The moving-platform is equipped with a pressure washer hose. In operation, pressurized water is projected onto the surface to clean, which creates a wrench onto the MP in the opposite direction of the water flow. Due to cable elasticity, this external wrench induces a small displacement of the MP along the normal to the plane of MP motion.

In the sequel, the ROWC prototype and its modeling are introduced in Sect. 2. Then, the SOTDA is described in Sect. 3. Experimental results on ROWC obtained with a control scheme integrating the SOTDA are presented in Sect. 4.

2 Prototype and Modeling

Figure 1 presents the prototype projecting water onto a facade. This CDPR uses a fully-constrained cable configuration with eight cables and six degrees of freedom. A video of the window cleaning application is available[1]. This prototype is actuated by B&R Automation synchronous motors of nominal speed and nominal torque equal to 2100 rpm and 2.6 Nm, respectively. The motors are coupled to planetary gearboxes with a reduction ratio of 5, and its winches have a diameter of 90 mm. A specific platform was designed for the window cleaning application, equipped with a water nozzle located at the extremity of a deployable mechanism. Due to its cable configuration, the prototype has an almost planar workspace and presents the lowest stiffness along the y_b axis, aligned with the direction of force due to water pressure. As a consequence, the SOTDA aims to select the set of cable tensions that maximizes the stiffness of the MP along the y_b axis.

Figure 1 depicts the main geometric parameters of the CDPR at hand. Cable exit points are denoted as A_i, while cable anchor points are denoted as B_i, with

[1] ROCKET window cleaning application video (link).

$i \in \{1, \dots, m\}$, m being the number of cables attached to the MP, $m = 8$ for ROWC. $\mathscr{F}_b$ is the robot base frame, and $\mathscr{F}_p$ is the MP frame. Vector ${}^b\mathbf{a}_i$ points from O to A_i and is expressed in frame $\mathscr{F}_b$. Vector ${}^p\mathbf{b}_i$ point from P to B_i and is expressed in frame $\mathscr{F}_p$. Vector ${}^b\mathbf{u}_i$ is the ith cable unit vector pointing from B_i to A_i. Vector ${}^b\mathbf{p}$ is the position vector of point P, the MP geometric center, expressed in $\mathscr{F}_b$. As the pulleys of the ROWC prototype are very small compared to the cable lengths, the following model is established from a geometric modeling neglecting the pulleys.

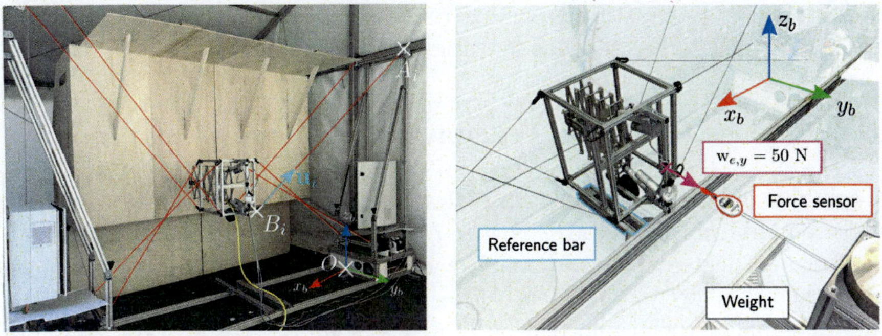

Fig. 1. (Left) ROWC prototype projecting pressurized water onto a boat model (Right) Experimental setup with lateral wrench for the static case.

The static equilibrium of the MP is given by

$$\mathbf{W}\mathbf{t} + \mathbf{w}_e + \mathbf{w}_g = 0 \tag{1}$$

where $\mathbf{t}$ is the cable tension vector, $\mathbf{w}_g$ the wrench applied to the MP due to gravity, $\mathbf{w}_e$ an external wrench expressed in frame $\mathscr{F}_b$, and $\mathbf{W}$ the wrench matrix of the robot, whose expression is given by:

$$\mathbf{W} = \begin{bmatrix} {}^b\mathbf{u}_1 & \dots & {}^b\mathbf{u}_i & \dots & {}^b\mathbf{u}_m \\ {}^b\mathbf{b}_1 \times {}^b\mathbf{u}_1 & \dots & {}^b\mathbf{b}_i \times {}^b\mathbf{u}_i & \dots & {}^b\mathbf{b}_m \times {}^b\mathbf{u}_m \end{bmatrix} \tag{2}$$

The stiffness matrix $\mathbf{K}$ of ROWC is expressed based on a linear cable elasticity model [2]. From [3] and [7], $\mathbf{K}$ is defined as:

$$\mathbf{K} = \frac{\delta \mathbf{w_e}}{\delta \mathbf{X}} = \mathbf{K}_p + \mathbf{K}_a \quad \text{with} \quad \delta\mathbf{X} = \left[\delta\mathbf{p}\ \delta\boldsymbol{\phi}\right]^{\mathrm{T}} \tag{3}$$

where a small change in the external wrench $\delta\mathbf{w}_e$ results in a small displacement screw $\delta\mathbf{X}$ of the MP. $\delta\mathbf{p}$ and $\delta\boldsymbol{\phi}$ are respectively the translational and rotational components of the displacement. $\mathbf{K}_p$ and $\mathbf{K}_a$ are named passive and active stiffness matrices, respectively, and take the form:

$$\mathbf{K}_p = \sum_{i=1}^{m} k_i \begin{bmatrix} {}^b\mathbf{u}_i {}^b\mathbf{u}_i^{\mathrm{T}} & {}^b\mathbf{u}_i {}^b\mathbf{u}_i^{\mathrm{T}} {}^b\hat{\mathbf{b}}_i^{\mathrm{T}} \\ {}^b\hat{\mathbf{b}}_i {}^b\mathbf{u}_i {}^b\mathbf{u}_i^{\mathrm{T}} & {}^b\hat{\mathbf{b}}_i {}^b\mathbf{u}_i {}^b\mathbf{u}_i^{\mathrm{T}} {}^b\hat{\mathbf{b}}_i^{\mathrm{T}} \end{bmatrix}$$

$$\mathbf{K}_a = -\sum_{i=1}^{m} t_i \begin{bmatrix} -\frac{1}{l_i}\left(\mathbf{I}_3 - {}^b\mathbf{u}_i {}^b\mathbf{u}_i^{\mathrm{T}}\right) & \frac{1}{l_i}\left(\mathbf{I}_3 - {}^b\mathbf{u}_i {}^b\mathbf{u}_i^{\mathrm{T}}\right) {}^b\hat{\mathbf{b}}_i \\ -\frac{1}{l_i} {}^b\hat{\mathbf{b}}_i \left(\mathbf{I}_3 - {}^b\mathbf{u}_i {}^b\mathbf{u}_i^{\mathrm{T}}\right) & \left[{}^b\hat{\mathbf{u}}_i + \frac{{}^b\hat{\mathbf{b}}_i}{l_i}\left(\mathbf{I}_3 - {}^b\mathbf{u}_i {}^b\mathbf{u}_i^{\mathrm{T}}\right)\right] {}^b\hat{\mathbf{b}}_i \end{bmatrix} \quad (4)$$

with k_i the ith cable stiffness and t_i its tension. $\mathbf{K}_p$ is a function of the cable elasticity, whereas $\mathbf{K}_a$ depends on cable tensions.

3 Stiffness Oriented Tension Distribution Algorithm

A solution to tension distribution is to find a set of cable tensions ensuring the static equilibrium of the MP. For over-constrained CDPRs with an actuation redundancy degree equal to two, like ROWC, the set of admissible solutions is characterized by the polygon of feasible tensions [10]. Let t_{min} and t_{max} be the lower and upper cable tension limits. A solution to the static equilibrium (Eq. (1)) can be written as:

$$\mathbf{t} = \mathbf{W}^{\dagger}\mathbf{f} + \mathbf{N}\lambda \quad \text{with} \quad \lambda = \left[\lambda_1 \; \lambda_2\right]^{\mathrm{T}} \quad (5)$$

where $\mathbf{N} = \text{null}(\mathbf{W})$ is the null space of the wrench matrix $\mathbf{W}$, and λ is an arbitrary vector of coordinates λ_1 and λ_2. Let Σ be the 2D space of solutions $\left[\lambda_1 \; \lambda_2\right]^{\mathrm{T}}$ obtained from Eq. (5), and Ω the hypercube of feasible cable tensions $\Omega = \left\{\mathbf{t} \mid 0 \leqslant t_{min} \leqslant t_i \leqslant t_{max}, i = 1, \ldots, m\right\}$. Their intersection $\Lambda = \Sigma \cap \Omega$ is a 2D convex polytope of feasible cable tensions sets, obtained from the affine map $\mathscr{A} = (\mathbf{N}, \mathbf{t}_0)$:

$$\mathscr{A}^{-1}(\Lambda) = \left\{\lambda \in \mathbb{R}^2 \mid \mathbf{t}_{min} \leqslant \mathbf{t}_0 + \mathbf{N}\lambda \leqslant \mathbf{t}_{max}\right\} \quad (6)$$

with $\mathbf{t}_0 = \mathbf{W}^{\dagger}\mathbf{f}$. $\mathscr{A}^{-1}(\Lambda)$ is also a 2D convex polytope, named feasible polygon.

The proposed TDA aims to minimize the moving-platform translational displacement $\delta\mathbf{p}$ due to an external wrench $\delta\mathbf{w}_e$ exerted onto the MP, along a given direction. From Eq. (3), if the goal is to maximize the MP stiffness along axis y_b, the SOTDA will have to minimize the following objective function:

$$\min_{\mathbf{t}} |\delta p_y(\mathbf{t})| \quad \text{with} \quad \delta\mathbf{X}(\mathbf{t}) = \mathbf{K}^{-1}(\mathbf{t})\delta\mathbf{w}_e \quad \text{and} \quad \delta\mathbf{p} = \left[\delta p_x \; \delta p_y \; \delta p_z\right]^{\mathrm{T}} \quad (7)$$

where the cable tension vector $\mathbf{t}$ is a function of λ_1 and λ_2, and $\delta\mathbf{w}_e$ is a small wrench in the same direction of the external wrench $\mathbf{w}_e$. The corresponding optimization problem can be solved by different methods, such as a nonlinear least square method.

Algorithm 1: Stiffness Oriented Tension Distribution Algorithm (SOTDA)

Input: Moving-platform pose $\mathbf{X}$, external wrench $\mathbf{w}_e$
Output: Stiffness oriented cable tensions $\mathbf{t}_{SO}$
Compute feasible polygon $\mathscr{A}^{-1}(\Lambda)$ at MP pose $\mathbf{X}$, the MP being subjected to external wrench $\mathbf{w}_e$
for $(\lambda_1,\ \lambda_2) \in \mathscr{A}^{-1}(\Lambda)$ **do**
  Compute corresponding cable tension vector $\mathbf{t} = \mathbf{t}(\lambda_1,\ \lambda_2)$
  Compute cable length vector $\mathbf{l} = \left[l_1\ \ldots\ l_m\right]^{\mathrm{T}}$
  Compute cable elasticity coefficients $k_i = \frac{ES}{l_i}$
  Compute the stiffness matrix $\mathbf{K} = \mathbf{K}_{\mathrm{p}} + \mathbf{K}_{\mathrm{a}}$
  Compute the MP displacement screw $\delta\mathbf{X} = \mathbf{K}^{-1}\delta\mathbf{w}_e$
  Select the second component of $\delta\mathbf{X}$, i.e., the MP displacement along axis y_b, $\delta p_y(\mathbf{t})$
end
Select and apply the set of cable tensions that minimizes δp_y : $\mathbf{t}_{SO} = \min\limits_{\mathbf{t}} |\delta p_y(\mathbf{t}_j)|$

Algorithm 1 describes the proposed SOTDA, $\mathbf{t}_{SO}$ being the resulting cable tension vector. For a real-time implementation, two assumptions are made to reduce the number of sets $(\lambda_1,\ \lambda_2)$ to be evaluated and the computing time: (*i*) the robot has the highest stiffness when at least one cable tension reaches the upper bound t_{max}; (*ii*) a second cable should either reach the lower cable tension limit t_{min} or the upper cable tension limit t_{max}. Under these assumptions, the optimal solution should then be located at the intersection of two vertices, that is at a corner of the feasible polygon. For an 8-cable CDPR with an actuation redundancy equal to two, these assumptions lead to a maximum of 16 sets (λ_1, λ_2) to be evaluated. It is noteworthy that even if the optimal solution is not amongst the set of solutions, the algorithm will consider the best of the computed ones.

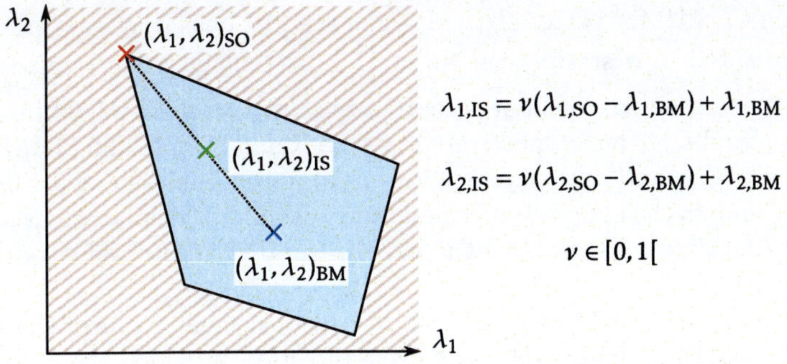

Fig. 2. Intermediate point (IS) inside the feasible polygon

In the sequel, an Intermediate Solution (IS) is introduced by selecting a new point between the barycenter of the feasible polygon and the stiffness oriented solution, based on the value of a coefficient ν bounded between 0 and 1. The implementation of ν defines a security margin with respect to the edges of the feasible polygon as shown in Fig. 2.

4 Experimental Results

Similarly to the implementation of classical Tension Distribution Algorithms (TDAs) [12,15], the SOTDA is located in the control scheme after the sum of the control signal $\boldsymbol{\tau}_c$ and feedforward term $\boldsymbol{\tau}_{da}$, as illustrated in Fig. 3. A proportional-derivative (PD) controller is chosen for its asymptotic stability, tuned according to the Ziegler-Nichols method.

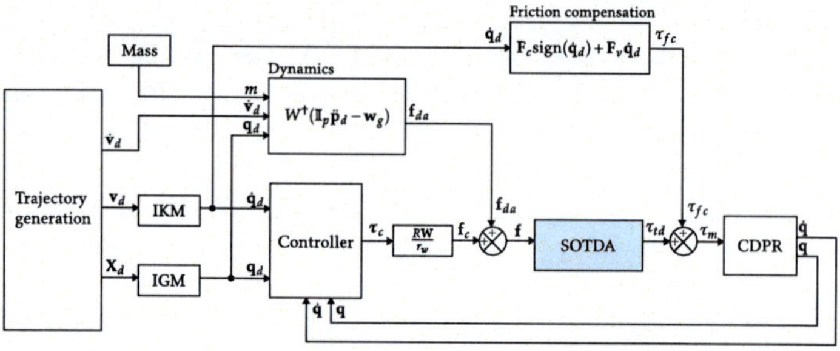

Fig. 3. Control scheme with SOTDA on controller and feedforward block outputs.

The force magnitude due to water jet in the actual cleaning application is about 12.5 N along the y_b axis. In the following, forces of 30 N and 50 N are applied on the MP for better observation of the MP behavior. First, the SOTDA is experimented in a static pose $\mathbf{X}_0 = \left[1.5\ 0.350\ 0.590\ 0\ 0\ 0\right]^{\mathrm{T}}$ to validate the stiffness gain under a constant lateral wrench $\mathbf{w}_e = \left[0\ 50\ 0\ 0\ 0\ 0\right]^{\mathrm{T}}$. Tensions limits t_{min} and t_{max} are respectively set to 10 N and 200 N. The Young modulus is $E = 80$ GPa and the cross-section area of the cables is $S = 0.75$ mm^2. Figure 4 presents the estimation in simulation of the MP displacement δp_y within the entire feasible polygon and the experimental results obtained on the real prototype.

From Eq. (7), the estimation of the MP displacement requires the definition of a small external wrench $\delta \mathbf{w}_e$ along the direction in which the stiffness must be maximized. Here, $\delta \mathbf{w}_e$ amounts to the total external wrench due to water pressure. Using MATLAB *meshgrid* function, the feasible polygon is discretized and the MP displacement along the y axis, δp_y, is computed for all points of the grid. The vertical colorbar presents the value of δp_y in mm within the feasible polygon.

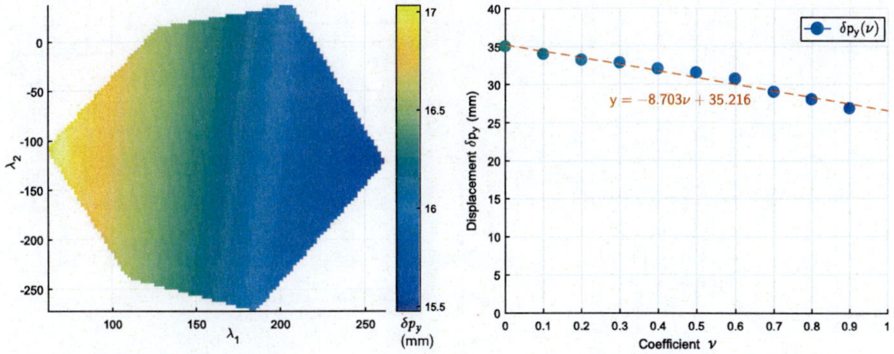

Fig. 4. (Left) Estimated MP displacement δp_y (mm) within the feasible polygon (Right) Measured MP displacement δp_y (mm) as a function of coefficient ν.

From the simulation results, the computed displacements show a difference of 1.5 mm between the lowest and highest values. It is noteworthy that the influence of the solution selection on the MP displacement is small compared to the minimal total error of 15.5 mm. It means that the passive part $\mathbf{K}_p$ of the stiffness matrix $\mathbf{K}$ might far outweighs its active part $\mathbf{K}_a$.

The experimental process in the static case is as follows:

1. The MP is positioned next to a fixed reference bar, as shown in Fig. 1;
2. Using a steel scribe, the resting position of the MP is marked on the reference bar when no external wrench is applied;
3. For each value of $\nu = \{0, 0.1, 0.2, \ldots, 0.9\}$, a lateral wrench of 30 N is applied onto the MP along y_b-axis, using a Kern HDB 10K10N force sensor. Then, the new steady position of the MP is marked on the bar.

The resulting MP displacements as a function of coefficient ν are plotted in Fig. 4. It is apparent that the higher ν, the smaller the MP displacement. The SOTDA leads a reduction of up to 24.8% of the displacement due to the lateral force compared to the solution obtained by selecting the feasible polygon barycenter to tune the cable tensions. With values bounded between 26 and 36 mm, the measured MP displacements are larger than the estimated ones. This difference is due to the modeling simplifications while neglecting the pulleys geometry and the choice of a linear elastic cable model. The simulation results also only consider the elasticity of the cable, while neglecting the compliance of the controller and the elasticity of the frame. Therefore, the real MP displacements are larger than the simulated ones.

In a second experiment, the MP follows the linear path shown in Fig. 5. An external force equal to 30 N is exerted onto the MP along the trajectory. The upper cable tension limit t_{max} is set to 150 N and the trajectory is performed for $\nu = 0$ and $\nu = 0.9$, respectively. The cable tension profiles obtained with the SOTDA for $\nu = 0$ and $\nu = 0.9$, respectively, are plotted in Fig. 5.

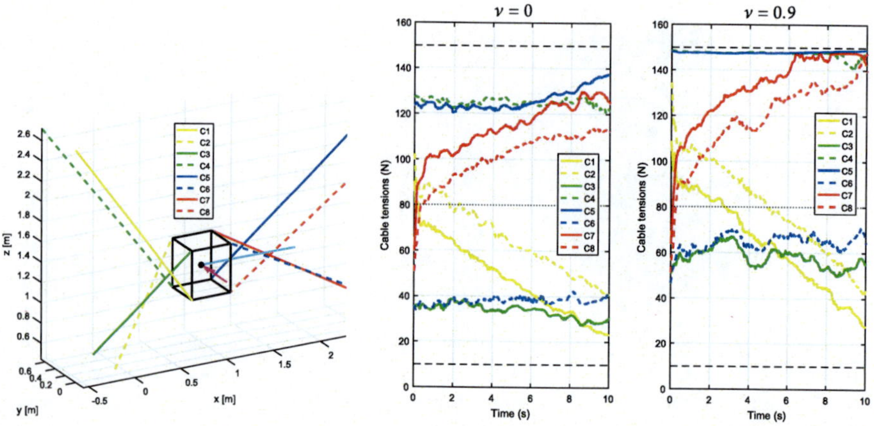

Fig. 5. (Left) Test path (cyan) (Right) Cable tensions profiles, i.e., SOTDA output, for $\nu = 0$ and $\nu = 0.9$, resp.

On the one hand, two cable tensions are close to t_{max} at all time when $\nu = 0.9$. Cables 4 and 5 are the most tensed ones along the trajectory since they oppose the external wrench and support the MP weight. As a result, their tensions are close to t_{max}, while maintaining the MP static equilibrium. On the other hand, cable tensions are around the average admissible cable tension of 80 N for $\nu = 0$. Besides, it is should be noted that cable tensions remain continuous with the SOTDA. The value of ν can then be selected to find a trade-off between the stiffness oriented solution and the one obtained with the barycenter approach.

5 Conclusion

In this paper, a stiffness oriented tension distribution algorithm (SOTDA) was proposed to select the set of cable tensions of a CDPR minimizing the displacement of its moving-platform, the latter being subject to a lateral force along a given direction. The algorithm was studied in simulation and tested experimentally on a fully-constrained CDPR. The experimental results show that the moving-platform displacement can be reduced by up to 24% with a good selection of cable tensions. It is expected that a higher upper tension limit could further improve the benefits of the SOTDA. The introduction of the stiffness coefficient ν for the selection of an intermediate point allows for the implementation of the SOTDA while maintaining a safety margin with respect to the maximum feasible tensions. Future work includes a sensitivity analysis of the SOTDA parameters such as tension limits and cable configuration, a finer identification of the CDPR elastic properties and experiments with a variable pressure water jet.

Acknowledgements. The authors wish to associate the industrial and academic partners of IRT Jules Verne in the framework of the ROCKET project, namely Chantiers de l'Atlantique, Clemessy, B&R Automation, Clarté and CNRS.

References

1. Albus, J., Bostelman, R., Dagalakis, N.: The NIST SPIDER, a robot crane. J. Res. Natl. Inst. Stand. Technol. **97**(3), 373 (1992)
2. Baklouti, S., Courteille, E., Caro, S., Dkhil, M.: Dynamic and oscillatory motions of cable-driven parallel robots based on a nonlinear cable tension model. J. Mech. Robot. **9**(6), 061, 014 (2017)
3. Behzadipour, S.: Stiffness of cable-based parallel manipulators with application to stability analysis. J. Mech. Des. **128**(6), 303–310 (2006)
4. Borgstrom, P.H., Jordan, B.L., Sukhatme, G.S., Batalin, M.A., Kaiser, W.J.: Rapid computation of optimally safe tension distributions for parallel cable-driven robots. IEEE Trans. Robot. **25**(6), 1271–1281 (2009)
5. Bruckmann, T., Pott, A., Hiller, M.: Calculating force distributions for redundantly actuated tendon-based stewart platforms. In: Lennarčič, J., Roth, B. (eds.) Advances in Robot Kinematics, pp. 403–412. Springer, Dordrecht (2006)
6. Fortin-Coté, A., Cardou, P., Gosselin, C.: An admittance control scheme for haptic interfaces based on cable-driven parallel mechanisms. In: Proceedings of the 2014 IEEE International Conference on Robotics and Automation (ICRA), Hong Kong, pp. 819–925 (2014)
7. Gagliardini, L., Caro, S., Gouttefarde, M., Girin, A.: Discrete reconfiguration planning for cable-driven parallel robots. Mech. Mach. Theory **100**, 313–337 (2016)
8. Gagliardini, L., Gouttefarde, M., Caro, S.: Determination of a dynamic feasible workspace for cable-driven parallel robots. In: Advances in Robot Kinematics 2016, vol. 4, 361–370 (2018)
9. Hassan, M., Khajepour, A.: Minimum-norm solution for the actuator forces in cable-based parallel manipulators based on convex optimization. In: Proceedings of the 2007 IEEE International Conference on Robotics and Automation, pp. 1498–1503 (2007)
10. Hassan, M., Khajepour, A.: Analysis of bounded cable tensions in cable-actuated parallel manipulators. IEEE Trans. Robot. **27**(5), 891–900 (2011)
11. Kawamura, S., Kino, H., Won, C., Kamawura, S., Kino, H., Won, C.: High-speed manipulation by using parallel wire-driven robots. Robotica **18**(1), 13–21 (2000)
12. Lamaury, J.: Contribution a la commande des robots paralleles a cables a redondance d'actionnement (2013)
13. Lamaury, J., Gouttefarde, M.: A tension distribution method with improved computational efficiency, pp. 71–85. Springer, Heidelberg (2013)
14. Merlet, J.P., Daney, D.: A portable, modular parallel wire crane for rescue operations. In: Proceedings of the 2010 IEEE International Conference on Robotics and Automation (ICRA), pp. 2834–2839 (2010)
15. Picard, E., Caro, S., Claveau, F., Plestan, F.: Pulleys and force sensors influence on payload estimation of cable-driven parallel robots. In: 2018 IEEE/RSJ International Conference on Intelligent Robots and Systems (IROS), pp. 1429–1436 (2018)
16. Pott, A.: An improved force distribution algorithm for over-constrained cable-driven parallel robots. In: Thomas, F., Perez Gracia, A. (eds.) Computational Kinematics, pp. 139–146. Springer, Dordrecht (2014)
17. Verhoeven, R.: Analysis of the workspace of tendon based Stewart platforms. Ph.D. thesis, Verlag nicht ermittelbar (2004)
18. Yao, R., Tang, X., Wang, J., Huang, P.: Dimensional optimization design of the four-cable-driven parallel manipulator in fast. IEEE/ASME Trans. Mechatron. **15**(6), 932–941 (2010)

A Forward Kinematic Code for Cable-Driven Parallel Robots Considering Cable Sagging and Pulleys

Marc Fabritius[1(✉)] and Andreas Pott[2]

[1] Fraunhofer Institute for Manufacturing Engineering and Automation IPA, Stuttgart, Germany
marc.fabritius@ipa.fraunhofer.de

[2] University of Stuttgart, Stuttgart, Germany
andreas.pott@isw.uni-stuttgart.de

Abstract. The large workspace of cable-driven parallel robots is one of their key benefits. Experiments have shown that in practice the reachable workspace is often smaller than the theoretically predicted one. To improve the workspace computation, a new forward kinematic code, which considers the previously neglected effects of cable sagging and pulleys, is introduced. For an exemplary robot with eight cables, the new forward kinematic predicts a 18.5% smaller workspace and a 27.8% lower platform stiffness than the standard geometric model.

The findings of this work show the importance of considering the effects of cable sagging and pulleys in the workspace computation and kinematic codes, especially for large cable-driven parallel robots.

Keywords: Cable-driven parallel robots · Forward kinematic code · Workspace analysis · Cable sagging · Pulley model

1 Introduction

The workspace of a cable-driven parallel robot (CDPR) is one of its most important properties, since its large size distinguishes it from other types of robots. Therefore, it is vital for the successful application of CDPR to both accurately compute their workspace and use kinematic codes which can exploit a large workspace in practice. Most previous works on the wrench-feasible workspace of CDPR, only consider the standard geometric model of CDPR, which is defined in [10, Section 3.2].

This model neglects the effects of cable sagging and pulleys for the benefit of simpler computations. Kraus [4] analyzes a CDPR running a kinematic code based on this model and finds that on line segments between the center of the workspace and its theoretically computed border, there is a critical point at roughly 80% of the distance, after which the cable tension limits are violated and the positioning accuracy of the platform decreases significantly. Kraus infers

J. Lenarčič and B. Siciliano (Eds.): ARK 2020, SPAR 15, pp. 218–225, 2021.
https://doi.org/10.1007/978-3-030-50975-0_27

that this is caused by the sagging of cables, which is not modeled in the kinematic code he uses [4, p. 91]. Riehl et al. [11] also conclude that the effects of cable sagging and pulleys should be modeled for large CDPR in order to achieve a good positioning accuracy.

Therefore, it is reasonable to expect that accounting for these effects in the workspace analysis and kinematic code will substantially improve both the theoretical estimate of the workspace boundary and the accuracy and cable tensions of CDPR in practice. In the field of CDPR, there are some works considering either a catenary cable model [2,6,7] or a pulley model [8,12]. To the best of the authors knowledge, only one other work combines these models. Gouttefarde et al. [1] do this, but are not considering the section of the cable which is wound on the pulley.

In this work, this section of the cable is accounted for and its unstretched length is denoted as l_0^{I}. Based on this catenary-pulley model, a new forward kinematic code is presented which provides substantially different results than kinematic codes for the standard geometric model.

The following work is structured as follows: First, the necessary definitions and assumptions for the modeling of CDPR are laid out in Sect. 2. Then, the equations of the catenary-pulley model are derived in Sect. 3. Section 4 presents a forward kinematic code for the catenary-pulley model. The definition of the wrench-feasible workspace for the catenary-pulley model is given in Sect. 5. In Sect. 6, this workspace and the forward kinematic code are evaluated on a fully-constrained CDPR and compared to the standard geometric model. Finally, the conclusion and outlook are given in Sect. 7.

2 Definitions and Assumptions

A CDPR consists of $m \in \mathbb{N}$ cables, which connect a mobile platform via pulleys to fixed winches. The pose of the platform $(\mathbf{r}, \mathbf{R}) \in \mathrm{SE}_3$, i.e. its position $\mathbf{r} \in \mathbb{R}^3$ and orientation $\mathbf{R} \in \mathrm{SO}_3$, can be manipulated through suitable changes of the unstretched cable lengths $\mathbf{l}_0 \in \mathbb{R}^m$. This pose $(\mathbf{r}, \mathbf{R})$ also characterizes the transformation from the world coordinate frame $\mathscr{K}_0$ to the platform frame $\mathscr{K}_p$.

The cables are modeled with the catenary-pulley model, which accounts for their elasticity, sagging, and wrapping around pulleys. Figure 1 illustrates all parameters and vectors associated with the modeling of each cable $i \in \{1, ..., m\}$. The point at which the pulley touches its panning axis $\mathbf{e}_i$ is called the proximal anchor point $\mathbf{a}_i \in \mathbb{R}^3$. The other end of the cable, which is connected to the mobile platform, is called distal anchor point $\mathbf{b}_i \in \mathbb{R}^3$. Its location is expressed as $\mathbf{r} + \mathbf{R}\mathbf{b}_i$ in the world coordinate frame $\mathscr{K}_0$.

For better readability, the index i is omitted from all symbols which describe the catenary-pulley model of cable i. In Fig. 1, these symbols and the force vectors acting on the cable are highlighted in blue.

The catenary-pulley model divides the cable into two sections. The first section of the cable is wound around a pulley of radius r from $\mathbf{a}_i$ up to the point $\mathbf{c}$, where the cable tangentially leaves the pulley. It is assumed that there

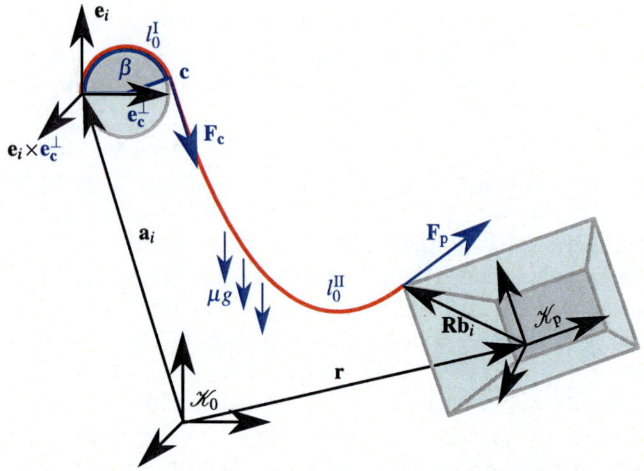

Fig. 1. Illustration of the catenary-pulley model for cable i

is no friction in the panning axis $\mathbf{e}_i$ of the pulley and between the cable and the pulley. Furthermore, it is assumed that the pulley is statically balanced around its panning axis $\mathbf{e}_i$.

The second section of the cable is suspended between $\mathbf{c}$ and $\mathbf{r} + \mathbf{Rb}_i$. It is modeled using Irvine's catenary model [3] with the cable weight force μg, the Young's modulus E, and the cable's cross-sectional area A. The unstretched length of the cable $l_0 = l_0^{\mathrm{I}} + l_0^{\mathrm{II}}$ is partitioned into the parts contained in the first l_0^{I} and second l_0^{II} cable section. This partition depends on the force $\mathbf{F}_{\mathbf{c}}$ in the cable at the boundary of the two cable sections $\mathbf{c}$.

3 Derivation of the Catenary-Pulley Model

The catenary-pulley model expresses the position of the distal anchor point $\mathbf{d}_i(\mathbf{F}_{\mathbf{c}}, l_0)$ of a cable i in $\mathcal{K}_0$ as a function of its force $\mathbf{F}_{\mathbf{c}}$ and unstretched cable length l_0.

The vector $\mathbf{F}_{\mathbf{c}}$ defines the direction in which the cable leaves the pulley at the point $\mathbf{c}$, and its norm is the force in the cable at this point. Furthermore, it is assumed that the pulley is statically balanced around its panning axis $\mathbf{e}_i$. Therefore the pulley and $\mathbf{F}_{\mathbf{c}}$ lie in the same plane. For further calculations, a local coordinate system in this plane is defined. Its first axis is the panning axis of the pulley $\mathbf{e}_i$ and its second axis $\mathbf{e}[^{\perp}_{\mathbf{c}}]$ points in the direction of the cable $\mathbf{e}_{\mathbf{c}}^{\perp} = \frac{\mathbf{F}_{\mathbf{c}} - \langle \mathbf{F}_{\mathbf{c}}, \mathbf{e}_i \rangle \mathbf{e}_i}{\| \mathbf{F}_{\mathbf{c}} - \langle \mathbf{F}_{\mathbf{c}}, \mathbf{e}_i \rangle \mathbf{e}_i \|}$. The point $\mathbf{c}$ on the pulley of radius r can be expressed in $\mathcal{K}_0$ using these axes

$$\mathbf{c} = \mathbf{a}_i + r\mathbf{e}_{\mathbf{c}}^{\perp} + \frac{r}{\|\mathbf{F}_{\mathbf{c}}\|} \left(\left\langle \mathbf{F}_{\mathbf{c}}, \mathbf{e}_{\mathbf{c}}^{\perp} \right\rangle \mathbf{e}_i - \langle \mathbf{F}_{\mathbf{c}}, \mathbf{e}_i \rangle \mathbf{e}_{\mathbf{c}}^{\perp} \right) . \tag{1}$$

The wrapping angle of the cable around the pulley is given by

$$\beta = \operatorname{atan2}\left(\langle \mathbf{F_c}, \mathbf{e_c^\perp}\rangle, \langle \mathbf{F_c}, \mathbf{e}_i\rangle\right) . \tag{2}$$

The unstretched cable lengths of the first l_0^{I} and second l_0^{II} section are calculated as

$$l_0^{\mathrm{I}} = \frac{\beta r}{1 + \frac{\|\mathbf{F_c}\|}{EA}} \quad \text{and} \quad l_0^{\mathrm{II}} = l_0 - l_0^{\mathrm{I}} . \tag{3}$$

The length l_0^{II} determines the weight of the catenary and therefore influences the cable force vector $\mathbf{F_p}$ at the distal anchor point of the cable on the platform

$$\mathbf{F_p} = \mathbf{F_c} + \mu g l_0^{\mathrm{II}} \mathbf{e}_z . \tag{4}$$

The horizontal components of the cable force are isolated in the vector $\mathbf{F}_{x,y} = \left[\langle \mathbf{F_c}, \mathbf{e}_x\rangle \ \langle \mathbf{F_c}, \mathbf{e}_y\rangle \ 0\right]^{\mathsf{T}}$. Hereby the vectors $\mathbf{e}_x$, $\mathbf{e}_y$ and $\mathbf{e}_z$ denote the axes of $\mathscr{K}_0$.

Combining previous calculations with Irvine's catenary model [3], the location of the distal anchor point $\mathbf{d}_i(\mathbf{F_c}, l_0)$ can be calculated in $\mathscr{K}_0$

$$\begin{aligned} \mathbf{d}_i(\mathbf{F_c}, l_0) = &\, \mathbf{c} + \frac{l_0^{\mathrm{II}}}{2EA}(\mathbf{F_p} + \mathbf{F_c}) \\ &+ \frac{1}{\mu g}\left(\log\left(\frac{\|\mathbf{F_p}\| + \langle \mathbf{F_p}, \mathbf{e}_z\rangle}{\|\mathbf{F_c}\| + \langle \mathbf{F_c}, \mathbf{e}_z\rangle}\right)\mathbf{F}_{x,y} + \left(\|\mathbf{F_p}\| - \|\mathbf{F_c}\|\right)\mathbf{e}_z\right) . \end{aligned} \tag{5}$$

With this formula, the catenary-pulley model equation for the i-th cable can be formulated as a nonlinear equation system in $\mathbf{F_c} \in \mathbb{R}^3$ and $l_0 \in \mathbb{R}$

$$\mathbf{d}_i(\mathbf{F_c}, l_0) - (\mathbf{r} + \mathbf{R}\mathbf{b}_i) = \mathbf{0} \in \mathbb{R}^3 . \tag{6}$$

For a given pose $(\mathbf{r}, \mathbf{R})$ and unstretched cable length l_0, this can be interpreted as a nonlinear least-square problem in $\mathbf{F_c}$. To solve this and other problems, the iterative Levenberg-Marquardt scheme [5] is used throughout this work since it has been proven to work robustly on a wide range of nonlinear problems.

4 A Forward Kinematic Code for the Catenary-Pulley Model

A forward kinematic code calculates a fitting platform pose $(\mathbf{r}, \mathbf{R})$ for given unstretched cable lengths $\mathbf{l}_0$. The kinematic code presented in the following, uses a Levenberg-Marquardt iteration scheme to minimize the scaled residual wrench $\mathbf{S}\mathbf{w}_{\mathrm{res}} \in \mathbb{R}^6$, which expresses the discrepancy between the wrenches applied to the platform by the cables, its mass, and external influences. Hereby $\mathbf{S}$ is a diagonal scaling matrix with the entries $\left[1\ 1\ 1\ \|\mathbf{b}_{\max}\|^{-1}\ \|\mathbf{b}_{\max}\|^{-1}\ \|\mathbf{b}_{\max}\|^{-1}\right]$, where $\|\mathbf{b}_{\max}\|$ is the largest distance between any of the distal anchor points $\mathbf{b}$ and their geometric center. This scaling of the moment components is done to ensure their comparability with the force components of the residual wrench. In each step k of the iteration scheme, the residual wrench is calculated as

$$\mathbf{w}_{\text{res}} = \mathbf{w}_{\text{p}} + \mathbf{w}_{\text{ext}} + \sum_{i=1}^{m} \mathbf{w}_i \ , \tag{7}$$

where $\mathbf{w}_{\text{p}}$ is the wrench due to the platform's mass m_{p} and gravity g.

Assuming, without loss of generality, that its center of mass is at the origin of the platform coordinate system $\mathcal{K}_p$, this wrench is the same for all iteration steps k and takes the form $\mathbf{w}_{\text{p}} = \begin{bmatrix} 0 \ 0 \ -m_{\text{p}}g \ 0 \ 0 \ 0 \end{bmatrix}^{\mathsf{T}}$. The external wrench $\mathbf{w}_{\text{ext}}$ is constant for all iteration steps k. However, each wrench $\mathbf{w}_i$ that is exerted on the platform by a cable i depends on the platform pose $(\mathbf{r}_k, \mathbf{R}_k)$ of the iteration step k. It is calculated using a nested Levenberg-Marquardt scheme which solves the catenary-pulley model from Eq. (6) for $\mathbf{F_c}$ with the input parameters $\mathbf{r}_k, \mathbf{R}_k, l_0$, where l_0 is the unstretched length of the corresponding cable i. The cable force at the platform $\mathbf{F}_{\text{p}}$ is obtained from Eq. (4) and used to calculate the wrench $\mathbf{w}_i$ of cable i on the platform

$$\mathbf{w}_i = -\begin{bmatrix} \mathbf{F}_{\text{p}} \\ \mathbf{R}_k\mathbf{b} \times \mathbf{F}_{\text{p}} \end{bmatrix} \ . \tag{8}$$

Once the Levenberg-Marquardt scheme finds a platform pose $(\mathbf{r}_k, \mathbf{R}_k)$ for which the residual wrench $\mathbf{w}_{\text{res}}$ from Eq. (7) vanishes, this pose together with its calculated cable forces, is considered the result of the forward kinematic code.

If the the iteration terminates without finding such a pose, the forward kinematic code has failed.

5 Workspace Definition for the Catenary-Pulley Model

The wrench-feasible workspace of the catenary-pulley model is the set of all platform poses $(\mathbf{r}, \mathbf{R}) \in \mathbf{SE}_3$ for which for any external wrench $\mathbf{w}_{\text{ext}} \in \mathrm{Q}$ in the wrench set $\mathrm{Q} \subseteq \mathbb{R}^6$, there exists a set of unstretched cable lengths $\mathbf{l}_0 \in \mathbb{R}^m$ such that the forward kinematic code from Sect. 4 returns a result with the following property: For each cable $i \in \{1, ..., m\}$, the norm of its cable force at the pulley $\mathbf{F_c}$ must be within the feasible interval $\|\mathbf{F_c}\| \in [f_{\text{min}}, f_{\text{max}}]$.

This condition is placed on the cable force at the pulley $\mathbf{F_c}$, since in practice the maximal cable force f_{max} of a CDPR is usually constrained by its winches.

6 Workspace Comparison

In this section, the previously introduced workspace of the catenary-pulley model is compared to the wrench-feasible workspace of the standard geometric model as defined in [10, see p. 167] for a redundantly-constrained CDPR with eight cables.

For this exemplary comparison, the set of external wrenches Q is chosen to only contain the zero wrench $Q = \{\mathbf{0}\}$. To simplify the visualization, only the translational workspace with the constant platform orientation $\mathbf{R} = \mathbf{I}$ is investigated.

For the standard geometric model, the workspace is computed using the improved closed-form force distribution algorithm from [9]. To compute the workspace of the catenary-pulley model, the workspace criterion from Sect. 5 is evaluated with one set of unstretched cable lengths $\mathbf{l}_0$ for each pose $(\mathbf{r}, \mathbf{R})$. These cable lengths are obtained by combining the force distribution from the standard geometric model with the pulley geometry calculations from [8]. The iterative forward kinematic code from Sect. 4 is started with the platform pose which was used to calculate the force distribution $(\mathbf{r}_0, \mathbf{R}_0) = (\mathbf{r}, \mathbf{R})$.

For the workspace comparison, a CDPR with $m = 8$ cables is considered. Its geometry parameters $\mathbf{a}_i$ and $\mathbf{b}$ are defined as all sign combinations of the vector sets $\mathbf{a}_i = \left[\pm 5.0\,\text{m}\ \pm 6.0\,\text{m}\ \pm 2.5\,\text{m}\right]^{\mathrm{T}}$ and $\mathbf{b}_\mathrm{i} = \left[\pm 0.5\,\text{m}\ \pm 0.3\,\text{m}\ \mp 0.4\,\text{m}\right]^{\mathrm{T}}$. The parameters of the catenary-pulley model for all cables are chosen to represent steel cables: $\mu g = 6\,\frac{\text{N}}{\text{m}}$, $EA = 7 \cdot 10^6\,\text{N}$, $r = 0.25\,\text{m}$. The panning axes of the pulleys are given by $e_i = \left[1\ 0\ 0\right]^{\mathrm{T}}$ for $i \in \{1, 2, 5, 6\}$ and $\mathbf{e}_i = \left[-1\ 0\ 0\right]^{\mathrm{T}}$ for $i \in \{3, 4, 7, 8\}$. The limits of the feasible cable forces are set to be $f_{\text{min}} = 1000\,\text{N}$ and $f_{\text{max}} = 10000\,\text{N}$. The gravitational force on the platform is $m_\text{p} g = 8000\,\text{N}$.

Figure 2 shows the comparison of wrench-feasible workspaces for the catenary-pulley and the standard geometric model through sliced views in the YX- and YZ-planes. The workspace of the catenary-pulley model (shown in green) has a volume of $131.6\,\text{m}^3$. It is contained in the larger workspace of the standard geometric model (shown in green and yellow), which has a volume of $161.6\,\text{m}^3$.

This shows that workspace calculations based on the standard geometric model can overestimate the actual workspace of a CDPR, which is more accurately modeled by the catenary-pulley model.

The stiffness of the platform calculated with the catenary-pulley model is also significantly lower than with the standard geometric model. For the catenary-pulley model, the homogenized stiffness matrix is the derivative of the scaled residual wrench $\mathbf{S}\mathbf{w}_{\text{res}}$ from Sect. 4 with respect to the platform's pose $(\mathbf{r}, \mathbf{R})$. Hereby the orientation $\mathbf{R}$ is parametrized through euler angles which are scaled with the factor $\|\mathbf{b}_{\text{max}}\|^{-1}$. At the central platform pose $(\mathbf{0}, \mathbf{I})$, the smallest singular value of the stiffness matrix, which is a lower bound for the overall stiffness, decreases by 27.8% when the effects of cable sagging and pulleys are considered.

The computation time of the workspace criterion for the catenary-pulley model is on average $4.0 \cdot 10^{-3}\,\text{s}$ per platform pose. The equivalent time for the standard geometric model is $7.3 \cdot 10^{-6}\,\text{s}$. This substantial difference in computation time stems from the nonlinearity of the catenary-pulley model, which is solved using two nested Levenberg-Marquardt schemes. All computations are implemented in C++ and measured on a laptop with an Intel® Core™ i5-7440HQ CPU running at 2.80 GHz.

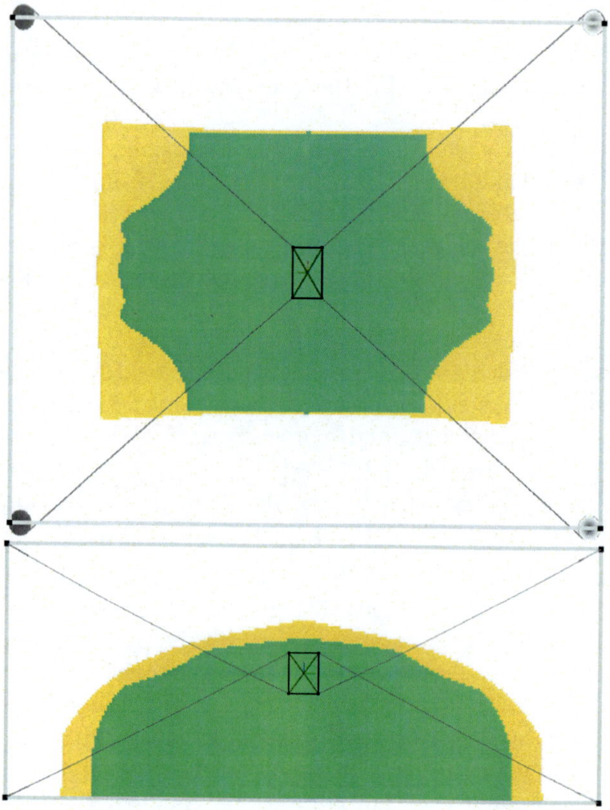

Fig. 2. Comparison of wrench-feasible workspaces

7 Conclusion and Outlook

In this work, a new forward kinematic code, which considers the effects of cable sagging and pulleys, is developed. The resulting wrench-feasible workspace is significantly smaller than the workspace obtained from the standard geometric model. The stiffness of the cable robot platform predicted by the new forward kinematic code, is also much lower than the stiffness obtained from the standard geometric model. These findings show the necessity of considering the effects of cable sagging and pulleys in the workspace calculation, especially for large CDPR.

Future research should validate the workspace predictions of the new forward kinematic code and also develop corresponding inverse kinematic codes to increase the workspace and improve the accuracy of CDPR in practice.

Acknowledgements. This work was supported by the European Union's H2020 Program (H2020/2014–2020) under the grant agreement No. 732513 (HEPHAESTUS project).

References

1. Gouttefarde, M., Nguyen, D.Q., Baradat, C.: Kinetostatic analysis of cable-driven parallel robots with consideration of sagging and pulleys. In: Advances in Robot Kinematics (ARK), pp. 213–221 (2014)
2. Hui, L.I.: A giant sagging-cable-driven parallel robot of fast telescope: its tension-feasible workspace of orientation and orientation planning. In: Proceedings of the 14th IFToMM World Congress, Taipei, pp. 373–381 (2015)
3. Irvine, H.: Cable Structures. MIT Press, Cambridge (1981)
4. Kraus, W.: Force control of cable-driven parallel robots. Ph.D. thesis, University of Stuttgart, Germany (2015)
5. Marquardt, D.W.: An algorithm for least-squares estimation of nonlinear parameters. J. Soc. Ind. Appl. Math. **11**(2), 431–441 (1963)
6. Merlet, J.P.: A new generic approach for the inverse kinematics of cable-driven parallel robot with 6 deformable cables. In: Lenarčič, J., Merlet, J.P. (eds.) Advances in Robotik Kinematics, pp. 209–216 (2016)
7. Merlet, J.P.: Some properties of the Irvine cable model and their use for the kinematic analysis of cable-driven parallel robots. In: Corves, B., Wenger, P., Hüsing, M. (eds.) Mechanisms and Machine Science, EuCoMeS 2018, vol. 59, pp. 409–416. Springer, Cham (2019)
8. Pott, A.: Influence of pulley kinematics on cable-driven parallel robots. In: Advances in Robot Kinematics (ARK), pp. 197–204 (2012)
9. Pott, A.: An improved force distribution algorithm for over-constrained cable-driven parallel robots. In: Computational Kinematics, vol. 15, pp. 139–146. Springer (2014)
10. Pott, A.: Cable-Driven Parallel Robots: Theory and Application. Springer Tracts in Advanced Robotics, vol. 120. Springer, Heidelberg (2018)
11. Riehl, N., Gouttefarde, M., Krut, S., Baradat, C., Pierrot, F.: Effects of non-negligible cable mass on the static behavior of large workspace cable-driven parallel mechanisms. In: 2009 IEEE International Conference on Robotics and Automation, pp. 2193–2198. IEEE (2009)
12. Schmidt, V.L.: Modeling techniques and reliable real-time implementation of kinematics for cable-driven parallel robots using polymer fiber cables. Dissertation, University of Stuttgart, Germany (2017). https://doi.org/10.18419/opus-9085

Singularity and Workspace Analysis of 3-SPS-U and 4-SPS-U Tensegrity Mechanisms

Swaminath Venkateswaran[1(✉)] and Damien Chablat[2]

[1] École Centrale de Nantes, Laboratoire des Sciences du Numérique de Nantes (LS2N), UMR CNRS 6004, 44300 Nantes, France
swaminath.venkateswaran@ls2n.fr

[2] Centre National de la Recherche Scientifique (CNRS), Laboratoire des Sciences du Numérique de Nantes (LS2N), UMR CNRS 6004, 44300 Nantes, France
damien.chablat@cnrs.fr

Abstract. This article analyzes the singularities and workspace of two tensegrity mechanisms that employ a passive universal joint and either three or four tension springs. These two architectures are correlated to 3-SPS-U and 4-SPS-U parallel mechanisms for determining their geometric equations. By fixing the limits of prismatic joints, the workspace for the mechanisms is generated and the parallel singularities are analyzed. Based on the singularity boundaries obtained from the workspace, the joint limits are modified to generate the maximal singularity free workspaces for both the architectures. A comparison is done based on the tilt limits obtained from the workspace of the mechanisms. The mechanism with the maximum tilt limits is implemented for a piping inspection robot to pass through pipe bends and junctions.

Keywords: Tensegrity mechanism · Singularities · Workspace analysis · 3-SPS-U, 4-SPS-U, Piping inspection robot

1 Introduction

A bio-inspired piping inspection robot was designed and developed at LS2N, France. This robot accomplishes the locomotion of a caterpillar in six steps to move inside a pipeline [1]. However, the prototype is a rigid model which restricts its application to straight pipelines. By the addition of articulation units, the robot can be made reconfigurable. Some interesting researches on piping inspection robots that pass through pipe bends include i) the robot of Chen et al. [2] where double hook joint is employed, ii) THES-I robot of Hirose et al. [3] where a universal joint actuated by DC-motor is used and iii) robot of Brunete et al. [4] that uses SMA spring with microcontrollers. Most of these researches have articulation units either in passive mode or active mode but not a combination of both. By analyzing key design issues namely passive compliance, active compliance and tilt limits, a tensegrity mechanism that uses a passive universal joint and three tension springs was proposed as a solution for the bio-inspired robot of LS2N [5]. This mechanism can work passively when the robot passes through

J. Lenarčič and B. Siciliano (Eds.): ARK 2020, SPAR 15, pp. 226–233, 2021.
https://doi.org/10.1007/978-3-030-50975-0_28

a pipe bend at $\pi/2$ radians. In the event of a T-union or junction, cable actuation can be performed on the tensegrity mechanism to follow a given path. In this article, the singularities and workspace analysis of the tensegrity mechanism is carried out for the Euler angles of the universal joint [6]. This analysis can help us to determine if the mechanism can work within the singularity boundaries under passive modes. Two types of tensegrity mechanisms are studied and compared where one configuration uses three tensions springs and the other uses four tensions springs. By using the SIROPA library in Maple, the workspace is generated by setting joint limits and parallel singularities. By analyzing these boundaries, the joint limits are modified to obtain singularity free workspaces. By comparing the tilt limits of both the architectures, the mechanism with maximum tilt range is implemented for the bio-inspired robot to overcome the issue of passive compliance.

The outline of the article is as follows. In the following section, the architectures and the geometric constraints of both the mechanisms are presented. The subsequent section deals with the derivation of singularities and joint limits equations. Followed by that, the workspace of the mechanisms are studied and a comparison is being made based on the tilt limits. The article then ends with conclusions.

2 Architecture of the Mechanisms

The tensegrity mechanisms under study uses a passive universal joint and three tensions springs for one configuration and four springs in the other. These mechanisms can be correlated to parallel manipulators of type 3-SPS-U [7] and 4-SPS-U [8]. Here, S indicates the spherical joints, U represents the universal joint and P represents the actuated prismatic joints. The architectures are represented in Fig. 1 in the home pose where the tilt angles α & β are equal to 0 radians.

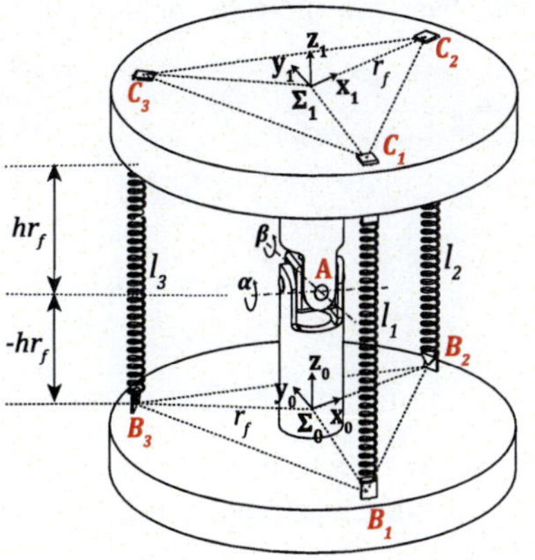

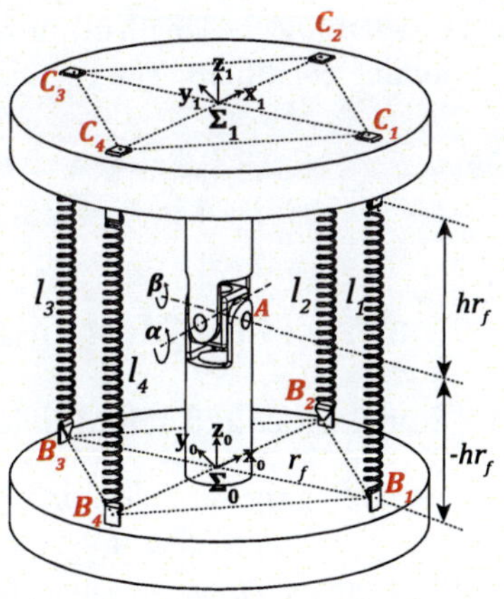

Fig. 1. The 3-SPS-U(left) and 4-SPS-U(right) tensegrity mechanisms at their home poses

The fixed co-ordinate frame of the base is represented by Σ_0 with the origin at B_0. The spring mounting points on the fixed base are represented by B_i $(i = 1,2,3)$ and they form the imaginary equilateral triangle of the manipulator base whose median is r_f for the 3-S$\underline{P}$S-U mechanism. The base mouting points B_i $(i = 1,2,3,4)$ forms an imaginary square for the 4-S$\underline{P}$S-U mechanism. The diagonal length for the square is $2r_f$. The vector co-ordinates for the base mounting points are given by

$$\mathbf{b}_i = \left[r_f \cos(\tfrac{i2\pi}{j}), r_f \sin(\tfrac{i2\pi}{j}), -r_f h\right]^T \tag{1}$$

where for $j = 3$, $i = 0$ to 2 for 3-S$\underline{P}$S-U & for $j = 4$, $i = 0$ to 3 for 4-S$\underline{P}$S-U

In Eq. 1, h represents a constant and a value of 1 is taken which corresponds to the inverse pendulum configuration [5]. The moving co-ordinate frame of the end-effector is represented by Σ_1 with its origin at C_0. The spring mounting points of the end-effector is represented by C_i $(i = 1,2,3)$ and C_i $(i = 1,2,3,4)$ for the 3-S$\underline{P}$S-U and 4-S$\underline{P}$S-U mechanisms. For estimating the vector co-ordinates of the end-effector mounting points, the XY Euler angle matrix is employed with respect to the central point A of the universal joint. The vector co-ordinates for the end-effector mounting points are given by

$$\mathbf{R} = \mathbf{R}_x(\alpha)\mathbf{R}_y(\beta) = \begin{bmatrix} c_\beta & 0 & s_\beta \\ s_\alpha s_\beta & c_\alpha & -s_\alpha c_\beta \\ -c_\alpha s_\beta & s_\alpha & c_\alpha c_\beta \end{bmatrix} \tag{2}$$

$$\mathbf{c}_i = \mathbf{R}\left[r_f \cos(\tfrac{i2\pi}{j}), r_f \sin(\tfrac{i2\pi}{j}), r_f h\right]^T \tag{3}$$

where for $j = 3$, $i = 0$ to 2 for 3-S$\underline{P}$S-U & for $j = 4$, $i = 0$ to 3 for 4-S$\underline{P}$S-U

In Eq. 2, $\mathbf{R} \in SE(3)$ represents the spatial transformation matrix obtained from the Euler angles of universal joint and is used to identify the end-effector co-ordinates as indicated in Eq. 3. The inverse geometric model for the mechanism is simpler and it is used to determine the length of springs between the base and end-effector at home-pose and working conditions. The equation is given by

$$l_i = \sqrt{(b_{ix} - c_{ix})^2 + (b_{iy} - c_{iy})^2 + (b_{iz} - c_{iz})^2} \tag{4}$$

with $i = 1$ to 3 for 3-S$\underline{P}$S-U & $i = 1$ to 4 for 4-S$\underline{P}$S-U

3 Analysis of the Mechanisms

3.1 Singularities and Joint Limits Equations

For determining the feasible workspace of the tensegrity mechanism, it is necessary to determine the singularities of the mechanism. For a parallel manipulator, the singularity equation is given by the well-known equation [9]

$$\mathbf{A}\mathbf{t} + \mathbf{B}\dot{\rho} = 0 \tag{5}$$

where $\mathbf{t}$ represents the angular velocity vector
and $\dot{\rho} = [\dot{l}_1, \dot{l}_2, \dot{l}_3]^T$ represents the joint velocity vector

Parallel singularities occur when the determinant of the direct kinematics matrix **A** of Eq. 5 vanishes. This occurs when the end-effector platform aligns itself with one of the springs of the mechanism. There exist no serial singularities for the mechanism as the determinant of the inverse kinematics matrix **B** does not vanish since the prismatic springs cannot be equal to 0 mm. The parallel singularities are analyzed by sub-dividing the manipulator into three and four sets of 2-SPS-U architecture for the 3-SPS-U and 4-SPS-U mechanisms. By using the Groebner base elimination technique, we can generate the joint limits equations with the help of the SIROPA library [10] in Maple. These equations can be generated using the *ConstraintEquations* syntax of the SIROPA library. The joint limit equations for the 3-SPS-U mechanism are given by

$$\mathscr{C}_{1+3i} : 2r_f^2 c_\alpha(c_\beta - s_\beta) - 2r_f^2(s_\beta + c_\beta) + 4r_f^2 - l_j^2 = 0 \tag{6}$$

$$\begin{aligned}\mathscr{C}_{2+(k-1)+3i} : -m(\left(c_\beta + \tfrac{s_\beta}{2} + 1\right) s_\alpha rf^2\sqrt{3} + 2\left(c_\beta + \tfrac{s_\beta}{2} - \tfrac{3}{4}\right) rf^2 c_\alpha + \\ r_f^2(s_\beta - \tfrac{c_\beta}{2} + 4) - l_j{}^2) = 0\end{aligned} \tag{7}$$

for $i = [0,1], \quad l_j = [l_{min}, l_{max}]$ & for $k = [1,2], \quad m = [-1,1]$ in Eq. 7

Similarly, the joint limit equations are generated for the 4-SPS-U mechanism and they are given by

$$\mathscr{C}_{1+2k+4i} : 2r_f^2 c_\beta(c_\alpha - 1) + m2r_f^2 s_\beta(c_\alpha + 1) + 4r_f^2 - l_j^2 = 0 \tag{8}$$

$$\mathscr{C}_{2+2k+4i} : 2r_f^2 c_\alpha(c_\beta - 1) - m2r_f^2 s_\alpha(c_\beta + 1) + 4r_f^2 - l_j^2 = 0 \tag{9}$$

for $i = [0,1], \quad l_j = [l_{min}, l_{max}]$ & for $k = [0,1], \quad m = [-1,1]$ in Eqs. 8 and 9

The singularity equations are generated for both the mechanisms to determine their workspace limits. The equations can be generated using *ParallelSingularities* syntax of the SIROPA library in Maple. Equations 10 to 11 and Eqs. 12 to 13 represents the singularity equations for the 3-SPS-U and 4-SPS-U mechanisms. These joint limits and singularity equations are employed to determine the singularity free workspace for the mechanisms.

By using the Cylindrical Algebraic Decomposition (CAD) technique, the real solutions are generated for the problem [11] by using the *CellDecompositionPlus* syntax of the SIROPA library. For isolating the aspects around home-pose, we transform the singularity and joint limits equations as inequalities [12] in Maple.

Singularity equations for the 3-SPS-U mechanism

$$\begin{aligned}\zeta_1 : 4s_\beta c_\beta c_\alpha - 14s_\beta c_\alpha^2 + 3c_\beta^2 c_\alpha + 7c_\beta c_\alpha^2 - 6s_\beta c_\beta + 2c_\alpha s_\beta + 8c_\beta^2 + 4c_\alpha c_\beta \\ + 6s_\beta - 3c_\beta + c_\alpha - 4 = 0\end{aligned} \tag{10}$$

$$\begin{aligned}\zeta_{1+k} : ((2c_\beta + 2s_\beta + 3)c_\alpha^2 + (3c_\beta^2 + (2 - s_\beta)c_\beta - 2s_\beta - 1)c_\alpha + c_\beta^2 + 3c_\beta s_\beta \\ - 2)\sqrt{3} + m(3s_\alpha((2 - s_\beta - c_\beta)c_\alpha + 3c_\beta^2 - (s_\beta + 1)c_\beta + s_\beta - 1)) = 0\end{aligned} \tag{11}$$

for $k = [1,2], \quad m = [-1,1]$ in Eq. 11

Singularity equations for the 4-SP̲S-U mechanism

$$\zeta_{1+2i}: s_\alpha s_\beta c_\beta + s_\alpha s_\beta c_\alpha + j(s_\alpha c_\beta^2 - s_\alpha c_\beta c_\alpha + s_\beta c_\beta c_\alpha - s_\beta c_\alpha^2 - s_\alpha c_\beta + s_\alpha c_\alpha - s_\beta c_\beta + c_\alpha s_\beta) + c_\beta^2 c_\alpha + c_\beta c_\alpha^2 - s_\alpha s_\beta + c_\beta^2 + c_\alpha c_\beta + c_\alpha^2 - 1 = 0 \quad (12)$$

$$\zeta_{2+2i}: j(s_\alpha s_\beta c_\beta + s_\alpha s_\beta c_\alpha - c_\beta^2 c_\alpha - c_\beta c_\alpha^2 - s_\alpha s_\beta - c_\beta^2 - c_\alpha c_\beta - c_\alpha^2 + 1) - s_\alpha c_\beta^2 + s_\alpha c_\beta c_\alpha + s_\beta c_\beta c_\alpha - s_\beta c_\alpha^2 + s_\alpha c_\beta - s_\alpha c_\alpha - s_\beta c_\beta + c_\alpha s_\beta = 0 \quad (13)$$

for $i = [0, 1], j = [-1, 1]$

3.2 Workspace Analysis

For generating the traces of workspace and singularities, we set limits of springs as $[l_{min}, l_{max}] = [10, 30]$ mm for the 3-SP̲S-U mechanism and [7, 30] mm for the 4-SP̲S-U mechanism. The lower limit l_{min} contributes mainly to the boundaries of the aspects centered in $\alpha = \beta = 0$ radians. A value of 11 mm is taken for the parameter r_f [5]. With these parameters being set, the workspace is estimated for both the mechanisms under study and a comparison is done based on their maximum tilt limits.

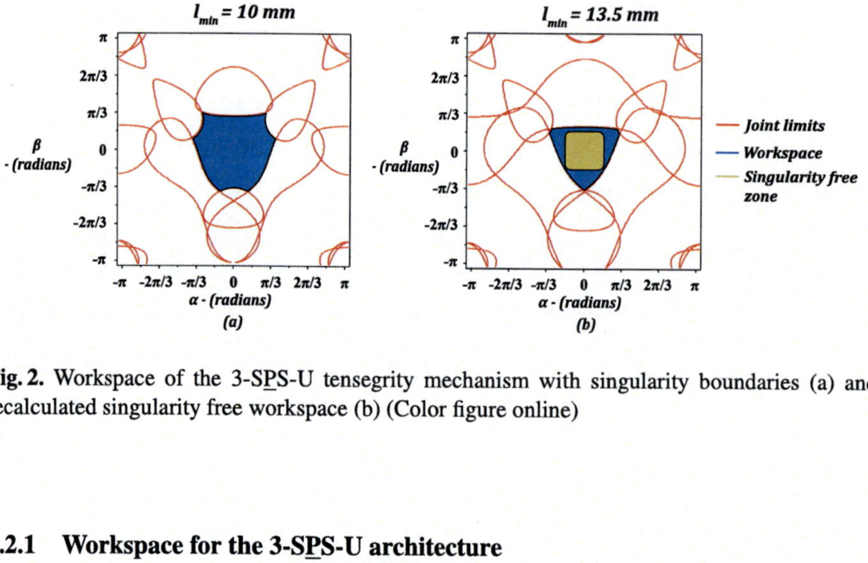

Fig. 2. Workspace of the 3-SP̲S-U tensegrity mechanism with singularity boundaries (a) and recalculated singularity free workspace (b) (Color figure online)

3.2.1 Workspace for the 3-SP̲S-U architecture

The workspace and singularity curves for the 3-SP̲S-U mechanism are represented in Fig. 2a. The blue regions indicate the workspace for the mechanism with the singularity boundaries. The joint limits for the three springs are represented in red lines and they appear superimposed on the plots. Parallel singularities can be observed for the mechanisms at the corners of the blue regions especially at $\alpha = \beta = \pi/3$ radians. The singularity free workspace for the mechanism is bounded by a triangular region which has to be extracted to determine the minimum limit for the springs. One edge of triangle is extracted at $\alpha = 0$ and $\beta = -\pi/3$ radians. At this position, one or two of the springs reach their minimum position with no singularities. A value of 13.5 mm is estimated for the lengths l_2 and l_3 at this position. This is the minimum limit for which the spring can go to avoid parallel singularities. For determining the remaining edges of the triangle, the minimum limits for the other length pairs ($l_1 - l_2$ & $l_1 - l_3$) are fixed as 13.5 mm. The values of α are taken as $\pm\pi/3$ radians from the workspace obtained in Fig. 2a. A value of 0.67 radians is found for β at these positions. With the modified lower limit for the springs, the workspace for the 3-SPS-U tensegrity mechanism is recalculated and plotted as shown in Fig. 2b. It could be seen that a singularity free workspace in the form of a Reuleaux triangle is obtained. The minimum square zone within this triangle is bounded by the limits $[\alpha, \beta] \in [-\pi/6, \pi/6]$ radians. Thus in order to avoid singularities during operation, the 3-SPS-U mechanism can tilt upto $\pm\pi/6$ radians.

3.2.2 Workspace for the 4-SPS-U architecture

The workspace and singularity surfaces are generated for the 4-SPS-U architecture and it is represented in Fig. 3a.

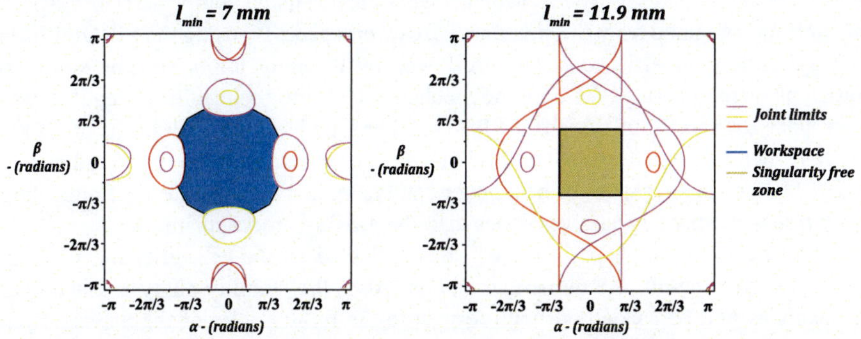

Fig. 3. Workspace of the 4-SPS-U tensegrity mechanism with singularity boundaries (a) and recalculated singularity free workspace (b)

In Fig. 3a, the joint limits are represented by the colored lines and the workspace is represented by the blue-colored region. Compared to the 3-SPS-U architecture, a wider workspace with singularity regions are observed for the 4-SPS-U mechanism. Based on the analysis, parallel singularities occur when both α and $\beta \in [\pm\pi/3, \pm\pi/3]$ radians. At these boundaries, two of the four legs reaches its minimum length. At α equal to 0 radians, β attains maximum tilt values of $\pm\pi/3$ radians. Using the joint limit Eqs. 8–9 and the minimum spring length value at the singularity boundary, the maximum tilt limit with singularity free workspace is estimated for α,β at $[0, \pi/3]$ radians. A tilt limit of $\pm 5\pi/18$ radians is obtained and the minimum spring length is estimated to be 11.9 mm. The recalculated workspace is in the form of a square for the 4-SPS-U mechanism and it is represented in Fig. 3b. The singularity free workspace for this mechanism is superimposed on the square with tilt limits of $\pm 5\pi/18$ radians.

3.3 Discussions

From the workspace analysis, it could be observed that the 3-SPS-U mechanism generates a higher tilt limit when compared to the 4-SPS-U mechanism. The articulation unit for the bio-inspired robot must be able to overcome pipe bends at $\pi/2$ radians in a passive mode. With the 3-SPS-U mechanism, it will be difficult to address this problem as there are possibilities that the mechanism might reach singular poses within a narrow tilt limit range. On the other hand, the 4-SPS-U mechanism offers higher tilt limits with a singularity free workspace. The possibilities of reaching singular positions by this mechanism is comparatively less than the 3-SPS-U mechanism. The tilt limit issues for the 3-SPS-U mechanism can be addressed by using stacked modules. A simulation of the 4-SPS-U mechanism coupled with the bio-inspired robot developed at LS2N was performed and it showed that the 4-SPS-U mechanism is capable of overcoming pipe bends at $\pi/2$ radians.

4 Conclusions

In this article, two tensegrity mechanisms were analyzed and compared with respect to the tilt limits obtained from the singularity free workspace. By using the SIROPA library in Maple, the geometric constraints followed by the joint limits and the singularity equations were generated for both the mechanisms. From the resulting singularity free workspaces, it was found that the 4-SPS-U mechanism has higher tilt limits of $\pm 5\pi/18$ radians when compared to the 3-SPS-U mechanism which provides $\pm\pi/6$ radians. The 4-SPS-U mechanism proves to be a better candidate, that can operate within the singularity free workspace in passive modes than the 3-SPS-U mechanism.

In future works, an experiment will be conducted on the tensegrity mechanism in order to correlate with the numerical results. Also, the stability analysis for both the mechanisms will be carried out with respect to the Euler angles of the universal joint. This analysis will be done by taking into account the weight of the robot for horizontal and vertical orientations of the pipeline.

Acknowledgements. We would like to thank the ANR Avineck project team for the support extended in the initial study of this mechanism.

References

1. Venkateswaran, S., Chablat, D., Boyer, F.: Numerical and experimental validation of the prototype of a bio-inspired piping inspection robot. Robotics **8**(2), 32 (2019)
2. Chen, J., Chen, T., Deng, Z.: Design method of modular units for articulated in-pipe robot inspecting system. In: 2011 Second International Conference on Digital Manufacturing & Automation, pp. 389–392. IEEE (2011)
3. Hirose, S., Ohno, H., Mitsui, T., Suyama, K.: Design of in-pipe inspection vehicles forpipes. In: Proceedings 1999 IEEE International Conference on Robotics and Automation (Cat. No. 99CH36288C), vol. 3, pp. 2309–2314. IEEE (1999)
4. Brunete, A., Hernando, M., Gambao, E.: Modular multi configurable architecture for low diameter pipe inspection microrobots. In: Proceedings of the 2005 IEEE International Conference on Robotics and Automation, pp. 490–495. IEEE (2005)

5. Venkateswaran, S., Furet, M., Chablat, D., Wenger, P.: Design and analysis of a tensegrity mechanism for a bio-inspired robot. In: Proceedings of the ASME 2019 International Design Engineering Technical Conferences & Computers and Information in Engineering Conference, vol. 5A, 43rd Mechanisms and Robotics Conference (2019). V05AT07A026
6. Shah, S., Saha, S., Dutt, J.: Denavit-hartenberg parameterization of Euler angles. J. Comput. Nonlinear Dyn. **7**(2), 021006 (2012)
7. Alici, G., Shirinzadeh, B.: Topology optimisation and singularity analysis of a 3-sps parallel manipulator with a passive constraining spherical joint. Mech. Mach. Theor. **39**(2), 215–235 (2004)
8. Pan, Y., Chen, Y., Li, L.: Analysis of kinematic dexterity and stiffness performance based on spring's wire-driven 4-sps/u rigid-flexible parallel trunk joint mechanism. Int. J. Struct. Integrity (2019)
9. Gosselin, C., Angeles, J.: Singularity analysis of closed-loop kinematic chains. IEEE Trans. Robot. Autom. **6**(3), 281–290 (1990)
10. Jha, R., Chablat, D., Baron, L., Rouillier, F., Moroz, G.: Workspace, joint space and singularities of a family of delta-like robot. Mech. Mach. Theory **127**, 73–95 (2018)
11. Chablat, D., Ottaviano, E., Moroz, G.: A comparative study of 4-cable planar manipulators based on cylindrical algebraic decomposition. In: Proceedings of the ASME 2011 International Design Engineering Technical Conferences & Computers and Information in Engineering Conference, pp. 1–10 (2011)
12. Chablat, D., Wenger, P.: Working modes and aspects in fully parallel manipulators. In: Proceedings of 1998 IEEE International Conference on Robotics and Automation (Cat. No. 98CH36146), vol. 3, pp. 1964–1969. IEEE (1998)

Check for updates

Degeneration to Infinity May Provide Information About Kinematics

Michel Coste[1(✉)] and Nestor Djintelbe[2]

[1] IRMAR, Université Rennes 1, Rennes, France
michel.coste@univ-rennes1.fr
[2] Université Assane Seck, Ziguinchor, Senegal
djin_nestor@yahoo.fr

Abstract. We propose to degenerate the kinematics of a mobile platform by letting the lengths of its legs tend to infinity. The rationale for this approach is that the degenerate situation is simpler to analyze and may provide relevant information for the kinematic analysis of the platform. We implement this approach in the framework of the Study model. We show its usefulness on the example of the general operation mode of the Tsai 3-UPU platform, whose kinematic analysis is rather difficult.

Keywords: Study quadric · Degeneration · 3-UPU platform

1 Introduction

The Study model has proved very useful for the algebraic modelization of problems in robotics and their resolution in computer algebra systems. It realizes the group of rigid motions as a subset of a 6-dimensional quadric (the Study quadric S) in the 7-dimensional real projective space. Precisely, the group of rigid motions is identified with the complement in S of a 3-plane E contained in S. For the Study model and its use in kinematics, one can see [3–5].

The exceptional plane E can be seen as the boundary of the group of rigid motions in S, but there is no useful interpretation of its points as degenerate rigid motions. So we modify the Study model in order to have a boundary of dimension 5 whose points can be seen as degenerate rigid motions with infinite translation part. This modification is described in Sect. 2. An alternative modification is described in [2], using the notion of blowing-up.

The purpose of the introduction of a meaningful boundary for the group of rigid motions is to obtain information on the kinematics of the robots as the lengths of their legs are large enough. We have chosen to illustrate this approach with the example of the Tsai 3-UPU platform; a thorough kinematic analysis of this parallel manipulator has been done in [6]. We present a short survey in Sect. 3, insisting on the so-called general operation mode which is rather mysterious. In Sect. 4, we show that the degeneration of this general mode produces new relevant informations on its kinematic behaviour.

J. Lenarčič and B. Siciliano (Eds.): ARK 2020, SPAR 15, pp. 234–241, 2021.
https://doi.org/10.1007/978-3-030-50975-0_29

2 The Study Model and Its Modification

We recall the Study model for the group of rigid motions and then define a modification of this model. This modification aims at obtaining a boundary which allows to degenerate rigid motions as the length of the translation vector tends to infinity.

2.1 The Study Model

The Study model of the group of rigid motions is the quadric S with equation

$$(\Sigma) \qquad x_0y_0 + x_1y_1 + x_2y_2 + x_3y_3 = 0 \tag{1}$$

where $[x_0 : \ldots : x_3 : y_0 : \ldots : y_3]$ are homogeneous coordinates in the real projective 7-space. We denote $\mathbf{x} = \left[x_0, x_1, x_2, x_3\right]^{\mathsf{T}}$ and $\mathbf{y} = \left[y_0, y_1, y_2, y_3\right]^{\mathsf{T}}$. We recall that to each point of this quadric outside of the exceptional 3-plane E with equation $\mathbf{x} = \mathbf{0}$, we associate the rigid motion with rotation matrix

$$\mathbf{R}(\mathbf{x}) = \frac{1}{\|\mathbf{x}\|^2}\begin{bmatrix} x_0^2+x_1^2-x_2^2-x_3^2 & 2(x_1x_2-x_0x_3) & 2(x_1x_3+x_0x_2) \\ 2(x_1x_2+x_0x_3) & x_0^2-x_1^2+x_2^2-x_3^2 & 2(x_2x_3-x_0x_1) \\ 2(x_1x_3-x_0x_2) & 2(x_2x_3+x_0x_1) & x_0^2-x_1^2-x_2^2+x_3^2 \end{bmatrix} \tag{2}$$

and translation vector

$$\mathbf{t}(\mathbf{x},\mathbf{y}) = \frac{2}{\|\mathbf{x}\|^2}\begin{bmatrix} x_0y_1 - x_1y_0 + x_2y_3 - x_3y_2 \\ x_0y_2 - x_1y_3 - x_2y_0 + x_3y_1 \\ x_0y_3 + x_1y_2 - x_2y_1 - x_3y_0 \end{bmatrix} \quad \text{where } \|\mathbf{x}\|^2 = x_0^2 + x_1^2 + x_2^2 + x_3^2 \,. \tag{3}$$

2.2 Modification of the Study Model

The modification of S with which we shall work is a subset $\widetilde{S}$ of the real 9-dimensional affine space with coordinates $w_0, w_1, w_2, w_3, s, y_0, y_1, y_2, y_3$. We denote $\mathbf{w} = \left[w_0, w_1, w_2, w_3\right]^{\mathsf{T}}$. This subset is defined by the equations and inequalities

$$\begin{array}{ll} \|\mathbf{w}\|^2 = w_0^2 + w_1^2 + w_2^2 + w_3^2 = 1 & \|\mathbf{y}\|^2 = y_0^2 + y_1^2 + y_2^2 + y_3^2 = 1 \\ \mathbf{w}\cdot\mathbf{y} = w_0y_0 + w_1y_1 + w_2y_2 + w_3y_3 = 0 \quad s \geq 0 & \end{array} \tag{4}$$

The modification $\widetilde{S}$ is related to S via the mapping $\pi : \widetilde{S} \to S$ defined by

$$(\mathbf{w}, s, \mathbf{y}) \longmapsto [w_0s : w_1s : w_2s : w_3s : y_0 : y_1 : y_1 : y_2 : y_3] \tag{5}$$

The image of π is the complement in S of the 3-plane $\mathbf{y} = \mathbf{0}$ corresponding to rigid motions whose translation vector is zero. Every rigid motion with a non zero translation vector is the image by π of two points $(\mathbf{w}, s, \mathbf{y})$ and $(-\mathbf{w}, s, -\mathbf{y})$ of $\widetilde{S}$ with $s > 0$; it is easily checked that s is the inverse of the length of the translation vector. The image by π of an element $(\mathbf{w}, 0, \mathbf{y}) \in \widetilde{S}$ is the element $[0 : 0 : 0 : 0 : y_0 : y_1 : y_1 : y_2 : y_3]$ of the exceptional 3-plane E. The element $(\mathbf{w}, 0, \mathbf{y}) \in \widetilde{S}$ may be seen as the degeneration of the rigid motion associated to $(\mathbf{w}, s, \mathbf{y}) \in \widetilde{S}$ as s tends to 0, i.e. as the length of the translation vector tends to infinity. It retains the information on the rotation part in $\mathbf{w}$ and on the direction of translation in $\mathbf{t}(\mathbf{w}, \mathbf{y})$.

3 Tsai 3-UPU

We recall the kinematic analysis of a mobile platform with three degrees of freedom known as the Tsai 3-UPU. We rely on [6], with slight changes in presentation.

3.1 Algebraic Modelization

The Tsai 3-UPU is a mobile platform with architecture illustrated in Fig. 1. The centers A_i and B_i of the universal joints on the base (resp. mobile platform) form an equilateral triangle. The rotation axes 1 and 5 in the kinematic chain for each leg are tangent to the circumscribed circles. The rotation axes 2 and 4 are parallel, and both orthogonal to (A_iB_i).

For the algebraic modelization, we work in the fixed frame attached to the base where the points A_i have coordinates

$$\mathbf{a}_1 : \left[0,\, h_1,\, 0\right]^\top \quad \mathbf{a}_2 : \left[0,\, -h_1/2,\, \sqrt{3}h_1/2\right]^\top \quad \mathbf{a}_3 : \left[0,\, -h_1/2,\, -\sqrt{3}h_1/2\right]^\top ,$$

where h_1 is the radius of the circle circumscribed to the base (in the computations, we shall fix $h_1 = 1$). We denote $\mathbf{v}_i$ the coordinates of unit vectors giving the direction of the rotation axes 1 for each leg.

We have a mobile frame attached to the platform where the points B_i have coordinates $\mathbf{b}_i$ similar to $\mathbf{a}_i$ with h_1 replaced by the radius h_2 of the circle circumscribed to the platform (we shall take for short $h_2 = h$ in the computations). The home pose of the mobile platform is when the mobile frame coincides with the fixed one. The coordinates of B_i in the fixed frame are $\mathbf{R}(\mathbf{x})\mathbf{b}_i + \mathbf{t}(\mathbf{x},\mathbf{y})$.

The constraint equation for each leg express that the axis 1, the line (A_i, B_i) and the axis 5 are coplanar. These equations are

$$\det(\mathbf{v}_i,\ \mathbf{R}(\mathbf{x})\mathbf{b}_i + \mathbf{t}(\mathbf{x},\mathbf{y}) - \mathbf{a}_i,\ \mathbf{R}(\mathbf{x})\mathbf{v}_i) = 0 \qquad \text{for } i = 1,2,3\,. \tag{6}$$

After chasing denominators, these constraint equations are homogeneous equations (C_i) of degree 4 in $\mathbf{x},\mathbf{y}$, the Study parameters. Together with the equation (Σ) of the Study quadric and the inequation $N_\mathbf{x} \neq 0$, the equations (C_i) describe the possible poses for the mobile platform.

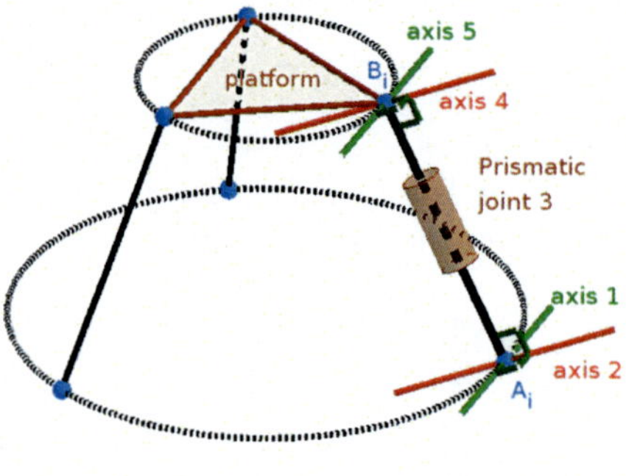

Fig. 1. A Tsai 3-UPU platform.

3.2 The Operation Modes

We form the ideal $\langle C_1, C_2, C_3, \Sigma\rangle$ generated by these four equations and saturate it with respect to $N_{\mathbf{x}}$ in order to get rid of parasitic components contained in the exceptional 3-plane (cf. [1] Chap. 4, § 4). We obtain a homogeneous ideal $\mathfrak{J}$ whose projective variety is an algebraic subset of dimension 3 of the Study quadric. The operation modes of the mobile platform are the irreducible components of this algebraic set. These irreducible components are obtained by computing the primary decomposition of $\mathfrak{J}$ (cf. [1] Chap. 4, § 8). We obtain:

- $\mathfrak{J}_1 = \langle y_0, x_1, x_2, x_3\rangle$: all poses obtained from the home pose by translation.
- $\mathfrak{J}_2 = \langle x_0, y_1, x_2, x_3\rangle$: all poses obtained by a half-turn with vertical axis followed by a translation.
- $\mathfrak{J}_3 = \langle y_0, y_1, x_2, x_3\rangle$: all poses obtained by rigid motion in the base plane.
- $\mathfrak{J}_4 = \langle x_0, x_1, y_2, y_3\rangle$: all poses obtained by a horizontal flip followed by motion in the base plane.
- A component whose kinematic analysis is difficult. We give only the first two and the last generators of the ideal produced by the computation:

$$\begin{aligned}\mathfrak{J}_5 = &\langle 2x_1y_1 + x_2y_2 + x_3y_3,\ 2x_0y_0 + x_2y_2 + x_3y_3,\ \ldots,\\ &x_3y_0(3x_2^2 - x_3^2)(h+1) + x_2y_1(3x_3^2 - x_2^2)(h-1) - (x_2^2 + x_3^2)y_0y_1\rangle\end{aligned}$$

We shall follow the terminology of [6] and call "general mode" the fifth mode of operation.

It is remarkable that a small change in the architecture of the 3-UPU platform can drastically affect its kinematic behaviour. The SNU 3-UPU differs from the Tsai one in the disposition of axes 1 and 5: they point towards the center of the circle circumscribed to the base (resp. platform) instead of being tangent to this circle. The analysis of the operation modes of the SNU 3-UPU is done in [7]. There is a purely rotational operation mode, the operation modes $\mathfrak{J}_1$ to $\mathfrak{J}_4$ are again present, but there is no general mode $\mathfrak{J}_5$. Instead, there are three components whose kinematic analysis is easy:

- $\mathfrak{J}_6 = \langle x_0, y_1, y_2, y_3\rangle$: all poses obtained by a half-turn with axis through origin, followed by a translation in the direction of the axis of half-turn.
- $\mathfrak{J}_7 = \langle y_0, x_1, y_2, y_3\rangle$: all poses obtained by the half-turn with vertical axis through origin, followed by a rigid motion of mode $\mathfrak{J}_6$; the rotation part is a rotation with horizontal axis.

plus a non-real component $\mathfrak{J}_8$ whose real points correspond to poses obtained by a translation along the vertical axis through origin, possibly composed with the half-turn around this axis. These singular poses belong also to other modes of operation.

4 Degeneration of the General Mode

The kinematic analysis of the general operation mode of the Tsai 3-UPU is difficult. We are going to compute the boundary of this operation mode in the modification of the Study model and obtain some relevant information from this degeneration.

4.1 The Boundary of the General Mode

The projective variety $V_5 = V(\mathfrak{J}_5) \subset S$ contains the poses of the general mode, plus some points of the exceptional 3-plane E. We compute its inverse image $\pi^{-1}(V_5)$ in the modification $\widetilde{S}$. This is done by substituting the variables x_i with $w_i s$ in the generators of $\mathfrak{J}_5$ and adding equations (4) to the list of generators. We obtain in this way an ideal in the ring of polynomials in variables $\mathbf{w}, s, \mathbf{y}$.

We then remove from $\pi^{-1}(V_5)$ the parasitic components entirely contained in the hyperplane $s = 0$. This is done by saturating the ideal obtained in the preceding step with respect to s. This new ideal $\widetilde{\mathfrak{J}_5}$ describes a subset $\widetilde{V_5} \subset \widetilde{S}$. The points $(\mathbf{w}, s, \mathbf{y}) \in \widetilde{V_5}$ with $s > 0$ correspond to poses in the general mode ($(\mathbf{w}, s, \mathbf{y})$ and $(-\mathbf{w}, s, -\mathbf{y})$ give the same pose). The points $(\mathbf{w}, 0, \mathbf{y}) \in \widetilde{V_5}$ may be seen as degenerate poses of the general mode, as the length of legs tends to infinity. The intersection of $\widetilde{V_5}$ with the hyperplane $s = 0$ forms the boundary of the general mode. This boundary is a 2-dimensional algebraic subset of the boundary of the group of rigid motions.

This boundary is computed by setting $s = 0$ in the ideal $\widetilde{\mathfrak{J}_5}$, thus obtaining an ideal in the ring of polynomials with variables $\mathbf{w}, \mathbf{y}$. The primary decomposition of this ideal reveals that the boundary of the general mode has several components (we dismiss those obtained by $\mathbf{w}, \mathbf{y} \mapsto -\mathbf{w}, -\mathbf{y}$):

- $\mathfrak{K}_6 = \langle w_0, y_1, y_2, y_3, y_0 - 1, w_1^2 + w_2^2 + w_3^2 - 1 \rangle$: all degenerate poses obtained by a half-turn with axis through origin, followed by an infinite translation in the direction of the axis of half-turn.
- $\mathfrak{K}_7 = \langle y_0, w_1, y_2, y_3, y_1 - 1, w_0^2 + w_2^2 + w_3^2 - 1 \rangle$: all degenerate poses obtained by the half-turn with vertical axis through origin, followed by a a half-turn with axis through origin and then an infinite translation in the direction of the axis of the latter half-turn; the rotation part is a rotation with horizontal axis.

There are also non-real components whose only real points are ($\mathbf{w} = [\pm 1, 0, 0, 0]^\top$, $\mathbf{y} = [0, 1, 0, 0]^\top$) and ($\mathbf{w} = [0, \pm 1, 0, 0]^\top$, $\mathbf{y} = [1, 0, 0, 0]^\top$), all singular, corresponding to degenerate poses obtained by infinite translation in the vertical direction, possibly composed with a half-turn with vertical axis.

We have found that the boundary of the general mode of the Tsai 3-UPU decomposes as the union of the boundaries of the modes $\mathfrak{J}_6$, $\mathfrak{J}_7$ and $\mathfrak{J}_8$ of the SNU 3-UPU and that the degenerate poses in this boundary have a simple kinematic description.

4.2 The Degenerate Direct Kinematic Problem for the General Mode

The actuated joints in the 3-UPU platform are the three prismatic joints. The inverse kinematic mapping (IKM) associates to each pose the lengths r_1, r_2, r_3 of the legs. We degenerate the IKM to the boundary of the set of poses for the general mode in the modification of the Study model. All lengths of legs tend to infinity, but the relevant information lies in the differences of the lengths of the legs.

For $(\mathbf{w}, s, \mathbf{y}) \in \widetilde{S}$ we obtain (recall $\mathbf{w}$, $\mathbf{y}$ and $\mathbf{t}(\mathbf{w}, \mathbf{y})$ are unit vectors):

$$\begin{aligned} r_i &= (1/s)\ \|s\,\mathbf{R}(\mathbf{w})\,\mathbf{b}_i + \mathbf{t}(\mathbf{w}, \mathbf{y}) - s\,\mathbf{a}_i\| \\ &= (1/s)\ \sqrt{1 + 2s\,\mathbf{t}(\mathbf{w}, \mathbf{y}) \cdot (\mathbf{R}(\mathbf{w})\,\mathbf{b}_i - \mathbf{a}_i) + s^2\ \|\mathbf{R}(\mathbf{w})\,\mathbf{b}_i - \mathbf{a}_i\|^2}\,. \end{aligned} \tag{7}$$

It follows that the limit of $r_i - r_3$ (for $i = 1,2$) as s tends to 0 is

$$d_i = \mathbf{t}(\mathbf{w},\mathbf{y}) \cdot \left(\mathbf{R}(\mathbf{w})\,(\mathbf{b}_i - \mathbf{b}_3) - (\mathbf{a}_i - \mathbf{a}_3)\right). \tag{8}$$

The remainders of the polynomials of Eq. (8) by a Gröbner basis (cf. [1] Chap. 2, § 6) for the ideals $\mathfrak{K}_6$ and $\mathfrak{K}_7$ give the formulation of the degenerate IKM for these two components of the boundary of the general mode.

For $\mathfrak{K}_6$, with $w_0 = y_1 = y_2 = y_3 = 0$, $y_0 = 1$ and $w_1^2 + w_2^2 + w_3^2 = 1$:

$$d_1 = (3/2)(h-1)\,w_2 + (\sqrt{3}/2)(h-1)\,w_3 \qquad d_2 = \sqrt{3}(h-1)\,w_3 \tag{9}$$

For $\mathfrak{K}_7$, with $y_0 = w_1 = y_2 = y_3 = 0$, $y_1 = 1$ and $w_0^2 + w_2^2 + w_3^2 = 1$:

$$d_1 = -(\sqrt{3}/2)(h+1)\,w_2 + (3/2)(h+1)\,w_3 \qquad d_2 = -\sqrt{3}(h+1)\,w_2 \tag{10}$$

For the two real singular points of $\mathfrak{K}_8$, we have $d_1 = d_2 = 0$.

The degenerate direct kinematic problem (DKP) for the general mode is now easily solved. For any (d_1, d_2) inside the ellipse

$$d_1^2 - d_1 d_2 + d_2^2 = \frac{9}{4}(h-1)^2 \quad (\text{resp.} \frac{9}{4}(h+1)^2) \tag{11}$$

there are two solutions in the mode $\mathfrak{K}_6$ (resp. $\mathfrak{K}_7$). For $d_1 = d_2 = 0$ there are solutions in $\mathfrak{K}_8$ which are singular degenerate poses.

4.3 Comparison with the Actual DKP

We fix $h = 1/2$ and compare the number of solutions to the DKP on lines $d_1 = k$, $d_2 = 2k$ in the degenerate joint space ($r_3 = \infty$) and $r_1 = r_3 + k$, $r_2 = r_3 + 2k$ for various values of r_3 in the actual joint space. One can see in Fig. 2 a striking stability, perturbed by the evolution of the four singular solutions for $k = 0$.

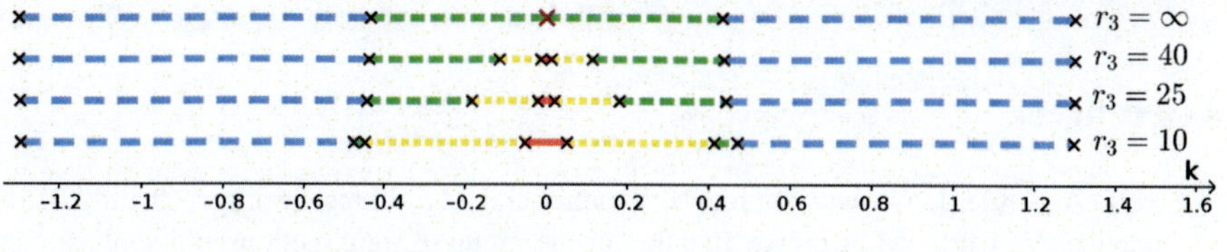

Fig. 2. Number of solutions to DKP. Red: 12, orange: 8, green: 4, blue: 2 (Color figure online)

The solutions of the DKP come in pairs obtained by reflection through the base plane. In Table 1, we compare the solutions to the DKP for $k = 0.3$ and various values of r_3 (including the degenerate case), retaining only one solution in a pair.

The solutions in the degenerate case evolve to solutions for large values of r_3. We notice the components $\mathfrak{K}_6$ on the first line and $\mathfrak{K}_7$ on the second line. The two extra solutions for $r_3 = 10$ come from the evolution of the singular solution ($\mathbf{w} = [0, 1, 0, 0]^\mathsf{T}, \mathbf{y} = [1, 0, 0, 0]^\mathsf{T}$) in $\mathfrak{K}_8$, responsible for the yellow part in Fig. 2.

Table 1. Solutions to DKP for $k = 0.3$ (component $\mathbf{y}$ normalized to $\widetilde{\mathbf{y}}$)

	$r_3 = \infty$	$r_3 = 40$	$r_3 = 25$	$r_3 = 10$		
x_0	0	−0.016	−0.022	−0.033	0.061	−0.015
x_1	0.721	0.684	0.651	0.551	0.846	0.981
x_2	−0.693	−0.717	−0.736	−0.780	−0.185	0.147
x_3	0.000	−0.134	−0.186	−0.294	0.497	−0.125
$\widetilde{y}_0$	1	0.998	0.996	0.985	0.966	0.957
$\widetilde{y}_1$	0	−0.023	−0.033	−0.060	0.069	−0.015
$\widetilde{y}_2$	0	−0.048	−0.068	−0.123	0.184	−0.046
$\widetilde{y}_3$	0	0.018	0.036	0.104	−0.167	−0.285
x_0	0.973	0.973	0.973	0.971		
x_1	0	0.003	0.004	0.011		
x_2	0.000	−0.008	−0.013	−0.031		
x_3	0.231	0.231	0.232	0.238		
$\widetilde{y}_0$	0	0.003	0.005	0.011		
$\widetilde{y}_1$	1	1.000	0.999	0.995		
$\widetilde{y}_2$	0	−0.001	−0.004	−0.021		
$\widetilde{y}_3$	0	−0.024	−0.038	−0.092		

5 Conclusion

We have shown with the example of the Tsai 3-UPU platform that the degeneration of an operation mode whose kinematic analysis is rather complex can give a boundary whose kinematic behaviour is very easy to understand. Moreover, the degenerate DKP for this boundary gives valuable information on the original DKP, for large enough lengths of legs. This approach can also be used for other examples of parallel manipulators with restricted degrees of freedom (3-RPS, 4-UPU ...).

References

1. Cox, D.A., Little, J., O'Shea, D.: Ideals, Varieties and Algorithms. Springer, Cham (2015)
2. Djintelbe, N., Coste, M.: Compactification of the group of rigid motions and applications to robotics. arXiv:1910.00319
3. Husty, M.L., Schröcker, H.P.: Algebraic geometry and kinematics. In: Emiris, I.Z., Sottile, F., Theobald, T. (eds.) Nonlinear Computational Geometry, pp. 85–107. Springer, New York (2010)
4. Selig, J.M.: Geometric Fundamentals of Robotics. Monographs in Computer Science. Springer, New York (2005)
5. Study, E.: Von den Bewegungen und Umlegungen. Math. Ann. **39**(4), 441–565 (1891)

6. Walter, D., Husty, M.: Kinematic analysis of the TSAI 3-UPU parallel manipulator using algebraic methods. In: The 13th IFToMM World Congress in Mechanism and Machine Science, pp. 1–10 (2011)
7. Walter, D.R., Husty, M.L., Pfurner, M.: A complete kinematic analysis of the SNU 3-UPU parallel robot. In: Bates, D.J., Besana, G., Di Rocco, S., Wampler, C.W. (eds.) Interactions of Classical and Numerical Algebraic Geometry, pp. 331–346. AMS, Providence (2009)

Kinematic Synthesis of a Modified Jansen Leg Mechanism

Kevin Chen and J. Michael McCarthy(✉)

Robotics and Automation Laboratory, University of California, Irvine, Irvine, CA, USA
{kevinc15,jmmccart}@uci.edu

Abstract. This paper modifies Jansen's leg mechanism so that it is formed by an RR serial chain with hip and knee joints driven by separate four-bar function generators with one input. Kinematic synthesis of the two function generators programs the joint movement of the RR leg to achieve a desired foot trajectory. The two input cranks can be connected by gears to form a one degree-of-freedom system that generalizes the Jansen leg mechanism. An example design is presented.

Keywords: Kinematic synthesis · Leg mechanism · Jansen linkage

1 Introduction

Theodorus Jansen designed an eight-bar leg mechanism that he used for a series of walking machines over the years since 1991 including the wind driven Strandbeest [7, 18]. The dimensions of his leg mechanism yield a distinctive foot trajectory, that has a straight-line movement along its interaction with the ground. This ensures that the leg does not expend energy lifting or lowering the walker body [6]. Figure 1 is a photograph of a model of his Strandbeest.

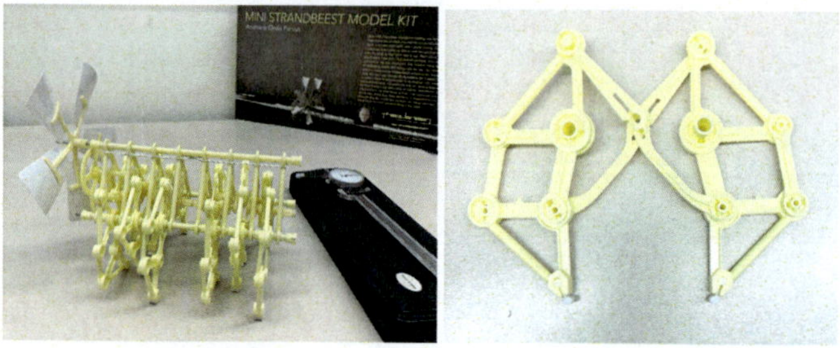

(a) Model (b) Leg Mechanism

Fig. 1. (a) A model of the Strandbeest developed by T. Jansen and (b) a closeup of the leg mechanisms.

In this paper, we design function generators to mechanically program the joint movement of the upper and lower leg elements of an RR serial chain around the hip

J. Lenarčič and B. Siciliano (Eds.): ARK 2020, SPAR 15, pp. 242–249, 2021.
https://doi.org/10.1007/978-3-030-50975-0_30

and knee joints. The end point of the RR serial chain, which is the foot of the leg mechanism, is positioned at five points around a required foot trajectory. This yields five sets of the hip and knee angles measured relative to a horizontal axis that are the outputs of the two function generators. The input angles are the same for both linkages so they can be driven by a single input. The result is one degree-of-freedom leg mechanism that has a structure similar to the Jansen linkage shown in Fig. 1.

2 Literature Review

Kaneko et al. (1985) [8] describe the separation of the leg movement of a walking machine into "body propelling motion" from the "terrain adapting motion", (also see Shigley (1960) [14]). They describe the benefits of a foot-trajectory that has straight-line movement while in contact with the ground, and therefore use the Chebyshev linkage to provide this foot trajectory, (also see Tchebyshev(1948) [16]).

Funabashi et al. (1985) [2] uses kinematic synthesis to design a six-bar linkage to guide the ankle-path of a biped walking machine. Shieh (1996) [12] presents the kinematic synthesis of six-bar and eight-bar linkages that provide useful one degree of freedom body propelling movement for walking machines, also see Shieh et al. (1998) [13].

Giesbrecht (2010) [4] and Ghassaei (2011) [3] used Jansen's linkage as a model for their design and optimization of eight-bar leg mechanisms for walking machines. Nansai et al. (2015) [10] studied variations of the foot trajectory obtained by varying link dimensions in order to obtain a reconfigurable Jansen's linkage. McCarthy (2019) [9] shows that versions of Jansen's and Ghassaei's leg mechanisms obtained by controlling an RR serial chain using four-bar function generators, and obtained new designs using graphical three-position-synthesis.

Here we show how five position synthesis of two four-bar function generators can be used to drive an RR serial chain along a specified foot trajectory, traced by the point **T**, in order to obtain a modified version of Jansen's linkage.

3 RR Leg and Foot Trajectory

A schematic version of Jansen's leg mechanism is shown in Fig. 2. The joints **A**, **C**, and **D** each connect three links. Separate these three joints so they connect two links each to find that Jansen's leg mechanism is an eight-bar linkage with 10 joints [17].

The measured dimensions of the physical model of the Jansen linkage shown in Fig. 2 do not match those identified by T. Jansen. This means the foot trajectory of our model does not match the well-known foot trajectory of Jansen's leg mechanism. However, our goal is to match the trajectory of the physical model with our modifications of the Jansen leg mechanism.

In order to define our modified Jansen linkage, we change the length of **DF** to match that of **CE**, so that the four-bar loop **CDFE** is a parallelogram linkage. In order to maintain a foot trajectory that includes an approximate straight-line, while in contact with the ground, we changed the lengths of **DT** and **FT** as shown in Table 1. Figure 3 shows

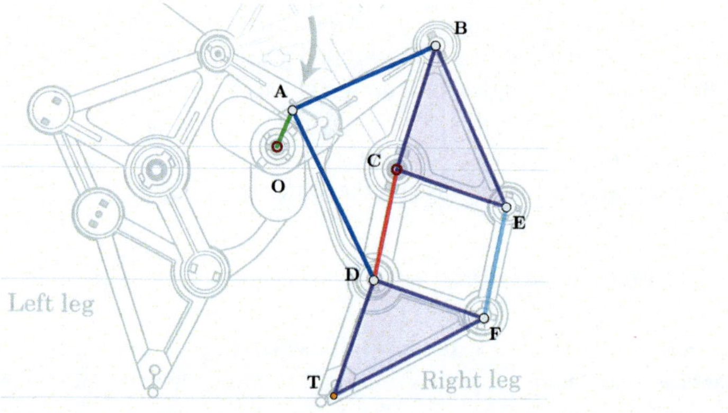

Fig. 2. The joints **A**, **C**, and **D** of Jansen's leg mechanism each connect three links, which means they are viewed as consisting of two joints. Thus, the leg mechanism consists of eight links and 10 joints.

Table 1. The dimensions of the physical leg mechanism and those of the mathematical model.

Link	Segment	Physical model (cm)	Mathematical model (cm)
1	|**OA**|	1.0	1.0
2	|**AB**|	3.45	3.45
3	|**CB**|	2.98	2.98
	|**BE**|	3.97	3.97
	|**CE**|	2.90	2.90
4	|**OC**|	2.75	2.75
5	|**CD**|	2.71	2.71
6	|**EF**|	2.71	2.71
7	|**AD**|	4.30	4.30
8	|**DF**|	2.52	2.90
	|**DT**|	2.50	3.12
	|**FT**|	3.78	4.64

the foot trajectory of the mathematical model obtained using these modified dimensions of the Jansen linkage.

Using this new set of dimensions for the Jansen leg mechanism, we identified five positions of **T** along its trajectory and determined the angles measured relative to the horizontal axis at the hip joint, **C**, and at the knee joint, **D**. See Fig. 4. The output angles at the hip and knee joints are listed in Table 2.

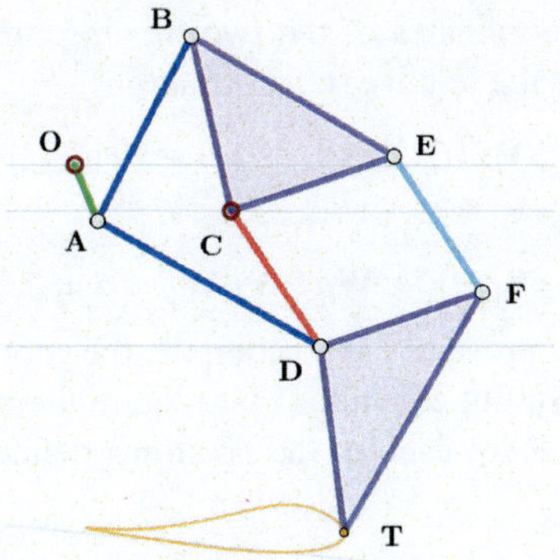

Fig. 3. The foot trajectory of this model of Jansen's leg mechanism, traced by the point **T**, has a linear trajectory while in contact with the ground.

Table 2. The input and output angles for the function generators that drive the upper leg around the hip joint and the lower leg around the knee joint of Jansen's linkage.

Position	Input angle	Upper leg (Hip joint)	Lower leg (Knee joint)
1	0.0°	240.9°	298.7°
2	70.0°	245.5°	244.6°
3	186.0°	283.0°	252.6°
4	240.0°	287.8°	276.2°
5	290.0°	278.5°	302.2°

4 Synthesis of a Four-Bar Function Generator

Here we provide the synthesis theory for one of these four-bar function generators. Our formulation differs from Freudenstein's approach [1,5], rather we use a version of finite-position synthesis [11].

This design procedure begins with the specification of the two fixed pivots, $\mathbf{O} = (o_x, o_y)$ and $\mathbf{C} = (c_x, c_y)$, of the four-bar function generator, and the five pairs of angles θ_i and ψ_i, $i = 1,\ldots,5$ that define the required functional relationship between the input and output links. See Fig. 5.

The positions of the input and output links relative to a fixed frame F are defined by the 3×3 homogeneous transformations, $\mathbf{S}(\theta_i, \mathbf{O})$ and $\mathbf{T}(\theta_i, \mathbf{C})$ given by,

$$\mathbf{S}(\theta_i,\mathbf{O}) = \begin{bmatrix} \cos\theta_i & -\sin\theta_i & o_x \\ \sin\theta_i & \cos\theta_i & o_y \\ 0 & 0 & 1 \end{bmatrix}, \mathbf{T}(\psi_i,\mathbf{C}) = \begin{bmatrix} \cos\psi_i & -\sin\psi_i & c_x \\ \sin\psi_i & \cos\psi_i & c_y \\ 0 & 0 & 1 \end{bmatrix}, \quad i = 1,\ldots,5. \quad (1)$$

Therefore, the positions of the pivots $\mathbf{V}_i$ and $\mathbf{W}_i$ in the input and output cranks are obtained as,

$$\mathbf{V}_i = \mathbf{S}(\theta_i,\mathbf{O})\mathbf{v}, \quad \mathbf{W}_i = \mathbf{T}(\psi_i,\mathbf{C})\mathbf{w}, \quad i = 1,\ldots,5, \quad (2)$$

where $\mathbf{v}$ and $\mathbf{w}$ are the coordinates of the two moving pivots in the local frame. For convenience we introduce the relative transformations,

$$\mathbf{S}_{1i} = \mathbf{S}(\theta_i, \mathbf{O})\mathbf{S}(\theta_1, \mathbf{O})^{-1}, \quad \mathbf{T}_{1i} = \mathbf{T}(\psi_i, \mathbf{C})\mathbf{T}(\psi_1, \mathbf{C})^{-1}, \tag{3}$$

so we have

$$\mathbf{V}_i = \mathbf{S}_{1i}\mathbf{V}_1, \quad \mathbf{W}_i = \mathbf{T}_{1i}\mathbf{W}_1, \quad i = 2, \ldots, 5, \tag{4}$$

The goal of our design procedure is to compute the coordinates of $\mathbf{V}_1$ and $\mathbf{W}_1$. This is achieved using the constraint equations that require the moving pivots $\mathbf{V}_i$ and $\mathbf{W}_i$ to maintain a fixed distance R for each of the positions defined by the angles $\theta_i, \psi_i), i = 1, \ldots, 5$, given by,

$$(\mathbf{W}_i - \mathbf{V}_i) \cdot (\mathbf{W}_i - \mathbf{V}_i) = R^2, \quad i = 1, \ldots, 5. \tag{5}$$

Substitute (4) into these constraint equations to obtain,

$$(\mathbf{T}_{1i}\mathbf{W}_1 - \mathbf{S}_{1i}\mathbf{V}_1) \cdot (\mathbf{T}_{1i}\mathbf{W}_1 - \mathbf{S}_{1i}\mathbf{V}_1) = R^2, \quad i = 1, \ldots, 5, \tag{6}$$

which yields five equations in the five unknowns $\mathbf{V}_1 = (u_1, v_1)$, $\mathbf{W}_1 = (x_1, y_1)$ and R. We solve these equations to obtain our four-bar function generator.

The synthesis of four-bar function generators can have order and branching defects [15], so we evaluate every design. We increase the number of designs by introducing to the required set of angles within tolerance zones as described in Plecnik and McCarthy (2011) [11]. The result is a variety of four-bar function generators that provide these coordinated rotations within the specified tolerances.

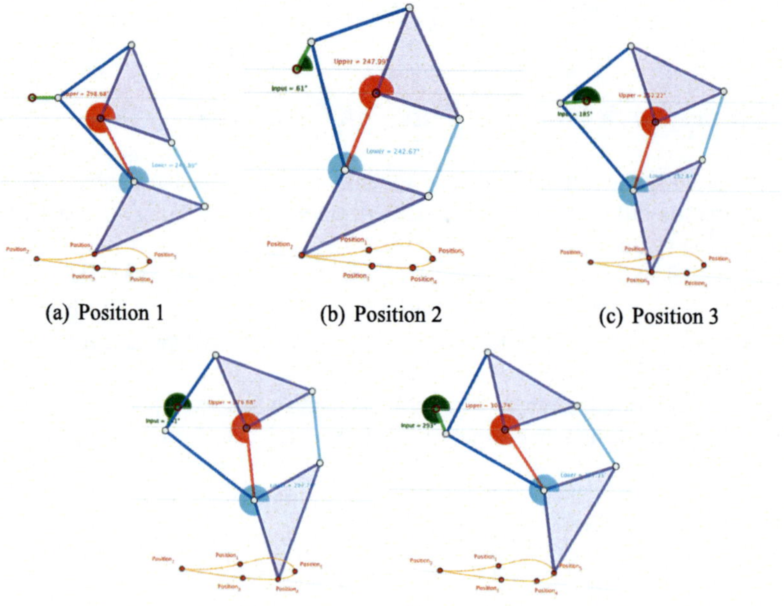

(a) Position 1 (b) Position 2 (c) Position 3

(d) Position 4 (e) Position 5

Fig. 4. Five positions of the Jansen leg model used for the design of four-bar function generators.

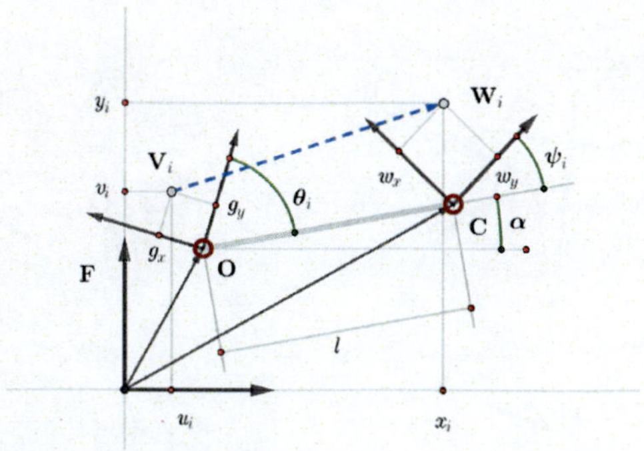

Fig. 5. The requirements for the four-bar function generator consists of two fixed pivots, **O** and **C**, and the set of five pairs of angles θ_i and ψ_i, $i = 1, \ldots, 5$ that define the coordinated rotations of the input and output links.

5 Example Result

The input and output angles for the lower leg were used to compute the dimensions of a function generator with input crank located at $\mathbf{O}_1$ and output crank located at **C**. The calculations yield the coordinates $\mathbf{A}_1$ and **B** that provide the required coordinates in the first task position. The rotation of **CB** is transmitted to the lower leg **DT** through the parallelogram **CEFD**. See Fig. 6.

Similarly, the input and output angles for the upper leg were used to define a function generator that has the input crank $\mathbf{O}_2$ and output crank **CG**. This yielded the coordinates for $\mathbf{A}_2$ and **G**. See Table 3.

A solid model of the leg mechanism is shown in Fig. 6. This shows separate input cranks driving the upper and lower legs, however, it is possible to design the two function generators so they use the same input crank.

Table 3. The coordinates for the joints of the two function generators $\mathbf{O}_1\mathbf{A}_1\mathbf{BC}$ and $\mathbf{O}_2\mathbf{A}_2\mathbf{GC}$ measured in the first task position.

Point	Coordinates	Point	Coordinates
$\mathbf{O}_1$	(0.00, 0.00)	$\mathbf{A}_1$	(0.61, −0.30)
B	(6.24, 0.65)	**C**	(3.10, −0.66)
$\mathbf{O}_2$	(1.37, 1.50)	$\mathbf{A}_2$	(2.20, 0.93)
G	(2.75, −3.36)	**C**	(3.10, −0.66)

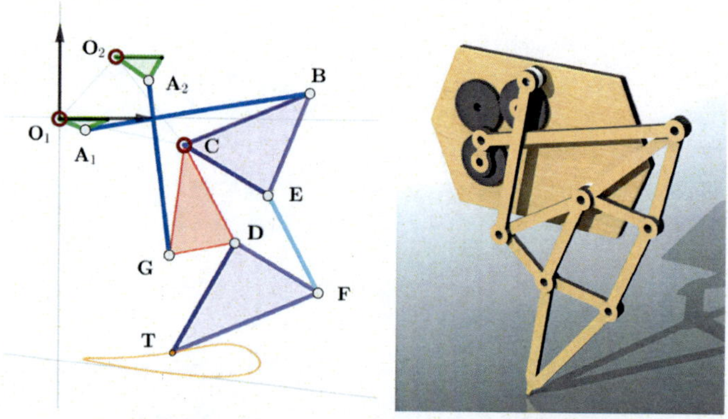

(a) Schematic (b) Solid Model

Fig. 6. (a) Schematic drawing of the modified Jansen leg mechanism with two coordinated input cranks $\mathbf{O}_1\mathbf{A}_1$ and $\mathbf{O}_2\mathbf{A}_2$, and (b) a solid model of the modified Jansen linkage with a gear driving the two input cranks.

6 Conclusion

This paper shows that the Jansen leg mechanism can be modified so that the movement of the upper and lower legs are controlled mechanically by separate four-bar function generators. This requires a minor modification of the lengths of the lower leg and a separation of the combined input crank into two separate but coordinated input cranks. We do not propose this modified Jansen leg mechanism as an improvement to the current design, rather as a way to design similar leg mechanisms as mechanically programmed RR serial chains using the principles of kinematic synthesis. This provides the opportunity to adjust the performance of the leg mechanism by modifying the performance of the function generators that control the upper and lower leg movements.

Acknowledgements. The authors gratefully acknowledge the assistance of Jeffrey Glabe and Angela Cardamone in the preparation of this paper, as well as reviewer comments that guided the final version of this paper.

References

1. Freudenstein, F.: Approximate synthesis of four-bar linkages. Trans. ASME **77**, 853–861 (1955)
2. Funabashi, H., Ogawa, K., Gotoh, Y., Kojima, F.: Synthesis of leg-mechanisms of biped walking machines. Bull. JSME **28**(237), 537–543 (1985)
3. Ghassaei, A.: The design and optimization of a crank-based leg mechanism. Technical report, Pomona College (2011)
4. Giesbrecht, D.: Design and optimization of a one-degree-of-freedom eight-bar leg mechanism for a walking machine. Master's thesis, The University of Manitoba (2010)
5. Hartenberg, R.S., Denavit, J.: Kinematic Synthesis of Linkages. McGraw-Hill Co., New York (1964)

6. Jansen, T.: Legsystem. https://youtu.be/ffs-2axfo1y. YouTube.com
7. Jansen, T. Strandbeest.com
8. Kaneko, M., Abe, M., Tanie, K.: A hexapod walking machine with decoupled freedoms. IEEE J. Robot. Autom. **RA–1**(4), 183–190 (1985)
9. McCarthy, J.M.: Kinematic Synthesis of Mechanisms: A Project-Based Approach. MDA Press, Brampton (2019)
10. Nansai, S., Rojas, N., Elara, M.R., Sosa, R., Iwase, M.: On a jansen leg with multiple gait patterns for reconfigurable walking platforms. In: Advances in Mechanical Engineering, pp. 1–18 (2015)
11. Plecnik, M.M., McCarthy, J.M.: Five position synthesis of a slider-crank function generator. In: Proceedings of the ASME 2011 International Design Engineering Technology Conferences, DETC2011-47581 (2011)
12. Shieh, W.B.: Design and optimization of planar leg mechanisms featuring symmetrical foot-point paths. Ph.D. thesis, The University of Maryland (1996)
13. Shieh, W.B., Tsai, L.W., Azarm, S., Tits, A.L.: A new class of six-bar mechanisms with symmetrical coupler curves. J. Mech. Des. **120**, 150–153 (1998)
14. Shigley, J.E.: The mechanics of walking vehicles. U.S, Army Ordnance Tank-Automotive Command (1960)
15. Singh, R., Chaudhary, H., Singh, A.K.: Defect-free optimal synthesis of crank-rocker linkage using nature-inspired optimization algorithms. Mech. Mach. Theory **116**, 105–122 (2017)
16. Tchebyshev, P.L.: On transformation of rotary movement into movement along some lines using joined systems. In: The Complete Works of P. L. Tchebyshev, vol IV. Theory of Mechanisms. AS USSR, Moscow-Leningrad (1948)
17. Tsai, L.W.: Mechanism Design: Enumeration of Kinematic Structures According to Function. CRC Press LLC, Boco Raton (2001)
18. Wikipedia: Jansen's linkage. https://en.wikipedia.org/wiki/jansen%27s_linkage

Exponential Displacement Coordinates by Means of the Adjoint Representation

Bertold Bongardt[1(✉)] and John J. Uicker[2]

[1] Institut für Robotik und Prozessinformatik, Technische Universität Braunschweig, 38102 Braunschweig, Germany
b.bongardt@tu-bs.de

[2] College of Engineering, University of Wisconsin-Madison, Madison, WI 53706, USA
uicker@engr.wisc.edu

Abstract. This paper introduces methods to determine the exponential coordinates of first kind and of second kind for an arbitrary spatial displacement given in terms of the left adjoint representation. Due to the algebraic properties of the (6×6)-matrix group, the obtained formulae are structure-preserving generalizations of their purely-rotative counterparts in coherence with the principle of transference. While the exponential coordinates of first kind represent the line-geometric parameters of a displacement (rotation and translation along a unit spear), the exponential coordinates of second kind coincide with dual Euler angles and with the parameters according to the kinematic convention by Sheth and Uicker. In either case, a spatial displacement is specified via six independent scalars, in form of a dual angle and a dual unit vector $(2+4=6)$ for the first kind and in form of three dual angles $(3\cdot2=6)$ along three sequentially orthogonal axes for the second kind. From a practical viewpoint, the parametrization method enables an automated parametrization of the kinematics of an arbitrary mechanism. From a theoretical viewpoint, the reported methods are relevant due to their structural simplicity and coherence.

Keywords: Adjoint representation · Sheth–Uicker parameters · Dual Euler angles · Line geometry · Screw theory · Principle of transference · Kinematic identification

1 Introduction

The symbolic notation described by Sheth and Uicker in 1971 represents a tool which permits specifying the kinematics of mechanisms of arbitrary topology and arbitrary geometry by means of a unified approach [12]. The notation has been developed as an extension to the notation described by Denavit and Hartenberg in 1955 [8], in particular, to cope with polynary links and with generic geometries of joint axes. The convention by Sheth and Uicker applies a decomposition of a particular displacement within a kinematic structure into three, consecutive helical displacements along standard coordinate axes. Due to this property, the

J. Lenarčič and B. Siciliano (Eds.): ARK 2020, SPAR 15, pp. 250–258, 2021.
https://doi.org/10.1007/978-3-030-50975-0_31

convention is able to parametrize any given displacement. The six parameters of a dual angle triplet $(\tilde{\gamma}, \tilde{\beta}, \tilde{\alpha}) \cong (\gamma, \mathring{\gamma}, \beta, \mathring{\beta}, \alpha, \mathring{\alpha})$ transmit relevant information about a mechanism's geometry: the location of a joint is maintained by the convention and the dual angle $\tilde{\beta} = \beta + \epsilon \cdot \mathring{\beta}$ specifies the translative and rotative distance between two consecutive joint axes [1]. Applied to the kinematics of a mechanism, a clear partition of the occurring displacements in time-invariant link displacements (given by design) and the time-variable joint displacements (given by application) is obtained [1,12]. In this document, a novel viewpoint on Sheth's and Uicker's convention is contributed by computing the six parameters from a displacement given in terms of the (6×6) left adjoint matrix [4,11]. It is demonstrated that the computation of Sheth–Uicker parameters from this matrix form is a straight-forward generalization of the computation of Euler angles from an orthogonal rotation matrix. This result corresponds to the identification of Sheth–Uicker parameters as the dual Euler angles of a displacement [1,10]. Due to the complementary relation of Sheth–Uicker parameters (exponential coordinates of second kind) and line-geometric parameters (exponential coordinates of first kind) of a displacement [1], the computation of the latter from an adjoint matrix is reported for sake of a self-contained analysis. The computation of the second exponential coordinates from the adjoint representation can also be interpreted as a simplification of the inverse kinematics solution to spatial 3C chains [3] for a specific design of axes which are consecutively intersecting orthogonally. The document is arranged as follows. As a basic means, two bivariate inverse tangent functions are introduced to invert exponential maps in Sect. 2. As a further prerequisite, the computation of exponential coordinates of first and second kind for rotations is summarized in Sect. 3. The generalized computation methods for displacements in the adjoint representation are stated for regular and singular geometries in Sect. 4. A brief discussion of applications and interrelations is provided in Sect. 5. The document is concluded by Sect. 6.

2 Complex and Dual-Complex Numbers

A complex number $z = x + \mathsf{i} \cdot y \in \mathbb{C}$ with $\mathsf{i}^2 = -1$ that features norm one $\|z\|^2 = z \cdot \overline{z} = 1$, with the complex conjugated $\overline{z} = x - \mathsf{i} \cdot y$, is characterized by Euler's identity

$$\hat{z} = \exp(\mathsf{i} \cdot \phi) = \cos(\phi) + \mathsf{i} \cdot \sin(\phi). \tag{1}$$

In the inverse direction, the (principal value of the) logarithm of a unit complex number $\hat{z}$ is determined via $\log(\hat{z}) = \mathsf{i} \cdot \phi = \mathsf{i} \cdot \arg(\hat{z})$, the 'argument' $\phi = \arg(\hat{z}) \in (\pi, \pi]$, an oriented angle. By means of the bivariate inverse tangent function

$$\operatorname{atan2}(y, x) = \begin{cases} \operatorname{atan}(y/x) & x > 0 \\ \operatorname{atan}(y/x) + \operatorname{sgn}(y) \cdot \pi & x < 0 \\ \operatorname{sgn}(y) \cdot \pi/2 & x = 0, \end{cases} \tag{2}$$

it holds that $\phi = \arg(x + \mathsf{i}\cdot y) = \mathrm{atan2}(y, x)$. Similarly, a dual complex number $\tilde{z} = \tilde{x} + \mathsf{i}\cdot\tilde{y} = x + \epsilon\cdot\mathring{x} + \mathsf{i}\cdot(y + \epsilon\cdot\mathring{y}) \in \tilde{\mathbb{C}}$ with $\epsilon^2 = 0$ that features norm one $\|\tilde{z}\|^2 = \tilde{z}\cdot\bar{\tilde{z}} = 1$ is characterized by the dual extension of Euler's identity

$$\hat{\tilde{z}} = \exp(\mathsf{i}\cdot\tilde{\phi}) = \widetilde{\cos}(\tilde{\phi}) + \mathsf{i}\cdot\widetilde{\sin}(\tilde{\phi}), \tag{3}$$

where $\widetilde{\cos}(\tilde{\phi}) = \cos(\phi) - \epsilon\cdot\mathring{\phi}\cdot\sin(\phi)$ and $\widetilde{\sin}(\tilde{\phi}) = \sin(\phi) + \epsilon\cdot\mathring{\phi}\cdot\cos(\phi)$. In the inverse direction, the complex logarithm $\log(\hat{\tilde{z}}) = \mathsf{i}\cdot\tilde{\phi} = \mathsf{i}\cdot\arg(\hat{\tilde{z}})$, It holds that $\tilde{\phi} = \arg(\tilde{x} + \mathsf{i}\cdot\tilde{y}) = \widetilde{\mathrm{atan2}}(\tilde{y}, \tilde{x})$ by means of the bivariate dual inverse tangent function [3]

$$\widetilde{\mathrm{atan2}}(\tilde{y}, \tilde{x}) = \mathrm{atan2}(y, x) + \epsilon\cdot\frac{x\mathring{y} - \mathring{x}y}{x^2 + y^2}. \tag{4}$$

While the set of unit complex numbers forms a unit circle $\mathcal{S}^1$, the set of unit dual-complex numbers forms the tangent bundle of a unit circle $\tilde{\mathcal{S}}^1 = \mathsf{T}\mathcal{S}^1 \cong \mathcal{S}^1 \times \mathbb{R} \subset \mathbb{R}^4$.

3 Rotation Coordinates

According to Euler's rotation theorem, a rotation $\boldsymbol{R} \in \mathsf{SO}(3)$ features an invariant axis, a unit vector $\hat{\boldsymbol{n}} \in \mathcal{S}^2$ and a rotation angle $\phi \in (-\pi, +\pi]$ about that axis. The trigonometric rotation formula by Rodrigues, Euler, Gauß, and Gelman reads

$$\boldsymbol{R} = \mathrm{rot}(\phi, \hat{\boldsymbol{n}}) = \exp(\phi\cdot\hat{\boldsymbol{n}}^{\otimes}) = \cos(\phi)\cdot\hat{\boldsymbol{n}}^{\circledcirc} + \sin(\phi)\cdot\hat{\boldsymbol{n}}^{\otimes} + \hat{\boldsymbol{n}}^{\oslash} \tag{5}$$

by means of three auxiliary tensors, the 'cross-matrix' $\boldsymbol{n}^{\otimes} = \begin{bmatrix} 0 & -n_3 & n_2 \\ n_3 & 0 & -n_1 \\ -n_2 & n_1 & 0 \end{bmatrix}$, the 'unit-matrix' $\boldsymbol{n}^{\circledcirc} = -(\boldsymbol{n}^{\otimes}\cdot\boldsymbol{n}^{\otimes})$, and the 'axial square-matrix' $\boldsymbol{n}^{\oslash} = (\boldsymbol{n} * \boldsymbol{n})\cdot\boldsymbol{I} - \hat{\boldsymbol{n}}^{\circledcirc}$, all associated to the axis of the rotation $\hat{\boldsymbol{n}}$ [3]. The operator $\oplus$ is defined via $(\boldsymbol{n}^{\otimes})^{\oplus} = \boldsymbol{n}$.

Coordinates of First Kind

Regular Geometries. In the inverse direction, the ('principal') matrix logarithm for a rotation matrix $\log(\boldsymbol{R}) = \phi\cdot\hat{\boldsymbol{n}}^{\otimes}$ is determined via the inversion of Eq. 5 as

$$\begin{aligned} \boldsymbol{n}^{\otimes} &:= \boldsymbol{R} - \boldsymbol{R}^T & \qquad c &:= \tfrac{1}{2}\cdot(\mathrm{trace}(\boldsymbol{R}) - 1) \\ \hat{\boldsymbol{n}}^{\otimes} &:= \mathrm{nrml}(\boldsymbol{n}^{\otimes}) & \qquad \phi &:= \mathrm{acos}(c). \end{aligned} \tag{6}$$

The normalization function $\mathrm{nrml}(\boldsymbol{n}^{\otimes}) = \|\boldsymbol{n}^{\otimes}\|^{-1}\cdot\boldsymbol{n}^{\otimes} = \hat{\boldsymbol{n}}^{\otimes}$ is induced by the matrix norm $\|\boldsymbol{n}^{\otimes}\| = (\tfrac{1}{2}\cdot\mathrm{trace}(\boldsymbol{n}^{\otimes} * \boldsymbol{n}^{\otimes}))^{1/2} = \|\boldsymbol{n}\|$.

Singular Geometries. In case of half-turns, rotations about π, the identities $c = -1$ and $\boldsymbol{R} = \mathrm{rot}(\pi, \hat{\boldsymbol{n}}) = \hat{\boldsymbol{n}}^{\oslash} - \hat{\boldsymbol{n}}^{\circledcirc} = 2\cdot\hat{\boldsymbol{n}}^{\oslash} - \boldsymbol{I} = \boldsymbol{R}^T$ permit to retrieve $\hat{\boldsymbol{n}}$ via $\boldsymbol{S} := \tfrac{1}{2}\cdot(\boldsymbol{R} + \boldsymbol{I})$ and

$$\hat{n}_+ := \left(\mathrm{diag}(\boldsymbol{S})\right)^{\circ\frac{1}{2}} \qquad \hat{\boldsymbol{n}} := \mathrm{diag}\Big(\mathrm{sgns}\big(\mathrm{diag}(\boldsymbol{S}\cdot(\hat{n}_+^{\oslash})^{+})\big)\Big)\cdot\hat{n}_+, \tag{7}$$

with elementwise root $(.)^{\circ\frac{1}{2}}$ and signum 'sgns', and generalized inverse $(.)^{+}$. For $\phi = 0$, it holds that $c = +1$ and $\boldsymbol{R} = \mathrm{rot}(0, \hat{\boldsymbol{n}}) = \boldsymbol{I}$ for which $\hat{\boldsymbol{n}} := \boldsymbol{0}$ is arranged.

Coordinates of Second Kind

According to Davenport, a rotation $\boldsymbol{R} \in \mathsf{SO}(3)$ can be factored into a sequence of rotations along three axes, if the second axis is orthogonal to the first and the third is orthogonal to the second [5]. The Bunge convention, the sequence of axes $(\hat{\boldsymbol{e}}_z,\ \hat{\boldsymbol{e}}_x,\ \hat{\boldsymbol{e}}_z)$, is a canonical choice that fulfills the requirement. Given three angles $\gamma, \beta, \alpha \in (-\pi, +\pi]$ with respect to the axes $(\hat{\boldsymbol{e}}_z,\ \hat{\boldsymbol{e}}_x,\ \hat{\boldsymbol{e}}_z)$, the rotation matrix $\boldsymbol{R} = \mathrm{rot}(\gamma, \hat{\boldsymbol{e}}_z) \cdot \mathrm{rot}(\beta, \hat{\boldsymbol{e}}_x) \cdot \mathrm{rot}(\alpha, \hat{\boldsymbol{e}}_z)$ features the nine elements

$$\boldsymbol{R} = \begin{bmatrix} \mathrm{c}_\gamma \mathrm{c}_\alpha - \mathrm{s}_\gamma \mathrm{c}_\beta \mathrm{s}_\alpha & -\mathrm{c}_\gamma \mathrm{s}_\alpha - \mathrm{s}_\gamma \mathrm{c}_\beta \mathrm{c}_\alpha & +\mathrm{s}_\gamma \mathrm{s}_\beta \\ \mathrm{s}_\gamma \mathrm{c}_\alpha + \mathrm{c}_\gamma \mathrm{c}_\beta \mathrm{s}_\alpha & -\mathrm{s}_\gamma \mathrm{s}_\alpha + \mathrm{c}_\gamma \mathrm{c}_\beta \mathrm{c}_\alpha & -\mathrm{c}_\gamma \mathrm{s}_\beta \\ \mathrm{s}_\beta \mathrm{s}_\alpha & \mathrm{s}_\beta \mathrm{c}_\alpha & \mathrm{c}_\beta \end{bmatrix}, \tag{8}$$

where 's' is shorthand for 'sine' and 'c' is shorthand for 'cosine'. In the inverse direction, the Euler angles are computed stepwise from the component representation in Eq. 8. Firstly, two values for the second angle β are determined by

$$\beta_+ := \mathrm{acos}\big(\boldsymbol{R}_{[3,3]}\big) \qquad\qquad \beta_- := -\beta_+. \tag{9}$$

By means of the auxiliary term $s_\beta := \sin(\beta_+)$, generic cases with two solutions for $s_\beta \neq 0$ are distinguished: from degenerate cases with infinite solutions for $s_\beta = 0$.

Regular Geometries. The values for the first γ and the third angle α are given by

$$\begin{aligned} \gamma_+ &:= \mathrm{atan2}\big(+\boldsymbol{R}_{[1,3]}, -\boldsymbol{R}_{[2,3]}\big) \qquad & \gamma_- &:= \mathrm{atan2}\big(-\boldsymbol{R}_{[1,3]}, +\boldsymbol{R}_{[2,3]}\big) \\ \alpha_+ &:= \mathrm{atan2}\big(+\boldsymbol{R}_{[3,1]}, +\boldsymbol{R}_{[3,2]}\big) \qquad & \alpha_- &:= \mathrm{atan2}\big(-\boldsymbol{R}_{[3,1]}, -\boldsymbol{R}_{[3,2]}\big) \end{aligned} \tag{10}$$

in a generic cases for $s \neq 0$, yielding two solution triplets $(\gamma_+, \beta_+, \alpha_+)$ and $(\gamma_-, \beta_-, \alpha_-)$.

Singular Geometries. The degenerate cases for $_\beta = 0$ feature infinite solution triplets and consist of half-turns ($\beta = \pi$) and zero-turns ($\beta = 0$) about the second axis. In case of $\beta = 0$, values for α and γ yield solutions as long as their sum $\alpha + \gamma$ remains constant. In case of $\beta = \pi$, the difference $\gamma - \alpha$ of α and γ needs to remain constant. By means of the auxiliary term $c_\beta = \cos(\beta) = \mathrm{sgn}(\cos(\beta)) = \mathrm{sgn}(\boldsymbol{R}_{[3,3]})$, two exemplary instances of the spectrum are given by

$$\begin{aligned} \gamma_- &:= \mathrm{atan2}\big(\boldsymbol{R}_{[2,1]}, \boldsymbol{R}_{[1,1]}\big) \qquad & \gamma_+ &:= 0 \\ \alpha_+ &:= c_\beta \cdot \mathrm{atan2}\big(\boldsymbol{R}_{[2,1]}, \boldsymbol{R}_{[1,1]}\big) \qquad & \alpha_- &:= 0. \end{aligned} \tag{11}$$

In the next section, the computations from Eqs. 6–7 and 9–11 are generalized from rotations to displacements in a structure-preserving manner employing the framework of (6×6) left adjoint matrices.

4 Displacement Coordinates

While the most popular representation of a displacement is the (4×4) matrix $\boldsymbol{D} = \left[\begin{smallmatrix} \boldsymbol{R} & \boldsymbol{t} \\ \boldsymbol{0} & 1 \end{smallmatrix}\right] \in \mathsf{SE}(3)$, it has been observed that the (6×6) matrices are a "more convenient" alternative [4].[1,2] In this document, the left adjoint representation [11] is indicated via $\grave{\mathsf{SE}}(3) = \mathrm{Ad}(\mathsf{SE}(3))$ and the matrix $\grave{\boldsymbol{D}} = \mathrm{Ad}(\boldsymbol{D})$ featuring the block structure

$$\grave{\boldsymbol{D}} = \begin{bmatrix} \boldsymbol{R} & \boldsymbol{0} \\ \boldsymbol{X} & \boldsymbol{R} \end{bmatrix} \in \grave{\mathsf{SE}}(3), \tag{12}$$

with $\boldsymbol{R} \in \mathsf{SO}(3)$ and left-bottom block $\boldsymbol{X} = \boldsymbol{t}^{\otimes} \cdot \boldsymbol{R}$. According to the displacement theorem by Mozzi and Chasles, any spatial displacement $\boldsymbol{D} \in \mathsf{SE}(3)$ features an invariant oriented line (spear), $\hat{\boldsymbol{\Lambda}} = \left[\begin{smallmatrix} \boldsymbol{n} \\ \boldsymbol{m} \end{smallmatrix}\right] \cong \hat{\boldsymbol{n}} + \epsilon \cdot (\boldsymbol{a} \times \hat{\boldsymbol{n}}) \in \tilde{\mathcal{S}}^2 = \mathsf{T}\mathcal{S}^2 \cong \mathcal{S}^2 \times \mathbb{R}^2 \subset \mathbb{R}^6$, in terms of Plücker coordinates. The 'line-geometric parameters', the tuple $(\tilde{\phi}, \hat{\boldsymbol{\Lambda}})$ of dual angle $\tilde{\phi} \in (-\pi, +\pi] \times \mathbb{R}$ and unit spear $\hat{\boldsymbol{\Lambda}}$ is also referred as 'screw displacement pair' [9]. The generalization of the rotation formula (5) to displacements reads

$$\grave{\boldsymbol{D}} = \mathrm{adsp}(\tilde{\phi}, \hat{\boldsymbol{\Lambda}}) = \exp(\tilde{\phi}^{\boxed{}} \cdot \hat{\boldsymbol{\Lambda}}^{\boxtimes}) = \widetilde{\cos}(\tilde{\phi})^{\boxed{}} \cdot \hat{\boldsymbol{\Lambda}}^{\boxdot} + \widetilde{\sin}(\tilde{\phi})^{\boxed{}} \cdot \hat{\boldsymbol{\Lambda}}^{\boxtimes} + \hat{\boldsymbol{\Lambda}}^{\boxslash}, \tag{13}$$

with dual number matrix $\tilde{x}^{\boxed{}} = (x + \epsilon \cdot \mathring{x})^{\boxed{}} = \left[\begin{smallmatrix} x\boldsymbol{I} & \boldsymbol{0} \\ \mathring{x}\boldsymbol{I} & x\boldsymbol{I} \end{smallmatrix}\right]$, and three auxiliary tensors, cross matrix $\boldsymbol{\Lambda}^{\boxtimes} = \left[\begin{smallmatrix} \boldsymbol{n}^{\otimes} & \boldsymbol{0} \\ \boldsymbol{m}^{\otimes} & \boldsymbol{n}^{\otimes} \end{smallmatrix}\right]$, unit matrix $\boldsymbol{\Lambda}^{\boxdot} = -(\boldsymbol{\Lambda}^{\boxtimes} \cdot \boldsymbol{\Lambda}^{\boxtimes})$, and axial square matrix $\boldsymbol{\Lambda}^{\boxslash} = (\boldsymbol{n} * \boldsymbol{n}) \cdot \boldsymbol{I} - \boldsymbol{\Lambda}^{\boxdot}$, all associated to the invariant spear $\boldsymbol{\Lambda}$ [3].

Coordinates of First Kind

Regular Geometries. In formal analogy to (6), the matrix logarithm for an adjoint displacement $\log(\grave{\boldsymbol{D}}) = \tilde{\phi}^{\boxed{}} \cdot \hat{\boldsymbol{\Lambda}}^{\boxtimes}$ is determined via the inversion of Eq. 13 as

$$\begin{aligned} \$^{\boxtimes} &:= \grave{\boldsymbol{D}} - \grave{\boldsymbol{D}}^{-1} & \qquad \tilde{c} &:= \tfrac{1}{2} \cdot (\widetilde{\mathrm{trace}}(\grave{\boldsymbol{D}}) - 1) \\ \hat{\boldsymbol{\Lambda}}^{\boxtimes} &:= \widetilde{\mathrm{nrml}}(\$^{\boxtimes}) & \qquad \tilde{\phi} &:= \widetilde{\mathrm{acos}}(\tilde{c}), \end{aligned} \tag{14}$$

via the dual normalization function $\widetilde{\mathrm{nrml}}(\$^{\boxtimes}) = (\|\$^{\boxtimes}\|^{-1})^{\boxed{}} \cdot \$^{\boxtimes} = \hat{\boldsymbol{\Lambda}}^{\boxtimes}$ with dual screw-matrix norm $\|\$^{\boxtimes}\| = \|\boldsymbol{n}\| + \epsilon \cdot \frac{\boldsymbol{n} * \boldsymbol{m}}{\|\boldsymbol{n}\|}$, via the dual inverse cosine $\widetilde{\mathrm{acos}}(\tilde{c}) = \mathrm{acos}(c) - \mathring{c} \cdot (\cos(\mathrm{asin}(c)))^{-1}$, and the dual trace function $\widetilde{\mathrm{trace}}(\grave{\boldsymbol{D}}) = \mathrm{trace}(\boldsymbol{R}) + \epsilon \cdot \mathrm{trace}(\boldsymbol{X})$.

[1] "*Algebraic manipulation for* (3×3) *matrices with dual-number elements is easier than that of screw algebra and dual quaternions since it, operations follow the same rules for* (3×3) *orthogonal real matrices.*" [13]. The same holds for (6×6) with real elements, since they are equivalent to (3×3) matrices with dual-number elements [3].

[2] Given a displacement in terms of a homogeneous (4×4) matrix, methods for computing the exponential coordinates of first kind [7] and of second kind [2] have been reported; a derivation of Denavit–Hartenberg parameters is described in [6]. A method to determine the matrix logarithm of an adjoint (6×6) displacement matrix is reported in [11].

Singular Geometries. In case of π-turns, the displacement matrix has the shape

$$\grave{\boldsymbol{D}} = \operatorname{adsp}(\pi + \epsilon \cdot s, \hat{\boldsymbol{\Lambda}}) = \begin{bmatrix} 2\hat{\boldsymbol{n}}^{\oslash} - \boldsymbol{I} & \boldsymbol{0} \\ \boldsymbol{X} & 2\hat{\boldsymbol{n}}^{\oslash} - \boldsymbol{I} \end{bmatrix},$$

with $\boldsymbol{X} = \boldsymbol{X}(\pi + \epsilon \cdot s, \hat{\boldsymbol{\Lambda}}) = 2 \cdot (\boldsymbol{m}^{\otimes} \cdot \hat{\boldsymbol{n}}^{\otimes} + \hat{\boldsymbol{n}}^{\otimes} \cdot \boldsymbol{m}^{\otimes}) - s \cdot \hat{\boldsymbol{n}}^{\otimes}$. In this case, the moment $\boldsymbol{m}$ of $\hat{\boldsymbol{\Lambda}}$ and the shift s are obtained via $\hat{\boldsymbol{n}}$ from Eq. 7 and with

$$\boldsymbol{m} := \tfrac{1}{2} \cdot \boldsymbol{X} \cdot \hat{\boldsymbol{n}} \quad \boldsymbol{S} := \tfrac{1}{2} \cdot (\boldsymbol{X} - \boldsymbol{X}^T) \quad s_+ := \|\boldsymbol{S}\| \quad s := -\operatorname{sgn}\big(\hat{\boldsymbol{n}} * (\boldsymbol{S}^{\oplus})\big) \cdot s_+ \,. \tag{15}$$

In case of $\tilde{\phi} = 0 + \epsilon \cdot s$, s and $\hat{\boldsymbol{\Lambda}}$ are obtained via $s := \|\boldsymbol{X}^{\oplus}\|$, $\hat{\boldsymbol{n}} := \boldsymbol{X}^{\oplus}/s$, and $\boldsymbol{m} := \boldsymbol{0}$.

Coordinates of Second Kind

Transferring Davenport's theorem to the general case, any displacement $\boldsymbol{D} \in \mathsf{SE}(3)$ can be factored into a sequence of helical displacements along three given spears, if the second spear is orthogonal to the first, and the third orthogonal to the second. The Bunge convention, the sequence of axes $(\hat{\boldsymbol{\Lambda}}_z, \hat{\boldsymbol{\Lambda}}_x, \hat{\boldsymbol{\Lambda}}_z)$, fulfills the requirement. Given dual angles $\tilde{\gamma}, \tilde{\beta}, \tilde{\alpha} \in (-\pi, \pi] \times \mathbb{R}$, the left adjoint displacement matrix, computed as

$$\grave{\boldsymbol{D}} = \operatorname{adsp}(\tilde{\gamma}, \hat{\boldsymbol{\Lambda}}_z) \cdot \operatorname{adsp}(\tilde{\beta}, \hat{\boldsymbol{\Lambda}}_x) \cdot \operatorname{adsp}(\tilde{\alpha}, \hat{\boldsymbol{\Lambda}}_z), \tag{16}$$

features the block structure of Eq. 12 with blocks $\boldsymbol{R} \in \mathsf{SO}(3)$ and $\boldsymbol{X} = \boldsymbol{t}^{\otimes} \cdot \boldsymbol{R}$. The 'primal' block $\boldsymbol{R}$ has the nine components listed in Eq. 8 and the 'dual' block $\boldsymbol{X}$ features the entries[3]

$$\boldsymbol{X} = \begin{bmatrix} x_{11} & x_{12} & \mathring{\gamma}(\mathrm{c}_\gamma \mathrm{s}_\beta) + \mathring{\beta}(\mathrm{s}_\gamma \mathrm{c}_\beta) \\ x_{21} & x_{22} & \mathring{\gamma}(\mathrm{s}_\gamma \mathrm{s}_\beta) - \mathring{\beta}(\mathrm{c}_\gamma \mathrm{c}_\beta) \\ \mathring{\beta}(\mathrm{c}_\beta \mathrm{s}_\alpha) + \mathring{\alpha}(\mathrm{s}_\beta \mathrm{c}_\alpha) & +\mathring{\beta}(\mathrm{c}_\beta \mathrm{c}_\alpha) - \mathring{\alpha}(\mathrm{s}_\beta \mathrm{s}_\alpha) & -\mathring{\beta}(\mathrm{s}_\beta) \end{bmatrix}.$$

For convenience, the auxiliary function $\widetilde{\Pi}(i, j; \grave{\boldsymbol{A}}) = \boldsymbol{A}_{[i,j]} + \epsilon \cdot \mathring{\boldsymbol{A}}_{[i,j]}$ is introduced to access a certain 'dual entry' of a matrix $\grave{\boldsymbol{A}} = \left[\begin{smallmatrix} \boldsymbol{A} & \boldsymbol{0} \\ \mathring{\boldsymbol{A}} & \boldsymbol{A} \end{smallmatrix}\right] \in \grave{\mathsf{SE}}(3)$. As in the rotational case, the decision variable $s_\beta := \sin(\operatorname{acos}(\boldsymbol{R}_{[3,3]}))$ is introduced to distinguish generic from degenerate geometries.

Regular Geometries. For a generic case with $_\beta \neq 0$, two values for the second angle are determined in generalization of Eq. 9 as

$$\tilde{\beta}_+ := \widetilde{\operatorname{acos}}(\widetilde{\Pi}(3, 3; \grave{\boldsymbol{D}})) \qquad \tilde{\beta}_- := -\beta_+. \tag{17}$$

---

[3] On the top-left, the entries are $x_{11} = \mathring{\gamma} \cdot (-\mathrm{s}_\gamma \mathrm{c}_\alpha - \mathrm{c}_\gamma \mathrm{c}_\beta \mathrm{s}_\alpha) + \mathring{\beta} \cdot (\mathrm{s}_\gamma \mathrm{s}_\beta \mathrm{s}_\alpha) + \mathring{\alpha} \cdot (-\mathrm{c}_\gamma \mathrm{s}_\alpha - \mathrm{s}_\gamma \mathrm{c}_\beta \mathrm{c}_\alpha)$, $x_{12} = \mathring{\gamma} \cdot (+\mathrm{s}_\gamma \mathrm{s}_\alpha - \mathrm{c}_\gamma \mathrm{c}_\beta \mathrm{c}_\alpha) + \mathring{\beta} \cdot (\mathrm{s}_\gamma \mathrm{s}_\beta \mathrm{c}_\alpha) + \mathring{\alpha} \cdot (-\mathrm{c}_\gamma \mathrm{c}_\alpha + \mathrm{s}_\gamma \mathrm{c}_\beta \mathrm{s}_\alpha)$, $x_{21} = \mathring{\gamma} \cdot (+\mathrm{c}_\gamma \mathrm{c}_\alpha - \mathrm{s}_\gamma \mathrm{c}_\beta \mathrm{s}_\alpha) - \mathring{\beta} \cdot (\mathrm{c}_\gamma \mathrm{s}_\beta \mathrm{s}_\alpha) + \mathring{\alpha} \cdot (-\mathrm{s}_\gamma \mathrm{s}_\alpha + \mathrm{c}_\gamma \mathrm{c}_\beta \mathrm{c}_\alpha)$, and $x_{22} = \mathring{\gamma} \cdot (-\mathrm{c}_\gamma \mathrm{s}_\alpha - \mathrm{s}_\gamma \mathrm{c}_\beta \mathrm{c}_\alpha) - \mathring{\beta} \cdot (\mathrm{c}_\gamma \mathrm{s}_\beta \mathrm{c}_\alpha) + \mathring{\alpha} \cdot (-\mathrm{s}_\gamma \mathrm{c}_\alpha - \mathrm{c}_\gamma \mathrm{c}_\beta \mathrm{s}_\alpha)$.

Then, the angles $\tilde{\gamma}$ and $\tilde{\alpha}$ are computed in generalization of Eq. 10 as

$$\begin{aligned}\tilde{\gamma}_+ &:= \widetilde{\text{atan2}}\big(+\tilde{\Pi}(1,3;\grave{\boldsymbol{D}}), -\tilde{\Pi}(2,3;\grave{\boldsymbol{D}})\big)\\ \tilde{\gamma}_- &:= \widetilde{\text{atan2}}\big(-\tilde{\Pi}(1,3;\grave{\boldsymbol{D}}), +\tilde{\Pi}(2,3;\grave{\boldsymbol{D}})\big)\\ \tilde{\alpha}_+ &:= \widetilde{\text{atan2}}\big(+\tilde{\Pi}(3,1;\grave{\boldsymbol{D}}), +\tilde{\Pi}(3,2;\grave{\boldsymbol{D}})\big)\\ \tilde{\alpha}_- &:= \widetilde{\text{atan2}}\big(-\tilde{\Pi}(3,1;\grave{\boldsymbol{D}}), -\tilde{\Pi}(3,2;\grave{\boldsymbol{D}})\big).\end{aligned} \tag{18}$$

These are two triplets of the dual Euler angles $(\tilde{\gamma}_+, \tilde{\beta}_+, \tilde{\alpha}_+)$ and $(\tilde{\gamma}_-, \tilde{\beta}_-, \tilde{\alpha}_-)$. With respect to Sheth–Uicker parameters, it is suitable to select one, for example, the first, solution triplet with $\beta \geq 0$, in order to deal with values in 'normal form'.

Singular Geometries. The exceptional cases for $\beta = 0$ feature infinite solutions. Here, the dual parts of $\tilde{\gamma}$ and $\tilde{\alpha}$ can be varied and the primal parts are fixed. For the second angle $\tilde{\beta}$, the values are

$$\beta := \text{acos}(\boldsymbol{R}_{[3,3]}) \qquad \mathring{\beta} := \sqrt{(\boldsymbol{X}_{[3,1]})^2 + (\boldsymbol{X}_{[3,2]})^2}. \tag{19}$$

The primal parts of $\tilde{\gamma}$ and $\tilde{\alpha}$ are either obtained by Eq. 11 or by

$$\gamma := \text{atan2}(\boldsymbol{X}_{[1,3]}, -\boldsymbol{X}_{[2,3]}) + \beta \quad \alpha := \text{atan2}(\boldsymbol{X}_{[3,1]}, \boldsymbol{X}_{[3,2]}) + \beta. \tag{20}$$

For the dual parts, the generalization to Eq. 11, with $c_\beta := \cos(\beta) = \text{sgn}(\cos(\beta)) = \text{sgn}(\tilde{\Pi}(3,3;\grave{\boldsymbol{D}})) = \text{sgn}(\boldsymbol{R}_{[3,3]})$, delivers

$$\begin{aligned}\mathring{\gamma}_- &:= \pi_\epsilon\big(\widetilde{\text{atan2}}\big(\tilde{\Pi}(2,1;\grave{\boldsymbol{D}}), \tilde{\Pi}(1,1;\grave{\boldsymbol{D}})\big)\big) & \mathring{\gamma}_+ &:= 0\\ \mathring{\alpha}_+ &:= \pi_\epsilon\big(c_\beta \cdot \widetilde{\text{atan2}}\big(\tilde{\Pi}(2,1;\grave{\boldsymbol{D}}), \tilde{\Pi}(1,1;\grave{\boldsymbol{D}})\big)\big) & \mathring{\alpha}_- &:= 0,\end{aligned} \tag{21}$$

where the auxiliary projection function $\pi_\epsilon(\tilde{x}) = \epsilon^{-1} \cdot (\tilde{x} - \underline{\tilde{x}})/2$ with dual conjugate $\underline{\tilde{x}} = x - \epsilon \cdot \mathring{x}$ is employed to retrieve the dual part $\mathring{x}$ of a dual number $\tilde{x}$.

5 Discussion

The sketch on the left hand side of Fig. 1 illustrates the application context of Sheth–Uicker parameters: An adjoint displacement matrix $\grave{\boldsymbol{D}}$ from Eq. 16 represents a time-invariant displacement between two joints J_{ij} and J_{jk} adjacent to a common link L_j, formally as

$$\grave{\boldsymbol{D}} = \grave{\boldsymbol{D}}_{ik} = \grave{\boldsymbol{P}}_{ij}^{-1} \cdot \grave{\boldsymbol{P}}_{jk}, \tag{22}$$

where $\grave{\boldsymbol{P}}_{ij}$ and $\grave{\boldsymbol{P}}_{jk}$ are the adjoint pose matrices of the Sheth-Uicker frames at the joints J_{ij} and J_{jk} in the mechanism's reference posture. The drawing on the right hand side of Fig. 1 illustrates the combinatorial point of view: Due to the partitioning into time invariant and time variable displacements, Sheth–Uicker

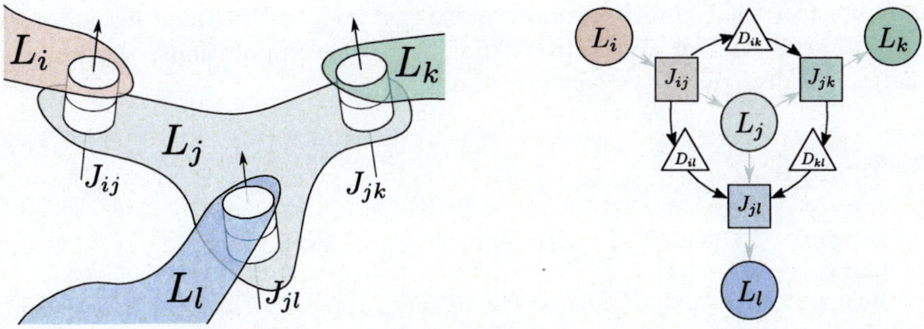

Fig. 1. A sketch of the geometry of a ternary link with skew joint axes (spears) in a mechanism and an augmentation of the mechanism's topological (sub-)graph by (squarish) joint vertices and (triangular) displacement vertices, as a tripartite directed graph.

parameters represent a tool for treating a mechanism of arbitrary complexity in a modular way in accordance with its topological graph. A k-nary link (branching or closed-loop) is treated in a canonical way that does not require manual intervention. Hence, a Sheth–Uicker parametrization provides a technical means to employ kinematic identification [6] of arbitrary mechanisms in an automated manner.

Table 1 provides an outline of the major formulae of the preceding sections for the four involved domains. By using the adjoint matrix representation for rendering a displacement, the formulae for inverting exponential maps feature strong structural similarity and coherence.

Table 1. Overview of four domains and their main formulae (for regular geometries).

Symbol	Domain	Exponential	Argument	First inversion	Second inversion
$\mathcal{S}^1$	Unit complex numbers	(1)	$\phi \cdot \mathsf{i}$	(2)	—
$\mathsf{SO}(3)$	Rotations	(5)	$\phi \cdot \hat{\boldsymbol{n}}^{\otimes}$	(6)	(9, 10)
$\mathcal{S}^1 \times \mathbb{R}$	Unit dual-complex numb	(3)	$\tilde{\phi} \cdot \mathsf{i}$	(4)	—
$\mathsf{SE}(3)$	Displacements	(13)	$\tilde{\phi}^{\boxdot} \cdot \hat{\boldsymbol{\Lambda}}^{\boxtimes}$	(14)	(17, 18)

6 Conclusions

The paper shows how an arbitrary spatial Euclidean displacement is parametrized by exponential coordinates of first kind and of second kind via elementary methods if the displacement is given in terms of a left adjoint (6×6) matrix. Based on the adjoint form, the procedures are structure-preserving extensions of their rotative counterparts that are well suitable for automated computations. As an application domain of the computations, the kinematic analysis of mechanical systems is outlined.

References

1. Bongardt, B.: Sheth-Uicker convention revisited. Mech. Mach. Theor. **69**, 200–229 (2013)
2. Bongardt, B.: Analytic approaches for design and operation of haptic human-machine interfaces. Ph.D. thesis, Universität Bremen (2015)
3. Bongardt, B.: The Adjoint trigonometric representation of displacements and a closed-form solution to the IKP of general 3C chains. J. Appl. Math. Mech. (accepted)
4. Borri, M., Trainelli, L., Bottasso, C.L.: On representations and parameterizations of motion. Multibody Syst. Dyn. **4**, 129–193 (2000)
5. Davenport, P.B.: Rotations about nonorthogonal axes. AIAA J. **11**(6), 853–857 (1973)
6. Faria, C., Vilaça, J.L., Monteiro, S., Erlhagen, W., Bicho, E.: Automatic Denavit-Hartenberg parameter identification for serial manipulators. In: Conference of the IEEE Industrial Electronics Society, pp. 610–617 (2019)
7. Gallier, J., Xu, D.: Computing exponentials of skew symmetric matrices and logarithms of orthogonal matrices. Int. J. Robot. Autom. **18**, 10–20 (2000)
8. Hartenberg, R.S., Denavit, J.: A kinematic notation for lower-pair mechanisms based on matrices. J. Appl. Mech. **22**, 215–221 (1955)
9. Hiller, M., Woernle, C.: A unified representation of spatial displacements. Mech. Mach. Theor. **19**(6), 477–486 (1984)
10. Rull, A., Thomas, F.: On generalized dual Euler angles. In: European Conference on Mechanism Science, pp. 61–68 (2014)
11. Selig, J.M.: Geometric Fundamentals of Robotics. Springer, Heidelberg (2005)
12. Sheth, P.N., Uicker, J.J.: A generalized symbolic notation for mechanisms. J. Eng. Ind. Ser. B **93**(70), 102–112 (1971)
13. Yang, A.T.: Displacement analysis of spatial five-link mechanisms using (3x3) matrices with dual-number elements. J. Eng. Ind. **91**(1), 152–157 (1968)

A Comparative Study on 2-DOF Variable Stiffness Mechanisms

Christoph Stoeffler[1(✉)], Shivesh Kumar[2], and Andreas Müller[3]

[1] AG Robotik, Universität Bremen, Bremen, Germany
stoeffle@uni-bremen.de
[2] Robotics Innovation Center, DFKI Bremen, Bremen, Germany
shivesh.kumar@dfki.de
[3] Johannes Kepler University, Linz, Austria
a.mueller@jku.at

Abstract. Based on the idea of *variable stiffness mechanisms*, a variety of such mechanisms is shown in this work. Specifically, 2-DOF *parallel kinematic machines* equipped with redundant actuators and non-linear springs in the actuated joints are presented and a comparative overview is drawn. Accordingly, a general stiffness formulation in task space of all mechanisms is given. Under fixed geometric parameters, optimization of task space stiffness is carried out on the designs comprising all kinematic solutions. Finally, a stiffness metric is introduced that allows a quantitative comparison of the given mechanism designs. This gives rise to design guidelines for engineers but also shows an interesting outline for future applications of variable stiffness mechanisms.

Keywords: Variable stiffness · Redundant actuation · Stiffness modeling · Optimal control

1 Introduction and Overview

Using *parallel kinematic machines* (PKMs) in modern robotic designs results in many advantages, due to the possibility of redundant actuation, higher load capacity and so on. Another recent tendency in robotics is the integration of flexible and soft links and joints to comply with higher safety standards, but also to increase the performance of robotic systems - may it be through energy storage or reduced mass, where for the latter flexibility is rather a consequence than an enhancement. Most notably, *serial elastic actuators* (SEAs) and *variable impedance actuators* (VIAs) find many applications in nowadays systems - see e.g [2]. Combining the conceptual ideas of PKMs and VIAs leads to a mechanism type that we call *variable stiffness mechanisms* (VSMs) and which we recently proposed in [10]. In general, VSMs allow simultaneous position and stiffness control of several end-effector DOFs and can thus be seen as a higher dimensional extension of VIAs. It is hoped that such designs lead to further enhancements of robotic systems, as they integrate compliance into multi-DOF mechanisms

J. Lenarčič and B. Siciliano (Eds.): ARK 2020, SPAR 15, pp. 259–267, 2021.
https://doi.org/10.1007/978-3-030-50975-0_32

and give rise to impact tolerance, energy storage and stiffness adaptations. In this work a comparative overview of 2-degree-of-freedom (2-DOF) VSMs based on stiffness bounds is presented. For this purpose various possible VSMs are presented in this section followed by a generic approach of *Cartesian stiffness* modeling of all presented VSMs in Sect. 2. A metric to quantify the stiffness in task space is presented and computed for all mechanisms in Sect. 4 that is used to derive some design guidelines. Finally, Sect. 5 draws conclusions and future perspectives of this work.

Like animal muscle groups, VSMs rely on antagonistic actuation within parallel structures that can be achieved in different ways. An early example of this approach can be found in [11], where stiffness control is opted via antagonistic actuation and later e.g. in [6], including the dynamics of the system. Figure 1 shows six different ways[1] to achieve a planar 2-DOF VSM, ordered by the degree of actuation and its implementation. Arrows in Fig. 1 indicate a module of non-linear spring combined with a drive. Pairs of those in one joint indicate pair of springs and drives what would typically be a VIA. For the sake of stiffness changes at the end-effector, non-linear springs are necessary - stiffness changes could be induced by the non-linearity and antagonism of the mechanism itself, like e.g. in [4,6], but we demand here, that the passive stiffness can also be changed. This in turn requires stiffness changes on joint level, what is expected to also enlarge the possible stiffnesses in the task space. A description of different redundancies in PKMs is given in [7] with the terms *activation of passive joints* and *increment of manipulator joint space* that apply to the centre box and right box of Fig. 1 respectively. But we can also exploit redundant actuation with VIAs in the base joints - shown in the left box of that figure.

2 Stiffness Modeling and Optimization

Stiffness is the coefficient in the first order approximation of the relation of force and displacement. It is thus determined by the first-order term of the *Taylor expansion* of a force $\mathbf{f}$ with respect to a displacement $\mathbf{s}$ whereas, usually higher-order terms are neglected and then stiffness generally writes

$$\mathbf{K}_s = \frac{d\mathbf{f}}{d\mathbf{s}}. \tag{1}$$

We can define stiffness in different spaces, but it is required that $\mathbf{f}$ and $d\mathbf{s}$ exist in a certain space and its *dual*. This naturally results in square stiffness matrices, where off-diagonal terms describe the coupling between the coordinates contained in $\mathbf{s}$. When force changes are decoupled from each other, diagonal stiffness matrices result. Those are e.g. present when the joint space stiffness of robotic systems is considered. A widely discussed problem is the conversion of stiffness

[1] For a 2-DOF mechanism, this is the number of reasonable designs, as we omit designs that are symmetric to others and that have higher degree of redundancy than end-effector DOFs.

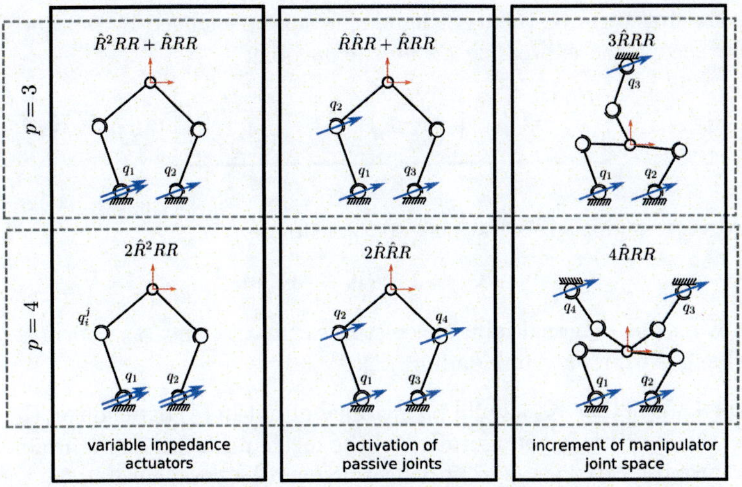

Fig. 1. Overview of six different planar 2-DOF VSM, distinguished by the degree of actuation p and the way it is achieved. Arrows indicate installed motor-spring combinations - single arrow: (nonlinear) SEA, double arrow: VIA. A mechanism can be described by full coordinates q_i^j (jth joint coordinate in ith leg) or by m generalized coordinates q_k respectively.

from joint space to task space e.g. $f : \mathbf{K}_q \mapsto \mathbf{K}_x$. A good overview that addresses this mapping is given in [8] encompassing different works on serial and parallel manipulators. One major difference in the accounted works is the treatment of external loading that is commonly said to be responsible for asymmetric stiffness matrices. It has been argued by Hoevenaars and others [3] that asymmetry is a modeling inconsistency, as the resulting matrices have to arise from *conservative* properties alone. Zefran and Kumar [12] show in a geometric setting that asymmetry is resolved by choosing an *affine connection* on the robots configuration - the Lie group $SE(3)$. In contrast to this, Chen [1] illustrates asymmetric matrices as a result of *non-coordinate transformations*. What follows, is a practical argumentation for a model of symmetric stiffness matrices of our designs, exemplified by the *conservative congruence transformation* (CCT) [9]. Mapping via Jacobian between joint space forces τ and task space forces $\mathbf{f}$ is generally

$$\boldsymbol{\tau} = \mathbf{J}^T \mathbf{f} \tag{2}$$

Differentiating and making use of the stiffness expression (1) yields

$$\begin{aligned} d\boldsymbol{\tau} &= \left(d\mathbf{J}^T\right)\mathbf{f} + \mathbf{J}^T\left(d\mathbf{f}\right) \\ \mathbf{K}_q d\mathbf{q} &= \left(\frac{\partial \mathbf{J}^T}{\partial \mathbf{q}} d\mathbf{q}\right)\mathbf{f} + \mathbf{J}^T \mathbf{K}_x d\mathbf{x} \end{aligned} \tag{3}$$

where the first term on the right hand side can be changed by reordering indices and the second via expressing $d\mathbf{x}$ by the Jacobian

$$\mathbf{K}_q d\mathbf{q} = \underbrace{\left[\left(\frac{\partial \mathbf{J}^T}{\partial q_1} \mathbf{f} \right) \left(\frac{\partial \mathbf{J}^T}{\partial q_2} \mathbf{f} \right) \quad \dots \quad \left(\frac{\partial \mathbf{J}^T}{\partial q_m} \mathbf{f} \right) \right]}_{\mathbf{K}_g} d\mathbf{q} + \mathbf{J}^T \mathbf{K}_x \mathbf{J} d\mathbf{q}$$

Finally, rearranging gives the task space stiffness

$$\mathbf{K}_x = \mathbf{J}^{-T}(\mathbf{K}_q - \mathbf{K}_g)\mathbf{J}^{-1} \tag{4}$$

where $\mathbf{K}_q$ is the diagonal joint space stiffness matrix and $\mathbf{K}_g$ captures the Jacobian changes from external loading.

Remark 1. Whether $\mathbf{K}_g$ should be included or not in (4) depends on the assumption whether a certain configuration of the mechanism can be maintained or not. In our proposed designs, spring deflections can be compensated by the serially attached motors. Thus external forces lead to stiffness changes on joint level, but the configuration of the mechanism is depending on commanded position $\mathbf{x}_c$ alone and $(d\mathbf{J}^T)\,\mathbf{f}$ in (3) vanishes. The stiffness mapping in (4) simplifies to

$$\mathbf{K}_x = \mathbf{J}^{-T}\mathbf{K}_q\mathbf{J}^{-1} \tag{5}$$

External loading can thus be seen as a problem that is solely reflected in $\mathbf{K}_q$ what is in accordance with [3].

2.1 Constraint Jacobian

The Jacobian matrix of a mechanism is given by differentiation of a constraint equation $\mathbf{g}(\mathbf{x}, \mathbf{q}) = \mathbf{0}$ such as

$$\frac{\partial \mathbf{g}(\mathbf{x}, \mathbf{q})}{\partial \mathbf{x}} \dot{\mathbf{x}} = -\frac{\partial \mathbf{g}(\mathbf{x}, \mathbf{q})}{\partial \mathbf{q}} \dot{\mathbf{q}} \tag{6}$$

$$\dot{\mathbf{x}} = -\mathbf{J}_x^{\dagger}\mathbf{J}_q\dot{\mathbf{q}} \tag{7}$$

$$\dot{\mathbf{x}} = \mathbf{J}\dot{\mathbf{q}} \tag{8}$$

where $\mathbf{J}_x^{\dagger}$ is the *pseudoinverse* in the case of $dim(\mathbf{g}) > dim(\mathbf{x})$ and otherwise the usual inverse of $\mathbf{J}_x$.

2.2 Stiffness Computation in Task Space

With the above equations and i legs attached to the end-effector, task space stiffness can be computed according to the scheme in Fig. 2, where the functions $f_i^{-1} : \mathbf{x} \mapsto q_i^j$ and $\partial f_i^{-1}/\partial \mathbf{x} : q_i^j \mapsto \mathbf{J}_i^{leg}$ are the inverse and differential kinematics of 2-DOF serial planar mechanisms, as to be found e.g. in [5]. It should be noted that the full state of the mechanism q_i^j is computed and the subset q_k is used to obtain the Jacobians of the parallel mechanism $\mathbf{J}$ to account for external loading. Finally, Cartesian stiffness is obtained by the sum of individual leg stiffnesses, where passive joints are accounted by zero joint stiffness.

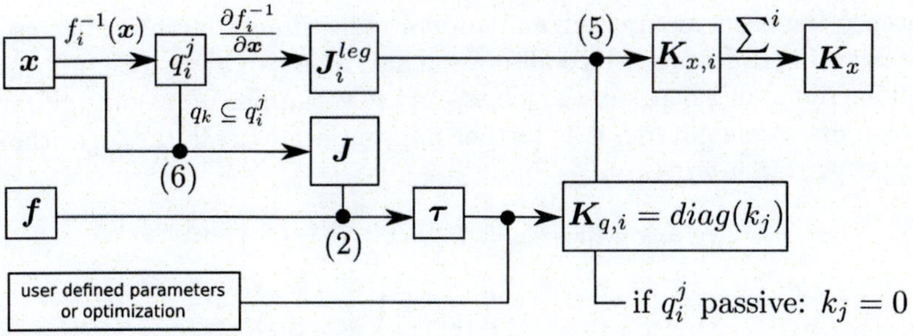

Fig. 2. Depiction of stiffness computation in task space with optimized or user defined stiffness. Connections denote the application of related functions to compute the boxed entities.

2.3 Jacobian Null Space and Joint Space Stiffness Optimization

Redundancy is reflected in the Jacobian matrix of (7) and can be directly used for stiffness optimization in task space. A general solution of (2) is presented through

$$\boldsymbol{\tau} = \mathbf{J}^T\mathbf{f} + \left(\mathbf{I}_n - \mathbf{J}^T\mathbf{J}^{\dagger T}\right)\boldsymbol{\tau}_n \tag{9}$$

with $\mathbf{I}$ as the identity matrix and $\boldsymbol{\tau}_n$ being an arbitrary vector in the space of $\boldsymbol{\tau}$. Since the second term of (9) represents the null space of $\mathbf{J}$ and thus $\boldsymbol{\tau}_n$ has no impact on $\mathbf{f}$, one could also write

$$\boldsymbol{\tau} = \mathbf{J}^T\mathbf{f} + \ker(\mathbf{J})\mathbf{u} \tag{10}$$

where $\ker(\mathbf{J})$ gives the basis of the null space of $\mathbf{J}$ and $\mathbf{u}$ is an arbitrary vector with dimension equal to the degree of redundancy. Optimizing Cartesian stiffness by spring deflection γ_i for a prescribed configuration then writes

$$\begin{aligned} \underset{\gamma_1,\dots,\gamma_m,\mathbf{u}}{\textbf{minimize}} \quad & \det(\mathbf{K}_x(\mathbf{x},\gamma_1,\dots,\gamma_m)) \\ \textbf{subject to} \quad & \tau(\gamma_1,\dots,\gamma_m) = \mathbf{J}^T\mathbf{f} + \ker(\mathbf{J})\mathbf{u} \\ & -\gamma_{max} < \gamma_i < \gamma_{max} \end{aligned}$$

In designs using VIAs (left box in Fig. 1), the objective function reduces to simple joint stiffness $k_j = d\tau^{s1}/d\gamma_1 + d\tau^{s2}/d\gamma_2$ where τ^{s1} and τ^{s2} are the spring forces in the VIA and the constraints reduce to a simple force equilibrium $\tau_i - \tau^{s1} + \tau^{s2} = 0$. The optimized joint space stiffness $\mathbf{K}_q$ is then mapped to task space as in the scheme above.

3 A Metric for Work Space Stiffness

By looking at the determinant of Cartesian stiffness matrices, a scalar value implying magnitude of the underlying transformation can be given. Taking the difference between maximized and minimized stiffness further informs about the achievable difference, since absolute magnitude can always be influenced by installing different springs in the joints. As the variations in magnitude can be of some orders of magnitude, it is further appropriate to take the logarithm, thus *stiffness variation* writes

$$sv = \log(\det(\mathbf{K}_x)^{max}) - \log(\det(\mathbf{K}_x)^{min}) = \log\left(\frac{\det(\mathbf{K}_x)^{max}}{\det(\mathbf{K}_x)^{min}}\right) \tag{11}$$

By integrating the variation over work space and normalizing, a configuration dependent *work space stiffness metric* is obtained

$$wssm = \frac{\int sv(\mathbf{x})dV}{\int dV} \tag{12}$$

4 Simulations and Discussion

To allow a comparison between all designs, location of the base joints is fixed as depicted in Fig. 3, leaving r, l and e as geometric parameters that are kept dimensionless for simulations. Springs are modeled by means of the point-symmetric force function $\tau^s = 5\sin(\gamma)\cdot\gamma^3$ in the interval $\gamma \in [-\pi, \pi]$ that results in a convex stiffness function $d\tau^s/d\gamma$ throughout this region. Even though, our formulations account for external forces, stiffness bounds of the different designs are computed without loading. In most cases and without external forces, stiffness minimization can simply be computed with $\mathbf{K}_q^{min} = f(\gamma_i = 0)$ and mapped to Cartesian space. Points in task space with a condition index $1/\text{cond}(\mathbf{J}\mathbf{J}^T) < 10^{-4}$ are discarded from computation to avoid optimization attempts close to singularities.

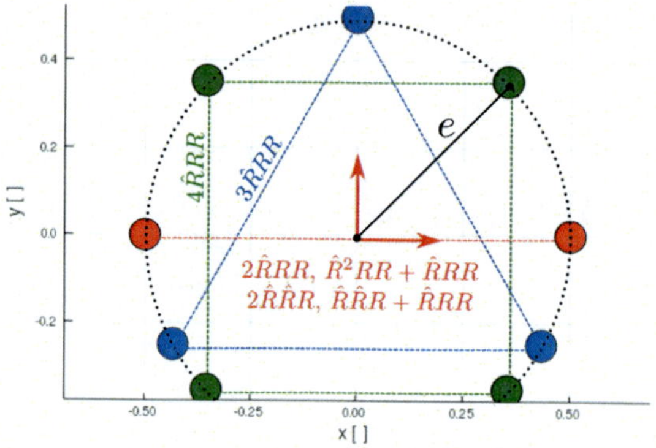

Fig. 3. Base joints (dots) lie on a circle with radius e in symmetric arrangements. Geometric parameters are normalized to the circles diameter. The coordinate system shows the work space origin.

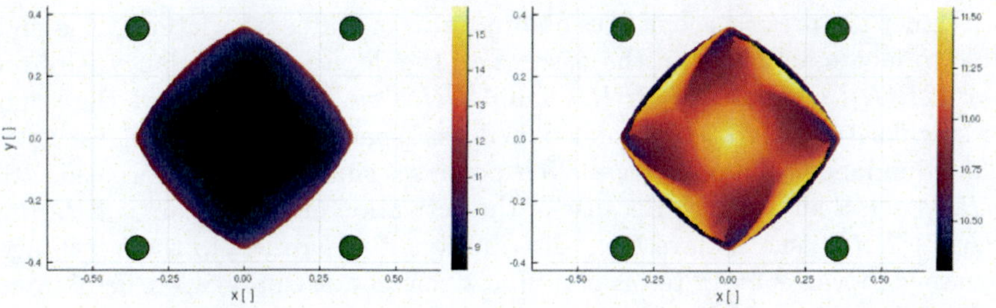

Fig. 4. Minimized (left) and maximized (right) $\log(\det(\mathbf{K}_x))$ of $4\hat{R}RR$: $r = 0.4$, $l = 0.4$, $e = 0.5$ of the first kinematic solution (out of 16)

Table 1. Work space stiffness metric for all designs and kinematic solutions. Numbers in the vectors indicate first or second inverse solution of the independent joint coordinates.

Mechanism	Kinematic solutions																
	$[1,1,1,1]^T$	$[2,1,1,1]^T$	$[1,2,1,1]^T$	$[2,2,1,1]^T$	$[1,1,2,1]^T$	$[2,1,2,1]^T$	$[1,2,2,1]^T$	$[2,2,2,1]^T$	$[1,1,1,2]^T$	$[2,1,1,2]^T$	$[1,2,1,2]^T$	$[2,2,1,2]^T$	$[1,1,2,2]^T$	$[2,1,2,2]^T$	$[1,2,2,2]^T$	$[2,2,2,2]^T$	$[q_1,q_2,q_3,q_4]^T$
Geometric parameters: $r_i = r = 0.5$, $l_i = l = 0.25$, $e = 0.5$																	
$2\hat{R}^2RR$	3.19	3.19	3.19	3.19	–	–	–	–	–	–	–	–	–	–	–	–	
$\hat{R}^2RR + \hat{R}RR$	1.59	1.59	1.59	1.59	–	–	–	–	–	–	–	–	–	–	–	–	
$2\hat{R}\hat{R}R$	2.61	2.60	2.60	2.61	–	–	–	–	–	–	–	–	–	–	–	–	
$\hat{R}\hat{R}R + \hat{R}RR$	2.04	2.02	2.02	2.04	–	–	–	–	–	–	–	–	–	–	–	–	
$3\hat{R}RR$	2.22	2.35	2.35	2.35	2.35	2.35	2.35	2.22	–	–	–	–	–	–	–	–	
$4\hat{R}RR$	2.63	2.54	2.54	2.63	2.54	2.62	2.63	2.54	2.54	2.63	2.62	2.54	2.63	2.54	2.54	2.63	
Geometric parameters: $r_i = r = 0.4$, $l_i = l = 0.4$, $e = 0.5$																	
$2\hat{R}^2RR$	3.19	3.19	3.19	3.19	–	–	–	–	–	–	–	–	–	–	–	–	
$\hat{R}^2RR + \hat{R}RR$	1.59	1.59	1.59	1.59	–	–	–	–	–	–	–	–	–	–	–	–	
$2\hat{R}\hat{R}R$	2.79	2.79	2.79	2.79	–	–	–	–	–	–	–	–	–	–	–	–	
$\hat{R}\hat{R}R + \hat{R}RR$	2.25	2.05	2.05	2.25	–	–	–	–	–	–	–	–	–	–	–	–	
$3\hat{R}RR$	2.35	2.13	2.13	2.13	2.13	2.13	2.13	2.35	–	–	–	–	–	–	–	–	
$4\hat{R}RR$	2.61	2.43	2.43	2.70	2.43	2.65	2.70	2.43	2.43	2.70	2.65	2.43	2.70	2.43	2.43	2.61	

In Fig. 4 a stiffness optimization of $4\hat{R}RR$ is shown. The values of the minimized solution increase at the boundaries, since one of the legs gets into a stretched position, possessing high stiffness in one direction. A similar phenomenon can be seen with the $3\hat{R}RR$. Table 1 contains all computed metrics of the investigated mechanisms for two arbitrary geometric settings. It can be observed, that the computed metric of designs involving VIAs ($2\hat{R}^2RR$ and $\hat{R}^2RR + \hat{R}RR$) in the base is independent of geometry and kinematic solution - the magnitude is solely reflected by the non-linear function of the springs. Moreover, the metric of mechanism $2\hat{R}^2RR$ with two VIAs, has twice the magnitude as that of $\hat{R}^2RR + \hat{R}RR$ which accounts that two eigenvalues of $\mathbf{K}_x$ can be changed independently instead of one.

Under the present geometries and in terms of stiffness variations, it is preferable to locate actuators in the base when the degree of redundancy is of order two ($2\hat{R}^2RR$ compared to $2\hat{R}\hat{R}R$ and $4\hat{R}RR$), while $4\hat{R}RR$ allows the removal of singularities in a then more confined workspace. This is in contrast to simple redundancy, where $\hat{R}^2RR + \hat{R}RR$ possesses smaller stiffness variations than $\hat{R}\hat{R}R + \hat{R}RR$ and $3\hat{R}RR$. Because of the activation of passive joints, $2\hat{R}\hat{R}R$ and $\hat{R}\hat{R}R + \hat{R}RR$ usually have mass distributions that are not to the advantage of dynamic movements. In terms of stiffness changes, it can however be of interest to use the design $\hat{R}\hat{R}R + \hat{R}RR$, since it allows bigger stiffness variations than $\hat{R}^2RR + \hat{R}RR$ but maintaining the same work space size.

5 Conclusion

In this work, a variety of planar 2-DOF VSMs has been introduced. The modeling applies to general PKM. The optimization of Cartesian stiffness in task space has been carried out. To this end, a metric has been introduced to assess and compare the different designs. We believe that the properties of such mechanisms should be further investigated, as their advantages can only be exploited with a thorough theoretical understanding.

Acknowledgements. Funded by the Deutsche Forschungsgemeinschaft (DFG, German Research Foundation) - 404971005. The third author acknowledges the support from the “LCM – K2 Center for Symbiotic Mechatronics” within the framework of the Austrian COMET-K2 program.

References

1. Chen, S.F.: The spatial conservative congruence transformation for manipulator stiffness modeling with coordinate and noncoordinate bases. J. Robot. Syst. **22**(1), 31–44 (2005)
2. Ham, R., Sugar, T., Vanderborght, B., Hollander, K., Lefeber, D.: Compliant actuator designs. IEEE Robot. Autom. Mag. **16**(3), 81–94 (2009)
3. Hoevenaars, A., Gosselin, C., Lambert, P., Herder, J.: Consistent modeling resolves asymmetry in stiffness matrices. Mech. Mach. Theor. **105**, 80–90 (2016)
4. Kock, S., Schumacher, W.: A parallel x-y manipulator with actuation redundancy for high-speed and active-stiffness applications. In: Proceedings. 1998 IEEE, ICRA (Cat. No.98CH36146), vol. 3, pp. 2295–2300. IEEE, Leuven (1998)
5. Lynch, K.M., Park, F.C.: Modern Robotics: Mechanics, Planning, and Control. Cambridge University Press, Cambridge (2017). OCLC: ocn983881868
6. Müller, A.: Stiffness control of redundantly actuated parallel manipulators. In: Proceedings 2006 IEEE, ICRA 2006, Orlando, FL, USA, pp. 1153–1158 (2006)
7. O'Brien, J., Wen, J.: Redundant actuation for improving kinematic manipulability. In: Proceedings 1999 IEEE International Conference on Robotics and Automation (Cat. No.99CH36288C), vol. 2, pp. 1520–1525. IEEE, Detroit (1999)
8. Pashkevich, A., Klimchik, A., Chablat, D.: Enhanced stiffness modeling of manipulators with passive joints. Mech. Mach. Theor. **46**(5), 662–679 (2011)

9. Chen, S.-F., Kao, I.: Simulation of conservative congruence transformation. Conservative properties in the joint and Cartesian spaces. In: Proceedings 2000 ICRA. Millennium Conference, IEEE International Conference on Robotics and Automation, Symposia Proceedings (Cat. No.00CH37065), vol. 2, pp. 1283–1288. IEEE, San Francisco (2000)
10. Stoeffler, C., Kumar, S., Peters, H., Brüls, O., Müller, A., Kirchner, F.: Conceptual design of a variable stiffness mechanism in a humanoid ankle using parallel redundant actuation. In: 2018 IEEE-RAS 18th International Conference on Humanoid Robots (Humanoids), pp. 462–468. IEEE, Beijing (2018)
11. Yi, B.J., Freeman, R., Tesar, D.: Open-loop stiffness control of over constrained mechanisms/robotic linkage systems. In: Proceedings. 1989 International Conference on Robotics and Automation, pp. 1340–1345. IEEE Computer Society Press, Scottsdale (1989)
12. Zefran, M., Kumar, R.V.: A geometric approach to the study of the Cartesian stiffness matrix. J. Mech. Des. **124**, 30–38 (2002)

Kinematics and Orientation Workspace of a 3-DOF Parallel Robotic Wrist Actuated by Spherical Four-Bar Linkages

Guanglei Wu[1,3](✉), Ning Zhang[1], Chuangchuang Cui[1], Huiping Shen[2], and Xuping Zhang[3]

[1] School of Mechanical Engineering, Dalian University of Technology, Dalian 116024, China
gwu@dlut.edu.cn
[2] School of Mechanical Engineering, Changzhou University, Changzhou 213164, China
shp65@126.com
[3] Department of Engineering, Aarhus University, 8000 Aarhus C, Denmark
xuzh@eng.au.dk

Abstract. This paper presents a three-degree-of-freedom fully parallel robotic wrist and its preliminary kinematics. This mechanism actuated by spherical four-bar linkages originates from a spherical parallel manipulator with co-axial inputs. The inequalities to determine the orientation workspace is derived and numerically illustrated, which shows that the manipulator can have large rotation angles, particularly, the twist angle of the end-effector. A 3D-printed prototype is built to validate the conceptual design.

Keywords: Parallel robotic wrist · Mirrored motion · Spherical four-bar linkage

1 Introduction

Three degree-of-freedom (3-DOF) spherical parallel manipulators (SPMs) are dedicated to camera-orientating device [6], minimally invasive surgical robots [8] and wrist joints [2] because of their large orientation workspace and high payload capacity. General SPM architectures usually generate limited end-effector rotation, e.g., the 'Agile Eye' [6] or agile wrist [4] with ±30° twist angle, whereas design of SPMs considers high end-effector rotational capability for extended application fields, for instance, the application of the asymmetrical SPMs with unlimited end-effector rotation as machine tool head [10]. When the SPMs function as orientating device, unlimited twist angle is unnecessary, while, high rotational capability will be an advantage amongst the counterparts. Thus, a fully parallel robotic wrist mechanism with large orientation angle is to be introduced in this work.

J. Lenarčič and B. Siciliano (Eds.): ARK 2020, SPAR 15, pp. 268–276, 2021.
https://doi.org/10.1007/978-3-030-50975-0_33

The proposed SPM originates from a special one with co-axial inputs [9], as depicted in Fig. 1. The original SPM is composed of three curved links sliding along a circular guide as the yellow base platform (BP) to support the pink mobile platform (MP), which can be mirrored about the symmetric plane of the BP. When the lower pink MP is fixed as the BP instead of the circular guide, the other MP orientation can be mirrored to the base about the symmetric plane, i.e., the geometric plane of BP. It is noted that the fixed BP eliminates the rotation around the normal direction of the BP, thus becoming a 2-DOF SPM, which means that every two curved links, sliding units, BP and circular guide become a spherical five-bar linkage. When the active revolute joint on the BP is replaced with a spherical four-bar linkage and an additional rotation around the vertical axis is attached to the base, the mobilities of this spherical mechanism include three pure rotations. This paper will present some preliminary kinematic analysis of the proposed robotic wrist.

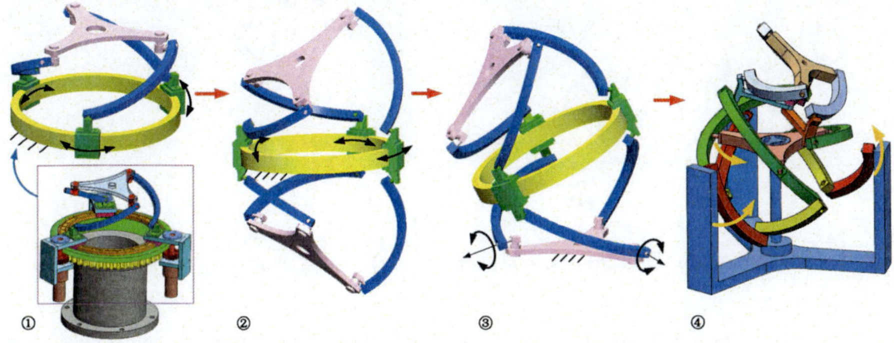

Fig. 1. The transition from the co-axial SPM to the robotic wrist mechanism actuated by spherical four-bar linkages.

2 Manipulator Architecture

The kinematic structure of the robotic wrist is described in Fig. 2, where the third leg is hidden for view. The reference coordinate system (x, y, z), with the origin located at the center of rotation (point O) of the BP, of which the z axis points toward the vertical direction and the x axis is parallel to unit vector $\mathbf{a}_1$. The local coordinate system (X, Y, Z) is built with the origin located at the geometric center P of the MP, namely, the center of rotation, the X-axis being parallel to unit vector $\mathbf{u}'_1$. The actuation unit is consisted of a deformable four-bar linkage, defined by geometric parameters of curve links α_1, α_2, α_3 and the torsion angle σ. The curved links have the same architecture, defined by β. Moreover, the orientations of the MP $G_1G_2G_3$ and the rotational base $D_1D_2D_3$ are symmetrical about the mirror plane ε that is normal to unit vector $\mathbf{n}$. The angles between the unit vector $\mathbf{v}_i$ ($\mathbf{v}_i'$) of passive joints at point E_i (F_i) have and $\mathbf{n}$ are denoted by γ.

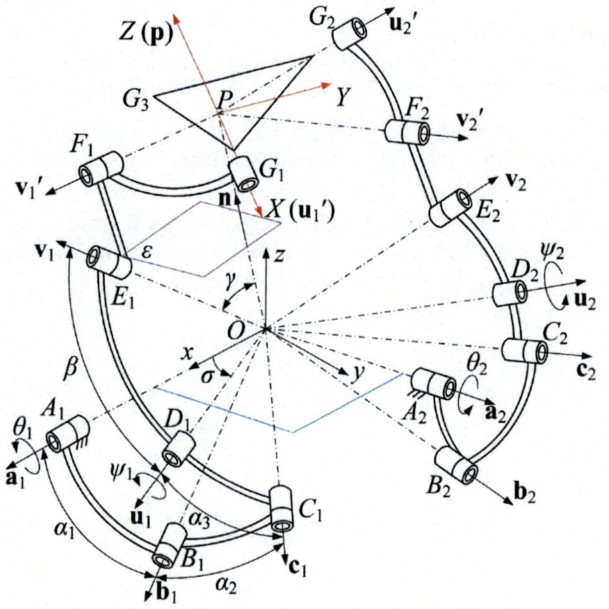

Fig. 2. The kinematic structure of the 3-DOF robotic wrist.

3 Inverse Geometry of the Wrist Mechanism

The orientation of the mobile platform is described by the *azimuth–tilt–torsion* angle (ϕ, θ, σ) [5], for which the rotation matrix is expressed as

$$\mathbf{Q} = \mathbf{R}_z(\phi)\mathbf{R}_y(\theta)\mathbf{R}_z(\sigma - \phi) \tag{1}$$

Thus, the unit vector $\mathbf{u}_i'$ of the revolute joint is a function of the MP orientation:

$$\mathbf{u}_i' = \mathbf{Q}\mathbf{w}_i;\ \mathbf{w}_i = \left[\cos\eta_i\ \sin\eta_i\ 0\right]^T,\ \eta_i = 2(i-1)\pi/3,\ i = 1,\ 2,\ 3 \tag{2}$$

where $\mathbf{w}_i$ is the unit vector in the home configuration.

Moreover, the normal vector $\mathbf{n}$ of the mirror plane ε is derived as

$$\mathbf{n} = \left[n_x\ n_y\ n_z\right]^T = \mathbf{R}_z(\phi)\mathbf{R}_y(\theta/2)\mathbf{k} \tag{3}$$

where $\mathbf{k}$ is the unit vector of z-axis. Consequently, the reflection matrix of plane ε is denoted by [7]

$$\mathbf{R}_e = \mathbf{I} - 2\left[n_x\mathbf{n}\ n_y\mathbf{n}\ n_z\mathbf{n}\right] = \mathbf{I} - 2\mathbf{n}\otimes\mathbf{n} \tag{4}$$

where $\mathbf{I}$ is the 3×3 identity matrix and $\otimes$ stands for the tensor product. Therefore, the unit vector $\mathbf{u}_i$ of the revolute joint on the rotational base is obtained from

$$\mathbf{u}_i = \mathbf{R}_e \mathbf{u}'_i = \left[\cos(\eta_i + \sigma) \ \sin(\eta_i + \sigma) \ 0\right]^T \tag{5}$$

which means that the angle of the rotational base with respect to the global coordinate frame is equal to the MP torsional angle as displayed in Fig. 2.

The input angle ψ_i of the revolute joint at point D_i can be solved from the following kinematic constraints equation:

$$\mathbf{v}_i \cdot \mathbf{n} - \cos\gamma = 0 \tag{6}$$

where unit vector $\mathbf{v}_i$ is obtained in terms of the angle ψ_i following the angle-axis representation [1], namely,

$$\mathbf{v}_i = \mathbf{R}_{\mathbf{u}_i}(\psi_i)\mathbf{v}_i^*; \ \mathbf{v}_i^* = \left[\cos(\eta_i + \sigma + \beta) \ \sin(\eta_i + \sigma + \beta) \ 0\right]^T \tag{7}$$

Solving Eq. (6), the angle ψ_i is expressed below:

$$\psi_i = 2\tan^{-1}\frac{I_i \pm \sqrt{I_i^2 + J_i^2 - K_i^2}}{K_i - J_i}; \ \begin{cases} I_i = \sin\beta\cos(\theta/2) \\ J_i = -\sin\beta\sin(\theta/2)\sin(\eta_i + \sigma - \phi) \\ K_i = \cos\beta\sin(\theta/2)\cos(\eta_i + \sigma - \phi) - \cos\gamma \end{cases} \tag{8}$$

With known angle ψ_i, the input angle θ_i of the actuated revolute joint on the base platform is resorting to the kinematic problem of the spherical four-bar linkage as displayed in Fig. 3, where the input/output (I/O) equation takes the form [3, 11]:

$$k_1 + k_2\cos\psi_i + k_3\cos\psi_i\cos\theta_i - k_4\cos\theta_i + k_5\sin\psi_i\sin\theta_i = 0 \tag{9}$$

with

$$k_1 \equiv \cos\sigma\cos\alpha_1\cos\alpha_3 - \cos\alpha_2, \ k_2 \equiv \sin\sigma\sin\alpha_1\cos\alpha_3$$
$$k_3 \equiv \cos\sigma\sin\alpha_1\sin\alpha_3, \ k_4 \equiv \sin\sigma\cos\alpha_1\sin\alpha_3, \ k_5 \equiv \sin\alpha_1\sin\alpha_3$$

To this end, the ith active input angle is solved as:

$$\theta_i = 2\tan^{-1}\frac{A_i \pm \sqrt{A_i^2 + B_i^2 - C_i^2}}{C_i - B_i}; \ \begin{cases} A_i = k_5\sin\psi_i \\ B_i = k_3\cos\psi_i - k_4 \\ C_i = k_1 + k_2\cos\psi_i \end{cases} \tag{11}$$

According to Eqs. (8) and (11), the inverse geometric problem of each limb has four solutions corresponding to the four working modes, which results in 64 working modes of the proposed manipulator.

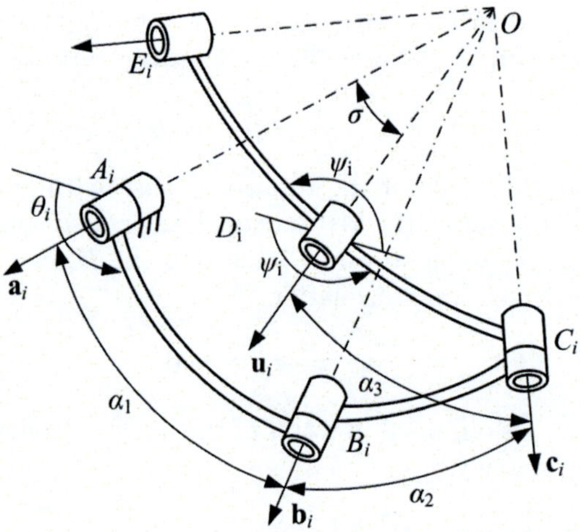

Fig. 3. The spherical four-bar linkage of the actuation unit.

4 Orientation Workspace

The workspace of the robotic wrist mechanism refers to the actual ranges of the *azimuth–tilt–torsion* angles, namely, the orientation workspace, subject to the condition that the solutions in Eqs. (8) and (11) are real numbers. This means that the following inequalities exist:

$$I_i^2 + J_i^2 - K_i^2 \geq 0,\ A_i^2 + B_i^2 - C_i^2 \geq 0;\ i = 1,\ 2,\ 3 \tag{12}$$

A tedious manipulation on the above equations yield

$$-\left[\cos(\eta_i - \phi + \sigma)\sin(\theta/2) - \cos\beta\cos\gamma\right]^2 + \sin^2\beta - \cos^2\gamma + \cos^2\beta\cos^2\gamma \geq 0 \tag{13a}$$

$$-\left(\sin\alpha_1\sin\sigma\cos\psi_i - \cos\alpha_2\cos\alpha_3 + \cos\alpha_1\cos\sigma\right)^2 + \sin^2\alpha_2\sin^2\alpha_3 \geq 0 \tag{13b}$$

Expanding the previous inequalities lead to

$$\cos(\beta + \gamma) \leq \cos(\eta_i - \phi + \sigma)\sin(\theta/2) \leq \cos(\beta - \gamma) \tag{14a}$$

$$\cos(\alpha_2 + \alpha_3) \leq \sin\alpha_1\sin\sigma\cos\psi_i + \cos\alpha_1\cos\sigma \leq \cos(\alpha_2 - \alpha_3) \tag{14b}$$

The reachable orientation workspace can be found from the six inequalities of Eqs. (14a) and (14b) that are subject to the link dimensions and geometric parameters. With the aid of the CAD model, the geometric parameters of the wrist mechanism are set to $\alpha_1 = 75°$, $\alpha_2 = 90°$, $\alpha_3 = 60°$, $\beta = 90°$, $\gamma = 60°$.

Figure 4 displays the orientation workspace of the wrist mechanism represented on a unit sphere with different torsional angle σ together with the interdependency of the angles of the orientation workspace in Fig. 5. The maximum

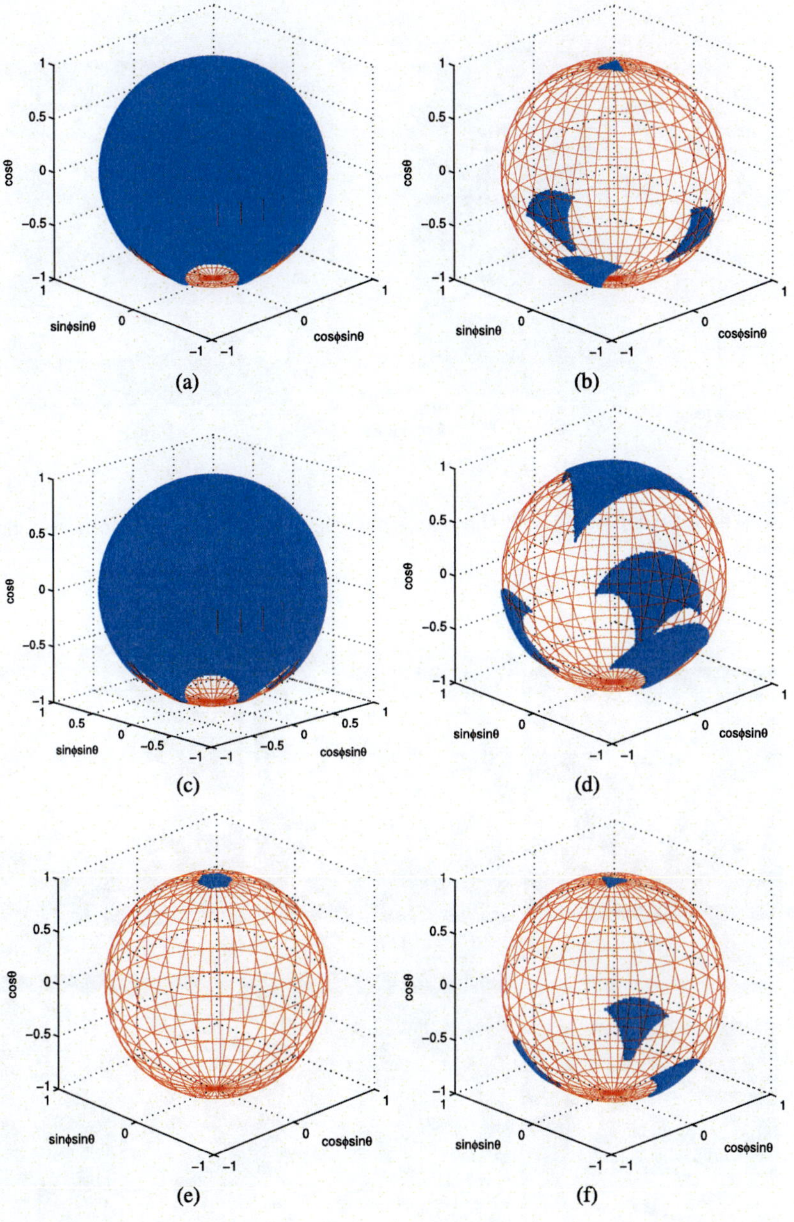

Fig. 4. The pan-tilt angle of the mobile platform with different constant twist angle: (a) $\sigma = -135°$; (b) $\sigma = -60°$; (c) $\sigma = 45°$; (d) $\sigma = 50°$; (e) $\sigma = 90°$; (f) $\sigma = 120°$.

torsional angle σ can reach up to $\pm 135°$. It is seen from Figs. 4(b), 4(d) and 4(f) that the workspace is not continuous, which means that difference workspace regions corresponds to different working modes. Figure 5 shows that the robotic

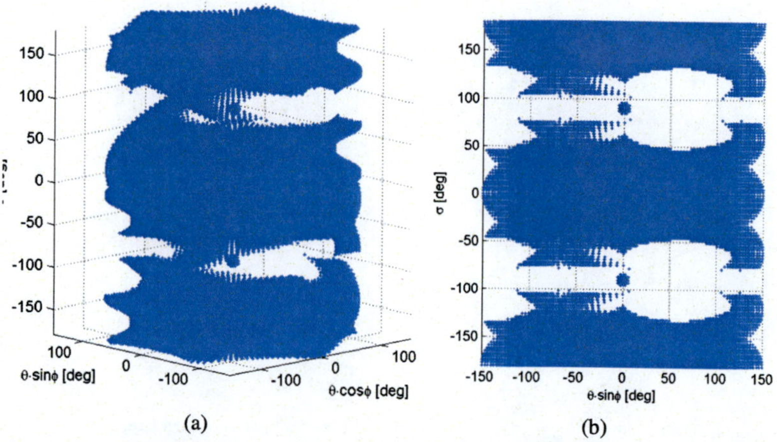

Fig. 5. The interdependency of the reachable orientation angle: (a) isometric view; (b) side view.

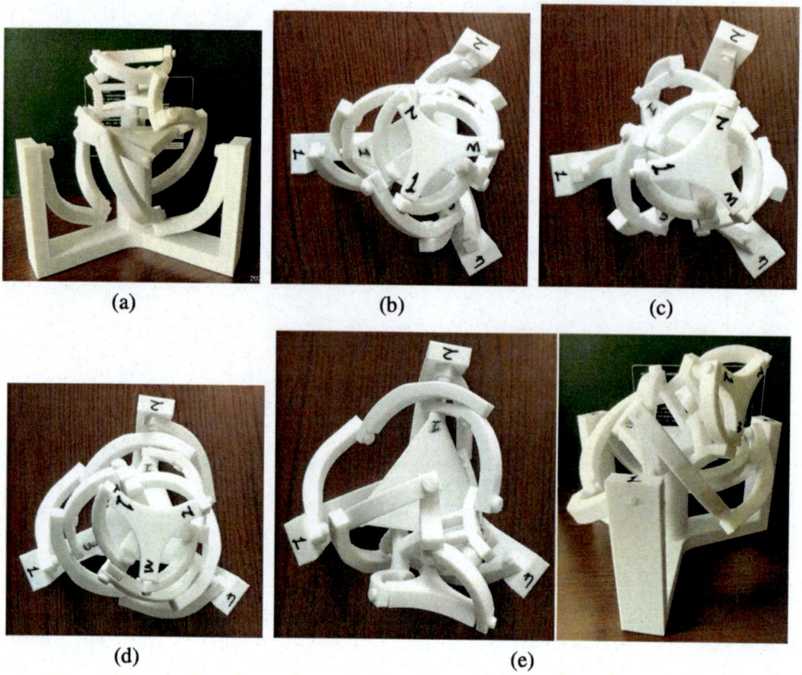

Fig. 6. A 3D-printed prototype of the parallel robotic wrist: (a) arbitrary configuration; (b) $\theta = \sigma = 0$; (c) $\theta = 0$, $\sigma = 60°$; (d) $\theta = 0$, $\sigma = 120°$; (e) $\theta = 90°$, $\sigma = 120°$ (top and side views).

wrist can have the orientation workspace with the range $\phi \in (-180°, 180°]$, $\theta \in [0, 150°]$, $\sigma \in [-50°, 50°]$ under one working mode.

Figure 6 depicts a 3D-printed prototype of the mechanism. This wrist mechanism can have relatively larger orientation angles compared to the most existing counterparts with symmetrical architectures.

5 Conclusion

In this paper, a three-degree-of-freedom fully parallel robotic wrist with spherical four-bar linkage based actuation units is proposed, which originates from a spherical parallel manipulator with co-axial inputs. The preliminary kinematics is studied to derive the geometric inequalities to determine the orientation workspace, subject to the link dimensions and geometric parameters. Numerical illustration is carried out to show that the manipulator can have large rotation angles, particularly, the twist angle of the end-effector. Moreover, the interdependency of the reachable orientation angles is graphically depicted. To this end, a 3D-printed prototype is built to validate the conceptual design.

Acknowledgments. The supports from the Fundamental Research Funds for the Central Universities (No. DUT19JC25), the Natural Science Foundation of Liaoning Province (No. 20180520028), and the China Scholarship Council (CSC No. 201906065026) are gratefully acknowledged.

References

1. Angeles, J.: Fundamentals of Robotic Mechanical Systems: Theory, Methods, and Algorithms. Springer, New York (2007)
2. Asada, H., Granito, J.: Kinematic and static characterization of wrist joints and their optimal design. In: IEEE International Conference on Robotics and Automation, pp. 244–250 (1985)
3. Bai, S., Angeles, J.: A unified input-output analysis of four-bar linkages. Mech. Mach. Theor. **43**(2), 240–251 (2008)
4. Bidault, F., Teng, C.P., Angeles, J.: Structural optimization of a spherical parallel manipulator using a two-level approach. In: ASME Design Engineering Technical Conferences and Computers and Information in Engineering Conference, Pittsburgh, Pennsylvania (2001)
5. Bonev, I., Gosselin, C.: Singularity loci of spherical parallel mechanisms. In: 2005 IEEE International Conference on Robotics and Automation, pp. 2957–2962 (2005)
6. Gosselin, C., Hamel, J.: The Agile Eye: a high-performance three-degree-of-freedom camera-orienting device. In: IEEE International Conference on Robotics and Automation, pp. 781–786 (1994)
7. Kovács, E.: Rotation about an arbitrary axis and reflection through an arbitrary plane. Ann. Math. Inform. **40**, 175–186 (2012)
8. Li, T., Payandeh, S.: Design of spherical parallel mechanisms for application to laparoscopic surgery. Robotica **20**(2), 133–138 (2002)
9. Wu, G., Caro, S., Bai, S., Kepler, J.: Dynamic modeling and design optimization of a 3- DOF spherical parallel manipulator. Robot. Auto. Syst. **62**, 1377–1386 (2014)

10. Wu, G., Zou, P.: Comparison of 3-dof asymmetrical spherical parallel manipulators with respect to motion/force transmission and stiffness. Mech. Mach. Theor. **105**, 369–387 (2016)
11. Yang, A., Freudenstein, F.: Application of dual-number quaternion algebra to the analysis of spatial mechanisms. ASME J. Appl. Mech. **31**, 300–307 (1964)

Analytical Determination of the Longest Cylinder Free of Gain-type Singularities Inside the Workspace of a 3-R$\underline{P}$S Spatial Manipulator

Argaja Deepak Shende, Bibekananda Patra, Prem Kumar Prasad, and Sandipan Bandyopadhyay(✉)

Department of Engineering Design, Indian Institute of Technology Madras, Chennai 600036, India
argaja10@gmail.com, bibeka.patra2@gmail.com, kumar.prasad.prem@gmail.com, sandipan@iitm.ac.in

Abstract. The knowledge of singularity-free regions (SFRs) inside the workspace of a manipulator is helpful in its path-planning and design. To identify such regions analytically, the singularity manifold needs to be determined first. This paper presents the derivation of the gain-type singularity manifold in the task-space of the 3-R$\underline{P}$S manipulator, which is further utilised to compute its SFR, in the form of a singularity-free cylinder (SFC), which is free of gain-type singularities. The problem of identifying the longest SFC for a given constant radius and a chosen base circle is posed as a constrained optimisation problem, which reduces to a 21-degree univariate polynomial in the length of the SFC. The formulation is illustrated via a numerical example.

Keywords: 3-R$\underline{P}$S manipulator · Gain-type singularity manifold · Singularity-free cylinder · Algebraic elimination

1 Introduction

Parallel manipulators, unlike serial manipulators, can *gain* one or more degree(s)-of-freedom (DoF) at certain singularities. These *gain-type singularities* [1,4], which cause uncontrollable motions of the end-effectors and/or certain links, occur *inside* the workspaces of parallel manipulators. The presence of such singularities within the workspace of a manipulator reduces its actual usable workspace, and complicates its shape. Hence, the identification of a convex-shaped SFR could be useful in path-planning, since any line segment contained inside this SFR will be free of singularities, while the SFR itself needs to be computed only once for a particular design of the manipulator.

Li et al. have computed the SFC of the planar 3-R$\underline{P}$R manipulator in [7] by discretising the translational workspace for a range of orientations. Jiang et al. have obtained a singularity-free orientation workspace for a given position of the Gough-Stewart platform manipulator in [5]. A method to compute

J. Lenarčič and B. Siciliano (Eds.): ARK 2020, SPAR 15, pp. 277–284, 2021.
https://doi.org/10.1007/978-3-030-50975-0_34

the singularity-free sphere (SFS) *analytically* in the translational workspace of a semi-regular Stewart platform manipulator is described in [9]. Prasad and Bandyopadhyay have reported a detailed study on the identification of the largest SFC in the translational workspace of the same manipulator in [10]. Karnam et al. [6] have found the SFRs as a subprocess of finding the *safe working zones* (SWZ) for the manipulators, planar 3-RRR and the spatial MaPaMan-I.

In this work, the SFR of the 3-RPS manipulator is computed. Sokolov and Xirouchakis in [12] have identified a singularity-free zone of the same manipulator in the form of a cylinder of infinite length, which has a singularity-free base circle centred at a constant height, in the translational workspace. Briot and Bonev have found a SFC in the plane of the Euler angles, azimuth and tilt, for a constant heave of the manipulator in [2]. Chablat et al. have reported in [3], a SFR in the space formed by two quaternion components and the heave variable, using the cylindrical algebraic decomposition (CAD) technique and inverse kinematics solutions of the manipulator, for both the operation modes.

In this paper, the derivation of the gain-type singularity manifold of the manipulator in the *task-space* is discussed in detail. Due to the nature of the DoF (i.e., two rotational and one translational) of the manipulator, it is natural to choose a cylinder as the shape of the SFR in the task-space, as explained in [6,8]. The analytical formulation using the Lagrangian method of constrained optimisation to determine the SFC is analogous to that presented in [10], in the context of the translational workspace of a semi-regular Stewart platform manipulator. The longest SFC is computed for a constant radius and a chosen base circle for a given architecture of the manipulator. The variations of the maximum length of the SFC are studied for different radii and centres of the chosen base circle.

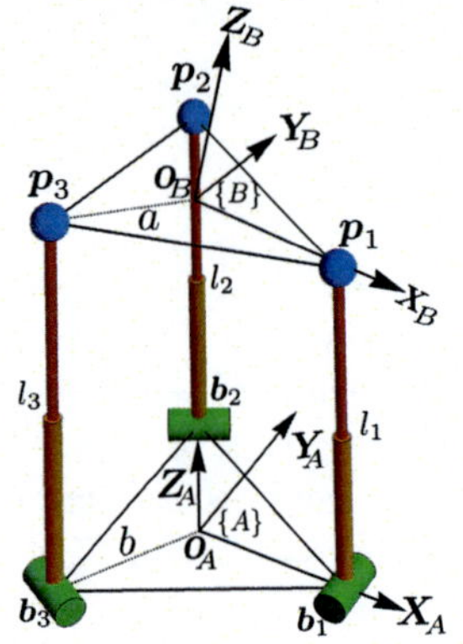

Fig. 1. Schematic of the 3-RPS manipulator.

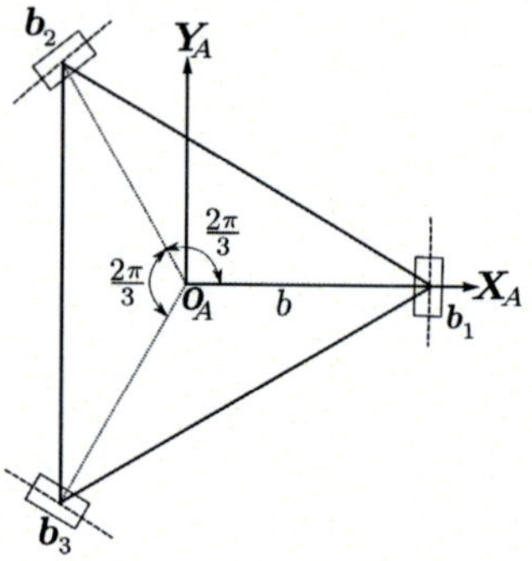

Fig. 2. Top-view of the planes perpendicular to the fixed platform containing one leg each.

The rest of the paper is organised as follows: the geometry of the 3-R$\underline{\text{P}}$S manipulator, formulation of the loop-closure constraint equations and derivation of the singularity manifold are described in Sect. 2, followed by the computation of the SFC in Sect. 3. A numerical example is presented in Sect. 4. The paper is concluded in Sect. 5.

2 Mathematical Formulation

A schematic of the 3-R$\underline{\text{P}}$S manipulator is shown in Fig. 1. The fixed and the moving platforms are equilateral triangles with circumradii b and a, respectively. The reference frames, namely, $\{A\}$ and $\{B\}$, are attached at the centroid of the fixed and the moving platform, respectively. The platforms are connected via the three R$\underline{\text{P}}$S limbs confined to three planes (perpendicular to the fixed platform) passing symmetrically through the axis $\boldsymbol{Z}_A$, as shown in Fig. 2. The vertices of the fixed platform, expressed in frame $\{A\}$, are: ${}^A\boldsymbol{b}_1 = [b, 0, 0]^\top$, ${}^A\boldsymbol{b}_2 = [-b/2, \sqrt{3}b/2, 0]^\top$, and ${}^A\boldsymbol{b}_3 = [-b/2, -\sqrt{3}b/2, 0]^\top$. Likewise, the vertices of the moving platform, expressed in frame $\{B\}$, are: ${}^B\boldsymbol{p}_1 = [a, 0, 0]^\top$, ${}^B\boldsymbol{p}_2 = [-a/2, \sqrt{3}a/2, 0]^\top$, ${}^B\boldsymbol{p}_3 = [-a/2, -\sqrt{3}a/2, 0]^\top$, which may be represented in frame $\{A\}$ as: ${}^A\boldsymbol{p}_i = {}^A\boldsymbol{O}_B + {}^A_B\boldsymbol{R}\,{}^B\boldsymbol{p}_i$, where $i = 1, 2, 3$; ${}^A\boldsymbol{O}_B = [x, y, z]^\top$ is the origin of frame $\{B\}$. Further, ${}^A_B\boldsymbol{R} \in \mathbb{SO}(3)$, parameterised in terms of *Rodrigues' parameters*, $\{c_1, c_2, c_3\} \in \mathbb{R}^3$, represents the orientation of the moving platform w.r.t. the fixed frame of reference $\{A\}$.

The configuration-space of the manipulator is described by $\boldsymbol{q} = [\boldsymbol{\theta}^\top, \boldsymbol{x}^\top]^\top$, where $\boldsymbol{\theta} = [l_1, l_2, l_3]^\top$ is designated as the *active/actuated* variable and the *task-space* variables are grouped into the vector $\boldsymbol{x} = [x, y, z, c_1, c_2, c_3]^\top$. The loop-closure equations are formed in terms of these variables, which are described below.

- The ith vertex of the moving platform, $\boldsymbol{p}_i$, lies on a sphere of radius l_i, centred at $\boldsymbol{b}_i$, at *every* instance, i.e.,

$$\eta_i \triangleq ({}^A\boldsymbol{p}_i - {}^A\boldsymbol{b}_i) \cdot ({}^A\boldsymbol{p}_i - {}^A\boldsymbol{b}_i) - l_i^2 = 0, \quad i = 1, 2, 3. \tag{1}$$

- Additional constraint equations arise from the fact that the vector along l_i is perpendicular to the vector ${}^A\boldsymbol{b}_{kj} = {}^A\boldsymbol{b}_k - {}^A\boldsymbol{b}_j$, parallel to the rotation axis at $\boldsymbol{b}_i$ (see Fig. 2), leading to:

$$\eta_{i+3} \triangleq ({}^A\boldsymbol{p}_i - {}^A\boldsymbol{b}_i) \cdot {}^A\boldsymbol{b}_{kj} = 0, \text{ where } i, j, k \in \{1, 2, 3\} \text{ permute cyclically.} \tag{2}$$

The expressions in the LHSs of Eqs. 1, 2 are assembled into a vector, $\boldsymbol{\eta}(\boldsymbol{q}) = [\eta_1, \ldots, \eta_6]^\top$. The constraint equations, $\boldsymbol{\eta}(\boldsymbol{q}) = \boldsymbol{0}$, are used to determine the gain-type singularity manifold of the manipulator.

Gain-type singularities occur when the manipulator gains one or more DoF at the moving platform. At such a singularity, one or more pairs of branches of forward kinematics of the manipulator merge. The mathematical condition for such a singularity can be obtained by invoking the *implicit function theorem* (IFT) (see [6]) as $\boldsymbol{S}_2(\boldsymbol{x}) = 0$, where:

$$\boldsymbol{S}_2(\boldsymbol{x}) = \det\left(\frac{\partial\boldsymbol{\eta}}{\partial\boldsymbol{x}}\right). \tag{3}$$

For the 3-R<u>P</u>S manipulator, $\boldsymbol{S}_2(\boldsymbol{x})$ is a polynomial of total degree 13 in $\boldsymbol{x}$, with a *size*[1] of 2.77 MB. The expression for the singularity manifold in the task-space variables, $\{c_1, c_2, z\}$, is obtained using Eqs. 1–3. By choosing to represent the task-space in the coordinates $\{c_1, c_2, z\}$, the analysis presented in this paper is applicable only to one of the *operation modes* of the manipulator described in [11], namely, the second. Rodrigues' parameters are not defined for the first mode, as it involves a *half-turn* of the moving platform about a horizontal axis. However, this theoretical incompleteness does not undermine the practical significance of the analysis, given that the half-turn typically causes a pair of legs to cross over, leading to interference among them, making the first mode impractical for standard operations of the manipulator. The equation describing the gain-type singularity manifold in terms of $\{c_1, c_2, z\}$ is derived in the following manner.

- The variables x, y are obtained in terms of $\{c_1, c_2, c_3\}$ by solving the Eqs. $\eta_4 = 0$ and $\eta_5 = 0$, linearly.
- Substituting these expressions in $\eta_6 = 0$ and simplifying leads to $c_3 = 0$, indicating that no *yaw motion* exists for the manipulator, as mentioned in [2].
- Using the above and simplifying, Eq. 3 reduces to $(1+c_1^2+c_2^2)^3 f(c_1, c_2, z) = 0$, i.e.,

$$f(c_1, c_2, z) = 0, \quad \text{as} \quad 1 + c_1^2 + c_2^2 \neq 0 \quad \forall c_1, c_2 \in \mathbb{R}. \tag{4}$$

  The manifold is described by Eq. 4, where f is of degree $\{8, 9, 3\}$ in $\{c_1, c_2, z\}$, respectively. The total degree and size of the polynomial f are found to be 11 and 25.26 KB, respectively.

The above computations have been done in `Mathematica` on a PC with an Intel® Core™ i7-4790 CPU running at a clock speed of 3.60 GHz. It takes 1.86 s (averaged over 50 runs) to find $f(c_1, c_2, z)$ in the closed-form. Using Eq. 4, the longest SFC for a given radius and base is determined in the following section.

[1] The term "size" refers to the amount of internal memory required to store the expression in the Computer Algebra System (CAS) `Mathematica` [13] used in this work.

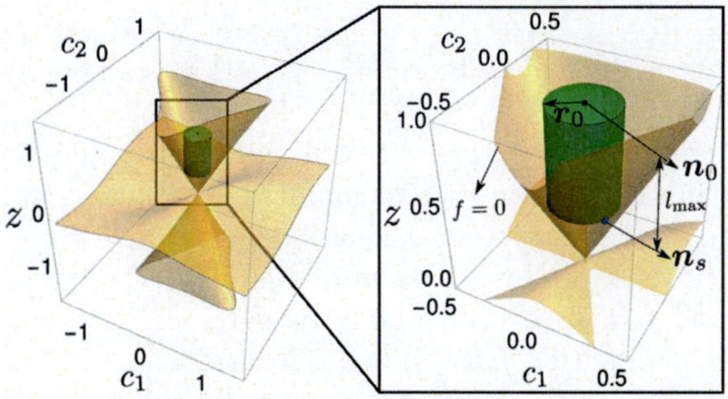

Fig. 3. Description of a singularity-free cylinder

3 Computation of the Longest SFC in the Task-Space

The objective is to determine the SFC of the maximum length (denoted by $l_{\max}$) for a given radius r_0 and the centre of the chosen base circle, $\boldsymbol{n}_0 = [c_{10}, c_{20}, z_0]^\top$ (see Fig. 3). The values of c_{10} and c_{20} need to be zero (see [8] for details) to afford a *metric* in the c_1c_2-plane, which is needed to define the radius of the cylinder, denoted by r_0. The circular cross-section of the SFC, computed at any constant $\boldsymbol{Z}$, i.e., in a c_1c_2-plane, is described as: $g(c_1, c_2) = c_1^2 + c_2^2 - r_0^2 = 0$.

To identify of the longest SFC, the *method of constrained Lagrangian optimisation* is used. The SFC intersects the singularity surface, described by Eq. 4, at a point $\boldsymbol{n}_s = [c_1, c_2, z_0 + l_{\max}]^\top$, as shown in Fig. 3. The equivalent *unconstrained* objective function can be written in terms of the Lagrange multipliers, λ_1 and λ_2, as $\mathcal{L} = l + \lambda_1 f(c_1, c_2, z_0 + l) + \lambda_2 g(c_1, c_2)$. The partial derivatives of $\mathcal{L}$ w.r.t. c_1, c_2, l, λ_1 and λ_2 are set to zero to construct the equations defining the *extremum* of $\mathcal{L}$:

$$\begin{aligned} h_1 &= \frac{\partial \mathcal{L}}{\partial c_1} = \lambda_1 \frac{\partial f}{\partial c_1} + \lambda_2 \frac{\partial g}{\partial c_1} = 0, \quad h_2 = \frac{\partial \mathcal{L}}{\partial c_2} = \lambda_1 \frac{\partial f}{\partial c_2} + \lambda_2 \frac{\partial g}{\partial c_2} = 0, \\ h_3 &= \frac{\partial \mathcal{L}}{\partial l} = 1 + \lambda_1 \frac{\partial f}{\partial l} = 0, \quad h_4 = \frac{\partial \mathcal{L}}{\partial \lambda_1} = f(c_1, c_2, z_0 + l) = 0, \\ h_5 &= \frac{\partial \mathcal{L}}{\partial \lambda_2} = g(c_1, c_2) = 0. \end{aligned} \tag{5}$$

These equations are reduced to a univariate polynomial equation in l through the following elimination steps.

- The multipliers λ_1 and λ_2 are eliminated *linearly* from the three equations $h_1(c_1, c_2, l, \lambda_1, \lambda_2) = 0$, $h_2(c_1, c_2, l, \lambda_1, \lambda_2) = 0$ and $h_3(c_1, c_2, l, \lambda_1) = 0$ resulting in $f_1(c_1, c_2, l) = 0$, where f_1 is of degrees $\{9, 8, 2\}$ in $\{c_1, c_2, l\}$, respectively.
- To eliminate c_1 from Eq. 5, the polynomial h_4 of degrees $\{8, 9, 3\}$ in $\{c_1, c_2, l\}$, is divided by the polynomial $h_5(c_1, c_2)$, which is quadratic in both $\{c_1, c_2\}$, w.r.t. the variable c_1. The original pair of polynomials $h_5(c_1, c_2)$ and $h_4(c_1, c_2, l)$ is now equivalent to the new pair consisting of $h_5(c_1, c_2)$ and $f_2(c_1, c_2, l)$, since,

$h_4(c_1, c_2, l) = \alpha(c_1, c_2, l)h_5(c_1, c_2) + f_2(c_1, c_2, l)$, where $\alpha(c_1, c_2, l)$ is the quotient in the above division. The reminder f_2 is a polynomial in $\{c_1, c_2, l\}$ of degrees $\{0, 6, 3\}$.

- Similarly, the polynomial $f_1(c_1, c_2, l)$ is divided by the polynomial $h_5(c_1, c_2)$, w.r.t. the variable c_1, resulting in a remainder polynomial $f_3(c_1, c_2, l)$ of degrees $\{1, 5, 2\}$ in $\{c_1, c_2, l\}$. The system of equations $f_1(c_1, c_2, l) = 0$ and $h_5(c_1, c_2) = 0$ is then reduced to the system of equations $f_3(c_1, c_2, l) = 0$ ans $h_5(c_1, c_2) = 0$. The polynomial $f_3(c_1, c_2, l)$ is of the form: $f_3(c_1, c_2, l) = c_1 f_4(c_2, l)$. Assuming $c_1 \neq 0$, a new equation $f_4(c_2, l) = 0$ is formed[2].
- The resultant of the two polynomials $f_2(c_2, l)$ and $f_4(c_2, l)$ w.r.t. c_2 is the desired univariate polynomial, $f_5(l)$, which is of degree 21.

The elimination process took 18.44 s (averaged over 50 runs) to obtain a 16.46 MB-sized polynomial, $f_5(l)$. The equation $f_5(l) = 0$ is solved for the given architecture parameters and the real solutions are extracted. The obtained solutions are validated by finding the common roots in c_2 of the equations $f_2 = 0$ and $f_4 = 0$, followed by solving $h_5 = 0$ for c_1 and thereafter, using the real solutions for $\{c_1, c_2, l\}$, the values for λ_1 and λ_2 are obtained. The solutions for which the *residue*[3] $\epsilon < 10^{-8}$ are considered to be accurate enough for further analysis. The least absolute value of l among the real ones gives $l_{\max}$, as the cylinder intersects the singularity surface at other points for the higher values of l.

4 Numerical Example

The above formulations are illustrated numerically for the architecture parameters[4] $b = 1$, and $a = 1/3$. Equation $f_5(l) = 0$ is solved using the in-built function `NSolve` of `Mathematica` for these parameters, along with $r_0 = 0.2$ and $z_0 = 1$. Only five solutions among the 21 are found to be real and accurate. The length of the SFC is computed to be $l_{\max} = -0.58$. Here, the negative sign of $l_{\max}$ signifies that $\boldsymbol{n}_s$ is measured from $\boldsymbol{n}_0$ along the negative $\boldsymbol{Z}$-axis, as shown in Fig. 3.

The variations of $l_{\max}$ w.r.t. z_0 and r_0 have been studied for the same architecture parameters. It has been observed that for $z_0 \leq 0.41$, there is no real cylinder which is free of singularity, for the chosen value of $r_0 = 0.2$. However, the absolute value of $l_{\max}$ monotonically increases for $z_0 > 0.41$, as shown in Fig. 4a. Figure 4b shows the reverse phenomenon at $z_0 = 1$ for increasing r_0. Figure 4c demonstrates a region for which a nonzero $l_{\max}$ of the SFC can be obtained in the range $r_0 \in [0.1, 0.2]$ and $z_0 \in [0.42, 1]$. The computation of such an SFC, with an associated accuracy of $\epsilon < 10^{-7}$, takes about 4 ms (averaged over 50 runs) on a single core of an Intel® Core™ i7-4790 CPU running at

[2] If c_1 is assumed to be zero, the chosen base circle converts to a point $[0, \pm r_0, z_0]^\top$; hence the SFC can not be found.

[3] The "residue", is defined as $\epsilon = ||(h_1, \ldots, h_5)||$, where h_i appear in Eq. 5, after substituting the solutions of $\{c_1, c_2, l, \lambda_1, \lambda_2\}$.

[4] All the link lengths and distances are scaled w.r.t. the base circumradius b; hence the length units are not specified.

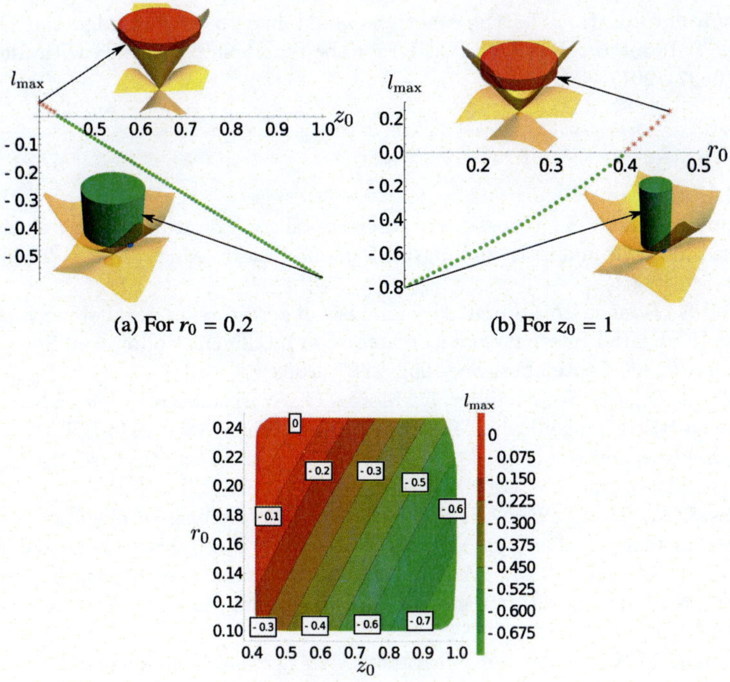

(a) For $r_0 = 0.2$

(b) For $z_0 = 1$

(c) For varying both z_0 and r_0

Fig. 4. Variation of l_{max} w.r.t. r_0 and z_0.

3.60 GHz. The algorithm is implemented in the `C` programming language, and compiled with `gcc`, version 7.4.0, with the `-Ofast` option.

5 Conclusions

In this paper, the closed-form expression of the gain-type singularity manifold of the 3-R$\underline{P}$S manipulator has been derived *analytically* in the task-space, represented by the variables $\{c_1, c_2, z\}$. The singularity surface, described by a polynomial equation of degree 11 in $\{c_1, c_2, z\}$, has been derived using the singularity condition and the constraint equations. Finding the longest SFC of the 3-R$\underline{P}$S manipulator has been demonstrated for the first time. The procedure to determine the SFC of the maximum length for a constant radius and a chosen base circle has been described in detail. A 21-degree univariate polynomial, $f_5(l)$, has been obtained, whose smallest absolute real root equals the derived maximum length, l_{max}. The computation of the singularity manifold and the SFC have been illustrated numerically. The variations of l_{max} of the SFC w.r.t. r_0 and z_0 have been studied numerically. These computations, being both fast and accurate, afford the means for in-depth studies of the design space leading to optimal designs of the manipulator in reasonable times.

Acknowledgements. The third author would like to acknowledge the Council of Scientific & Industrial Research (CSIR) for the scholarship provided with file number: 09/084(0697)/2017-EMR-I.

References

1. Bandyopadhyay, S., Ghosal, A.: Analysis of configuration space singularities of closed-loop mechanisms and parallel manipulators. Mech. Mach. Theor. **39**(5), 519–544 (2004)
2. Briot, S., Bonev, I.A.: Singularity analysis of zero-torsion parallel mechanisms. In: 2008 IEEE/RSJ International Conference on Intelligent Robots and Systems, Nice, France, 22–26 September, pp. 1952–1957 (2008)
3. Chablat, D., Jha, R., Rouillier, F., Moroz, G.: Workspace and Joint Space Analysis of the 3-RPS Parallel Robot. In: Proceedings of the ASME 2014 IDETC/CIE, 38th Mechanisms and Robotics Conference, Buffalo, New York, USA, 17–20 August, vol. 5A (2014). https://doi.org/10.1115/DETC2014-34593
4. Ghosal, A., Ravani, B.: A differential-geometric analysis of singularities of point trajectories of serial and parallel manipulators. J. Mech. Des. **123**(1), 80–89 (1998)
5. Jiang, Q., Gosselin, C.M., Wang, Y., Fang, C.: Maximal singularity-free orientation workspace over a position region of Gough-Stewart platform. Adv. Robot. **29**(22), 1427–1436 (2015)
6. Karnam, M.K., Baskar, A., Srivatsan, R.A., Bandyopadhyay, S.: Computation of the safe working zones of planar and spatial parallel manipulators. Robotica **38**(25), 861–885 (2020)
7. Li, H., Gosselin, C.M., Richard, M.J.: Determination of maximal singularity-free zones in the workspace of planar three-degree-of-freedom parallel mechanisms. Mech. Mach. Theor. **41**(10), 1157–1167 (2006)
8. Muralidharan, V., Bose, A., Chatra, K., Bandyopadhyay, S.: Methods for dimensional design of parallel manipulators for optimal dynamic performance over a given safe working zone. Mech. Mach. Theor. **147**, 103721 (2020)
9. Nag, A., Reddy, V., Agarwal, S., Bandyopadhyay, S.: Identifying singularity-free spheres in the position workspace of semi-regular Stewart platform manipulators. In: Lenarčič, J., Merlet, J.P. (eds.) ARK 2016. Springer Proceedings in Advanced Robotics, vol. 4, pp. 421–430. Springer, Cham (2018)
10. Prasad, P.K., Bandyopadhyay, S.: Identification of the largest singularity-free cylinders in the translational workspace of the semi-regular Stewart platform manipulator. In: Uhl, T. (ed.) Advances in Mechanism and Machine Science. IFToMM WC 2019. Mechanisms and Machine Science, vol. 73, pp. 569–578. Springer, Cham (2019)
11. Schadlbauer, J., Walter, D., Husty, M.: The 3-RPS parallel manipulator from an algebraic viewpoint. Mech. Mach. Theor. **75**, 161–176 (2014)
12. Sokolov, A., Xirouchakis, P.: Singularity analysis of a 3-DOF parallel manipulator with R-P-S joint structure. Robotica **24**(1), 131–142 (2006)
13. Wolfram Research, Inc.: Mathematica, Version 12.0. Champaign, IL (2019)

Clifford's Identity and Generalized Cayley-Menger Determinants

Federico Thomas(✉) and Josep M. Porta

Institut de Robòtica i Informàtica Industrial (CSIC-UPC), Llorens Artigas 4-6, 08028 Barcelona, Spain
{fthomas,porta}@iri.upc.es

Abstract. Distance geometry is usually defined as the characterization and study of point sets in $\mathbb{R}^k$, the k-dimensional Euclidean space, based on the pairwise distances between their points. In this paper, we use Clifford's identity to extend this kind of characterization to sets of n hyperspheres embedded in $\mathbb{S}^{n-3}$ or $\mathbb{R}^{n-3}$ where the role of the Euclidean distance between two points is replaced by the so-called *power* between two hyperspheres. By properly choosing the value of n and the radii of these hyperspheres, Clifford's identity reduces to conditions in terms of generalized Cayley-Menger determinants which has been previously obtained on the basis of a case-by-case analysis.

Keywords: Clifford's identity · Cayley-Menger determinants · Distance geometry

1 Introduction

In general, when using distance geometry to characterize point sets based on the pairwise distances between the points, no more constrains are enforced than the mere requirement that all distances are positive [14]. Therefore, this characterization is said to be performed in a semimetric space because the triangle inequality, the tetrangular inequality, and any other higher-order inequality, are not necessarily considered. Nevertheless, since distance geometry performs all characterizations in terms of Cayley-Menger determinants, these constraints, and those related to the orientation of simplices, can be incorporated in a simple way [18]. This generalization permits to formulate and solve such relevant problems in Kinematics as the inverse kinematics of serial robots and the forward kinematics of parallel ones (see [20] and the references therein).

Another important generalization of distance geometry consists in considering other geometric elements than points. To this end, Cayley-Menger determinants must be extended to involve the *distance* between these other geometric elements. In this paper, we will see that this generalization is possible by adopting the concept of *power* instead of the standard Euclidean distance. As a result, the generalizations of Cayley-Menger determinants presented, for example, in [15], naturally arise as particular cases of Clifford's identity, instead of relying on a case-by-case analysis.

This paper is organized as follows. Section 2 gives the definition of the power of two circles, a generalization of Steiner's power of a point and a circle. The limit cases (those

J. Lenarčič and B. Siciliano (Eds.): ARK 2020, SPAR 15, pp. 285–292, 2021.
https://doi.org/10.1007/978-3-030-50975-0_35

involving at least one cicle with zero or infinite radius) are treated in Sect. 3, where the straightforward generalization to hyperspheres is also given. Some further generalizations are detailed in Sect. 4. Section 5 gives the most general form of Clifford's identity and explains how generalized Cayley-Menger determinants—involving points, hyperplanes and hyperspheres—can be deduced from it. Finally, Sect. 6 identifies some points deserving further research efforts.

2 The Power of Two Circles

In 1826, Steiner defined the power of a point P with respect to a circle c of radius r as $h = d^2 - r^2$, where d is the distance between P and the center O of the circle [9,21]. According to this definition, points inside the circle have negative power, points outside have positive power, and points on the circle have zero power. For external points, the power equals the square of the length of a tangent from the point to the circle. Steiner proved that for any line through P intersecting c in points P' and P'', the power of the point with respect to the circle coincides, up to sign, with the product of the lengths of the segments from P to P' and P to P''.

In 1872, Darboux extended the idea of the power of a point with respect to a circle to the power of two circles [8]. Clifford also gave the same definition in two papers that appeared in his collected works [3,4].

Let c_i denote a circle with center at O_i and radius r_i, and let $d_{i,j}$ denote the distance between the centers of the circles c_i and c_j. Then, the mutual power of the circles c_i and c_j is defined as:

$$(c_i c_j) = d_{i,j}^2 - r_i^2 - r_j^2. \tag{1}$$

When both circles intersect, the power thus defined reduces to

$$(c_i c_j) = -2 r_i r_j \cos\theta_{i,j}, \tag{2}$$

where $\theta_{i,j}$ is the angle of intersection of the two circles. As a consequence, when they cut at right angles, their power vanishes and vice versa.

3 Limit Cases

In terms of Cartesian coordinates, c_i with center coordinates (x_i, y_i) can be expressed as:

$$c_i : (x - x_i)^2 + (y - y_i)^2 - r_i^2 = x^2 + y^2 + 2 f_i x + 2 g_i y + h_i = 0, \tag{3}$$

where $f_i = -x_i$, $g_i = -y_i$, and $h_i = x_i^2 + y_i^2 - r_i^2$. Then, the power of two circles given in (1) can be rewritten as:

$$(c_i c_j) = h_i + h_j - 2 f_i f_j - 2 g_i g_j, \tag{4}$$

which is sometimes also referred to as the product of c_i and c_j [7]. This alternative formula is useful to analyze the limit cases in which one circle degenerates into a line possibly at infinity. Next, we analyze three relevant cases. All others follow directly.

Table 1. The powers between points, hyperspheres, and hyperplanes ($\delta \to \infty$).

$c_i \setminus c_j$	Point	Hypersphere	Hyperplane	Hyperplane at ∞
Point	$d_{i,j}^2$	$d_{i,j}^2 - r_j^2$	$2\delta p_{j,i}$	$-\delta^2$
Hypersphere	$d_{i,j}^2 - r_i^2$	$d_{i,j}^2 - r_i^2 - r_j^2$	$2\delta p_{j,i}$	$-\delta^2$
Hyperplane	$2\delta p_{i,j}$	$2\delta p_{i,j}$	$-2\delta^2 \cos\theta_{i,j}$	0
Hyperplane at ∞	$-\delta^2$	$-\delta^2$	0	NA

1. *One circle degenerates into a line.* Considering a line as a circle of infinite radius, let us set $r_i = \delta \to \infty$. Then, using (1), we have that

$$\lim_{\delta \to \infty} (c_i c_j) = \lim_{\delta \to \infty} \left[(p_{i,j} + \delta)^2 - \delta^2 - r_j^2 \right] = 2 p_{i,j} \delta, \tag{5}$$

where $p_{i,j}$ is the perpendicular distance from the center of c_j to the line. The side of the line on which $p_{i,j}$ is positive is regarded as the outside of the infinite circle.
2. *One circle degenerates into a line at infinity.* The equation of the line at infinity can be written as

$$x^2 + y^2 + 2 f_i x + 2 g_i y - \delta^2 = 0, \tag{6}$$

where $\delta \to \infty$. Identifying (6) and (3), we have that $h_i = -\delta^2$. Then, the power of c_i and c_j, c_i being the line at infinity, can be expressed as:

$$(c_i c_j) = \lim_{\delta \to \infty} (-\delta^2 + h_j - 2 f_i f_j - 2 g_i g_j) = -\delta^2. \tag{7}$$

Observe that the negative sign is consistent with the previous case because the line at infinite *wraps* all finite points.
3. *Both circles degenerate into lines.* If we set $r_i = r_j = \delta \to \infty$ in (2), then

$$\lim_{\delta \to \infty} (c_i c_j) = -2\delta^2 \cos\theta_{i,j}. \tag{8}$$

where $\theta_{i,j}$ is the angle between both lines. Observe that the power of two coincident lines is $-2\delta^2$.

The cases in which at least one of the radii is zero follow directly. Moreover, observe that the definition of power of two circles can be directly extended to the power of two hyperspheres, and that its value for the limit cases obtained above also apply to this generalization. The resulting powers for all possible combinations are summarized in Table 1. To deal with powers involving points, hyperplanes and hyperspheres at the same time, it is important, for dimensional consistency, to keep the factors δ and δ^2, contrarily to what is done, for example, in [11]. As a result, all powers have the dimension of length squared as δ has length dimension.

4 Further Generalizations

Let us suppose that c_i and c_j are two circles, on a sphere of radius R, with spherical radii r_i and r_j, respectively. Let $\omega_{i,j}$ denote the length of the arc joining their centers.

Then, the power of c_i and c_j is defined as [2, §105]

$$(c_i c_j) = \cos(\kappa \omega_{i,j}) - \cos(\kappa r_i)\cos(\kappa r_j), \tag{9}$$

where $\kappa = 1/R$ denotes the curvature of the sphere.

If c_i and c_j intersect and the angle of intersection between both circles is denoted by $\theta_{i,j}$, then we have that

$$(c_i c_j) = \sin(\kappa r_i)\sin(\kappa r_j)\cos\theta_{i,j}, \tag{10}$$

which is zero when the circles cut orthogonally.

If $\kappa \to 0$ (*i.e.*, the sphere becomes a plane), $\cos(\kappa\alpha) \to 1 - \kappa^2\alpha^2$. Then, it is easy to verify that

$$\lim_{\kappa \to 0}(c_i c_j) = -\kappa^2(\omega_{i,j}^2 - r_i^2 - r_j^2), \tag{11}$$

which is consistent with the definition of power of two circles on a plane given by (1) up to the factor $-\kappa^2$.

Other generalization of the concept of power, that have received little attention, can be found in the literature. For example, in [12], Lachlan extended the concept of power to certain pairs of conics following the work of Casey in [1]. Laguerre also defined the power of a point with respect to any algebraic curve of degree n to be the product of the distances from the point to the intersections of a circle through the point with the curve, divided by the nth power of the diameter d [13]. Laguerre showed that this number is independent of the diameter. In the case when the algebraic curve is a circle the result does not coincide with the power of a point with respect to a circle defined by Steiner, but differs from it by a factor of d^2. The topic of the power of a point with respect to an algebraic curve has recently be revisited in [22] following the work of Neville [16, 17], which was influenced by that of Laguerre.

5 Clifford's Identity

Clifford's identity states that, given two sets of five circles on the unit sphere $\mathbb{S}^2$, their mutual powers are related through the following relationship [3]:

$$\begin{vmatrix} (c_1c_6) & (c_1c_7) & (c_1c_8) & (c_1c_9) & (c_1c_{10}) \\ (c_2c_6) & (c_2c_7) & (c_2c_8) & (c_2c_9) & (c_2c_{10}) \\ (c_3c_6) & (c_3c_7) & (c_3c_8) & (c_3c_9) & (c_3c_{10}) \\ (c_4c_6) & (c_4c_7) & (c_4c_8) & (c_4c_9) & (c_4c_{10}) \\ (c_5c_6) & (c_5c_7) & (c_5c_8) & (c_5c_9) & (c_5c_{10}) \end{vmatrix} = 0, \tag{12}$$

where powers have to be computed according to Eq. (9).

The great mathematics historian Julian Coolidge pointed out in [5, p. 135] that this identity was first presented by Darboux in [8] and that Frobenius announced that he had long been familiar with it and proceeded to publish his results in [10]. Nevertheless, Coolidge himself later realized [6, p. 168] that Clifford presented this result for the first time in [3]. In this later reference, Clifford's identity is presented in an alternative form:

that involving two sets of six spheres in $\mathbb{R}^3$, in which case powers have to be computed according to Eq. (1).

Although Clifford's identity has been presented for $\mathbb{S}^2$ or $\mathbb{R}^3$, thus involving five or six hyperspheres, respectively, both versions are easily stepped up or down to as many dimensions as required. Indeed, we can present Clifford's identity in all its generality as

$$\begin{vmatrix} (c_1c_{n+1}) & (c_1c_{n+2}) & \cdots & (c_1c_{2n}) \\ (c_2c_{n+2}) & (c_2c_{n+2}) & \cdots & (c_2c_{2n}) \\ \vdots & \vdots & \ddots & \vdots \\ (c_nc_{2n}) & (c_nc_{n+2}) & \cdots & (c_nc_{2n}) \end{vmatrix} = 0, \tag{13}$$

where $\{c_1,\dots,c_n\}$ and $\{c_{n+1},\dots,c_{2n}\}$ are two sets of $n > 4$ hyperspheres (which includes points and hyperplanes as degenerate cases) embedded in $\mathbb{S}^{n-3}$ or $\mathbb{R}^{n-3}$.

Many interesting relationships can be obtained as particular cases of (13). Due to space limitations, next we only analyze different cases in $\mathbb{R}^2$ and, to better appreciate the advantages of the presented approach, this analysis follows the same sequence of cases presented in [15].

In $\mathbb{R}^2$, Clifford's identity involves two sets of five circles. If both sets are equal, and their radii are zero, Clifford's identity reduces to

$$\begin{vmatrix} 0 & d_{12}^2 & d_{13}^2 & d_{14}^2 & d_{15}^2 \\ d_{21}^2 & 0 & d_{23}^2 & d_{24}^2 & d_{25}^2 \\ d_{31}^2 & d_{32}^2 & 0 & d_{34}^2 & d_{35}^2 \\ d_{41}^2 & d_{42}^2 & d_{43}^2 & 0 & d_{45}^2 \\ d_{51}^2 & d_{52}^2 & d_{53}^2 & d_{54}^2 & 0 \end{vmatrix} = 0. \tag{14}$$

Let us remind that $d_{i,j} = d_{j,i}$ is the distance between points c_i and c_j. If $c_1 = c_6$ is a line instead of a point, after taking the common factor δ of the first row and column out of the resulting determinant, we have that

$$\begin{vmatrix} 0 & 2p_{12} & 2p_{13} & 2p_{14} & 2p_{15} \\ 2p_{12} & 0 & d_{23}^2 & d_{24}^2 & d_{25}^2 \\ 2p_{13} & d_{32}^2 & 0 & d_{34}^2 & d_{35}^2 \\ 2p_{14} & d_{42}^2 & d_{43}^2 & 0 & d_{45}^2 \\ 2p_{15} & d_{52}^2 & d_{53}^2 & d_{54}^2 & 0 \end{vmatrix} = 0. \tag{15}$$

In this case $p_{1,j}$ is the oriented distance between the line c_1 and the point c_j. If c_1 is a line at infinity, after taking the common factor $-\delta^2$ of the first row and column out of the resulting determinant, the identity (15) becomes

$$\begin{vmatrix} 0 & 1 & 1 & 1 & 1 \\ 1 & 0 & d_{23}^2 & d_{24}^2 & d_{25}^2 \\ 1 & d_{32}^2 & 0 & d_{34}^2 & d_{35}^2 \\ 1 & d_{42}^2 & d_{43}^2 & 0 & d_{45}^2 \\ 1 & d_{52}^2 & d_{53}^2 & d_{54}^2 & 0 \end{vmatrix} = 0, \tag{16}$$

that is, the standard Cayley-Menger determinant for four points on a plane. Now, let us assume that $c_5 = c_{10}$ is also a line instead of a point. Then, after taking the common factor δ of the last row and column out of the resulting determinant, we have that

$$\begin{vmatrix} 0 & 1 & 1 & 1 & 0 \\ 1 & 0 & d_{23}^2 & d_{24}^2 & 2p_{25} \\ 1 & d_{32}^2 & 0 & d_{34}^2 & 2p_{35} \\ 1 & d_{42}^2 & d_{43}^2 & 0 & 2p_{45} \\ 0 & 2p_{52} & 2p_{53} & 2p_{54} & -2 \end{vmatrix} = 0, \tag{17}$$

which concurs with the result presented in [15]. If $c_4 = c_9$ is also a line instead of a point, after taking the common factor δ of the third row and column out of the determinant, we have that

$$\begin{vmatrix} 0 & 1 & 1 & 0 & 0 \\ 1 & 0 & d_{23}^2 & 2p_{24} & 2p_{25} \\ 1 & d_{32}^2 & 0 & 2p_{34} & 2p_{35} \\ 0 & 2p_{42} & 2p_{43} & -2 & -2\cos\theta_{45} \\ 0 & 2p_{52} & 2p_{53} & -2\cos\theta_{54} & -2 \end{vmatrix} = 0, \tag{18}$$

which coincides with the one derived in [15] up to a factor of -2 in the last two rows, which does not alter the result. Finally, if $c_3 = c_8$ is also a line, it is straightforward to show that

$$\begin{vmatrix} 0 & 1 & 0 & 0 & 0 \\ 1 & 0 & 2p_{23} & 2p_{24} & 2p_{25} \\ 0 & 2p_{32} & -2 & -2\cos\theta_{34} & -2\cos\theta_{35} \\ 0 & 2p_{42} & -2\cos\theta_{43} & -2 & -2\cos\theta_{45} \\ 0 & 2p_{52} & -2\cos\theta_{53} & -2\cos\theta_{54} & -2 \end{vmatrix} = 0, \tag{19}$$

from which we can conclude that

$$\begin{vmatrix} 1 & \cos\theta_{34} & \cos\theta_{35} \\ \cos\theta_{43} & 1 & \cos\theta_{45} \\ \cos\theta_{53} & \cos\theta_{54} & 1 \end{vmatrix} = 0. \tag{20}$$

In other words, the location of point $c_2 = c_7$ is irrelevant and only the angles between the three finite lines, $c_3 = c_8$, $c_4 = c_9$, and $c_5 = c_{10}$, are mutually constrained. This concurs with the result presented in [15] without having to analyze separately this particular case.

Finally, it is interesting to observe that the determinant in (20) can be interpreted as the Gramian of three points on the unit circle. Gramians can actually be seen as the counterpart in $\mathbb{S}^n$ of Cayley-Menger determinants in $\mathbb{R}^n$ [19,23].

6 Conclusion

In the past, the identities in Sect. 5 involving generalized Cayley-Menger determinants have been individually proved. Nevertheless, we have shown how Clifford's identity, together with Table 1, is enough to derive them in a simple and unified way. The analysis presented, as an example, is valid for $\mathbb{R}^2$, but its generalization to higher dimensions does not offer any extra complexity than that derived from the increment in the number of cases.

The analysis of Clifford's identity in $\mathbb{S}^n$ deserves further attention because it should be possibly to obtain the results for $\mathbb{R}^n$ from equivalent expressions in $\mathbb{S}^n$ as a limit in which the curvature of the embedding sphere tends to zero.

Another venue for further investigation concerns the generalization of Clifford's identity to conics and, in general, to arbitrary algebraic curves following Laguerre's ideas.

Acknowledgements. This work was partially supported by the Spanish Ministry of Economy and Competitiveness through the projects DPI2017-88282-P and MDM-2016-0656.

References

1. Casey, J.: On bicircular quartics. Trans. Irish Acad. **24**, 457–569 (1871)
2. Casey, J.: A Treatise on Spherical Trigonometry, and Its Application to Geodesy and Astronomy, with Numerous Examples. Longmans, Green & Co., London (1889)
3. Clifford, W.K.: On the power-coordinates in general. In: Tucker, R. (ed.) Mathematical Papers, pp. 546–555. Macmillan and Co., London (1882)
4. Clifford, W.K.: On the power of spheres. In: Tucker, R. (ed.) Mathematical Papers, pp. 332–336. Macmillan and Co., London (1882)
5. Coolidge, J.L.: A Treatise On The Circle And The Sphere. Oxford University Press, Oxford (1916)
6. Coolidge, J.L.: A History of Geometrical Methods. Oxford University Press, Oxford (1940)
7. Cox, H.: On systems of circles and bicircular quartics. Q. J. Math. Pure Appl. Math. **19**, 74–124 (1883)
8. Darboux, G.: Sur les relations entre les groupes de points, de cercles et de sphéres dans le plan et dans l'espace. Annales Scientifiques de l'École Normale Supérieure **1**, 323–392 (1872)
9. Fried, M.: Mathematics as the science of patterns - Jacob Steiner and the power of a point. Convergence (2010)
10. Frobenius, F.G.: Anwendungen der determinantentheorie auf die geometrie des maasses. Journal für die reine und angewandte Mathematik **79**, 185–247 (1875)
11. Johnson, W.: Some theorems relating to groups of circles and spheres. Am. J. Math. **14**(2), 97–114 (1892)
12. Lachlan, R.: On systems of circles and spheres. Proc. R. S. Lond. Philos. Trans. R. Soc. **177**, 481–625 (1886)
13. Laguerre, E.N.: Sur les courbes planes algébriques. Comptes rendus des séances de l'Académie des sciences **60**, 18–22 (1865)
14. Menger, K.: New foundation of Euclidean geometry. Am. J. Math. **53**(4), 721–745 (1931)
15. Michelucci, D., Foufou, S.: Using Cayley-Menger determinants for geometric constraint solving. In: Elber, G., Patrikalakis, N., Brunet, P. (eds.) Proceedings of the Ninth ACM Symposium on Solid modeling and Applications, Genoa, Italy, pp. 285–290 (2004)
16. Neville, E.H.: Products of lengths of tangents and of normals. Math. Gazette **38**(325), 192–193 (1954)
17. Neville, E.H.: The power of a point for a curve. Math. Gazette **40**(331), 11–14 (1956)
18. Porta, J.M., Rull, A., Thomas, F.: Sensor localization from distance and orientation constraints. Sensors **16**(7), 1–19 (2016)
19. Porta, J.M., Sarabandi, S., Thomas, F.: Angle-bound smoothing with applications in kinematics. In: IFToMM Asian Mechanism and Machine Science (Asian MMS 2018), Bengaluru, India (2018)

20. Porta, J.M., Thomas, F.: Yet another approach to the Gough-Stewart platform forward kinematics. In: Proceedings of the IEEE International Conference in Robotics and Automation, Brisbane, Australia (2018)
21. Steiner, J.: Einige geometrische betrachtungen. J. für die reine und angewandte Mathematik **1**, 161–184 (1826)
22. Suceava, B.D., Vajiac, A.I., Vajiac, M.B.: The power of a point for some real algebraic curves. Math. Gazette **92**(523), 22–28 (2008)
23. Thomas, F., Pérez-Gracia, A.: A new insight into the coupler curves of the RCCC four-bar linkage. In: 7th IFToMM International Workshop on Computational Kinematics, Poitiers, France (2017)

A New Approach for Continuous Wrapping of a Thick Strand on a Surface — The Planar Case with Constant Length and Free Ends

Katharina Müller(✉) and Andres Kecskemethy

Chair of Mechanics and Robotics, University of Duisburg-Essen, Lotharstr. 1, 47057 Duisburg, Germany
katharina.mueller@uni-due.de

Abstract. In musculoskeletal simulations, it is standard practice to model muscles or strands wrapping on a surface as thin massless lines. When the strand has non-negligible and in particular non-constant cross section thickness, the assumptions of infinitely thin lines does not apply anymore, and current methods rely on discretizations of the strands as chains of spherical beads, which however are discontinuous and thus produce jerky motions which lead to unrealistic forces when coupled to dynamic muscle models. In this work, we present a novel continuous approach which solves the problem using smooth differential equations as the limit of infinitesimally close beads. The present contribution is an extension of our previous work in which the equations were first derived for the case of a planar conical piece of strand lying on a given arc of a surface. The paper further develops these equations for the case of varying strand thickness and prescribed motion of the free ends under the constraint of constant strand length. The results are compared with the bead method, showing their superiority both in terms of smoothness and computational efficiency. Based on this approach, the 3D and stretchable-strand case can be tackled.

Keywords: Muscle wrapping · Thick strands · Musculoskeletal modelling

1 Introduction

In biomechanical musculoskeletal simulations, it is standard practice to model the muscle as a thin massless line. [7] was one of the first to use simple geometries like spheres and cylinders to fit them into bone geometries and wrap the muscle path onto them. Further improvements to this approach were made by [3], calling it obstacle-set method. [1] and [4] considered the whole muscle path between origin and insertion treating it as an optimization problem. [6] presented an analytical solution to this problem using natural geodesic variations and achieved

J. Lenarčič and B. Siciliano (Eds.): ARK 2020, SPAR 15, pp. 293–302, 2021.
https://doi.org/10.1007/978-3-030-50975-0_36

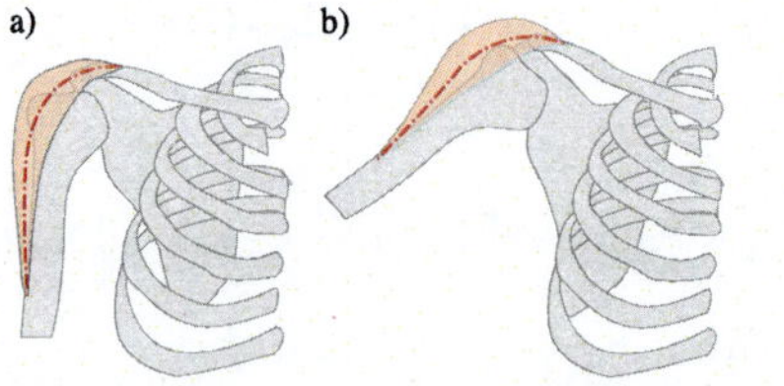

Fig. 1. Example: Musculus deltoideus pars acromialis for a) adduction b) abduction

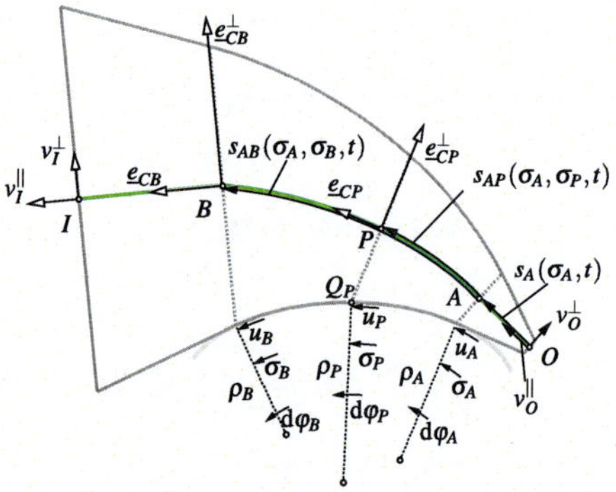

Fig. 2. Thick wrapping between O and I

high computational efficiency. All of these methods need to have the wrapping surfaces carefully placed to obtain a muscle path that is close to the centerline of the muscle, and fail when the muscle itself has varying thickness, in which case it is not possible to adapt the surface so that it follows the centerline while the muscle slides. An example of such a case is illustrated in Fig. 1 for the musculus deltoideus pars acromialis, which slides and stretches on the humerus when flexing the shoulder joint. To deal which such a problem, [2] present a discrete method using spheres ("beads") connected by springs to overcome this shortness. While such a bead chain method can be used generally, it induces intricate contact problems, high computation times and discontinuities.

The aim of this paper is to present a novel continuous approach where the thickness of the strand can be mathematically described along its centerline and the wrapping of this thick strand over the surface leads to a centerline that is automatically computed according to the local thickness and the surface curvature. The paper is a generalization of the method described in [5], where the new continuous approach was presented for the first time for the planar case for wrapping a thick, conical muscle on an arbitrary convex surface over a given arc length. This approach is further developed in this paper for a nonlinearly-variable, convex strand-thickness distribution as well as free end points, under the assumptions again of constant length and perfectly slack (no bending moments) strands on a frictionless convex surface, and that cross sections remain planar and undeformed. Due to the complex differential geometric relationships involved in the derivation, the approach is presented here for the two-dimensional case on the assumption that the muscle length is constant. While this does not solve the full problem shown in Fig. 1, as stretching is excluded, it tackles already the main geometric issue of wrapping a nonlinear thickness distribution on a surface. Later developments will treat the nonconstant length as well as the 3D case. The paper contains the main derivations as well as a comparison with the bead method, showing its advantages in computation time and smoothness.

2 Description of the Method

a) Strand on Surface

Let a thick strand be described by a wrapped part between the points A and B on the surface and two free ends O and I on both sides (Fig. 2). Let s denote the centerline path length, s_A the length of the straight line connecting points O and A, s_{AB} the length of the centerline of the wrapped strand segment between the points A and B and s_{BI} the length of the straight line segment between the points B and I. Moreover, let $b(s)$ describe the half thickness of the strand as a function of the undeformed centerline path length, and $\tan\alpha = \partial b/\partial s = b'$ the local half "conicity" angle $\alpha(s)$ over the centerline (also arising between the surface normal and the cross section).

Regard an arbitrary point P on the wrapped centerline of the strand between A and B. On the surface, let Q_P describe the contact point, $\underline{x}_{QP}$ its position vector, u_P the arc coordinate, σ_P a corresponding natural surface coordinate, ρ_P the local curvature radius, and $\mathrm{d}\varphi_P$ an infinitesimal rotation increment of the surface normal when progressing on the surface by an infinitesimal arc increment $\mathrm{d}u$. Then, it holds

$$\mathrm{d}u_P = \left\| \frac{\partial \underline{x}_{QP}}{\partial \sigma} \right\| \mathrm{d}\sigma_P \;\widehat{=}\; \mathrm{J}_{u\sigma}\big|_P \,\mathrm{d}\sigma_P; \;\; \mathrm{d}\varphi_P = \frac{\mathrm{d}u_P}{\rho_P} \;\widehat{=}\; \mathrm{J}_{\varphi\sigma}\big|_P \,\mathrm{d}\sigma_P. \tag{1}$$

where $\mathrm{J}_{\varphi\sigma}\big|_P$ and $\mathrm{J}_{u\sigma}\big|_P$ denote the surface Jacobians evaluated at point Q_P.

As the strand can slide on the surface, the centerline path length s_P at point P is a function of the "wrapping" surface coordinate σ_P for fixed time, and the change of strand location for fixed surface coordinate as a function of time (termed "pulling" according to increasing the value of s_A). An infinitesimal change $\mathrm{d}s_P$ of P is thus composed of two independent increments, denoted in shortcut as "wrap" and "pull":

$$\mathrm{d}s_P = \frac{\partial s}{\partial \sigma}\bigg|_{t=\mathrm{const}} \mathrm{d}\sigma_P + \frac{\partial s}{\partial t}\bigg|_{\sigma_P=\mathrm{const}} \mathrm{d}t \;\widehat{=}\; \mathrm{d}s^{\mathrm{wrap}} + \mathrm{d}s^{\mathrm{pull}}. \tag{2}$$

The same dependency holds for increments of angle α, thickness b, and vector $\underline{r}_P$:

$$\mathrm{d}\alpha_P = \frac{\partial \alpha}{\partial \sigma}\bigg|_{t=\mathrm{const}} \mathrm{d}\sigma_P + \frac{\partial \alpha}{\partial t}\bigg|_{\sigma_P=\mathrm{const}} \mathrm{d}t \;\widehat{=}\; \mathrm{d}\alpha^{\mathrm{wrap}} + \mathrm{d}\alpha^{\mathrm{pull}}, \tag{3}$$

$$\mathrm{d}b_P = \frac{\partial b}{\partial \sigma}\bigg|_{t=\mathrm{const}} \mathrm{d}\sigma_P + \frac{\partial b}{\partial t}\bigg|_{\sigma_P=\mathrm{const}} \mathrm{d}t \;\widehat{=}\; \mathrm{d}b^{\mathrm{wrap}} + \mathrm{d}b^{\mathrm{pull}}, \tag{4}$$

$$\mathrm{d}\underline{r}_P = \frac{\partial \underline{r}_P}{\partial \sigma}\bigg|_{t=\mathrm{const}} \mathrm{d}\sigma_P + \frac{\partial \underline{r}_P}{\partial t}\bigg|_{\sigma_P=\mathrm{const}} \mathrm{d}t \;\widehat{=}\; \mathrm{d}\underline{r}_P^{\mathrm{wrap}} + \mathrm{d}\underline{r}_P^{\mathrm{pull}} \;. \tag{5}$$

For α and b it holds via the chain rule

$$\mathrm{d}\alpha^{\mathrm{wrap}} = \frac{\partial\alpha}{\partial s}\cdot\left.\frac{\partial s}{\partial\sigma}\right|_{t=\mathrm{const}} \mathrm{d}\sigma_P \;\widehat{=}\; \mathrm{J}_{\alpha s}\,\mathrm{d}s^{\mathrm{wrap}} \quad ; \quad \mathrm{d}\alpha^{\mathrm{pull}} = \frac{\partial\alpha}{\partial s}\cdot\left.\frac{\partial s}{\partial t}\right|_{\sigma_P=\mathrm{const}} \mathrm{d}t \;\widehat{=}\; \mathrm{J}_{\alpha s}\,\mathrm{d}s^{\mathrm{pull}} \tag{6}$$

$$\mathrm{d}b^{\mathrm{wrap}} = \frac{\partial b}{\partial s}\cdot\left.\frac{\partial s}{\partial\sigma}\right|_{t=\mathrm{const}} \mathrm{d}\sigma_P \;\widehat{=}\; b'\,\mathrm{d}s^{\mathrm{wrap}} \quad ; \quad \mathrm{d}b^{\mathrm{pull}} = \frac{\partial b}{\partial s}\cdot\left.\frac{\partial s}{\partial t}\right|_{\sigma=\mathrm{const}} \mathrm{d}t \;\widehat{=}\; b'\,\mathrm{d}s^{\mathrm{pull}}. \tag{7}$$

with the angle-Jacobian $\mathrm{J}_{\alpha s} = \partial\alpha/\partial s$, while the increment $\mathrm{d}\underline{r}_P$ of point P can be described by the (linear) sum of the increment at the surface, the rotation of the segment $\overline{Q_P\,P}$, and the increment of the length of the segment $\overline{Q_P\,P}$ as

$$\mathrm{d}\underline{r}_P = \rho_P\,\underline{e}_{tP}\,\mathrm{d}\varphi_P + b_P\,(\mathrm{d}\varphi_P - \mathrm{d}\alpha_P)\,\underline{e}_{CP} + \mathrm{d}b_P\,\underline{e}_{CP}^{\perp} \tag{8}$$

with $\underline{e}_{tP}$ the unit tangent at the surface, $\underline{e}_{CP}$ the unit vector normal to the cross section, and $\underline{e}_{CP}^{\perp}$ the unit vector normal to $\underline{e}_{CP}$. With Eqs. (1) to (7) this gives

$$\begin{aligned}\mathrm{d}\underline{r}_P = &\Big[\rho_P\,\mathrm{J}_{\varphi\sigma}\big|_P\,\mathrm{d}\sigma_P\,\underline{e}_{tP} + \big(b_P\,\mathrm{J}_{\varphi\sigma}\big|_P\,\mathrm{d}\sigma_P - b_P\,\mathrm{J}_{\alpha s}\big|_P\,\mathrm{d}s_P^{\mathrm{wrap}}\big)\,\underline{e}_{CP} + b_P'\,\mathrm{d}s_P^{\mathrm{wrap}}\,\underline{e}_{CP}^{\perp}\Big]\\ &+\Big[-b_P\,\mathrm{J}_{\alpha s}\big|_P\,\mathrm{d}s_P^{\mathrm{pull}}\,\underline{e}_{CP} + b_P'\,\mathrm{d}s_P^{\mathrm{pull}}\,\underline{e}_{CP}^{\perp}\Big]\,.\end{aligned} \tag{9}$$

Projecting Eq. (9) onto $\underline{e}_{CP}$ for fixed time ($\mathrm{d}s_P^{\mathrm{pull}} = 0$) gives on the left side the path length increment $\mathrm{d}s_P^{\mathrm{wrap}}$ along the wrapping path from A to B. As the latter also occurs on the right hand side, the resulting path length increment is obtained as

$$\mathrm{d}s_P^{\mathrm{wrap}} = \frac{\rho_P\,\cos\alpha_P\,\mathrm{J}_{\varphi\sigma}\big|_P + b_P\,\mathrm{J}_{\varphi\sigma}\big|_P}{1 + b_P\,\mathrm{J}_{\alpha s}\big|_P}\,\mathrm{d}\sigma_P \;\widehat{=}\; \mathrm{J}_{s\sigma}\big|_P\,\mathrm{d}\sigma_P \tag{10}$$

where the Jacobian $\mathrm{J}_{s\sigma}\big|_P$ is used in the following as a shortcut. The length s_{AB} of the centerline from A to B thus results as:

$$s_{AB} = \int_{\sigma_A}^{\sigma_B} \mathrm{J}_{s\sigma}(\sigma,t)\big|_{t=\mathrm{const}}\,\mathrm{d}\sigma. \tag{11}$$

b) Constraints

The constraints arise from the boundary conditions at points A and B and the condition of constant total length. At both end points, there are two boundary conditions: one in longitudinal and one in orthogonal direction. For simplicity reasons, the constraints are expressed in terms of velocities which correspond to the infinitesimal expressions derived above divided by $\mathrm{d}t$.

Let the velocity of the end point O be split into a longitudinal component $v_O^{||}$ in direction of the unit vector $\underline{e}_{CA}$ normal to the strand cross section in A, and a component $v_O^{\perp}$ in the perpendicular $\underline{e}_{CA}^{\perp}$ to this direction. The same applies to $v_I^{||}$ and $v_I^{\perp}$ at the other end of the strand. The longitudinal constraint at point A states the difference of velocities of points A and O projected on the cross-section normal must be equal to the total change of the path coordinate at point A:

$$(\underline{v}_A - \underline{v}_O)\,\underline{e}_{CA} = \dot{s}_A \;\widehat{=}\; \dot{s}_A^{wrap} + \dot{s}_A^{pull}. \tag{12}$$

With Eq. (9) and noting that by Eq. (10) the terms with $\dot{s}_A^{\text{wrap}}$ cancel each other, one obtains

$$\dot{s}_A^{\text{pull}} = -\frac{v_O^{||}}{1 + b_A\,\mathrm{J}_{\alpha s}\big|_A}. \tag{13}$$

The second condition is that the free line at A is orthogonal to the strand cross section, i.e. $(\underline{r}_A - \underline{r}_O)\;\underline{e}_{CA}^{\perp} = 0$, leading to the kinematic condition

$$(\underline{v}_A - \underline{v}_O)\;\underline{e}_{CA}^{\perp} + (\underline{r}_A - \underline{r}_O)\;\underline{\dot{e}}_{CA}^{\perp} = 0. \tag{14}$$

For the rate of $\underline{\dot{e}}_{CA}^{\perp}$ it holds

$$\underline{\dot{e}}_{CA}^{\perp} = (\dot{\varphi}_A - \dot{\alpha}_A)\,\underline{e}_{CA} = \left[\mathrm{J}_{\varphi\sigma}\big|_A\,\dot{\sigma}_A - \mathrm{J}_{\alpha s}\big|_A\,\dot{s}_A^{wrap} - \mathrm{J}_{\alpha s}\big|_A\,\dot{s}_A^{pull}\right]\,\underline{e}_{CA}. \tag{15}$$

Inserting Eq. (15), Eq. (9) and $s_A = \underline{r}_A - \underline{r}_O$ into Eq. (14) and solving for $\dot{\sigma}_A$ yields

$$\dot{\sigma}_A = \frac{v_O^{\perp} - \left(b_A' - s_A \mathrm{J}_{\alpha s}\big|_A\right)\,\dot{s}_A^{pull}}{b_A'\mathrm{J}_{s\sigma}\big|_A + s_A\left(\mathrm{J}_{\varphi\sigma}\big|_A - \mathrm{J}_{\alpha s}\big|_A \mathrm{J}_{s\sigma}\big|_A\right) - \rho_A \mathrm{J}_{\varphi\sigma}\big|_A\,\mathrm{sin}\alpha_A}. \tag{16}$$

The constraint defining the longitudinal direction at point B arises from the condition that the overall length of the strand must be constant, i.e., $s_A + s_{AB} + s_{BI} = \ell = \text{const}$. This leads to the corresponding time derivative

$$\dot{s}_A + \dot{s}_{AB} + \dot{s}_{BI} = 0. \tag{17}$$

The time derivative of $\dot{s}_{AB}$ is given by

$$\dot{s}_{AB} = \left.\frac{\partial s_{AB}}{\partial \sigma_A}\right|_{\sigma_B, t=\text{const}} \dot{\sigma}_A + \left.\frac{\partial s_{AB}}{\partial \sigma_B}\right|_{\sigma_A, t=\text{const}} \dot{\sigma}_B + \left.\frac{\partial s_{AB}}{\partial t}\right|_{\sigma_A, \sigma_B=\text{const}} \tag{18}$$

for which with Eq. (11) one obtains

$$\left.\frac{\partial s_{AB}}{\partial \sigma_A}\right|_{\sigma_B, t=\text{const}} \dot{\sigma}_A = -\left.\frac{\partial s}{\partial \sigma}\right|_{t=\text{const}, A} \dot{\sigma}_A = -\dot{s}_A^{wrap}, \quad \left.\frac{\partial s_{AB}}{\partial \sigma_B}\right|_{\sigma_A, t=\text{const}} \dot{\sigma}_B = \left.\frac{\partial s}{\partial \sigma}\right|_{t=\text{const}, B} \dot{\sigma}_B = \dot{s}_B^{wrap} \tag{19}$$

$$\left.\frac{\partial s_{AB}}{\partial t}\right|_{\sigma_A, \sigma_B=\text{const}} = \dot{s}_B^{pull} - \dot{s}_A^{pull}. \tag{20}$$

Inserting $\dot{s}_A$ and $\dot{s}_{AB}$ into Eq. (17) yields for the length change $\dot{s}_{BI}$

$$\dot{s}_{BI} = -\dot{s}_B^{wrap} - \dot{s}_B^{pull} \tag{21}$$

Analogously to point A, one obtains for the relationship between $v_I^{||}$ and $\dot{s}_B^{pull}$

$$v_I^{||} = -\left[1 + b_B\, \mathrm{J}_{\alpha s}\big|_B\right] \dot{s}_B^{pull}. \tag{22}$$

On the other hand, $\dot{s}_B^{pull}$ is related to $\dot{s}_A^{pull}$ by the velocity transmission factor $\hat{\kappa}_{AB}$ as

$$\hat{\kappa}_{AB} = \frac{\dot{s}_B^{\mathrm{pull}}}{\dot{s}_A^{\mathrm{pull}}} = \frac{(\rho_B\ \cos\alpha_B + b_B)\rho_A}{(\rho_A\ \cos\alpha_A + b_A)\rho_B}\, e^{\kappa} \tag{23}$$

where e^{κ} is the quotient of output to input velocity on surface level defined by

$$e^{\kappa} = \frac{\dot{u}_B}{\dot{u}_A} = \exp\left(\int_{\sigma_A}^{\sigma_B} \frac{r(s(\sigma))}{\rho(\sigma)\,[\,\rho(\sigma) + r(s(\sigma))\,]}\, \rho'\, \mathrm{d}\sigma\right) \tag{24}$$

(see[1] below). Thus, it holds for $\dot{s}_B^{pull}$

$$\dot{s}_B^{pull} = \hat{\kappa}_{AB}\, \dot{s}_A^{pull} = -\hat{\kappa}_{AB}\, \frac{v_O^{||}}{1 + b_A\, \mathrm{J}_{\alpha s}\big|_A}. \tag{25}$$

and a global longitudinal input-output velocity transmission is obtained as

$$v_I^{||} = \hat{\kappa}_{AB}\, \frac{1 + b_B\, \mathrm{J}_{\alpha s}\big|_B}{1 + b_A\, \mathrm{J}_{\alpha s}\big|_A}\, v_O^{||}. \tag{26}$$

The orthogonality condition at point B finally yields analogously to point A

$$\dot{\sigma}_B = -\frac{v_I^{\perp} - \left(b_B' + \ell_{BI}\, \mathrm{J}_{\alpha s}\big|_B\right) \dot{s}_B^{pull}}{-b_B'\mathrm{J}_{s\sigma}\big|_B + \ell_{BI}\left[\mathrm{J}_{\varphi\sigma}\big|_B - \mathrm{J}_{\alpha s}\big|_B\, \mathrm{J}_{s\sigma}\big|_B\right] + \rho_B\, \mathrm{J}_{\varphi\sigma}\big|_B\ \sin\alpha_B}. \tag{27}$$

[1] This equation results from the limit of pair-wise velocity transmission between beads as firstly derived in [5]. If ρ_i is the surface curvature at bead "i", r_i is its radius, and $\Delta\rho_i$ is the difference of surface curvature from bead "i" to bead "i+1", then the total velocity transmission from first to last bead is the product of these pair-wise transmissions. However, in [5], for the limit of infinitesimally close beads, in [5] we erroneously neglected higher-order terms of infinitesimal quantities. This is wrong and the correct limit condition is

$$\lim_{\Delta\rho_i \to 0} (1 + \eta_1\, \Delta\rho_1) \ldots (1 + \eta_{N-1}\, \Delta\rho_{N-1}) = \lim_{\Delta\rho_i \to 0} \exp\left(\sum_{i=1}^{N-1} \eta_i\, \Delta\rho_i\right) \quad \text{with } \eta_i = \frac{r_i}{\rho_i\,(\rho_i + r_i)}.$$

which for $\Delta\rho \to \mathrm{d}\rho = (\partial\rho/\partial\sigma)\, \mathrm{d}\sigma = \rho'\, \mathrm{d}\sigma$ yields the integral form. Unfortunately we cannot include the proof here due to lack of space, so we refer here to this equation as a correction of [5].

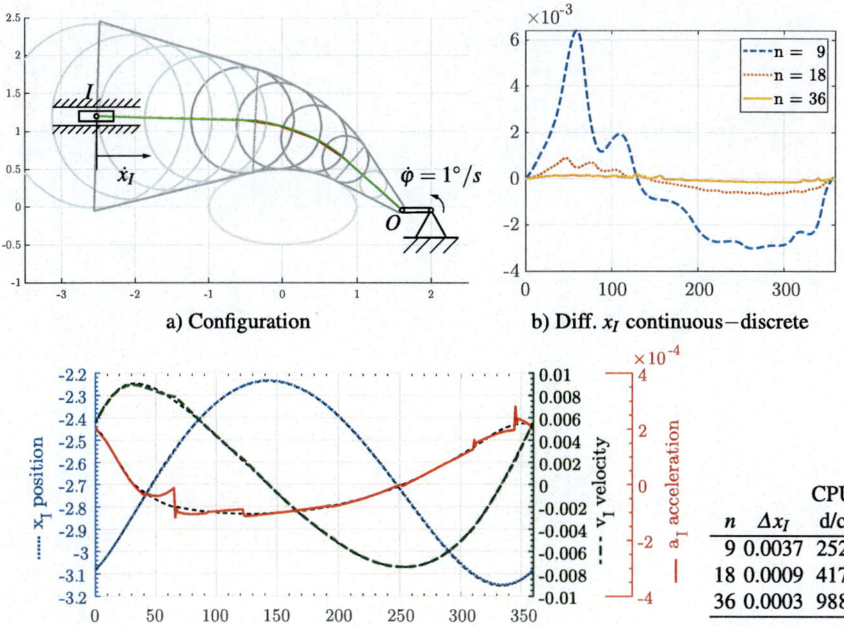

a) Configuration   b) Diff. x_I continuous−discrete

c) Kinematics of x_I: discrete ($n = 9$, colored) vs. continuous (dashed black)

n	Δx_I	CPU d/c
9	0.0037	252
18	0.0009	417
36	0.0003	988

d) Comparison

Fig. 3. Comparison of thick strand: discrete vs. continuous case for constant angle α

3 Computational Results

The continuous solution was obtained by numerical integration of equations (13), (16), (26), and (27) using the Matlab Runge-Kutta routine `ode45` with `reltol`$= 10^{-6}$. For each evaluation of the differential equations at a new point, an inner integration over s_{AB} according to Eq. (11) as well as that of the input-output velocity transmission factor e^κ according to Eq. (24) is necessary, which were again carried out again using `ode45` and the precision of the outer integration.

The regarded example is a stretch of thick strand wrapped on an ellipse with semi-major axis $a = 1$ and semi-minor axis $b = 0.5$ with dimensionless coordinates and as described in [5] (Fig. 3). The end O is rotated on a circle of radius $r = 0.4$ with constant angular velocity $\Omega = 1°/\text{s}$, while the end I can slide in horizontal direction. This resembles a slider-crank mechanism with the thick muscle acting as a "coupler" and the crank angle acting as input, while the slider motion x_I is the dependent variable. For the numerical integration, the relative maximum error in all constraints lied below 10^{-6}, showing good convergence and precision.

For comparison purposes, the continuous muscle was also approximated by a chain of inscribed circles with equidistant centers so that the continuous muscle becomes their hull. The "fineness' of the discretization is described by the number of circles within the total stretch of the continuous muscle. The contact between

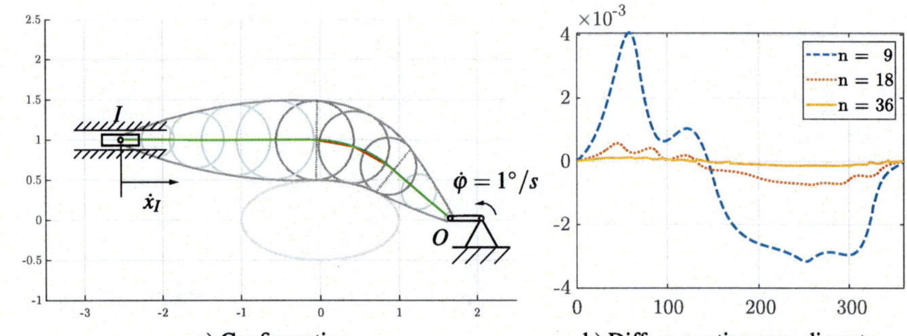

a) Configuration

b) Diff. x_I continuous−discrete

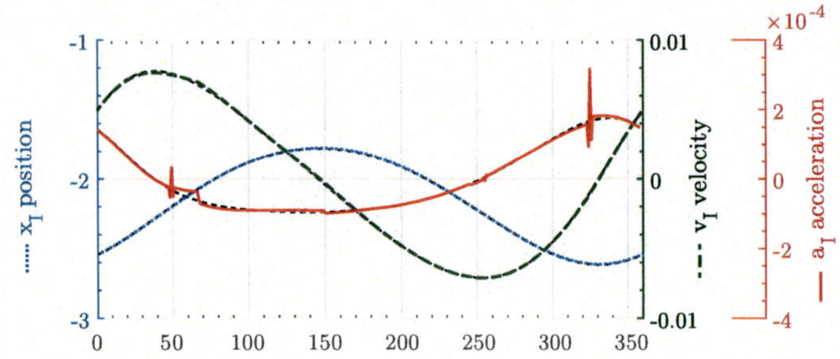

c) Kinematics of x_I: discrete ($n = 9$, colored) vs. continuous (dashed black)

n	Δx_I	CPU d/c
9	0.0040	239
18	0.0007	461
36	0.0001	887

d) Comparison

Fig. 4. Comparison of thick strand: discrete vs. continuous case for non-constant angle α

the circles ("beads") and the ellipse was solved by Newton iterations using the Matlab solver `fsolve` with constant rotation increments of 1° at the crank. If a bead entered or left the contact with the ellipse at a step, the corresponding bead location was corrected, i.e., no interval nesting was performed to determine the exact crank angle value at which touch-down or lift-off occurs. Details are described in [5].

Two cases of thick strands are regarded: one with constant angle α (termed constant "conicity"), and one with parabolic change of strand thickness over strand length. Figures 3 and 4 show the corresponding configurations (a) (boundary cross sections of the continuous strand shown as dotted lines) and the results of the difference of the computed value for the dependent slider variable x_I between the continuous method and the bead discretization for three different degrees of resolution (b).

It can be seen that the bead case approaches the continuous method for finer discretizations, proving the consistency of the latter. Moreover, it can also be seen that the discretized approach leads to unrealistic "pulsating" motions, with frequency and amplitude inversely proportional to the bead-pair distance. In figure (c), position, velocity and acceleration of point I are plotted for the case of $n = 9$ beads (colored curves) and compared to the smooth curves of the continuous method (dashed thin black lines). While the position graph seems smooth, the acceleration graph shows a jerky behavior including "spikes" due to sudden bead contact. This is disadvantageous for applications where the strand force is a function of strand length and strand length rate of change such as e.g. in the Hill muscle model, as then the force will the vary correspondingly discontinuously, slowing down numerical integration schemes. In the table (d), the maximal difference between displacement x_I of the discrete and the continuous method as well as the relative CPU time as factor discrete/continuous ("CPU d/c") are shown for a sample of discretization fineness ("n"). Visibly, the continuous approach leads to a much faster, more precise, and more easily implementable approach.

4 Conclusions

By the proposed continuous approach, the problem of determining the wrapping line of constant-length strands with varying cross section is shown to be more efficient, more precise, and smoother that the existing approach using discretizations by chains of spherical beads. The paper is a generalization of [5], where only a "conical" muscle geometry with constant aperture angle α and fixed location along the surface was considered, and regards non-linear, convex distributions of strand thickness over its (constant) length, as well as prescribed or depended motion of its free ends, solving also the sliding case. The results are verified by comparison with the bead method for different levels of discretization fineness. While the derivation of the equations is more complex than for the discrete bead method, the resulting equations are much easier to implement than the discrete case and lead to computational savings in the order of factors of 200–800. As future work, the method is planned to be extended to elongation with compressible cross sections, wrapping over multiple surfaces, and the spatial contact surface case.

References

1. Audenaert, A., Audenaert, E.: Global optimization method for combined spherical-cylindrical wrapping in musculoskeletal upper limb modelling. Comput. Methods Programs Biomed. **92**(1), 8–19 (2008). https://doi.org/10.1016/j.cmpb.2008.05.005
2. Franci, R., Parenti-Castelli, V.: A new tool to investigate the interactions between elastic fibers and rigid bodies. In: Proceedings of IFToMM 2007 (2007)
3. Garner, B.A., Pandy, M.G.: The obstacle-set method for representing muscle paths in musculoskeletal models. Comput. Methods Biomech. Biomed. Eng. **3**(1), 1–30 (2000). https://doi.org/10.1080/10255840008915251
4. Marsden, S.P., Swailes, D.C., Johnson, G.R.: Algorithms for exact multi-object muscle wrapping and application to the deltoid muscle wrapping around the humerus. Proc. Inst. Mech. Eng. Part H, J. Eng. Med. **222**(7), 1081–1095 (2008). https://doi.org/10.1243/09544119JEIM378
5. Müller, K., Kecskeméthy, A.: A continuous and computationally efficient method for wrapping a "thick" strand over a surface – the planar single-surface case. In: Uhl, T. (ed.) Advances in Mechanism and Machine Science, pp. 709–718. Springer, Cham (2019)

6. Scholz, A., Sherman, M., Stavness, I., Delp, S., Kecskeméthy, A.: A fast multi-obstacle muscle wrapping method using natural geodesic variations. Multibody Syst. Dyn. **36**(2), 195–219 (2016). https://doi.org/10.1007/s11044-015-9451-1
7. van der Helm, F.: A finite element musculoskeletal model of the shoulder mechanism. J. Biomech. **27**(5), 551–569 (1994). https://doi.org/10.1016/0021-9290(94)90065-5

Higher Order Path Synthesis of Four-Bar Mechanisms Using Polynomial Continuation

Aravind Baskar and Mark Plecnik(✉)

University of Notre Dame, Notre Dame, IN 46556, USA
{abaskar,plecnikmark}@nd.edu

Abstract. The path synthesis of a four-bar linkage is commonly framed so as to size links such that a coupler curve passes through specified planar precision points. It may instead be framed with consideration of higher order derivative information at each precision point. In this paper, we enumerate all 30 combinations of higher order path synthesis that lead to square polynomial systems for the four-bar. We present in detail the synthesis equations for (1) the case when one precision point is specified along with derivative information up to its eighth derivative, and (2) when three precision points are specified along with derivative information up to the second derivative at each point. For each case, 224 and 227 possible Roberts' cognate triplets were computed, respectively.

Keywords: Path synthesis · Numerical continuation · Higher order synthesis

1 Introduction

The path synthesis of a four-bar linkage is classically set up so that a point attached to the coupler link traces through specified planar precision points. When nine precision points are specified, the number of synthesis equations and unknowns are equal (called a square system), suggesting that solutions (linkage dimensions) are isolated points in the design space. The number of linkage designs for four-bar path synthesis for nine positions can be no larger than 1,442 Roberts' cognates as determined by Wampler et al. [12]. However, root counts for higher order four-bar path synthesis have not been characterized. Higher order synthesis specifies precision points and derivative information at each precision point. Higher order derivative information is often available through a parametric version of the plane curve to synthesize, and can increase the range over which the plane curve behaves as intended around the precision point. For each precision point, zero or several derivatives can be specified. These problems are generally referred as *mixed-order* problems. Figure 1a shows a mixed-order

J. Lenarčič and B. Siciliano (Eds.): ARK 2020, SPAR 15, pp. 303–310, 2021.
https://doi.org/10.1007/978-3-030-50975-0_37

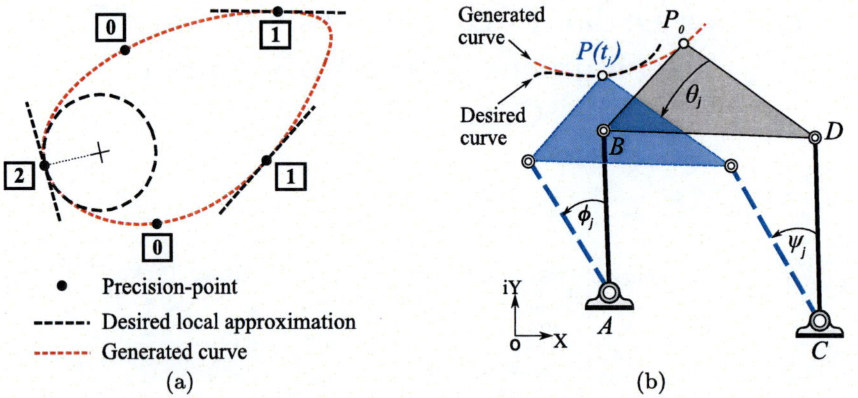

Fig. 1. (a) A mixed-order synthesis problem involving five precision points, boxed numbers indicate the order of derivative information specified at each point. (b) Schematic for higher order path synthesis of a four-bar linkage.

problem involving five precision points, of which two points have only positional requirements, two with first order derivatives (tangents) and one with up to second order derivatives (osculating circle) specified, respectively. On one extreme, nine precision points can be specified with no derivative information, of which the complete solution was obtained in [12]. On the other extreme, one precision point can be specified along with derivative information up to the eighth order. As a middle case, three precision points can be specified with derivative information specified up to the second order for each. In this paper, we present in detail the latter two cases and enumerate all others.

Early investigations of higher order synthesis for motion generation (which considers both position and orientation of a moving body) were completed by Müller. Tesar and Sparks recounted Müller's contribution and investigated all combinations of finitely and infinitesimally separated positions related to finding the Burmester points for five multiply separated positions [11]. Major contributions to the theory of higher order path generation came from Freudenstein [7] who discovered loci of moving points with prescribed curvature ratios, and Bottema and Roth [5,10] on instantaneous invariants and canonical coordinate systems. Cera and Pennestrì [6] give a more detailed recount of the literature on this subject.

Our current work formulates the equations of higher order path synthesis with the tools of numerical homotopy continuation in mind. In doing so, we obtain a large numbers of roots to the nonlinear system. In this sense, our work is an extension of Wampler et al.'s result [12]. Here we look at two problems in detail: (1) one position synthesis with derivatives up to the eighth order specified, and (2) three position synthesis with derivative information up to the second order specified at each point. We numerically compute a maximum of 224 and 227 Roberts' cognate triplets for each case, respectively.

2 Mathematical Model

Let us consider a parametric form of a plane curve in complex representation, $P(t) = x(t) + iy(t)$ where x and y are Cartesian coordinates and t is any parameter of the curve. Then, $t = t_j$ corresponds to the point $P(t_j) = x(t_j) + iy(t_j)$ about which the plane curve is to be approximated locally using a four-bar coupler curve. The n-th order derivative of the plane curve at t_j is given by $P^{(n)}(t_j) = x^{(n)}(t_j) + iy^{(n)}(t_j)$.

Consider a schematic of a planar four-bar mechanism as shown in Fig. 1b. The ground pivots are A & C, and the moving pivots are B & D. The linkage can be synthesized such that the coupler point P traverses a curve that locally approximates $P(t)$ at $t = t_j$. The same parameter t can be retained as the motion parameter of the four-bar mechanism for convenience. Let ϕ, θ and ψ represent the rotation angles of the links AB, BD and CD, respectively, measured relative to any arbitrary configuration of the mechanism. Typically, the reference configuration is assumed to correspond to one of the precision points $P(t_0) = P_0$ without loss of generality. At this reference configuration, $\phi(t_0) = \theta(t_0) = \psi(t_0) = 0$.

The vector loop closure equations of the system can be framed as:

$$A + (B - A)e^{i\phi(t)} + (P_0 - B)e^{i\theta(t)} = P(t), \tag{1}$$

$$C + (D - C)e^{i\psi(t)} + (P_0 - D)e^{i\theta(t)} = P(t). \tag{2}$$

The derivatives of Eqs.(1–2) w.r.t. t form higher order constraints. For example, Eq. (1) results in:

$$(B - A)\frac{\partial^n}{\partial t^n}(e^{i\phi(t)}) + (P_0 - B)\frac{\partial^n}{\partial t^n}(e^{i\theta(t)}) = P^{(n)}(t). \tag{3}$$

The derivatives of exponential rotations can be expanded using Faà di Bruno's Formula [14], e.g.:

$$\frac{\partial^n}{\partial t^n}(e^{i\theta(t)}) = e^{i\theta(t)} \sum_{k=1}^{n} i^k \boldsymbol{B}_{n,k}\left(\theta'(t), \theta''(t), ..., \theta^{(n)}(t)\right), \tag{4}$$

where $\boldsymbol{B}_{n,k}\left(\theta'(t), \theta''(t), ..., \theta^{(n)}(t)\right)$ for $k = 1, ..., n$ represent the partial Bell polynomials. Note that these polynomials are only a function of the higher order derivatives, $\theta^{(k)}(t), k = 1, .., n$ and not an explicit function of $\theta(t)$. Since the vector loop equations are written in terms of complex numbers, it is necessary to consider the conjugate equations alongside to complete the formulation [13].

It is possible to eliminate the rotation variables before considering higher order derivative constraints. This step would reduce the number of variables but result in equations of high degree. Hence, during homotopy path tracking, these equations yield a stiff system that requires greater numerical precision and slows computations down. Furthermore, the computational advantage provided

Table 1. Enumeration of all the higher order path synthesis problems for the four-bar linkage that lead to square systems of equations.

Sets of higher order precision point specification† of the form $\{O_j\}_{j=1}^N$		
9: {0,0,0,0,0,0,0,0,0}	**8**: {1,0,0,0,0,0,0,0}	**7**: {2,0,0,0,0,0,0}, {1,1,0,0,0,0,0}
6: {3,0,0,0,0,0}, {2,1,0,0,0,0}, {1,1,1,0,0,0}		
5: {4,0,0,0,0}, {3,1,0,0,0}, {2,2,0,0,0}, {2,1,1,0,0}, {1,1,1,1,0}		
4: {5,0,0,0}, {4,1,0,0}, {3,2,0,0}, {3,1,1,0}, {2,2,1,0}, {2,1,1,1}		
3: {6,0,0}, {5,1,0}, {4,2,0}, {4,1,1}, {3,3,0}, {3,2,1}, {2,2,2}		
2: {7,0}, {6,1}, {5,2}, {4,3}	**1**: {8}	

†For example, a set of $N = 4$ precision points consisting of one 3rd order point, two 1st order points, and one 0th order point is written as $\{3, 1, 1, 0\}$.

by the continuation technique that we employ does not gain its benefit from the discovery of a reduced Bézout's bound. Hence, we work with the original system without eliminating any variables.

A spectrum of mixed-order synthesis problems can be formulated in this framework. Table 1 lists all 30 combinations that lead to square systems of synthesis equations. Each mixed-order problem is denoted as $\{O_j\}_{j=1}^N$, where O_j is the order of the jth precision point in a set of N points. For each system to be square, $\sum_{j=1}^N O_j = 9 - N$. Only three mixed-order problems, namely, $\{0, 0, 0, 0, 0, 0, 0, 0, 0\}$, $\{2, 2, 2\}$ and $\{8\}$, distribute constraints evenly across precision points. The first one is Alt's problem which was solved in [12]. In this work, we report the complete solution to the other two unsolved problems using numerical continuation.

2.1 Eighth Order Path Synthesis of the Four-Bar Linkage

In this case, all the desired requirements are higher order constraints up to the order 8 at a lone precision point. Without loss of generality, this precision point is assumed to be the origin, $P(t_0) = P_0 = 0$. Since the configuration at this precision point is the reference configuration, the rotation angles of the moving links are, $\phi(t_0) = \theta(t_0) = \psi(t_0) = 0$. It must be noted that this does not imply that their derivatives are zero. Under these assumptions, the position equations Eqs. (1–2) vanish identically. The higher order constraints follow Eqs. (3) and (4). The resulting set of equations and their conjugates form a system of 32 polynomials whose roots represent the dimensions of four-bar mechanisms that approximate the desired curve up to the eighth order. The variables of this system are $\{A, B, C, D, A^*, B^*, C^*, D^*, \phi^{(n)}(t_0), \theta^{(n)}(t_0), \psi^{(n)}(t_0)\}$, $n = 1, ..., 8$, which are also 32 in number. Thus, a square system of polynomials is formed. The derivative specifications $P^{(n)}(t_0)$ and $P^{(n)}(t_0)^*$, $n = 1, ..., 8$, are computed from a parametric form of the desired curve.

2.2 Three Position Second Order Path Synthesis of the Four-Bar Linkage

The design requirements for this problem are defined by three distinct precision points with a second order local approximation in each one's neighborhood. Let t_0, t_1 and t_2 represent the parametric values of t that correspond to the precision points P_0, P_1 and P_2, respectively. Angles are measured with respect to the reference configuration corresponding to P_0, which is chosen to be the origin. Similar to the earlier case, a system of 32 polynomials in 32 variables can be assembled considering both position and higher order constraints. By considering the exponential rotations, e.g., $e^{i\theta(t_j)}$, as unknowns, the synthesis equations are rendered as polynomials.

3 Numerical Section

The non-linear systems of polynomial equations that arise in kinematic synthesis are difficult to handle using symbolic toolboxes. In several works in the recent decade [2,8,9], it has been shown that root generation algorithms that follow a probabilistic model are well suited to solve these problems numerically. We use the algorithm of random monodromy loops [3,8] with the `Bertini` path tracker [4] to solve the two problems formulated in Sect. 2. The algorithm works primarily by employing the following homotopy:

$$\mathbf{H}(\mathbf{z}, s) := \mathbf{F}(\mathbf{z}, \mathbf{p})(1 + e^{is}) + \mathbf{F}(\mathbf{z}, \mathbf{p}_\gamma)(1 - e^{is}) = \mathbf{0}, \quad s \in [0, 2\pi]. \tag{5}$$

Here, $\mathbf{F}$ represents the system of polynomials in the variable set $\mathbf{z}$. The parameter set $\mathbf{p}$ represents the target system while $\mathbf{p}_\gamma$ defines an auxiliary system that is randomly initialized before each iteration. As the homotopy variable s varies from 0 to 2π, a known set of solutions is randomly mapped into the total set of solutions in a one-to-one manner, thereby finding more solutions in the process. A seed solution is necessary to initiate the iterations. The algorithm terminates when no new solutions are obtained in the subsequent iteration.

A root set was first obtained for a numerically general set of design parameters to find the maximum number of finite roots that the monomial structure of the system admits. This strategy is referred to as an *ab initio* computation [4]. The solutions to path synthesis problems occur in groups of six due to the occurrence of Roberts' cognates and the inherent topological symmetry in the problem formulation [12]. Exploiting this structure reduces the amount of computation. The numerical run for the one position, eight derivative problem found 224 Roberts' triplets. Adaptive precision path tracking up to 1,024 bits was used and had about 10% path failure rate. The computation for the three position, two derivative problem found 227 Roberts' triplets. Path tracking with 128 bit precision lead to a nearly 0% failure rate.

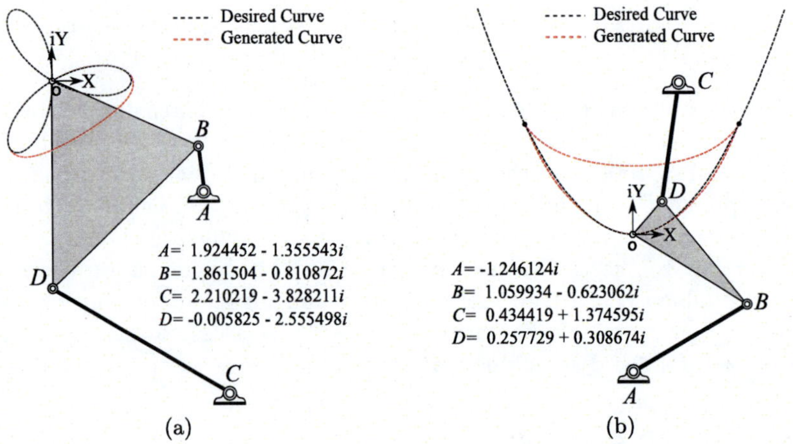

Fig. 2. (a) A four-bar mechanism that approximates the trifolium at the origin locally up to the eighth order. (b) A four-bar design for generating parabola, computed through three position second order synthesis.

3.1 Eighth Order Approximation of the Trifolium

We use parameter homotopy to synthesize four-bar linkages that generate an eighth-order approximation of the trifolium. Its parametric form is:

$$(x(t), y(t)) = (\cos(t)\cos(3t), \sin(t)\cos(3t)), t \in (-\pi, \pi]. \tag{6}$$

The precision point is chosen at the parameter value $t = \frac{\pi}{6}$ that corresponds to the origin. All derivative values $P^{(n)}(t)$ are straightforwardly computed from Eq. (6), defining a target polynomial system to solve. Next, the ab initio roots from above are deformed into roots of the target system using a procedure known as parameter homotopy [4]. The numerical computation yielded 134 groups of solutions out of the 224 paths. The remaining paths failed due to numerical errors when the computations were carried out with 128 bit precision. The number of physical solution groups was 19.

Typically, a large majority of the solutions in any exact synthesis suffer from kinematic defects such as circuit and order defects [1]. Circuit defects arise due to the occurrence of disconnected coupler curve components (called *circuits*) wherein not all of the specified points occur in the same circuit. Order defects occur when the specified points do not occur in the same order as intended in a circuit. For the eighth order synthesis, the solutions are free of these defects inherently. However, only a few of them generated the trifolium approximately over a desirable range. One such solution is shown in Fig. 2a.

3.2 Three Position Second Order Synthesis of a Parabola

Three position second order synthesis is demonstrated with a parabola. The parametric form of the chosen curve is $(x(t), y(t)) = (t, t^2), t \in \mathbb{R}$. The three precision points are chosen to correspond to $t = 0, 1$, and -1. The numerical computation of the parameter homotopy for this problem resulted in 123 successful solutions out of the 227 solution paths. Of these solutions, only six of them corresponded to physical mechanisms. Unlike the earlier case, the three position synthesis solutions can be affected by circuit defects. Order defects are not a concern in problems with up to three precision points because the desired order can be achieved by reversing the direction of motion, if needed. Figure 2b shows one of the only two solutions which are free of the circuit defect.

4 Summary

In this work, we presented a mathematical framework to solve the path synthesis of four-bar linkages for higher order requirements. General formulae for mixed-order synthesis are given and two cases were carried out in depth: eighth order synthesis at one position and second order synthesis at three positions. Complete solutions to these problems are computed using the method of random monodromy loops. The finite root count of these problems are estimated to consist of 224 and 227 Roberts' cognate triplets, respectively. Two numerical examples of four-bar mechanisms are given: generation of the trifolium and a parabola. For the eighth order example, we found a mechanism that approximates the trifolium curve at relatively far distances from the specified precision point. We also note that eighth order solutions do not suffer from kinematic defects, unlike other mixed-order solutions.

Acknowledgements. The authors would like to acknowledge the helpful discussions with Prof. Michael Stanisic and Prof. J. M. McCarthy in conceiving this work.

References

1. Balli, S.S., Chand, S.: Defects in link mechanisms and solution rectification. Mech. Mach. Theory **37**(9), 851–876 (2002)
2. Baskar, A., Bandyopadhyay, S.: An algorithm to compute the finite roots of large systems of polynomial equations arising in kinematic synthesis. Mech. Mach. Theory **133**, 493–513 (2019)
3. Baskar, A., Plecnik, M.: Synthesis of Stephenson III timed curve generators using a probabilistic continuation method. In: ASME 2019 International Design Engineering Technical Conferences and Computers and Information in Engineering Conference, 18–21 August, Anaheim, California, USA. American Society of Mechanical Engineers Digital Collection (2019)
4. Bates, D.J., Hauenstein, J.D., Sommese, A.J., Wampler, C.W.: Numerically solving polynomial systems with Bertini. SIAM (2013)
5. Bottema, O., Roth, B.: Theoretical Kinematics. North Holland Publishing Company, Amsterdam, Netherlands (1979)
6. Cera, M., Pennestrì, E.: Generalized Burmester points computation by means of Bottema's instantaneous invariants and intrinsic geometry. Mech. Mach. Theory **129**, 316–335 (2018)

7. Freudenstein, F.: Higher path-curvature analysis in plane kinematics. J. Eng. Ind. **87**(2), 184–190 (1965)
8. Hauenstein, J.D., Oeding, L., Ottaviani, G., Sommese, A.J.: Homotopy techniques for tensor decomposition and perfect identifiability. Journal für die reine und angewandte Math. (Crelles Journal) **753**, 1–22 (2019)
9. Plecnik, M.M., Fearing, R.S.: Finding only finite roots to large kinematic synthesis systems. J. Mech. Robot. **9**(2), 021005 (2017)
10. Roth, B.: On the advantages of instantaneous invariants and geometric kinematics. Mech. Mach. Theory **89**, 5–13 (2015)
11. Tesar, D., Sparks, J.W.: The generalized concept of five multiply separated positions in coplanar motion. J. Mech. **3**(1), 25–33 (1968)
12. Wampler, C., Morgan, A., Sommese, A.: Complete solution of the nine-point path synthesis problem for four-bar linkages. J. Mech. Des. **114**(1), 153–159 (1992)
13. Wampler, C.W.: Isotropic coordinates, circularity, and Bézout numbers: planar kinematics from a new perspective. In: Proceedings of ASME Design Engineering Technical Conference and Computers in Engineering Conference, Irvine, California, August, pp. 18–22 (1996)
14. Weisstein, E.W.: Faà di bruno's formula. From MathWorld–A Wolfram Web Resource. http://mathworld.wolfram.com/FaadiBrunosFormula.html

Development of a Reconfigurable Four-Bar Mechanism for a Human Robot Collaborative Gripper

Keerthi Sagar[1(✉)], Vishal Ramadoss[1], Michal Jilich[1], Matteo Zoppi[1], Dimiter Zlatanov[1], and Alessandro Zanella[2]

[1] PMAR Robotics, University of Genova, Genova, Italy
{keerthi,zoppi,zlatanov}@dimec.unige.it,
{vishal.ramadoss, michal.jilich}@edu.unige.it

[2] Centro Ricerche FIAT SCpA, Strada Torino 50, 10043 Orbassano, Italy
alessandro.zanella@crf.it

Abstract. With the rise of collaborative robots in industries, this paper proposes a human robot collaborative gripper for a windshield assembly and visual inspection application. The collaborative interface which acts as a haptic feedback device is mounted on the gripper using a deployable mechanism. The kinematics of a reconfigurable mechanism are analyzed to illustrate the advantages for using it as an unit mechanism and the concept is extended to a parallelogram based deployable four bar mechanism. A novel threefold reconfigurable four bar mechanism is developed by creating adjacent units orthogonally and the connection between such units are investigated. The proposed mechanism can be deployed and stowed in three directions. Locking of the mechanism is proposed using mechanism singularity. Kinematic simulations are performed to validate the proposed designs and analyses.

Keywords: Kinematics · Orthogonal · Deployable mechanism · Human-robot collaboration · Gripper

1 Introduction

Robots have long been a vital part of the everyday industries and the future of manufacturing portray robots and people working in a cooperative manner to complete collaborative tasks. The collaborative robots (co-bots) are designed to work alongside humans assisting them with various tasks. In this paper, the main collaborative application in focus is towards an EU H2020 CoLLaboratE usecase [5], visual inspection and windshield assembly in *Centro Ricerche Fiat, Italy.* Assembly of windshield on the car chassis is usually an automated procedure due

K. Sagar and V. Ramadoss—Equal contribution.

J. Lenarčič and B. Siciliano (Eds.): ARK 2020, SPAR 15, pp. 311–318, 2021.
https://doi.org/10.1007/978-3-030-50975-0_38

to strong tolerances in positioning and safety measures. In the existing workcell setup, the operator picks the windshield using a zero-gravity manipulator and places it on a rotating table for assembling the rear view mirror and sensors on the windshield. The operator also simultaneously checks for visual cracks on the outer surface. After the consequent assembly, a robot picks and places the windshield on the car chassis. With such a layout, the operator has to manually orient the gravity aiding manipulator to place the windshield on the assembly table and make more logistic steps in picking the assembly components from the material handling line. In the proposed collaborative workcell, the robot performs the operation of picking the product variant from the rack and presents it to the operator. The operator performs all the operations directly on the windshield mounted on the robot. The gripper acts as the collaborative interface between the robot and the human operator. The paper is organized as follows: Sect. 2 provides a brief overview of the collaborative gripper system. The kinematics of the reconfigurable mechanism are formulated in Sect. 3. The working modes and the singularity analysis of the mechanism are illustrated in the Sect. 4. Section 5 concludes the paper and presents areas of future work.

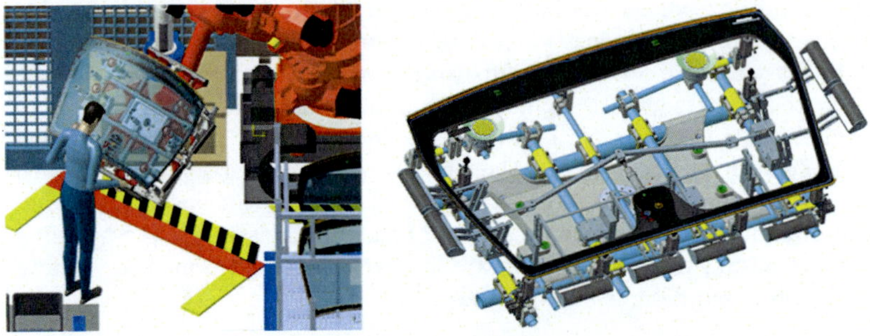

Fig. 1. a) Conceptual design of the collaborative gripper system to place robot skin for human interaction b) The proposed threefold reconfigurable four bar mechanism for the deployment of tactile handles in three different directions

2 The Collaborative Gripper System

The conceptual design of the collaborative gripper system is shown in Fig. 1(a). The gripper is provided with a handle equipped with tactile feedback sensors (robot skin). The operator can use the tactile handle to manipulate (position and orient) the robot and send signals to trigger specific actions of the robot [1] which is illustrated in Fig. 1. The tactile handles should be deployed only during the human interaction phase and stay in a retracted position. This is a requirement, since a static structure of the tactile handle can cause interference during picking operation from the rack and assembly on the car chassis. This article focuses on the development of a threefold reconfigurable four bar mechanism for the deployment of tactile handles.

3 The Reconfigurable Mechanism

The operator has to access the collaborative gripper in three directions. This is elucidated in Fig. 1(a). Most of the research that has been made on reconfigurable mechanisms has focused on structures that consist of scissor-like elements (SLEs)[8], Hoeken linkage [4], Sarrus linkage [7] and metamorphic mechanisms [2]. The four bar linkage is more suitable as a unit mechanism as it can achieve the desired behavior for our application with limited number of actuators. First, the desired motion is limited as we require nominal magnification ratio. Second, parallelogram based four bar is stable linkage and can withstand heavy loads. For these reasons, we propose a novel solution by leveraging the principles of reconfigurablity and four bar linkage.

3.1 Motion Analysis of the Reconfigurable Mechanism

The mobility using Gruebler-Kutzbach [3] criterion was predicted incorrectly. Instead of using the modified GK criterion, the constraint and free motion analysis are investigated through reciprocal screw theory [3]. Because of the symmetry of the structure of the proposed system, the motion of limb 1, in one branch is the same as that of limb 2 in the other branch. What follows will only analyze the proposed mechanism to figure out the mobility. The mechanism has 1 DOF. For modeling purposes, we consider a triangular structure as representation which is actuated by a prismatic joint. The revolute joint at D_1 can be considered as virtually two revolute joints with a point in between for the purpose of analysis, thereby forming symmetrical limbs, which is represented in Fig. 2(a). The mechanism is analyzed as a parallel manipulator by considering E as an end-effector, hence forming a three limb architecture.

The limb 1 (L_1) is a 4R chain, 2-system of screws with a constraint force in the direction of line formed by the intersection of two planes π_4 and π_1 and a couple represented by a solid pyramid arrow in red color perpendicular to the screw axis of joint C_1 and E_1. The limb 2 (L_2) is also a 2-system of screws with a constraint force in the direction of line formed by the intersection of two planes π_5 and π_2 and a couple represented by a solid pyramid arrow in blue color, which is perpendicular to the screw axis of joint C_2 and E_2. The limb 3 (L_3) is a RR chain, 4-system with two constraint couples, a constraint force passing through the joint axes and a planar pencil of screws (from which any one of the screws in the set can be a reciprocal screw), all represented in yellow color. We choose the basis of screw system according to the choice of end-effector. The dimension of the constraint wrench system is defined using the formula $dim(U + V) = dim(U) + dim(V) - dim(U \cap V)$. The dimension of the constraint wrench system is $4 + 2 + 2 - 3 = 5$-system of constraint wrenches, where $dim(U \cap V)$ is three forces (two intersecting forces at point I) and another force parallel to the pencil of screws of M_3E_3, passing through point I. Therefore, dimension of the twist system is 1 with translational degrees of freedom which depends on the configuration and moves in the symmetric plane π_s. M_3E_3, E_3D_1 and the projection of D_1 on the midpoint of M_1M_2 forms a parallelogram that

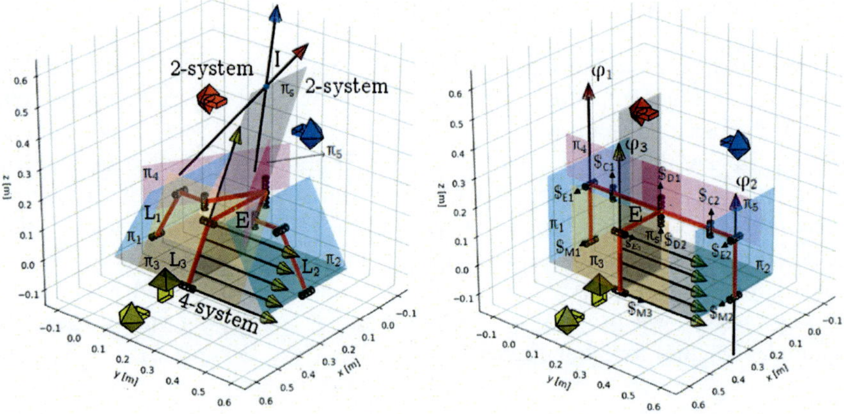

(a) (b)

Fig. 2. Constraint Analysis using screw theoretic representation

acts like a four-bar linkage as depicted in Fig. 3. Thus, the link E_3D_1 acts as an end-effector for the four bar linkage and thereby 1 DOF translational motion.

3.2 Kinematic Modeling

The kinematics of the reconfigurable mechanism which is shown in Fig. 3, is described in this section. Firstly, the position analysis is performed in which the position of E_1, E_2 and E_3 are determined. This is crucial as the path traced by these points determine the folding and unfolding pattern. A_i, B_i and E_i for $i = 1, 2, 3$ forms a triangle in XZ and YZ planes. $\triangle A_1B_1E_1$ and $\triangle A_2B_2E_2$ are in XZ plane for the configuration shown in the Fig. 3. $\triangle C_1D_1C_2$ is in XY plane. l_{ki} for $k, i = 1, 2, 3$ denote the length of the links for corresponding triangles. M_1, M_2 and M_3 represent the projection of the points E_1, E_2 and E_3 on the XY plane respectively. δ_1, δ_2 and δ_3 are the corresponding lengths of A_1M_1, A_2M_2 and A_3M_3. A piston is connected between the E_3 and D_1 with a length L. The links are connected using revolute joints for the defined points. The orientation of revolute joints are shown in the figure. θ_1 and θ_2 are angles between π_1(shifted from XZ) and π_2(shifted from XY) plane defined at points B_1 and B_2 rotating in counter-clockwise direction. θ_3 is the angle between YZ and XY plane defined at point B_3. Since E_1 and C_1, E_2 and C_2 are collinear, θ_4 and θ_6 are considered as zero in the mechanism. θ_5 and θ_7 are the angles between YZ and XZ plane defined at points C_1 and C_2.

The coordinates of A_i are defined as $[x_i, y_i, z_i]$ for $i = 1, 2, 3$. The coordinates of B_i (for $i = 1, 2$) are given below:

$$\begin{aligned} B_1 &= [x_1 + l_{13}, y_1, z_1] \\ B_2 &= [x_2 + l_{23}, y_2, z_2] \end{aligned} \tag{1}$$

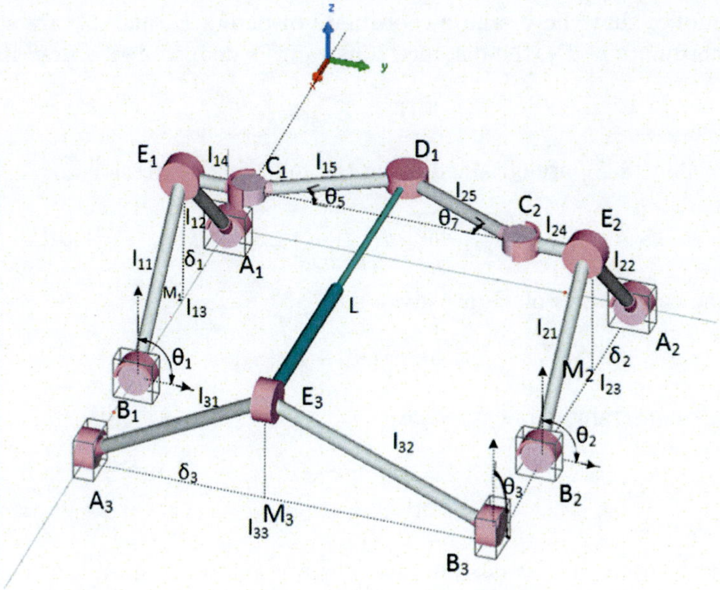

Fig. 3. Reconfigurable mechanism description

The coordinates of E_i (for $i = 1, 2$) are given below:

$$\begin{aligned} E_1 &= [x_1 + \delta_1, y_1 + \sqrt{(l_{11}^2 - \delta_1^2)}c\theta_1, z_1 + \sqrt{(l_{11}^2 - \delta_1^2)}s\theta_1] \\ E_2 &= [x_2 + \delta_2, y_2 + \sqrt{(l_{21}^2 - \delta_2^2)}c\theta_2, z_2 + \sqrt{(l_{21}^2 - \delta_2^2)}s\theta_2] \end{aligned} \tag{2}$$

The coordinates of C_1, C_2 and D_1 are:

$$\begin{aligned} C_1 &= E_1 + [0, l_{14}, 0] \\ C_2 &= E_2 + [0, -l_{24}, 0] \\ D_1 &= C_1 + [l_{15}s\theta_5, l_{15}c\theta_5, 0] \end{aligned} \tag{3}$$

where $c(.)$ and $s(.)$ notations are used for cosine and sine of the arguments respectively.

Expanding Eq. 3, it is given as:

$$\begin{aligned} C_1 &= [x_1 + \delta_1, y_1 + \sqrt{(l_{11}^2 - \delta_1^2)}c\theta_1 + l_{14}, z_1 + \sqrt{(l_{11}^2 - \delta_1^2)}s\theta_1] \\ C_2 &= [x_2 + \delta_2, y_2 + \sqrt{(l_{21}^2 - \delta_2^2)}c\theta_2 - l_{24}, z_2 + \sqrt{(l_{21}^2 - \delta_2^2)}s\theta_2] \\ D_1 &= [x_1 + \delta_1 + l_{15}s\theta_5, y_1 + \sqrt{(l_{11}^2 - \delta_1^2)}c\theta_1 + l_{14} + l_{15}c\theta_5, z_1 + \sqrt{(l_{11}^2 - \delta_1^2)}s\theta_1] \end{aligned} \tag{4}$$

The equality constraints are expressed as:

$$\begin{aligned} E_3[y] &= D_1[y] \\ E_3[z] &= D_1[z] \end{aligned} \tag{5}$$

which implies that the y- and z-coordinate of points E_3 and D_1 are same. For the x-coordinate of E_3, the distance constraint is defined and is described as:

$$\| D_1 - E_3 \| = L \tag{6}$$

The coordinates E_3 are obtained using the following expression:

$$E_3 = [x_3 + \sqrt{(l_{31}^2 - \delta_3^2)}c\theta_1, y_3 + \delta_3, z_3 + \sqrt{(l_{31}^2 - \delta_3^2)}s\theta_3] \tag{7}$$

where the coordinates of B_3 are given as:

$$B_3 = [x_3, y_3 + l_{33}, z_3] \tag{8}$$

The angle constraints are expressed as:

$$\begin{aligned} \theta_2 &= \pi - \theta_1 \\ 0^\circ &< \theta_1 < 90^\circ \end{aligned} \tag{9}$$

For a given input θ_1, the position analysis was performed and the coordinates, path traced by the points E_1, E_2 and E_3 are simulated respectively.

4 Kinematic Simulation

A mathematical mock-up of the reconfigurable mechanism was programmed and simulated using Python 2.7. The simulations were also validated with the educational tool for kinematic analysis of mechanism, GIM Software [6]. The deployment process is illustrated with different planes shown in Fig. 4(b). The results of both simulations are presented in Fig. 4.

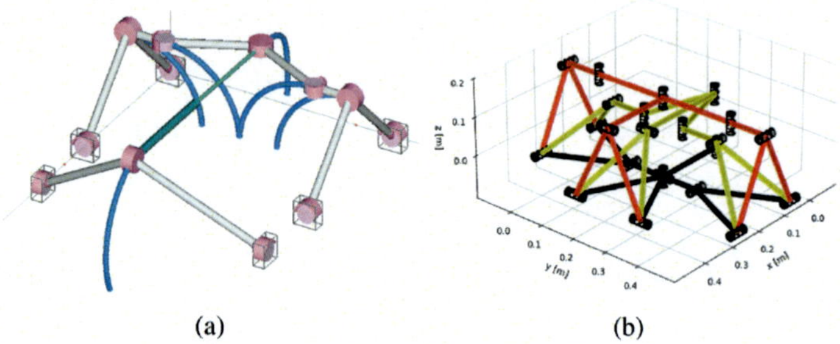

Fig. 4. (a) The path traced by the reconfigurable mechanism (b) Simulation of the deployment process: fully folded state(black color), partially folded state(yellow color) and fully deployed state(red color)

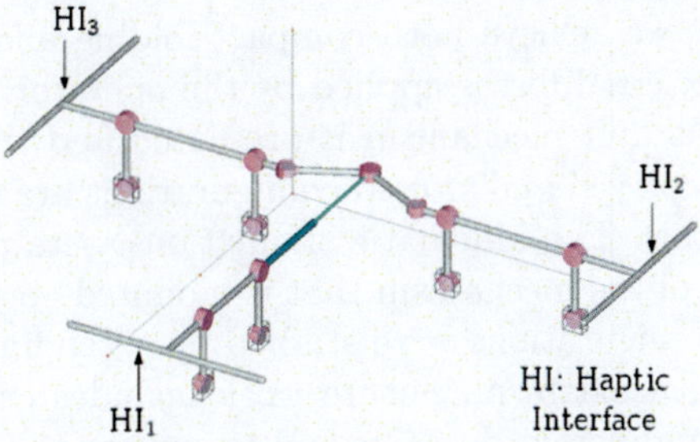

Fig. 5. Reconfigurable mechanism with three orthogonal four bar linkages

4.1 Singularity Analysis

The singularity analysis [9] is performed and the singular configurations for proposed mechanism is shown in Fig. 6. The fully deployed configuration, shown in Fig. 6(a), is a singular configuration as the three lines intersect at a common point in a plane. As shown in Fig. 2(b), the constraint forces φ_1, φ_2 and φ_3 present in limbs 1, 2 and 3 becomes parallel to each other, hence are linearly dependent. The dimension of total constraint wrench systems becomes six and hence results in 0 DOF. This singular configuration of the reconfigurable mechanism based on four bar linkage can be exploited in a way to perform certain tasks such as opposing human operative loads. The singular configuration shown in Fig. 6(b) can be avoided by allowing the motion within joint limits.

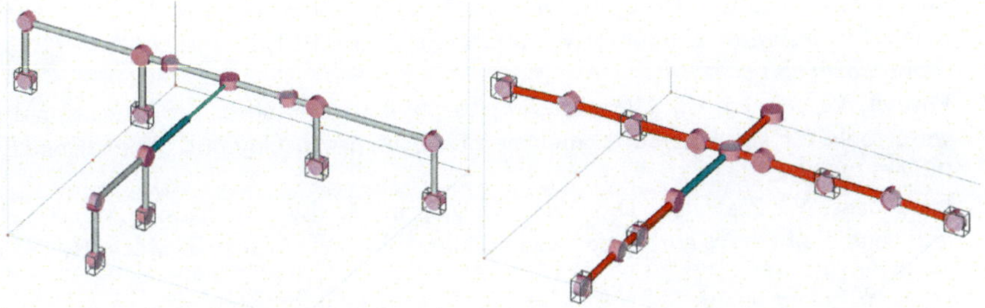

(a) Fully deployed configuration (b) Fully folded configuration

Fig. 6. Singular configurations of the reconfigurable mechanism

5 Conclusion

This paper introduced a reconfigurable robotic solution for effective mounting of human robot collaborative gripper which the operator utilizes as a tactile handle to manipulate the robot. By developing a novel reconfigurable mechanism based on four bar linkage, we achieve both compact folding and maximum expansion and to withstand external forces applied by the operator. The kinematic model of the reconfigurable unit mechanism is presented and then the working mode is simulated. The deployed and stowed configurations are modeled and analyzed using simulation tools. The kinematic simulations were presented to show the capability and ease of the mechanism that is mounted on the collaborative gripper. The singular configurations were studied by extending the unit mechanism to a four bar based mechanism. Future work includes employing the reconfigurable mechanism in different domains and to explore the possibility of stacking the mechanisms.

Acknowledgements. The research leading to these results has received funding by the EU Horizon 2020 Research and Innovation Programme under grant agreement No. 820767, CoLLaboratE project.

References

1. Cannata, G., Maggiali, M., Metta, G., Sandini, G.: An embedded artificial skin for humanoid robots. In: 2008 IEEE International Conference on Multisensor Fusion and Integration for Intelligent Systems, pp. 434–438. IEEE (2008)
2. Dai, J.S., Rees Jones, J.: Mobility in metamorphic mechanisms of foldable/erectable kinds. J. Mech. Des. **121**(3), 375–382 (1999)
3. Hunt, K.H.: Kinematic Geometry of Mechanisms, vol. 7. Oxford University Press, USA (1978)
4. Lu, S., Zlatanov, D., Ding, X., Molfino, R.: A new family of deployable mechanisms based on the Hoekens linkage. Mech. Mach. Theory **73**, 130–153 (2014)
5. Macovetchi, Kirstein, Doulgeri, Dimeas: A methodology to formulate user requirements for designing collaborative robots. Zenodo (2019). https://doi.org/10.5281/zenodo.3266850
6. Petuya, V., Macho, E., Altuzarra, O., Pinto, C., Hernandez, A.: Educational software tools for the kinematic analysis of mechanisms. Comput. Appl. Eng. Educ. **22**(1), 72–86 (2014)
7. Ramadoss, V., Zlatanov, D., Ding, X., Zoppi, M., Lyu, S.: Design, construction, and control of curves and surfaces via deployable mechanisms. J. Mech. Rob. **11**(6), 1–10 (2019)
8. Yang, Y., Peng, Y., Pu, H., Chen, H., Ding, X., Chirikjian, G.S., Lyu, S.: Deployable parallel lower-mobility manipulators with scissor-like elements. Mech. Mach. Theory **135**, 226–250 (2019)
9. Zlatanov, D., Bonev, I.A., Gosselin, C.M.: Constraint singularities as c-space singularities. In: Lenarčič, J., Thomas, F. (eds.) Advances in Robot Kinematics, pp. 183–192. Springer, Dordrecht (2002)

Kinematic Analysis of a Planar Manipulator with Anti-parallelogram Joints and Offsets

Philippe Wenger[1(✉)] and Matthieu Furet[1,2]

[1] Laboratoire des Sciences du Numérique de Nantes (LS2N), CNRS, 44321 Nantes, France
Philippe.Wenger@ls2n.fr

[2] Ecole Centrale de Nantes, 44321 Nantes, France

Abstract. Industrial robotic manipulators mostly use revolute or prismatic joints. In this paper, an anti-parallelogram mechanism, referred to as *X-joint*, is considered as an alternative choice to a revolute joint in the kinematic design of robots. This choice lends itself well to a remote actuation with cables, which contributes to lower inertia. We show that replacing revolute joints with X-joints in a planar 2-dof manipulator improves drastically the size of the workspace in the presence of joint limits. On the other hand, the kinematic analysis becomes significantly more difficult, owing to the much more complicated algebra involved in the input/output equations. The inverse kinematics, the singularities and the workspace optimization are investigated. A tentative design of a 2-X planar linkage is proposed and compared to its 2-R counterpart.

Keywords: Anti-parallelogram joint · Inverse kinematics · Singularity · Workspace · Optimal design

1 Introduction

An anti-parallelogram mechanism, referred to as *X-joint* in this paper, has two opposite pairs of equal-length bars like in the parallelogram joint, but contrary to this latter, the bars of one pair cross each other. The X-joint is less popular than the parallelogram joint and, as far as we know, it has never been used in industrial robots. It can be found in knee [1] and bird neck models [2], in gear trains [3] or in tensegrity mechanisms [4–8]. In [2,7] and [8], the X-joints were arranged in series and adjacent to each other to form planar manipulators, as shown in Fig. 1, left. In this paper, an offset is added between two successive X-joints and between the last X-joint and the reference point P, like in Fig. 1, right. The use of X-joints is suitable for a remote actuation with cables [2,8], an interesting choice for a lightweight design. As it will be shown further, moreover, X-joints can generate much larger workspaces than R joints. On the other hand, it turns out that the inverse kinematics of such a manipulator becomes much

J. Lenarčič and B. Siciliano (Eds.): ARK 2020, SPAR 15, pp. 319–326, 2021.
https://doi.org/10.1007/978-3-030-50975-0_39

more complicated. When two X-joints are adjacent as in Fig. 1, left, it was shown in [9] that the inverse kinematics can be solved with a quartic polynomial and admits up to four solutions. When offsets are added, it turns out that the degree of the inverse kinematics polynomial is twice higher, as shown further. In the next section, the inverse kinematics of a 2-dof planar manipulator with X-joints and offsets is derived and the singularities as well as the workspace boundaries in the absence of joint limits are determined. Then, the effect of joint limits is analyzed. Finally, a comparative workspace-based study is carried out between a classical 2-R planar manipulator and a 2-X manipulator with offsets.

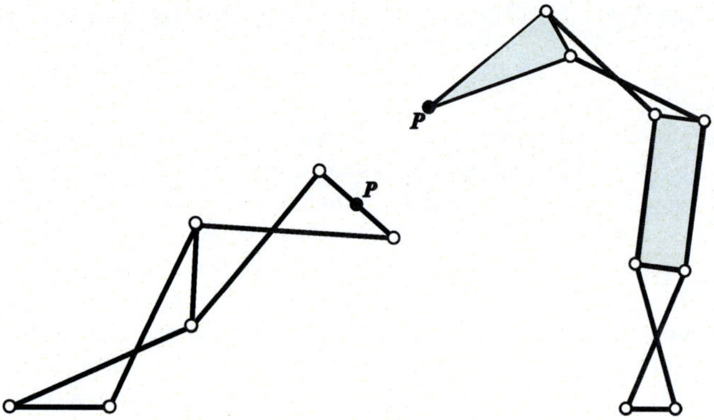

Fig. 1. A 2-X manipulator with no offsets *(left)* and with offsets *(right)*.

2 Manipulator Studied

The manipulator studied consists of two identical X-joints and two offset links in series as shown in Fig. 2. Each X-joint i has a base bar and an upper bar of length b and two crossed bars of length L with $L > b$. The two links are defined as a rectangle and a isoceles triangle, respectively. The rectangle, of height h, links the two X-joints while the triangle, of altitude h, links the second X-joint and the reference point $P = (x, y)$. The reference frame is centered at the middle of the base bar with the x-axis horizontal and pointing to the right. Let α_i define the orientation of the upper bar of the ith X-joint w.r.t. its base bar. For a matter of completeness in the kinematic analysis, each X-joint is assumed to move within its full range. In this case, $-2\pi < \alpha_i < 2\pi$. Thus, α_i cannot define the mechanism configuration unambiguously. Let l_i be the length of the line segment that links the middle points of the top and base bars of each X-joint i (shown in red dotted line in Fig. 2). The angle between this line and the direction orthogonal to the base bar is referred to as θ_i. It is easy to verify that $\alpha_i = 2\theta_i$. When $-\pi < \theta_i < \pi$, thus, the mechanism makes a full turn and the manipulator configuration can be defined by (θ_1, θ_2). Two more angle parameters are introduced, namely, ϕ_1

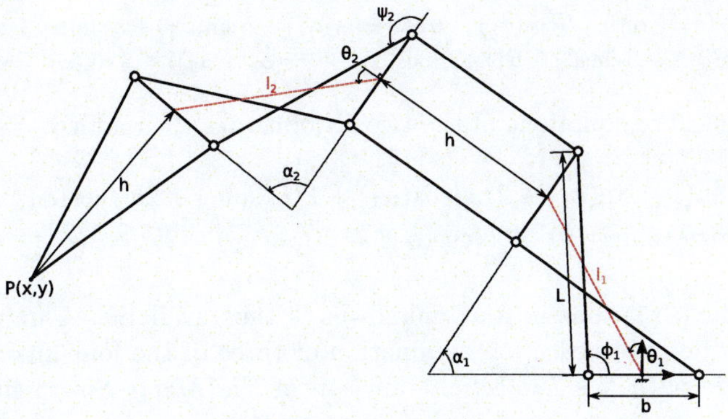

Fig. 2. Manipulator description.

and ψ_2, which define the angle of the left crossed-bar of the first X-joint and the angle of the right crossed-bar of the second X-joint, respectively (see Fig. 2). These angles are useful for the derivation of the inverse kinematics. The length l_i of the line segment that links the middle points of the top and base bars can be expressed as follows [2]:

$$l_i(\theta_i) = \sqrt{L^2 - b^2\cos^2(\theta_i)} \tag{1}$$

3 Direct and Inverse Kinematics

The direct kinematic equations of the manipulator studied are obtained upon expressing the Cartesian coordinates (x, y) of the reference point in the base frame:

$$\begin{cases} x = -h(\sin(2\theta_1) + \sin(2\theta_1 + \theta_2)) - l_1(\theta_1)\sin(\theta_1) - l_2(\theta_2)\sin(2\theta_1 + \theta_2) \\ y = h(\cos(2\theta_1) + \cos(2\theta_1 + \theta_2)) + l_1(\theta_1)\cos(\theta_1) + l_2(\theta_2)\cos(2\theta_1 + \theta_2) \end{cases} \tag{2}$$

where $l_1(\theta_1)$ and $l_2(\theta_2)$ are defined in (1).

The inverse kinematics is usually established after deriving the so-called characteristic polynomial, a univariate polynomial in one of the input variables that is obtained from the direct kinematics equations. It is not convenient here to combine the two equations in (2) to obtain this polynomial because of the presence of square roots. These square roots would have to be first cleared out, thus doubling the degree of the resulting polynomial and yielding extraneous roots. Instead, we proceed like in [9] by writing x and y as functions of the crossed-bar angles ϕ_1 and ψ_2 as follows:

$$\begin{cases} x = \frac{b}{2}\cos(\alpha_1+\alpha_2) + L\cos(\phi_1) + L\cos(\psi_2+\alpha_1) - h\sin(\alpha_1) + h\sin(\alpha_1+\alpha_2) - \frac{b}{2} \\ y = L\sin(\phi_1) + L\sin(\psi_2+\alpha_1) + h(\cos(\alpha_1)+\cos(\alpha_1+\alpha_2)) + \frac{b}{2}\sin(\alpha_1+\alpha_2) \end{cases} \quad (3)$$

The loop closure equations of the two X-joints are then added:

$$\begin{cases} -2Lb\sin(\alpha_1)\sin(\phi_1) - 2Lb(\cos(\alpha_1)+1)\cos(\phi_1) + 2b^2(\cos(\alpha_1)+1) = 0 \\ 2Lb\sin(\alpha_2)\sin(\psi_2) + 2Lb(\cos(\alpha_2)+1)\cos(\psi_2) + 2b^2(\cos(\alpha_2)+1) = 0 \end{cases} \quad (4)$$

A set of four equations in four unknowns is thus available. A univariate polynomial can be obtained upon elimination of three of the four unknowns, which is done here with the *Projection* function of the *Maple Siropa* library [9,10]. This function projects the system of four equations in order to obtain one single equation in one single variable, chosen here as ϕ_1. The half-tangent substitution yields a factored polynomial, one of which defines the characteristic polynomial. Interestingly, the characteristic polynomial obtained turns out to be of degree 8 in $t = \tan(\phi_1/2)$, leading to the possible existence of eight inverse kinematic solutions. Note that this degree was only four for the 2-X manipulator with no offsets and a maximum of four inverse kinematic solutions were found [9]. The expression of the 8-degree polynomial is very large and cannot be displayed here. It can be found in a technical report [11]. The inverse kinematic solutions are obtained by solving in cascade the first equation in (4) and the two equations in (3), following the same as process as in [9].

The inverse kinematics is solved for a 2-X manipulator of dimensions $L = 1/2, h = 1$ and $b = 1/3$ at $x = 0$ and $y = 9/10$. Eight solutions are found (see Fig. 3). It is worth noting that these eight solutions exist as long as no restriction is imposed on the rotation range of the X-joints, i.e. $-\pi < \theta_i < \pi$. When $-\pi/2 < \theta_i < \pi/2$, for example, it can be verified that the only solutions are those two symmetric configurations furthest to the right and left (in purple and pink in Fig. 3).

4 Singularity and Workspace Analysis

The Jacobian matrix of the manipulator can be derived from system (2). Its determinant is calculated to plot the singularity curves in the joint space. After a careful inspection of a great number of plots, it turns out that there are six, four or two curves depending on the link dimensions. Figure 4 shows the singularity plots when $b = 1/3$, $h = 1$, $L = 1$ (left) and $L = 2$ (right).

In the absence of joint limits, the workspace boundaries are defined by the graph of the discriminant of the characteristic polynomial. A huge polynomial containing 2905 monomials and of degree 32 in x and y is obtained. Figure 5 shows the workspace boundaries for a manipulator with link dimensions $b = 1/3, h = 1, L = 1/2$ (left) and $L = 1$ (right). Internal boundaries divide the workspace into constant accessibility regions. Accessibility can be obtained in each region upon choosing an arbitrary point and solving the inverse kinematics. The number of solutions is indicated with colors as explained in the figure legend.

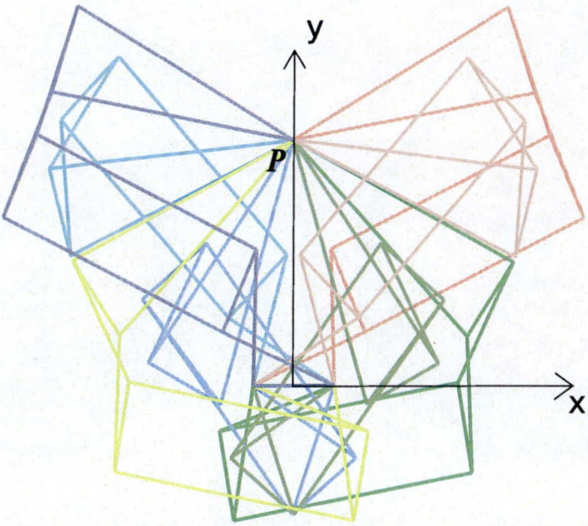

Fig. 3. The eight solutions at $x = 0$, $y = 9/10$ ($L = 1/2, h = 1$ and $b = 1/3$).

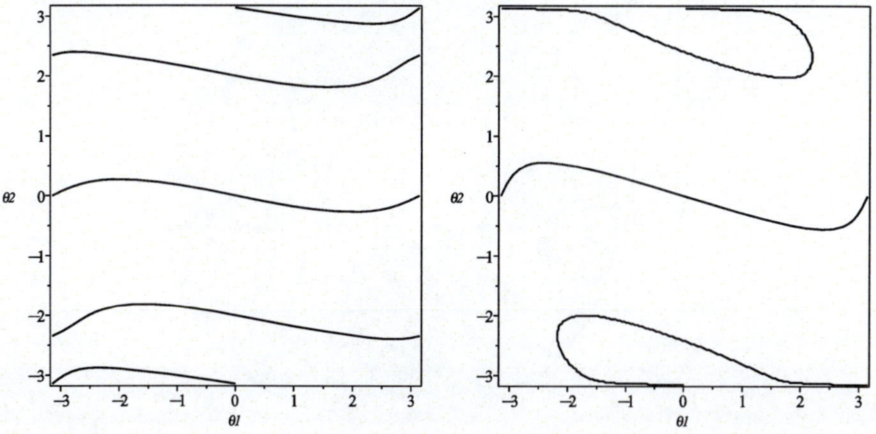

Fig. 4. Singularity curves in the joint space when $h = 1$, $b = 1/3$: there are four curves when $L = 1$ *(left)* and two curves when $L = 2$ *(right)*.

We now study the workspace when the rotation range of each X-joint is defined by $-\pi/2 < \theta_i < \pi/2, i = 1, 2$. These joint limits are necessary to avoid the flat singularities of the anti-parallelograms and to allow a remote actuation with cables. This time, the workspace boundaries are plotted upon mapping into the (x, y) plane the joint space singularity curves as well as the joint space boundaries. Figure 6 shows the resulting workspace boundaries when $b = 1/3$, $h = 1$, $L = 1$ and $-\pi/2 < \theta_i < \pi/2$.

The manipulator has one (resp. two) solutions in the light-grey (resp. dark-grey) areas, respectively. The height of the void Y_{vh} can be determined easily as

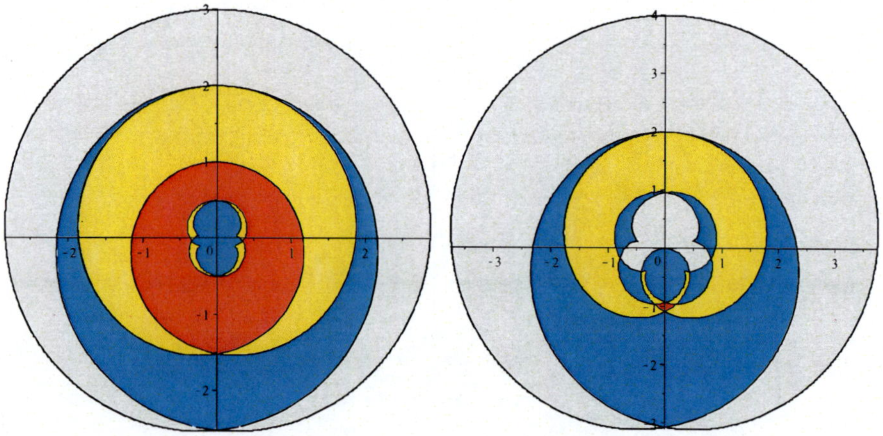

Fig. 5. Workspace when $h = 1$, $b = 1/3$ and $L = 1/2$ *(left)*, $L = 1$ *(right)*. Number of solutions is 2 (resp. 4, 6, 8) in the areas filled in grey (resp. blue, yellow, red).

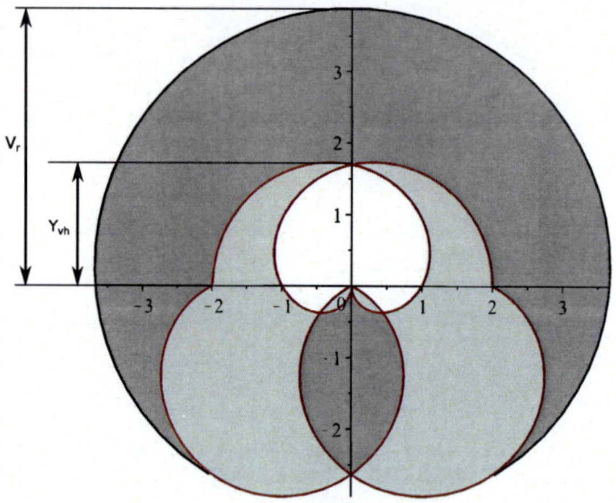

Fig. 6. Workspace when $b = 1/3$, $h = 1$, $L = 1$ and $-\pi/2 < \theta_i < \pi/2$. Light-grey (resp. dark grey) areas have 1 (resp. 2) inverse kinematic solutions.

it is bounded by the curves defined by the joint limits. More precisely, the upper point of the void is reached when $\theta_2 = \pi/2$ and $x = 0$. Solving $x = 0$ for θ_1 in the first equation of (2) and replacing the obtained solution into the second equation of (2) yields $Y_{vh} = \sqrt{3L^2 - b^2}$. Interestingly, the height of the void is independent of h. The void cannot be removed since b must be lower than L, but it can be minimized when L and b are small and b is small w.r.t. L. The vertical maximal reach V_r of the manipulator is equal to $V_r = 2h + 2\sqrt{L^2 - b^2}$. So to have a large maximal reach, knowing that L and b should be small, h should be chosen high. Suppose that the desired maximal reach is $V_r = 2$. Figure 7, left, shows the workspace of a suitable design ($b = 0.05$, $h = 0.91$, $L = 0.1$), exhibiting a void of height $Y_{vh} = 0.166$, which is twelve times less than the maximal reach. Note that, in addition, the accessibility is equal to two almost everywhere. It drops to one only in the two symmetric curved strips located in the negative values of y and around the void. For a matter of comparison, the workspace of a 2-R manipulator with the same maximal reach $V_r = 2$ (link lengths $L_1 = L_2 = 1$) and the same joint limits ($-\pi/2 < \theta_i < \pi/2$) is shown in Fig. 7, right. The chosen joint limits would allow a remote actuation with cables for both manipulators. It is apparent that the workspace of the 2-R manipulator is much smaller and the accessibility is equal to two only in the upper region of the workspace.

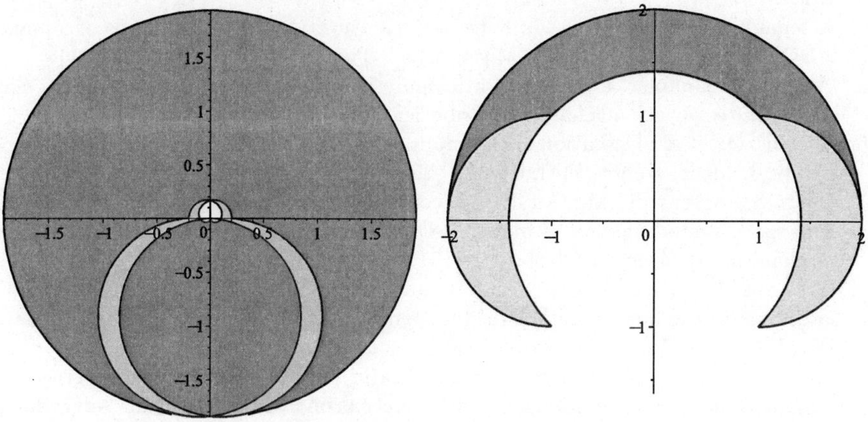

Fig. 7. Workspace of a 2-X (*left*, $b = 0.05$, $h = 0.91$, $L = 0.1$) and 2-R (*right*, $L_1 = L_2 = 1$) manipulator designed for a maximal reach equal to two.

5 Conclusion

A new family of manipulators with two anti-parallelogram joints and offset links has been considered as an alternative design to classical 2-revolute-jointed linkages. At the price of more complex input/output equations and singularity expressions, 2-X manipulators have been shown to offer a much larger workspace in the presence of joint limits. Moreover, they lend themselves well to a remote actuation with cables, which make them good candidates to lightweight designs suitable for safe interactions. In future work, a comparative study between 2-R and 2-X manipulators actuated with cables will be performed on the basis of their wrench-feasible workspace.

Acknowledgement. This work was conducted with the support of the French National Research Agency (AVINECK Project ANR-16-CE33-0025).

References

1. Hamon, A., Aoustin, Y., Caro, S.: Two walking gaits for a planar bipedal robot equipped with a four-bar mechanism for the knee joint. Multibody Syst. Dyn. **31**(3), 283–307 (2014)
2. Furet, M., Wenger, P.: Kinetostatic analysis and actuation strategy of a planar tensegrity 2-X manipulator. ASME J. Mech. Robot. 1–19 (2019). https://doi.org/10.1115/1.4044209
3. Zhang, X., Deng, H., et al.: Use of anti-parallelogram linkage mechanism and ordinary gear train for power transmission on a rotary engine. In: Proceedings of Third International Conference on Mechanic Automation and Control Engineering, July 2012
4. Arsenault, M., Gosselin, C.M.: Kinematic, static and dynamic analysis of a planar 2-DOF tensegrity mechanism. Mech. Mach. Theory **41**(9), 1072–1089 (2006)
5. Wenger, P., Chablat, D.: Kinetostatic analysis and solution classification of a class of planar tensegrity mechanisms. Robotica **37**, 1214–1224 (2019)
6. Boehler, Q., et al.: Definition and computation of tensegrity mechanism workspace. ASME J. Mech. Robot. **7**(4), 044502 (2015)
7. Chen, S., Arsenault, M.: Analytical computation of the actuator and cartesian workspace boundaries for a planar 2-Degree-of-Freedom translational tensegrity mechanism. J. Mech. and Rob. **4**, 011010 (2012)
8. Bakker, D.L., et al.: Design of an environmentally interactive continuum manipulator. In: Proceedings of 14th World Congress in Mechanism and Machine Science, IFToMM'2015, Taipei, Taiwan (2015)
9. Furet, M., Lettl, M., Wenger, P.: Kinematic analysis of planar tensegrity 2-X manipulators. In: Proceedings of 16th International Symposium on Advances in Robot Kinematics, Bologna, Italie (2018)
10. Rouillier, F., et al.: Siropa Library V1, IDDN.FR.001.140015.000.S.P.2017.000.20600
11. Furet, M., Wenger, P.: Inverse kinematics of 2-X manipulators with offsets. Technical report, LS2N-CNRS (2019)

On Singularity and Instability in a Planar Parallel Continuum Mechanism

Oscar Altuzarra(✉) and Francisco J. Campa

University of the Basque Country UPV/EHU, 48013 Bilbao, Spain
{oscar.altuzarra,fran.campa}@ehu.es

Abstract. Parallel Continuum Mechanisms are closed-loop mechanical systems formed by flexible rods connected to a rigid end-effector and actuated from the base attachments of such rods. It was shown in previous works that they have analogous kinematic features to rigid-link parallel mechanisms. Position analysis is a force equilibrium problem with a multiplicity of solutions. These correspond to distinct aspects of workspaces, delimited by singularity curves. Equilibrium poses need further analysis to verify their stability, i.e. if the stationary solution is a local minima of the potential energy of the device. This conference paper investigates singularity and instability issues to understand the behavior of the mechanism in such poses.

Keywords: Parallel Continuum Mechanisms · Singularity · Instability

1 Introduction

Compliant mechanisms with closed loop kinematic chains [7] do not need an specific kinematic analysis method on the hypothesis that notches devised to play the role of kinematic pairs allow a pseudo-rigid body model whose kinematic analysis is analogous to that of classical rigid-link mechanisms. On the contrary, Continuum mechanisms, in which mobility is due to large nonlinear deformation of slender elements, require the numerical integration of a set of differential equations to solve its kinematic analysis.

Rucker et al. [5] coined the term parallel continuum robot. They have developed an intensive series of research aiming a real-time control of an hexapod where limbs are flexible rods whose length is controlled to pose a rigid end-effector. In [9,10] the real-time forward kinematic problem is stated using the Cosserat rod model. Stability of such static equilibrium is assessed in [11] using the similarity of the problem with one of optimal control. And recently, they present the dynamic analysis for real-time control [8]. These seminal works are essentially based on the direct numerical integration of the system of differential equations that define the quasi-static equilibrium (lately the dynamic one as well) subjected to some geometrical constraints. The approach taken is to solve a boundary value problem with a shooting method that uses a guess solution obtained starting from a home position.

Our research is focused on exploring the analogies between the classical full kinematic analysis performed in multi degree of freedom rigid-body parallel mechanisms, and that of parallel continuum robots. For the sake of simplicity, planar mechanisms

J. Lenarčič and B. Siciliano (Eds.): ARK 2020, SPAR 15, pp. 327–334, 2021.
https://doi.org/10.1007/978-3-030-50975-0_40

are the case of study for the time being. Aspects to tackle are: the multiplicity of solutions of position problems, assembly modes and working modes, kinematic singularities, workspace and joint-space associated to assembly modes and working modes, and workspace analysis. Regarding the position analysis, we contributed a numerical method for the full position analysis in [1], and proved those results with a certified procedure that makes use of interval analysis in [3]. As a result of those previous works, authors have a tool to find all the multiple solutions of the forward or inverse position problem of those mechanisms subjected to given loads. Consequently, we can search iteratively the different solutions in the workspace or the joint space. In a preliminary work [2], while analysing multiplicity of solutions, rods' buckling modes were investigated and we concluded that there was no issue in changing the buckling mode while moving. Also, Kinematic singularities were evaluated, and it was showed that they effectively play the role of frontiers between distinct areas.

The stability issue of equilibrium configurations is also important. Elastic devices are prone to instabilities under certain loads [6]. In [11] it was proved that assembled systems of flexible rods, deformed by changing boundary conditions, could became unstable even under no external loading. Hence, here there is an issue that must be considered at the kinematic analysis, and that it was not present in rigid-link systems.

The purpose of this paper is to show the kinematic and stability characteristics of the different type of poses of a parallel continuum mechanism. In order to do that, several poses found in the workspace are analysed with indication of velocity ellipsoids and stability of its static equilibrium. From the latter, we explore the snap motions generated out of unstable poses. A deeper attention is given to singular poses.

2 Jacobian and Stiffness Matrices

Parallel continuum manipulators are essentially mechanisms and therefore they are subject to classical Jacobian analysis. In rigid-link parallel manipulators algebraic finding of velocity equations provides a set of linear equations that define Jacobians analytically. In continuum parallel manipulators this is not the case, Jacobians are found numerically at every configuration evaluating the effect of small changes of inputs $\Delta\boldsymbol{\alpha}$ in outputs $\mathbf{p}_{\mathscr{P}}$, or viceversa, namely

$$\Delta\mathbf{p}_{\mathscr{P}} = \mathbf{J}_{FK} \cdot \Delta\boldsymbol{\alpha} \tag{1}$$

The planar parallel continuum mechanism considered in this paper is shown in Fig. 1 at a regular pose. It is denoted by $2\underline{R}FR$ where $\underline{R}$ stands for the rotational actuator that controls the slope α_i at rod's clamped extreme O_i, F stands for flexible rods, and the final R stands for the revolute kinematic pair that joins both rods at end-effector $\mathscr{P}$. The workspace shown corresponds to the plotted working mode, divided by singularity curves into different aspects shown separately on the right [1]. Once Jacobian is found we can get its condition number, velocity ellipsoid and the like. In this regard, singularity analysis is performed iteratively analysing each of the configurations obtained in the position analysis.

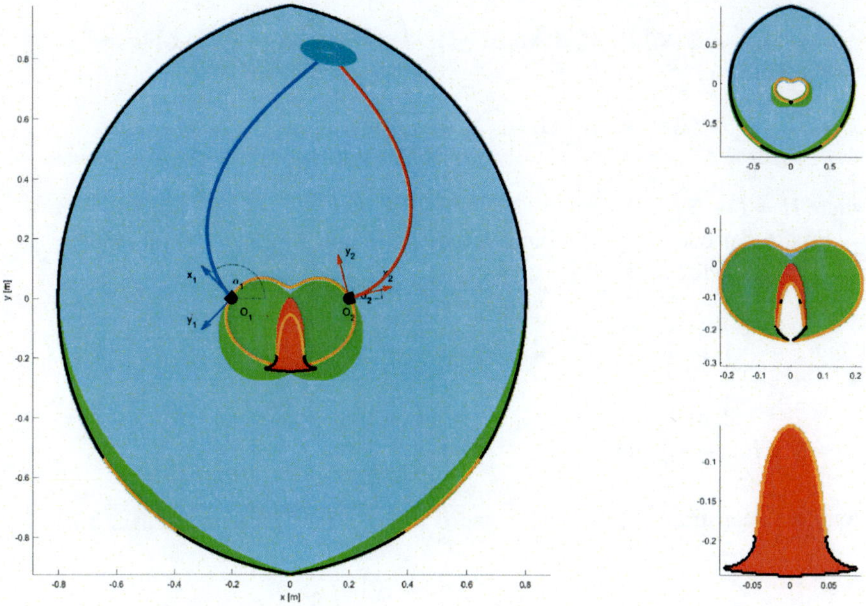

Fig. 1. Regular Stable Pose of $2\underline{R}FR$. Velocity ellipsoid and workspace plot.

The solutions obtained to the position problem fulfil the condition of static equilibrium, and correspond to the minimisation of the energy of the system. The following natural step is to check if that equilibrium is stable or not. Flexible rods may exhibit multiple equilibrium states whose stability can be investigated on the positive definiteness of the Hessian matrix of the total potential energy [6]. Variational approaches can be considered for such analysis [4]. Also, in [11], it can be found a procedure to use the similarity of the static equilibrium problem with a classical optimal control problem. In doing that analogy, it is possible to use the second-order sufficient conditions to find the stability of such elastic systems. The procedure is numerical and quite computational intensive. Let's show it here in brief for the mechanism under analysis.

A vector $\mathbf{x} = [\mathbf{x}_1\ \mathbf{x}_2]^T$ encompasses pose variables $\mathbf{x}_i = \{x_i(s)\ y_i(s)\ \theta_i(s)\}$ for each rod i, i.e.point coordinates in a fixed frame and slope of rod along arc length s. The energy to be minimised encloses the elastic energy of Kirchhoff rods Γ and the potential energy of external load ϕ:

$$J = -\{F_x\ F_y\ 0\} \begin{Bmatrix} x_{1_L} \\ y_{1_L} \\ \theta_{1_L} \end{Bmatrix} + \int_0^L \frac{1}{2} EI \sum_{i=1}^{2} u_{z_i}^2(s) ds = \phi(\mathbf{x}_{1_L}) + \int_0^L \Gamma(u_{z_1} u_{z_2} s)\, ds \tag{2}$$

where u_{z_i} is the curvature along the arc length of rod i.

Constraints at $s = 0$ are $\{x_{i_0},\ y_{i_0},\ \theta_{i_0}\}$; constraints at distal ends $s = L$ are stated in equations $\boldsymbol{\beta}(\mathbf{x}_{1L},\ \mathbf{x}_{2L}) = \mathbf{0}$; and a set of differential equations $\frac{d\mathbf{x}_i}{ds} = \mathbf{f}_i$ rules the change of $\mathbf{x}_i$. The augmented cost function is formed with Lagrange multipliers $\boldsymbol{\lambda}$ and $\boldsymbol{\nu}$ to enforce differential constraints $\frac{d\mathbf{x}_i}{ds} = \mathbf{f}_i$ and terminal constraints $\boldsymbol{\beta} = \mathbf{0}$:

$$J' = \phi(\mathbf{x}_{1L}) + \mathbf{v}^T \boldsymbol{\beta}(\mathbf{x}_{1L},\, \mathbf{x}_{2L}) + \int_0^L (\Gamma + \sum_{i=1}^{2} \boldsymbol{\lambda}_i^T \left[\frac{d\mathbf{x}_i}{ds} - \mathbf{f}_i\right])ds =$$

$$= G(\mathbf{x}_L, \mathbf{v}) + \int_0^L (H + \sum_{i=1}^{2} \boldsymbol{\lambda}_i^T \frac{d\mathbf{x}_i}{ds})ds \tag{3}$$

where $H = \Gamma - \sum_{i=1}^{2} \boldsymbol{\lambda}_i^T \mathbf{f}_i$.

First-order necessary conditions are obtained applying $\delta J' = 0$ for any change $\partial \mathbf{x}$, $\partial \mathbf{u}_z$, $\partial \mathbf{v}$ or $\partial \boldsymbol{\lambda}$ for each rod i, namely

$$\begin{Bmatrix} \frac{dx_i}{ds} \\ \frac{dy_i}{ds} \\ \frac{d\theta_i}{ds} \end{Bmatrix} = \begin{Bmatrix} \cos\theta_i \\ \sin\theta_i \\ u_{zi} \end{Bmatrix} \qquad \begin{Bmatrix} \frac{d\lambda_{i1}}{ds} \\ \frac{d\lambda_{i2}}{ds} \\ \frac{d\lambda_{i3}}{ds} \end{Bmatrix} = \begin{Bmatrix} 0 \\ 0 \\ \lambda_{i1}\sin\theta_i - \lambda_{i2}\cos\theta_i \end{Bmatrix}$$
$$EI\, u_{zi} - \lambda_{i3} = 0 \tag{4}$$

along with constraints $\boldsymbol{\lambda}_L = -\frac{\partial G}{\partial \mathbf{x}_L}^T$ and $\boldsymbol{\beta} = \mathbf{0}$. For the $2RFR$ mechanism $\boldsymbol{\beta}$ is:

$$\boldsymbol{\beta} = \begin{Bmatrix} x_{2_L} - x_{1_L} \\ y_{2_L} - y_{1_L} \end{Bmatrix} = \mathbf{0} \tag{5}$$

then:

$$G = -F_x x_{1_L} - F_y y_{1_L} + v_1(x_{2_L} - x_{1_L}) + v_2(y_{2_L} - y_{1_L}) \tag{6}$$

and we can get $\boldsymbol{\lambda}_L = -\frac{\partial G}{\partial \mathbf{x}_L}^T$, with elimination of Lagrange multipliers v_1 and v_2:

$$\begin{Bmatrix} \lambda_{11} + \lambda_{21} - F_x \\ \lambda_{12} + \lambda_{22} - F_y \\ \lambda_{13} \\ \lambda_{23} \end{Bmatrix} = \mathbf{0} \tag{7}$$

where Lagrange multipliers λ_{ij} represent the actions of each rod's end-tip that enforce end-effector conditions, i.e. the static equilibrium of the end-effector and the hinged condition of end-tips.

Second-order conditions for stability are: the strong Legrende-Clebsch condition

$$\frac{\partial^2 H}{\partial^2 u_{z_i}} = EI > 0 \tag{8}$$

and the nonexistence of conjugate points s_{cp} inside the interval $[s = 0,\ s = L)$. This can be determined by examining a certain matrix $\bar{\mathbf{S}}(s)$ that relates any infinitesimal change of $\mathbf{x}$ at s with the small change in $\boldsymbol{\lambda}$ required to satisfy the terminal boundary constraints:

$$\delta\boldsymbol{\lambda}(s) = \bar{\mathbf{S}}(s)\delta\mathbf{x}(s) \tag{9}$$

A conjugate point occurs at s_{cp} if $\bar{\mathbf{S}}$ becomes infinite. Matrix $\bar{\mathbf{S}}(s)$ is formulated from terminal boundary conditions $\mathbf{b} = \left\{\boldsymbol{\lambda}_L + \frac{\partial G}{\partial \mathbf{x}_L}^T \boldsymbol{\beta}(\mathbf{x}_L)\right\}^T = \mathbf{0}$ where any change of $\mathbf{x}$ or $\boldsymbol{\lambda}$ should not alter the fulfilment of the conditions, that is

$$\delta \mathbf{b} = \frac{\partial \mathbf{b}}{\partial \mathbf{x}} \delta \mathbf{x} + \frac{\partial \mathbf{b}}{\partial \boldsymbol{\lambda}} \delta \boldsymbol{\lambda} = \mathbf{0} \tag{10}$$

and hence, rearranging we get matrix $\bar{\mathbf{S}}(s)$ in

$$\delta \boldsymbol{\lambda} = \left[-\left(\frac{\partial \mathbf{b}}{\partial \boldsymbol{\lambda}}\right)^{-1} \frac{\partial \mathbf{b}}{\partial \mathbf{x}} \right] \delta \mathbf{x} \tag{11}$$

A conjugate point exists at s_{cp} when $\bar{\mathbf{S}}$ becomes infinite, conversely if $\frac{\partial \mathbf{b}}{\partial \boldsymbol{\lambda}}$ is singular. For each value of s considered inside the interval $[s = 0,\ s = L)$, we need to evaluate the singularity of the matrix:

$$\frac{\partial \mathbf{b}}{\partial \boldsymbol{\lambda}} = \frac{\partial \mathbf{b}}{\partial \mathbf{x}_L} \frac{\partial \mathbf{x}_L}{\partial \boldsymbol{\lambda}} + \frac{\partial \mathbf{b}}{\partial \boldsymbol{\lambda}_L} \frac{\partial \boldsymbol{\lambda}_L}{\partial \boldsymbol{\lambda}} = \frac{\partial \mathbf{b}}{\partial \mathbf{x}_L} \boldsymbol{\Phi}_{12}^{-1} + \frac{\partial \mathbf{b}}{\partial \boldsymbol{\lambda}_L} \boldsymbol{\Phi}_{22}^{-1} \tag{12}$$

where $\frac{\partial \mathbf{b}}{\partial \mathbf{x}_L}$ and $\frac{\partial \mathbf{b}}{\partial \boldsymbol{\lambda}_L}$ can be obtained analytically from $\mathbf{b}$, while finding $\frac{\partial \mathbf{x}_L}{\partial \boldsymbol{\lambda}}$ and $\frac{\partial \boldsymbol{\lambda}_L}{\partial \boldsymbol{\lambda}}$ requires the numerical evaluation of a so-called transition matrix $\boldsymbol{\Phi}$ defined as:

$$\begin{Bmatrix} \delta \mathbf{x} \\ \delta \boldsymbol{\lambda} \end{Bmatrix} = \boldsymbol{\Phi} \begin{Bmatrix} \delta \mathbf{x}_L \\ \delta \boldsymbol{\lambda}_L \end{Bmatrix} \tag{13}$$

Moreover, since each set of rod's equations is decoupled, we can get individual transition matrices $\boldsymbol{\Phi}_i(6 \times 6)$ to compute $\frac{\partial \mathbf{b}}{\partial \boldsymbol{\lambda}_i}(6 \times 3)$, and assemble them into:

$$\frac{\partial \mathbf{b}}{\partial \boldsymbol{\lambda}} = \left[\frac{\partial \mathbf{b}}{\partial \boldsymbol{\lambda}_1} \ \frac{\partial \mathbf{b}}{\partial \boldsymbol{\lambda}_2} \right] \tag{14}$$

For each equilibrium pose to be analysed, in order to evaluate matrix $\boldsymbol{\Phi}_i(s_j)$, we introduce an increment Δ in component x_{i_L}, solve the new equilibrium of the rod, retrieve the results $\mathbf{x}_i(s_j)$ and $\boldsymbol{\lambda}_i(s_j)$ at every s_j, find the change with respect to the original situation, and divide it by Δ to get the column of $\boldsymbol{\Phi}_i(s_j)$ corresponding to x_{i_L}, and proceed with the rest of components of $\mathbf{x}_{i_L}$ and $\boldsymbol{\lambda}_{i_L}$. Once matrix $\boldsymbol{\Phi}_i(s_j)$ is completed, solve the inverse and extract blocks $\boldsymbol{\Phi}_{i_{12}}^{-1}(3 \times 3)$ and $\boldsymbol{\Phi}_{i_{22}}^{-1}(3 \times 3)$ to get matrices $\frac{\partial \mathbf{b}}{\partial \boldsymbol{\lambda}_i}$, compose matrix $\frac{\partial \mathbf{b}}{\partial \boldsymbol{\lambda}}(6 \times 6)$ at s_j, and evaluate its determinant.

An alternative approach is to evaluate the stability of the mechanism under a perturbation of the end-effector load. Such condition is condensed in the stiffness matrix $\mathbf{K}$ found solving the position problem under fixed actuators and additional unit load in different directions, getting the deformation at $\mathscr{P}$:

$$\Delta \mathbf{F}_{\mathscr{P}} = \mathbf{K} \cdot \Delta \mathbf{p}_{\mathscr{P}} \tag{15}$$

Stability of a configuration can be assessed automatically evaluating the iff $det\mathbf{K} > 0$ and $K_{11} > 0$. For the different aspects in the workspace, see right of Fig. 1, abovementioned analyses showed that some are stable and some are not. For example, any pose inside the second aspect as in Fig. 2, showed a velocity ellipsoid indicating a regular pose, but existence of a conjugate point indicating instability. Its potential energy, plotted left of Fig. 1, has a saddle point.

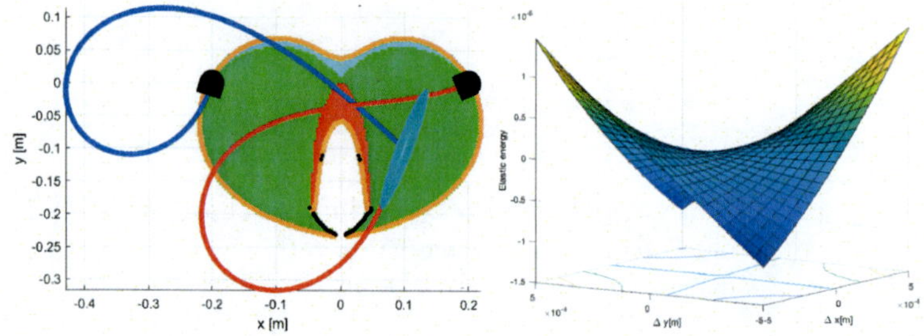

Fig. 2. Regular Unstable Pose along with velocity ellipsoid and Saddle Point of Potential Energy.

It was found that singular poses of the Direct Kinematic problem (Fig. 3) presented a degenerate velocity ellipsoid whose only axis indicates the direction of uncontrolled motion, and a saddle point in their potential energy plot indicating instability. This means that when the mechanism is actuated towards a DKP singularity, as it encounters such singular pose becomes unstable. If actuators are blocked then, the mechanism snaps towards a stable pose corresponding with those inputs.

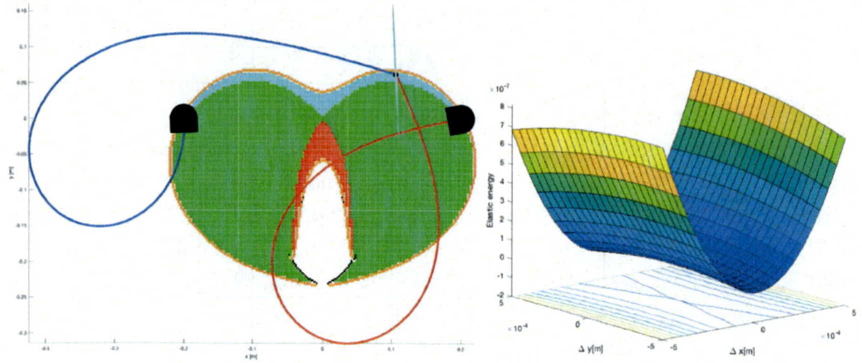

Fig. 3. DKP Singularity with ellipsoid and Saddle Point.

Such uncontrolled motion can be solved, in a first approximation, as a quasi-static problem in which the end-effector will move following the minimum available values of energy at each pose. In Fig. 4 there is a plot of the DKP singular pose from where the motion starts, the path followed by the end-effector, and the final pose reached by the mechanism. As observed in the right part of Fig. 4, that pose corresponds to a singular pose of the Inverse Kinematic problem, i.e. an outer edge of the workspace.

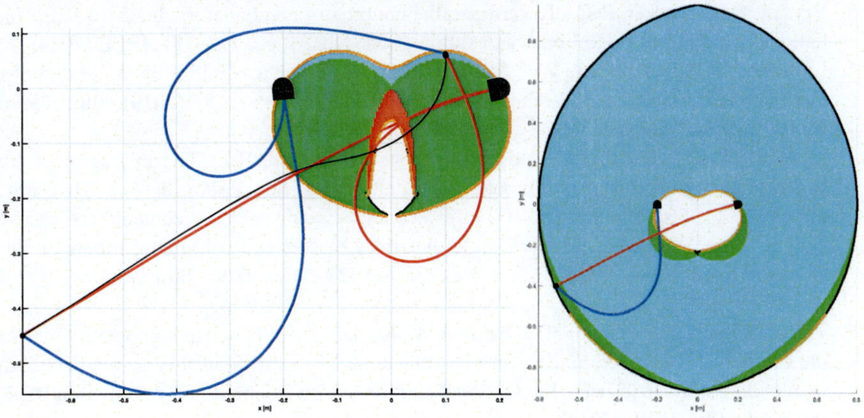

Fig. 4. Transition form DKP to IKP Singularity.

3 Conclusion

Instability is a major issue in workspace analysis of parallel continuum mechanisms. It can be assessed exhaustively with [11], or in a simplified manner evaluating the stiffness matrix. We have not found discrepancies between results of both methods. All poses in some aspects of a workspace are unstable. Kinematic singularities can be analysed with a numerical Jacobian, ellipsoids and condition numbers can be used. DKP singularities are unstable. From those poses the mechanism snaps to a pose that is an IKP singularity. A deeper thought on this phenomenon is required.

Acknowledgements. Authors wish to acknowledge the contribution of Diego Caballero and Unai Urrutia with MATLAB code, and financial support from Spanish Government (DPI2015-64450-R) and Regional Government of the Basque Country (Project IT949-16).

References

1. Altuzarra, O., Caballero, D., Campa, F.J., Pinto, C.: Position analysis in planar parallel continuum mechanisms. Mech. Mach. Theory **132**, 13–29 (2019). https://doi.org/10.1016/j.mechmachtheory.2018.10.014. http://www.sciencedirect.com/science/article/pii/S0094114X18307481
2. Altuzarra, O., Caballero, D., Zhang, Q., Campa, F.J.: Kinematic characteristics of parallel continuum mechanisms. In: Lenarcic, J., Parenti-Castelli, V. (eds.) Advances in Robot Kinematics 2018, pp. 293–301. Springer International Publishing, Cham (2019)

3. Altuzarra, O., Merlet, J.P.: Certified kinematics solution of 2-dof planar parallel continuum mechanisms. In: Uhl, T. (ed.) Advances in Mechanism and Machine Science, pp. 197–208. Springer International Publishing, Cham (2019)
4. Bretl, T., McCarthy, Z.: Quasi-static manipulation of a Kirchhoff elastic rod based on a geometric analysis of equilibrium configurations. Int. J. Robot. Res. **33**(1), 48–68 (2014)
5. Bryson, C.E., Rucker, D.C.: Toward parallel continuum manipulators. In: 2014 IEEE International Conference on Robotics and Automation (ICRA), pp. 778–785. IEEE (2014)
6. Levyakov, S.V., Kuznetsov, V.V.: Stability analysis of planar equilibrium configurations of elastic rods subjected to end loads. Acta Mechanica **211**(1), 73–87 (2010). https://doi.org/10.1007/s00707-009-0213-0.
7. McClintock, H., Temel, F.Z., Doshi, N., Koh, J.S., Wood, R.J.: The millidelta: A high-bandwidth, high-precision, millimeter-scale delta robot. Sci. Robot. **3**(14) (2018). https://doi.org/10.1126/scirobotics.aar3018. https://robotics.sciencemag.org/content/3/14/eaar3018
8. Till, J., Aloi, V., Rucker, C.: Real-time dynamics of soft and continuum robots based on Cosserat rod models. Int. J. Robot. Res. **38**(6), 723–746 (2019). https://doi.org/10.1177/0278364919842269.
9. Till, J., Bryson, C.E., Chung, S., Orekhov, A., Rucker, D.C.: Efficient computation of multiple coupled Cosserat rod models for real-time simulation and control of parallel continuum manipulators. In: 2015 IEEE International Conference on Robotics and Automation (ICRA), pp. 5067–5074. IEEE (2015)
10. Till, J., Rucker, D.C.: Elastic rod dynamics: Validation of a real-time implicit approach. In: 2017 IEEE/RSJ International Conference on Intelligent Robots and Systems (IROS), pp. 3013–3019 (2017). https://doi.org/10.1109/IROS.2017.8206139
11. Till, J., Rucker, D.C.: Elastic stability of cosserat rods and parallel continuum robots. IEEE Trans. Robot. **33**(3), 718–733 (2017). https://doi.org/10.1109/TRO.2017.2664879

Modeling and Simulation of Hybrid Soft Robots Using Finite Element Methods: Brief Overview and Benefits

Stanislao Grazioso[1(✉)], Giuseppe Di Gironimo[1], Luciano Rosati[2], and Bruno Siciliano[3]

[1] Dipartimento di Ingegneria Industriale, Università degli Studi di Napoli Federico II, Napoli, Italy
{stanislao.grazioso,giuseppe.gironimo}@unina.it

[2] Dipartimento di Strutture per l'Ingegneria e l'Architettura, Università degli Studi di Napoli Federico II, Napoli, Italy
luciano.rosati@unina.it

[3] Dipartimento di Ingegneria Elettrica E Tecnologie Dell'Informazione, Università degli Studi di Napoli Federico II, Napoli, Italy
bruno.siciliano@unina.it

Abstract. Mathematical modeling of hybrid soft robots is complicated by the description of the complex shape that they undergone when subject to actuation and external loads. It might be noticed that several approaches have been used so far in robotics, and the problem is not yet fully solved. This short paper aims at presenting an overview of modeling and simulation approaches for soft robots based on finite element methods. Benefits and perspectives of future directions are also discussed.

Keywords: Soft robotics · Modeling and simulation · Finite element method

1 Introduction

In the recent years there is a growing interest in building robotic systems with internal compliance, which can be either concentrated at the joint level or distributed along the manipulator's structure. The literature refers to the first class of systems as soft articulated robots [1], while to the second category as soft–bodied robots [2]. Soft robots are becoming pervasive in several applications, including medicine, rehabilitation, manufacturing, inspection and maintenance, remote explorations. However, despite the recent progresses, most of existing applications are limited to laboratories. In order to further advance the use of soft robots towards real-world applications and the development of commercially available solutions, it is important to fully understand and control their

J. Lenarčič and B. Siciliano (Eds.): ARK 2020, SPAR 15, pp. 335–340, 2021.
https://doi.org/10.1007/978-3-030-50975-0_41

mechanical behavior. As a matter of fact, modeling and simulation of soft robots is still an hot research topic in the community. Existing mathematical formulations are well–established for soft articulated robots [3]. Indeed, literature about soft–bodied robots and multi–link soft–bodied robots is less mature. With this respect, the current most accurate approaches are those based on continuum mechanics for finite deformations [4]. Such models require a numerical technique, as the finite element method, for space discretization and thus, numerical solution. In this brief work we consider four main aspects related to the use of finite element methods in soft robotics: the most used approaches for modeling; the available simulators; the benefits of using finite element–based modeling techniques and simulation tools; a generic discussion about the open issues. To the best of the authors' knowledge, the present paper is the first attempt to provide a survey on the topic.

2 Finite Element Modeling for Soft Robotics

The mechanics of a soft robot is formulated as a nonlinear structural problem. In structural analysis, two major sources generate nonlinear behaviors: geometry and material. Geometric nonlinearity refers to a nonlinear relationship between displacements and strains. Indeed, material nonlinearity arises when the constitutive equations between internal stresses and strains are nonlinear: this means that the material behavior depends on the current deformation state and past history.

Finite element methods foresee the discretization of the mechanical structure using mono–dimensional, bi–dimensional and three–dimensional finite elements. By referring to classic robotic structures, the finite element discretization process involve the use of: (1) beam elements, when one dimension of the geometry is predominant over the two others (as in the case of slender continuum robots for minimally invasive surgery [5]); (2) shell elements, when two dimensions are comparable and the third is negligible (as in the case of robotic fishes with tiny structures [6]); (3) solid elements, when the three dimensions are comparable.

Regarding the mechanical approaches used for description of large displacements and deformations, two approaches are mainly adopted in the literature: (i) corotational frame (CF) formulation [7,8]; (ii) inertial frame (IF) formulation [9,10]. The approaches differ from the coordinate frames used to formulate the equations of motion. In CF formulations, a corotational frame is defined for each finite element, and it follows a mean rigid body motion of the component. Here, strains and stresses are measured from the corotated configuration whereas a fixed configuration is used as reference for measuring rigid body motions. Indeed, in IF formulations (also called geometrically exact finite element formulations) a global inertial frame is used as reference frame for all motions. Here, the exact representation of the rigid body motions is achieved by deriving frame invariant deformation measures from continuum mechanics. Since the motion of a component is viewed as a whole, without a priori decomposition into a mean rigid body and a superimposed flexible motion, the formulations based on the geometrically exact theories are the most general and accurate to account for flexibility [11].

Examples of using the CF formulation for soft robotics applications can be found in the works by Duriez et al. [12,13]. Indeed, the use of IF formulations for soft robots is available in [14].

3 Finite Element Simulators for Soft Robotics

Simulators in robotics provide assistance for robot analysis, testing of control algorithms, motion planning and interaction with the environments. General platforms such as V–Rep [15] and Gazebo [16] mainly provide simulation of articulated rigid robots. They are based on physics engine such as ODE [17] and Bullet [18], which are not currently optimized for deformable bodies. The gold standard approach for analysis of soft robots is the use of commercial finite element solvers for nonlinear structural analysis; in this respect, the current most adopted software is ABAQUS[1] (Dassault Systèmes, Vélizy–Villacoublay, FR). As an example, ABAQUS has been largely used in modeling and simulation of soft pneumatic actuators [19,20]. One simulator which is gaining attention is SOFA [21,22], an open source framework for physics simulation of deformable bodies. Initially devoted mainly to medical simulation, SOFA is today one of the most used simulators for soft robots, also thanks to the development of the SoftRobots plugin,[2] used for interactive simulation and control of soft robots. SOFA allows developing meshes with mono–dimensional and three–dimensional elements, and it offers multiple material models and solvers. Recently, the SimSOFT simulator [14] has been developed for dynamic simulation of soft–bodied and soft articulated robots. It describes a soft–bodied robot as a nonlinear beam element and it implements kinematic joints for their multi–link connection.

4 Advantages

According to the authors' point of view, finite element modeling techniques and simulation tools bring the following advantages for soft robotic applications.

Serial/Parallel Kinematic Chains. Finite element approaches allow modeling and simulation of serial and parallel kinematic topologies within the same framework, since after discretization the model comprises a series of elements connected through nodes with special boundary conditions. This is of great interest for the soft robotics community, where also parallel structures are showing great potentials. As an example, for surgical applications, parallel robots offer a design which is easy to miniaturize and can achieve multi degrees–of–freedom motion in confined spaces [23].

Library with Rigid and Soft Bodies. Finite element solvers use mathematical formulations that can accommodate the simulation of rigid elements, considered as special cases of elements with zero internal deformations. Regarding continuum

[1] https://www.3ds.com/.

[2] https://project.inria.fr/softrobot/.

bodies, finite element simulators offer the possibility to mesh a robotic structure with beam, shell or solid elements.

Library with Multiple Kinematic Joints. Finite element methods treat also connection between bodies. Existing commercial software for nonlinear structural analysis (i.e. ABAQUS) are connected with multibody dynamics packages (as SIMPACK, always from Dassault Systèmes) for performing analyses with multiple bodies. Different finite element solutions as SOFA or SimSOFT, indeed, have built–in joint models for connecting multiple bodies.

5 Discussion and Open Issues

Advancements in the domain of finite element modeling and solvers for soft robots have been relatively slow. Two major reasons can be identified: (i) computational power of the current machines; (ii) lack of reliable and universal recognized theoretical models. From this literature analysis, we can state that, in order to facilitate the design and development of effective soft robotic systems, three grand challenges remain in the development of: (1) conceptual design tools; (2) model–based controllers; (3) integrated tools for design, analysis, control and virtual simulations.

1. Conceptual Design Tools. The available simulators for soft robots usually deal with analysis and verification of systems, when a detailed computer–aided–design (CAD) model is already available. This strategy can result in wasted time and effort for detailed designs which may not work. A special effort is worth to be made towards the development of computer aided conceptual design tools specifics for soft robots. Indeed, model–based conceptual design tools might speed up the development of effective soft robots, by providing non–experts key design parameters of the systems according to their specific application.

2. Model–Based Controllers. Model–based control is barely popular for soft robotic systems. Few applications are related to simple models for continuum robots [24] or model order reduction techniques, which use heavy offline stages [13]. A great effort should thus be pursued in the direction of purely invertible models for soft–bodied robots, such that all the model–based controllers used for rigid robotics can be applied also for soft robots.

3. Integrated Tools for Design, Analysis, Control, Virtual Simulations. Simulators for soft robots should foresee a unique software environment where the integration of geometry and analysis of soft mechanisms is natural. As a soft robot usually works in its deformed configuration, it is necessary to have an environment where the CAD geometry follows the real physics behavior of the manipulator during the working trajectories. Such integrated environment should provide multiple material models, multibody dynamics, computational mechanics and (eventually) computational fluid dynamics capabilities (the latter is required for flying and underwater soft robots). Furthermore, effective virtual prototyping tools for soft robots should foresee the possibility to test control algorithms and different input trajectories. To this end, models specifically developed for control purposes and not only for simulation purposes are required. Interactive and real–time virtual simulation tools, together with advanced interfaces might allow engineers and users to moving closer to a better model viewing, manipulation and feedback on the design and practical usage of soft robots. To do that, the following features are needed: (i) rapid computational mechanics tools based on efficient dynamic models; (ii) integration of graphics and mechanics; (iii) interfaces with mapping algorithms able to control a distributed system as a soft robot with a limited set of inputs. Another topic which is worth to be investigated is the development of a unique product design representation of soft robots.

As a lesson learned from this preliminary literature review, it is worth highlighting the fact that we are still far away from an integrated software environment for conceptual design, analysis and control of hybrid soft robots.

References

1. Della Santina, C., Piazza, C., Gasparri, G.M., Bonilla, M., Catalano, M.G., Grioli, G., Garabini, M., Bicchi, A.: The quest for natural machine motion: an open platform to fast-prototyping articulated soft robots. IEEE Robot. Autom. Mag. **24**(1), 48–56 (2017)
2. Laschi, C., Mazzolai, B., Cianchetti, M.: Soft robotics: technologies and systems pushing the boundaries of robot abilities. Sci. Robot. **1**(1), 3690 (2016)
3. Albu-Schaffer, A., Eiberger, O., Grebenstein, M., Haddadin, S., Ott, C., Wimbock, T., Wolf, S., Hirzinger, G.: Soft robotics. IEEE Robot. Autom. Mag. **15**(3), 20–30 (2008)
4. Antman, S.S.: Nonlinear Problems of Elasticity, vol. 107. Springer, New York (2005)
5. Burgner-Kahrs, J., Rucker, D.C., Choset, H.: Continuum robots for medical applications: a survey. IEEE Trans. Robot. **31**(6), 1261–1280 (2015)
6. Renda, F., Giorgio-Serchi, F., Boyer, F., Laschi, C.: Locomotion and elastodynamics model of an underwater shell-like soft robot. In: 2015 IEEE International Conference on Robotics and Automation (ICRA), pp. 1158–1165. IEEE (2015)
7. Crisfield, M.A., Remmers, J.J., Verhoosel, C.V., et al.: Nonlinear Finite Element Analysis of Solids and Structures. Wiley, Hoboken (2012)
8. Belytschko, T., Liu, W.K., Moran, B., Elkhodary, K.: Nonlinear Finite Elements for Continua and Structures. Wiley, Hoboken (2013)
9. Simo, J.C., Vu-Quoc, L.: A three-dimensional finite-strain rod model. Part ii: computational aspects. Comput. Meth. Appl. Mech. Eng. **58**(1), 79–116 (1986)
10. Simo, J., Vu-Quoc, L.: On the dynamics of flexible beams under large overall motions–the plane case: part ii. J. Appl. Mech. **53**(4), 855–863 (1986)
11. Bauchau, O.A., Betsch, P., Cardona, A., Gerstmayr, J., Jonker, B., Masarati, P., Sonneville, V.: Validation of flexible multibody dynamics beam formulations using benchmark problems. Multibody Sys. Dyn. **37**(1), 29–48 (2016)
12. Koehler, M., Okamura, A.M., Duriez, C.: Stiffness control of deformable robots using finite element modeling. IEEE Robot. Autom. Lett. **4**(2), 469–476 (2019)
13. Goury, O., Duriez, C.: Fast, generic, and reliable control and simulation of soft robots using model order reduction. IEEE Trans. Robot. **34**(6), 1565–1576 (2018)
14. Grazioso, S., Di Gironimo, G., Siciliano, B.: A geometrically exact model for soft continuum robots: the finite element deformation space formulation. Soft Robot. **6**(6), 790–811 (2019)

15. Freese, M., Singh, S., Ozaki, F., Matsuhira, N.: Virtual robot experimentation platform v-rep: a versatile 3d robot simulator. In: International Conference on Simulation, Modeling, and Programming for Autonomous Robots, pp. 51–62. Springer (2010)
16. Koenig, N., Howard, A.: Design and use paradigms for gazebo, an open-source multi-robot simulator. In: 2004 IEEE/RSJ International Conference on Intelligent Robots and Systems, vol. 3, pp. 2149–2154. IEEE (2004)
17. Smith, R., et al.: Open dynamics engine (2005)
18. Coumans, E.: Bullet physics engine. Open Source Software: http://bulletphysics.org 1(3), 84 (2010)
19. Elsayed, Y., Vincensi, A., Lekakou, C., Geng, T., Saaj, C., Ranzani, T., Cianchetti, M., Menciassi, A.: Finite element analysis and design optimization of a pneumatically actuating silicone module for robotic surgery applications. Soft Robot. **1**(4), 255–262 (2014)
20. Moseley, P., Florez, J.M., Sonar, H.A., Agarwal, G., Curtin, W., Paik, J.: Modeling, design, and development of soft pneumatic actuators with finite element method. Adv. Eng. Mater. **18**(6), 978–988 (2016)
21. Allard, J., et al.: Sofa-an open source framework for medical simulation (2007)
22. Faure, F., Duriez, C., Delingette, H., Allard, J., Gilles, B., Marchesseau, S., Talbot, H., Courtecuisse, H., Bousquet, G., Peterlik, I., et al.: Sofa: a multi-model framework for interactive physical simulation. In: Soft Tissue Biomechanical Modeling for Computer Assisted Surgery, pp. 283–321. Springer (2012)
23. Anderson, P.L., Mahoney, A.W., Webster, R.J.: Continuum reconfigurable parallel robots for surgery: shape sensing and state estimation with uncertainty. IEEE Robot. Autom. Lett. **2**(3), 1617–1624 (2017)
24. Della Santina, C., Katzschmann, R.K., Bicchi, A., Rus, D.: Model-based dynamic feedback control of a planar soft robot: trajectory tracking and interaction with the environment. Int. J. Robot. Res., 0278364919897292 (2019)

Exoskeleton Control Based on Network of Stable Heteroclinic Channels (SHC) Combined with Gaussian Mixture Models (GMM)

Tadej Petrič[1(✉)], Marko Jamšek[2], and Jan Babič[2]

[1] Department for Automation, Biocybernetics and Robotics, Jožef Stefan Institute, Ljubljana, Slovenia
tadej.petric@ijs.si

[2] Laboratory of Neuromechanics and Biorobotics, Department for Automation, Biocybernetics and Robotics, Jožef Stefan Institute, Ljubljana, Slovenia
{marko.jamsek,jan.babic}@ijs.si

Abstract. One of the major causes of disability and sick day leaves is lower back pain. Hence it can result in a decreased life quality and lower industrial productivity. One of the possible solutions to lower back pain could be the use of exoskeletons, which would reduce the spinal loading. One of such solutions is a quasi-passive spinal exoskeleton that engages and disengages the passive support depending on the movements performed by the user. This enables the spinal support for the user when lifting a heavy load and for all other tasks, the user motion is unobstructed. To achieve autonomous clutch activation, the main challenge is to properly classify the beginning of each motion. In this paper, we proposed a novel control method that uses Gaussian Mixture Models (GMM) for movement classifiers and a network of Stable Heteroclinic Channels (SHC) for designing a phase-state-machine. Integrating GMM into the SHC network enables a fast and reliable control of the clutch mechanism of the quasi-passive spinal exoskeleton. The control system capabilities were demonstrated in an experiment with a male subject wearing the quasi-passive exoskeleton while executing three different movements representative for an industrial working environment: walking, standing, and lifting.

Keywords: Spinal exoskeleton · Stable Heteroclinic Channels · Gaussian mixture models

1 Introduction

Studies have shown that there is a strong correlation between physically demanding jobs and prevalence of lower back pain (LBP) [13], therefore typical commercially available exoskeletons for lower-back pain prevention are targeting

J. Lenarčič and B. Siciliano (Eds.): ARK 2020, SPAR 15, pp. 341–348, 2021.
https://doi.org/10.1007/978-3-030-50975-0_42

blue-collar workers. It has been shown that such devices can effectively reduce the muscular activity of the lower back [10]. However, since the commercially available devices are completely passive they suffer from a limited versatility [3] as they will hinder some of the typical user movements, for example walking. To improve the usability of such devices we have developed a controller for a quasi-passive spinal exoskeleton device, with the ability to disengage the support when needed. Since a quasi-passive exoskeleton system stores energy during downward motion, it is crucial to design control systems in a way that it acts at the very beginning of the user's motion. This makes the control design specifically challenging since the control system needs to identify the user intentions (at the very beginning), where most of the motions have common characteristics. A possible solution, which was proposed for active exoskeletons [2] is to use a threshold-based approach to identify the type of a lifting movement. Their main assumption was also that the lifting starts from a squat-like posse, which made it easier to classify movement. While this is not feasible with the quasi-passive system due to the mechanical limitations of the clutch. For a quasi-passive system, we proposed in [1] a control method based on Gaussian Mixture Models (GMM) combined with a finite-state machine. Here the main limitation was the immediate discrete switching between sates, which could potentially lead the system into unwanted states, due to the similarities at the very beginning of the movements.

In this paper we improve the reliability of the control approach by merging a network of Stable Heteroclinic Channels (SHC) [8] with GMM. The novelty of our approach is the use of GMM to drive the input signals of the SHC network. This results in a continuous path through state-space that visits a succession of fixed points. SHC consist of sequences of saddle equilibrium points (nodes) [8]. Note that here the SHC is a particular form of trajectory that revisits a sequence of fixed points [7]. SHC was already introduced as a successful tool for human-robot interaction in [4], where they introduced the ability of the control system to reverse and hold transitions towards the next successor state. The novelty in this paper is the use of the GMM classifiers to drive the transitions.

2 Control Architecture

The foundation of the proposed control method is a network of SHC [8,11]. Such network can have an arbitrary number of saddle points, which can be interpreted as states. The SHC network is driven with GMM which results in a hybrid state machine system, where transitions between states are continuous. Note that the transitions are also nonexclusive and reversible.

Network of Stable Heteroclinic Channels (SHC): The dynamics of the SHC network, where $\boldsymbol{x}$ is a n-dimensional state vector, as introduced in [4], is governed by

$$\dot{\boldsymbol{x}} = \boldsymbol{x} \circ (\boldsymbol{\alpha} + \boldsymbol{x}^{\gamma} (\boldsymbol{\rho}_0 + \boldsymbol{\rho}_{\Delta} \circ (\mathbf{T} + \mathbf{G}))) \, \eta(t) + \dot{\boldsymbol{\delta}}(t), \tag{1}$$

where the element-wise multiplication is denoted by $\circ$, γ is a positive scalar exponent, $\eta(t)$ is a scalar multiplier that adjusts the SHC network evolution speed, and $\dot{\boldsymbol{\delta}}(t)$ is used to push the system away from saddle points and modulate transitions. The SHC network state transitions are defined with a transition matrix $\mathbf{T}$, where transitions from i to j exist if $T_{j,i} = 1$, otherwise $T_{j,i} = 0$, and the matrix $\mathbf{G}$ which weights competing transitions. The $n \times n$ matrix $\boldsymbol{\rho}_0$ is selected such that the SHC network has the n mutually exclusive saddle points placed on its exclusive coordinate axis. The SHC is created with the $n \times n$ matrix $\boldsymbol{\rho}_\Delta$ multiplied with $\mathbf{T}$. Based on n-dimensional parameter vectors [8]: growth rate $\boldsymbol{\alpha}$, saddle point locations $\boldsymbol{\beta}$ and saddle point shapes $\boldsymbol{\nu}$, we can compute the matrices $\boldsymbol{\rho}_0$ and $\boldsymbol{\rho}_\Delta$

$$\boldsymbol{\rho}_0 = \left(\boldsymbol{\alpha}(\boldsymbol{\beta}^{-1})^T\right) \circ \left(\mathbf{I} - \mathbf{1} - (\boldsymbol{\alpha}\boldsymbol{\alpha}^T)^{-1}\right), \tag{2}$$

$$\boldsymbol{\rho}_\Delta = \left(\boldsymbol{\alpha} \circ (\mathbf{1} + \boldsymbol{\nu}^{-1})\right)(\boldsymbol{\beta}^{-1})^T. \tag{3}$$

The advantage of this formulation is that $\boldsymbol{\rho}_0$ and $\boldsymbol{\rho}_\Delta$ are constant, while $\mathbf{T}$ and $\mathbf{G}$ can change arbitrarily.

To express each state and each possible transition, a function introduced in [4] that computes activation for all possible transitions is given by

$$\boldsymbol{\Lambda}^t = \frac{16\boldsymbol{x}\boldsymbol{x}^T|\boldsymbol{x}^2|}{(\boldsymbol{x}\mathbf{1}^T + \mathbf{1}\boldsymbol{x}^T)^2 + |\boldsymbol{x}|^2} \circ \mathbf{T}, \tag{4}$$

where $\mathbf{1}$ is n-dimensional vector of ones. The function ensures that the transitions are invariant to scaling of $\boldsymbol{x}$ and hence, bounded to the interval of $[0, 1]$. Further on, the state can be computed from the residual of the transition matrix $\boldsymbol{\Lambda}^t$. The state is given by

$$\boldsymbol{\lambda}^s = \frac{\boldsymbol{x}^2}{\sum \boldsymbol{x}^2}\left(\mathbf{1} - \sum \boldsymbol{\Lambda}^t\right). \tag{5}$$

Since the diagonal elements of Λ^t are meaningless, [4] proposed to combine all transitions and state activations into one activation matrix given by

$$\boldsymbol{\Lambda}_{ji} = \begin{cases} \boldsymbol{\Lambda}^t_{ji}, & j \neq i \\ \boldsymbol{\lambda}^s_i, & j = i \end{cases}. \tag{6}$$

Since all SHC are located on a two-dimensional plane spanned between their predecessor and successor state coordinate axis it is possible to compute the phase variable. The phase for an active transition from state i to state j is given by

$$\phi_{j,i} = \frac{|\boldsymbol{x}_j|}{|\boldsymbol{x}_j| + |\boldsymbol{x}_i|}. \tag{7}$$

As pointed out in [4], in terms of control, states can be associated with constant control goals, i.e. they can be seen as states in state-machine. While in typical state-machine the transition between states is discrete, here the transitions have a phase, which can be further-on used to drive any phase-related trajectory generator like DMP [9], CMP [5], etc.

The system described with Eq. 1, has four main inputs, which are used to modulate the network of SHC. The first one is the matrix $\mathbf{T}$ that defines which transitions exist. Note that it can be modified at any time. The second one is the greediness matrix $\mathbf{G}$, which is used to modify the transition direction, especially in cases when we have multiple competing transitions. Note that the greediness matrix $\mathbf{G}$ is also used to stop and/or reverse transitions. The third one is the vector $\boldsymbol{\delta}$, which modulates the time spent in a state and determines which transitions are active to leave a state. It is governed by

$$\dot{\boldsymbol{\delta}} = (\boldsymbol{\Lambda} \circ \mathbf{B}) + \epsilon \boldsymbol{W}(t), \tag{8}$$

where elements $\mathbf{B_{ji}}$ of matrix $\mathbf{B}$ corresponds to bias towards state j in state i. It is used to weigh individual states, i.e. to select the next successor state out of several ones if transitions exist. To ensure that the system does not stay in a saddle point forever a parametric ϵ stochastic noise $\boldsymbol{W}(t)$ is added, which ensures that the system takes a random transition after a random amount of time.

The fourth driving parameter of the SHC network is the transition velocity of η, which is used to speed up or slow down the system. It is given by

$$\eta = 2^{\sum \boldsymbol{\Lambda} \circ \mathbf{A}}, \tag{9}$$

where matrix A is used to define transition speeds. Note that if $\mathbf{A} = 0$, then the transition speeds are defined by $\boldsymbol{\alpha}$, i.e. unmodified system behavior. For more in-depth details please see [4].

Gaussian Mixture Models (GMM): Gaussian mixture models can be used either as an unsupervised or supervised classification algorithm. In supervised learning of GMM the expectation-maximization algorithm is used to fit a mixture of Gaussians to the training dataset and create a model. For each class an arbitrary number of Gaussians can be specified for fitting the data, i.e. each class is represented with a multivariate normal distribution. Typically we can compute a probability density function P_m for a multivariate normal distribution with the flowing equation

$$P_m = f(\boldsymbol{\chi}) = \frac{e^{-\frac{1}{2}(\boldsymbol{\chi}-\boldsymbol{\mu})^T \boldsymbol{\Sigma}^{-1}(\boldsymbol{\chi}-\boldsymbol{\mu})}}{\sqrt{(2\pi)^k |\boldsymbol{\Sigma}|}}. \tag{10}$$

Here $\boldsymbol{\mu}$ is the mean vector and $\boldsymbol{\Sigma}$ is the covariance matrix of the Gaussian mixtures. The vector $\boldsymbol{\chi}$ denotes a feature space input. In this notation, the equation yields P_m as a scalar value. Note that each state (cluster of data) has it's own probability density function P_m^s, denoted with superscript s. To represent all states we define a probability density vector $\boldsymbol{P_m} = [P_m^1, P_m^2, ..., P_m^s]$. For the sake of completeness, a summary of Gaussian Mixture Models (GMM) is given in [6].

Integration of GMM into SHC Network: We propose to use GMM probability density functions P_m^s to modulate the network of SHC. Hence, the resulting system will act as probability-phase-state-machine. We assume that the number of states n in the SHC network equals to the number of probability density functions P_m^s of GMM, i.e. $n = s$. Note that each state in the SHC network is associated with one probability density function P_m of GMM. By using the

GMM we propose to modulate the SHC network by changing on-line the transition matrix **T**, greediness matrix **G** and the bias matrix **B**. The transition matrix **T** is modulated only when saddle points are reached, i.e. when the transition phase is finished. It is given by

$$\mathbf{T} = \mathbf{T_0} \circ \boldsymbol{\lambda}^s, \tag{11}$$

where $\mathbf{T_0}$ is the transition matrix with all possible state transitions. This formulation ensures that from a current state only transitions towards the next successor states can be active, hence ensuring the system will go only towards possible successor states. Note that this guarantees the system will not result in a cyclic behavior, which is otherwise known as SHC cycles [8].

As proposed in [4], the greediness matrix **G** is computed with

$$\mathbf{G} = \left(\mathbf{T} \circ \overrightarrow{\mathbf{G}} - \overrightarrow{\mathbf{G}}^T \circ \mathbf{T}^T\right) + \left(\mathbf{T}\mathbf{T}^T \circ (1 - \mathbf{I})\right) \circ \overleftrightarrow{\mathbf{G}}, \tag{12}$$

where $\overrightarrow{\mathbf{G}}$ is the greediness of a successor with respect to the predecessor state and $\overleftrightarrow{\mathbf{G}}$ is the competitive greediness between competing successor states [4]. They are given by

$$\overrightarrow{\mathbf{G}}_{ji} = \begin{cases} -1 & g_j < -1 \\ 0.5(g_j - 1) & else \\ 0 & g_j > 1 \end{cases} \tag{13}$$

$$\overleftrightarrow{\mathbf{G}}ji = 0.5\left(1 - g_i + \kappa\left(g_j - g_i\right)\right), \tag{14}$$

where $\kappa = 0.25$, if not stated otherwise. While the greediness vector $\boldsymbol{g}$ can be constant as in [4], we proposed to compute it using the GMM probability density values. greediness vector $\boldsymbol{g}$ is now given with

$$\boldsymbol{g} = \begin{cases} \frac{\boldsymbol{P_m}}{|\boldsymbol{P_m}|_\infty} - \frac{\boldsymbol{\lambda}^s}{|\boldsymbol{\lambda}^s|_\infty} & , \Upsilon(\boldsymbol{P_m}) = \Upsilon(\sum_{i=1}^n(\mathbf{T}_{ij})) \\ \frac{\boldsymbol{P_m}}{|\boldsymbol{P_m}|_\infty} + \sum_{i=1}^n(\mathbf{T}_{ij}) & , else \end{cases}, \tag{15}$$

where function $\Upsilon(.)$ returns arguments of the maxima function (*argmax*). The proposed modulation of g ensures that the system stays in the current state or the transition can be reversed if the current state has the highest probability density among others. Otherwise, the greediness will be modulated based on the current probability density vector $\boldsymbol{P_m}$ and the vector (row sum of transition matrix **T**) which ensures that the system has the ability to leave the current state. When competing transitions exist the system will favor the one with the largest probability. Similarly, we propose to compute the bias matrix **B** such that the transitions are weighted based on the probability density, hence it will favor the transition with the largest probability. It is governed by

$$\mathbf{B} = \xi * \mathbf{1}^T * \left(\frac{\boldsymbol{P_m}}{|\boldsymbol{P_m}|_\infty} - 1\right), \tag{16}$$

where **1** is a n-dimensional vector of ones and ξ is a scalar constant.

The prosed combination of transition matrix **T**, greediness matrix **G** and the bias matrix **B** enables the use of the probability-phase-state machine system in human-robot interaction scenarios.

3 Exoskeleton Control

The proposed method aimed to control the clutch mechanism of the bilateral powered exoskeleton. To provide support to the user the exoskeleton employs a spring mechanism that generates an extension torque profile around the hip joints. The original mechanism [12] was upgraded with a clutch system that provides the possibility to enable or disable the support provided to the user. Due to the mechanical design the clutch can only be activated close to the normal standing pose.

To detect and provide information about the current pose of the user the exoskeleton is equipped with a sensory system. An inertial measurement unit (IMU) is mounted on the chest straps of the exoskeleton that provides a measure of the trunk inclination relative to the world coordinate system. While the IMU together with rotational encoders embedded in the hip joints is used to monitor left and right hip flexion/extension. The angles that are used for computing GMM are absolute trunk angle in the sagittal plane of the human and the left and right hip angles. The GMM were computed using the method reported in [1].

To control the clutch of the quasi-passive spinal exoskeleton we propose to exploit the internal state $\boldsymbol{x}$ of the phase-state-machine system. To demonstrate the behavior we selected three most typical actions for the use of the quasi-passive spinal exoskeleton: standing, walking and lifting. Note that here, the exoskeleton should engage the clutch only during the lifting task. To acquire the GMM the subject was asked to perform a sequence of actions, i.e. standing, walking and lifting a 3 kg box. The sequence of actions was random, and instructions were given out verbally after every completed action. Subject performed in total of 20 walks, 20 box lifts, 20s of standing. The subject was not instructed to perform any particular lifting technique. The acquired measured data was then also hand labeled to train GMM for each action in a supervised way. Once the GMM were obtained we used them for driving the phase-state-machine and hence to control the clutch of the quasi-passive spinal exoskeleton.

Here, by exploiting the internal state $\boldsymbol{x}$, we proposed that the clutch is activated when the internal coordinate for the lifting task x_3 of the phase-state-machine reaches a predefined threshold ι. The control algorithm of the clutch is given by

$$u = \begin{cases} 1 & x_3 > \iota \\ 0 & else \end{cases} \tag{17}$$

If not stated otherwise ι was experimentally set to $\iota = 0.35$.

To verify the approach we asked a subject to perform random actions, i.e. walking, standing and lifting, by wearing an exoskeleton with the proposed controller. One example session is shown in Fig. 1. Here we show a 200s time section of the experiment. The preliminary result shows that we were able to identify the lifting movements and activate the clutch on demand. By comparing the hand-labeled ground truth with the actual clutch activation, we can see that they are well matched. Even more, these results also showed that we eliminated unwanted behavior, i.e. short clutch activation signals, when the clutch should not be engaged, which was a limitation of our previous proposed controller [1].

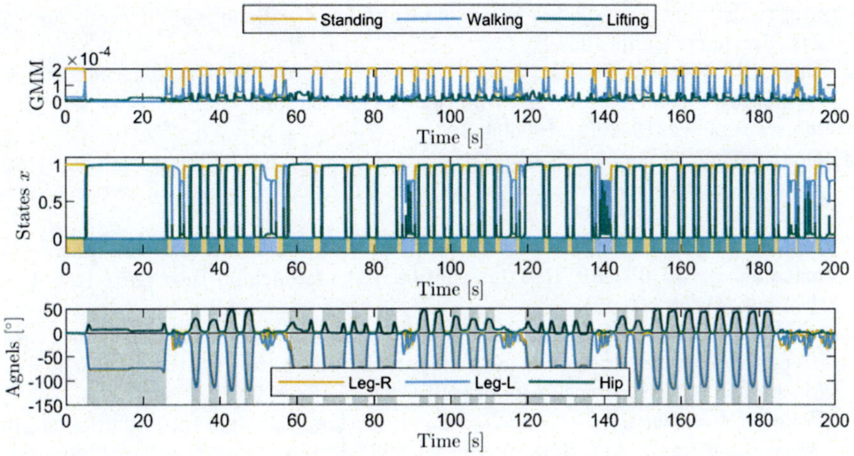

Fig. 1. Experimental validation of the proposed control system. The top plot shows all three GMMs. The middle plot shows the internal states together with the hand-labeled ground. Note that hand-label ground truth is indicated with the shaded colors below 0. The bottom plot shows the measured angles and the clutch activation. Here the clutch activation is illustrated with the gray shaded areas.

4 Discussion and Concussions

In this paper, we have introduced a novel control algorithm for human-robot physical interaction that is based on a phase-state machined driven with probability models. As such the system can autonomously and intuitively adapt and predict user behavior, which is important for exoskeleton systems. Further-on the internal states of the phase-state machine can be used to control actuators of the exoskeleton system. Which results in faster and more robust performance of the exoskeleton system.

Despite the limitations of the quasi-passive exoskeleton system, we have shown the conceptual results for the phase-state machine driven with the probability models. We have also shown that the system can be effectively applied to a real quasi-passive exoskeleton system. These preliminary experimental results have confirmed that a control approach based on the network of SHC combined with GMMs is an effective tool for tasks where human-robot interaction is present.

References

1. Cevzar, M., Petrič, T., Jamšek, M., Babič, J.: Real-time control of quasi-active hip exoskeleton based on gaussian mixture model approach. In: Carrozza, M.C., Micera, S., Pons, J.L. (eds.) Wearable Robotics: Challenges and Trends, pp. 244–248. Springer, Cham (2019)
2. Chen, B., Lanotte, F., Grazi, L., Vitiello, N., Crea, S.: Classification of lifting techniques for application of a robotic hip exoskeleton. Sensors **19**(4), 963 (2019). https://doi.org/10.3390/s19040963
3. De Rossi, S.M.M., et al.: Sensing pressure distribution on a lower-limb exoskeleton physical human-machine interface. Sensors **11**(1), 207–227 (2010). https://doi.org/10.3390/s110100207
4. Deimel, R.: Reactive interaction through body motion and the phase-state-machine. In: 2019 IEEE/RSJ International Conference on Intelligent Robots and Systems (IROS) (2019)
5. Denisa, M., Gams, A., Ude, A., Petric, T.: Learning compliant movement primitives through demonstration and statistical generalization. IEEE/ASME Trans. Mechatron. (2016). https://doi.org/10.1109/TMECH.2015.2510165
6. Fraley, C., Raftery, A.E.: Model-based clustering, discriminant analysis, and density estimation. J. Am. Stat. Assoc. **97**(458), 611–631 (2002). https://doi.org/10.1198/016214502760047131
7. Friston, K., Mattout, J., Kilner, J.: Action understanding and active inference. Biol. Cybern. **104**(1–2), 137–160 (2011). https://doi.org/10.1007/s00422-011-0424-z
8. Horchler, A.D., Daltorio, K.A., Chiel, H.J., Quinn, R.D.: Designing responsive pattern generators: stable heteroclinic channel cycles for modeling and control. Bioinspiration Biomimetics **10**(2), 1–16 (2015). https://doi.org/10.1088/1748-3190/10/2/026001
9. Ijspeert, A.J., Nakanishi, J., Hoffmann, H., Pastor, P., Schaal, S.: Dynamical movement primitives: learning attractor models for motor behaviors. Neural Comput. **25**(2), 328–373 (2013)
10. Koopman, A.S., Kingma, I., Faber, G.S., de Looze, M.P., van Dieën, J.H.: Effects of a passive exoskeleton on the mechanical loading of the low back in static holding tasks. J. Biomech. **83**, 97–103 (2019). https://doi.org/10.1016/j.jbiomech.2018.11.033
11. Laurent, G., Stopfer, M., Friedrich, R.W., Rabinovich, M.I., Volkovskii, A., Abarbanel, H.D.: Odor encoding as an active, dynamical process: experiments, computation, and theory. Ann. Rev. Neurosci. **24**(1), 263–297 (2001). https://doi.org/10.1146/annurev.neuro.24.1.263
12. Näf, M.B., Koopman, A.S., Baltrusch, S., Rodriguez-Guerrero, C., Vanderborght, B., Lefeber, D.: Passive back support exoskeleton improves range of motion using flexible beams. Front. Robot. AI **5**(June), 1–16 (2018). https://doi.org/10.3389/frobt.2018.00072
13. Waddell, G., Burton, A.K.: Occupational health guidelines for the management of low back pain at work: evidence review. Occup. Med. **51**(2), 124–135 (2001). https://doi.org/10.1093/occmed/51.2.124

Model Predictive Controller for a Planar Tensegrity Mechanism with Decoupled Position and Stiffness Control

JR Jurado Realpe(✉), Salih Abdelaziz, and Philippe Poignet

LIRMM, Université de Montpellier,
CNRS, Montpellier, France
{juradoreal,abdelaziz,poignet}@lirmm.fr

Abstract. Precise trajectory tracking and stiffness modulation for tensegrity mechanisms are a challenging topic that can open new horizon of applications for this type of systems. This paper presents a new control strategy of tensegrity mechanisms using a model predictive controller (MPC). Based on a dynamic model, the proposed approach allows to track trajectories with low and relatively high dynamics as well as to modulate the mechanism stiffness by changing only the controller's parameters. Trajectories of 30 s, 5 s and 1 s are performed showing a trajectory tracking improvement of up to 64% in the root mean square error when compared to literature results.

Keywords: Dynamic modeling · MPC controller · Stiffness modulation · Tensegrity mechanism · Trajectory tracking

1 Introduction

The word tensegrity comes from the contraction of tensional and integrity [1]. A tensegrity system is composed by compressive and tensile elements that interacts to define a stable volume [2]. The integrity of the structure is then ensured by only the internal forces produced by the tensile elements (cables, springs). Their large workspace to size ratio, their light weight and their ability to modulate stiffness, make tensegrity systems an interesting solution for structural applications [3,4], deployable systems [5,6], mobile robots [7,8] or manipulators [9,10].

Different control approaches have been proposed to control the motion of tensegrity mechanisms. The authors are commonly focused in the generation of gait for locomotion using open loop controllers [11,12], neuronal networks [13,14], kinematic models [10] or Lyapunov function [15]. Nevertheless, the trajectory tracking has not been a priority. To our knowledge, only two works [10,16] propose strategies to control finely trajectory tracking. On one hand, a PID controller and a tension distribution algorithm [10], inspired from cable-driven robots controllers, has been proposed to control the position and the stiffness of a 1 DOF planar tensegrity mechanism. This controller presents good

J. Lenarčič and B. Siciliano (Eds.): ARK 2020, SPAR 15, pp. 349–358, 2021.
https://doi.org/10.1007/978-3-030-50975-0_43

position tracking for trajectories with low dynamic but loose performance when tracking relatively high dynamics trajectories. On the other hand, a Model Predictive Controller (MPC) with inverse static optimization has been introduced to realize the trajectory tracking of a tensegrity spine robot. Nevertheless, stiffness modulation is not addressed.

Our interest here is to fully exploit a tensegrity system by controlling its position and stiffness. A dynamic-based control strategy would be pertinent to deliver good trajectory tracking performance. A MPC using a dynamic model is here proposed to modulate the stiffness and trajectory tracking of the mechanism in an asynchronous way by modifying the control parameters of the controller. The proposed strategy is experimentally validated on a 1 DOF planar cable-driven tensegrity mechanism.

This paper is organized as follow. The static, stiffness and dynamic models of the planar tensegrity mechanism is developed in Sect. 2. Then a MPC strategy is proposed in Sect. 3. Position tracking and stiffness modulation results are discussed in Sect. 4. Finally conclusions and future work are presented in Sect. 5.

2 System Modelling

The system studied here is a 1 DOF planar tensegrity mechanism, whose end effector is considered as the bar bc (Fig. 1). The end effector orientation is denoted θ. The structure is composed by an articulated equilateral parallelogram, of length L, and is driven by two cables. Each cable is attached to one corner of the parallelogram at one end and to a linear spring, of stiffness k, on the other end. The mechanism is actuated by two actuators. The angular position for each motor i is denoted α_i. Pulleys, of radius R, are mounted on the actuators so that $\rho_i = R\alpha_i$ describes the displacement of the cable i. This redundancy of actuation is performed so as to allow stiffness modulation of the system.

2.1 Static and Stiffness Models

The static model of the system is computed using the virtual work approach. It is described as:

$$\mathbf{W}\boldsymbol{\tau} = \Gamma \tag{1}$$

where $\mathbf{W} = [L\sin(\theta/2),\ -L\cos(\theta/2)]$. Γ denotes the torque generated by the two cables on the end effector. The components of the vector $\boldsymbol{\tau} = [\tau_1, \tau_2]^T$ describe the cables tensions, computed as:

$$\begin{cases} \tau_1 = k\Delta l_1 = k\big(\rho_1 + 2L\cos(\theta/2) - L\sqrt{2}\big) \\ \tau_2 = k\Delta l_2 = k\big(\rho_2 + 2L\sin(\theta/2) - L\sqrt{2}\big) \end{cases} \tag{2}$$

where Δl_i denotes the elongation of the spring i. The potential energy U of the system is assumed to be only the elastic energy stored in the springs:

$$U = \frac{1}{2}k(\Delta l_1^2 + \Delta l_2^2) \tag{3}$$

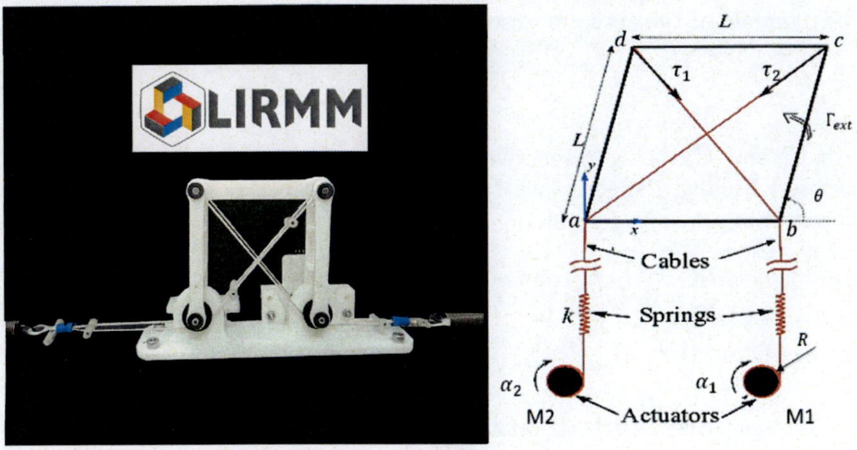

Fig. 1. Planar mechanism.

The static equilibrium of the platform is obtained by solving $\frac{\partial U}{\partial \theta} = 0$:

$$\theta = 2\arctan\left(\frac{L\sqrt{2} - \rho_2}{L\sqrt{2} - \rho_1}\right) \tag{4}$$

The angular stiffness K_θ at the equilibrium configuration is obtained by:

$$K_\theta = \frac{\partial^2 U}{\partial \theta^2} = -\frac{\partial \Gamma}{\partial \theta} = -\frac{L}{2}\Big(\cos(\theta/2)\tau_1 + \sin(\theta/2)\tau_2\Big) + kL^2 \tag{5}$$

2.2 Dynamic Model

The dynamic model of the structure is calculated using Euler-Lagrange formulation. The total kinetic energy E_k of the system is defined by:

$$E_k = \frac{5}{6} m L^2 \dot{\theta}^2 \tag{6}$$

where m denotes the mass of one bar of the parallelogram. The equation of motion is therefore given by:

$$\mathrm{f_d} + \mathrm{f_q} + \Gamma_{ext} = \frac{\mathrm{d}}{\mathrm{dt}}\frac{\partial \mathscr{L}}{\partial \dot{\theta}} - \frac{\partial \mathscr{L}}{\partial \theta} \tag{7}$$

where Γ_{ext} denotes the external torque applied on the end-effector and $\mathscr{L} = E_T - U$. In this study, the friction and the gravity forces, respectively, $\mathrm{f_d}$ and $\mathrm{f_q}$, are neglected. The structure dynamics is therefore described as:

$$\Gamma_{ext} = \frac{5}{3} m L^2 \ddot{\theta} + \frac{\partial U}{\partial \theta} \tag{8}$$

The dynamic of the actuator i is defined using Newton's law:

$$\mathrm{C_i} - \mathrm{C_v}\dot{\alpha}_i - R\tau_i = \mathrm{I_m}\frac{\mathrm{d}\dot{\alpha}_i}{\mathrm{dt}} \tag{9}$$

with $\mathrm{C_i}$ and $\mathrm{C_v}$ being respectively the torque and viscous coefficient of the actuator. Finally, the complete dynamic model of the tensegrity mechanism is expressed in a matrix formulation as:

$$\begin{pmatrix} R\mathrm{C}_1 \\ R\mathrm{C}_2 \\ \Gamma_{ext} \end{pmatrix} = \begin{pmatrix} \mathrm{I_m} & 0 & 0 \\ 0 & \mathrm{I_m} & 0 \\ 0 & 0 & \frac{5mL^2}{3} \end{pmatrix} \begin{pmatrix} \ddot{\rho_1} \\ \ddot{\rho_2} \\ \ddot{\theta} \end{pmatrix} + \begin{pmatrix} \mathrm{C_v}\dot{\rho_1} \\ \mathrm{C_v}\dot{\rho_2} \\ 0 \end{pmatrix} + \begin{pmatrix} R^2\tau_1 \\ R^2\tau_2 \\ \frac{\partial U}{\partial \theta} \end{pmatrix} \tag{10}$$

3 Proposed Control Strategy

A control approach that allows to asynchronously track low/high dynamic trajectories and modulate the stiffness of the tensegrity mechanism is hereafter presented. It consists on a linear MPC based on a discrete-time state-space model.

3.1 Dynamic Model Linearization

The linearization of (10) is performed around an equilibrium point $(\rho_{10}, \rho_{20}, \theta_0)$. Only $\partial U/\partial\theta$, τ_1 and τ_2 present nonlinear terms. The linearized terms are computed using the first-order Taylor expansion as:

$$\begin{pmatrix} \tau_1 \\ \tau_2 \end{pmatrix} = \begin{pmatrix} k\rho_1 - f_1\theta + f_3 \\ k\rho_2 + f_2\theta + f_4 \end{pmatrix} \tag{11}$$

$$\frac{\partial U}{\partial \theta} = -f_1\rho_1 + f_2\rho_2 - f_5\theta - f_6 \tag{12}$$

such that:

$$\begin{aligned}
f_1 &= kL\sin(\theta_0/2) \\
f_2 &= kL\cos(\theta_0/2) \\
f_3 &= k\rho_{10} + 2f_2 - kL\sqrt{2} \\
f_4 &= k\rho_{20} + 2f_1 - kL\sqrt{2} \\
f_5 &= -L\sqrt{2}/2(f_1 + f_2) + f_1\rho_{20}/2 + f_2\rho_{10}/2 \\
f_6 &= L\sqrt{2}(f_2 - f_1) - f_1\rho_{10} + f_2\rho_{20}
\end{aligned}$$

By defining the state vector $\mathbf{x} = [\rho_1, \rho_2, \theta, \dot{\rho}_1, \dot{\rho}_2, \dot{\theta}]^T$ and the control vector as the torque of the motors $\mathbf{u} = [\mathrm{C}_1, \mathrm{C}_2]^T$ and by using the linearized expressions, the continuous-time state-space is expressed as:

$$\dot{\mathbf{x}} = \mathbf{A}_c\mathbf{x} + \mathbf{B}_c\mathbf{u} + \mathbf{d}_c \tag{13}$$

with:

$$\mathbf{A}_c = \begin{pmatrix} 0 & 0 & 0 & 1 & 0 & 0 \\ 0 & 0 & 0 & 0 & 1 & 0 \\ 0 & 0 & 0 & 0 & 0 & 1 \\ \frac{-kR^2}{\mathrm{I_m}} & 0 & \frac{R^2 f_1}{\mathrm{I_m}} & \frac{-\mathrm{C_v}}{\mathrm{I_m}} & 0 & 0 \\ 0 & \frac{-kR^2}{\mathrm{I_m}} & \frac{-R^2 f_2}{\mathrm{I_m}} & 0 & \frac{-\mathrm{C_v}}{\mathrm{I_m}} & 0 \\ \frac{3f_1}{5mL^2} & \frac{-3f_2}{5mL^2} & \frac{3f_5}{5mL^2} & 0 & 0 & 0 \end{pmatrix} \mathbf{B}_c = \begin{pmatrix} 0 & 0 \\ 0 & 0 \\ 0 & 0 \\ \frac{R}{\mathrm{I_m}} & 0 \\ 0 & \frac{R}{\mathrm{I_m}} \\ 0 & 0 \end{pmatrix} \mathbf{d}_c = \begin{pmatrix} 0 \\ 0 \\ 0 \\ \frac{R^2 f_3}{\mathrm{I_m}} \\ \frac{R^2 f_4}{\mathrm{I_m}} \\ \frac{3f_6}{5mL^2} \end{pmatrix} \quad (14)$$

The linear discrete-time model is defined as:

$$\mathbf{x}_{k+1} = \mathbf{A}_k \mathbf{x}_k + \mathbf{B}_k \mathbf{u}_k + \mathbf{d}_k \quad (15)$$

3.2 MPC Formulation

The objective of the MPC controller is to minimize the error $\mathbf{E}_k$ over an horizon $\mathrm{H_p}$ between the desired trajectory $\mathbf{T}_k = [\mathbf{x}^d_{k+1}, \mathbf{x}^d_{k+2}, \ldots, \mathbf{x}^d_{k+\mathrm{H_p}}]^T$ and the predicted output $\mathbf{Y}_k = [\mathbf{x}_{k+1}, \mathbf{x}_{k+2}, \ldots, \mathbf{x}_{k+\mathrm{H_p}}]^T$. The cost function is defined as [17]:

$$\mathbf{V}_k = \mathbf{E}_k^T \mathbf{Q} \mathbf{E}_k + \boldsymbol{\Delta}\mathbf{U}_k^T \mathbf{R} \boldsymbol{\Delta}\mathbf{U}_k \quad (16)$$

with $\mathbf{E}_k = \mathbf{T}_k - \mathbf{Y}_k$. $\mathbf{Q}$ and $\mathbf{R}$ are diagonal weight matrices.

In order to solve the predictive control problem, it is necessary to predict the future states $\mathbf{x}_{k+i}$, $i = 1, ..., \mathrm{H_p}$. To do so, we consider that over a horizon $\mathrm{H_p}$ the linear time varying system (15) can be approximated by a linear invariant system. This yields to consider the matrices $\mathbf{A}_k$, $\mathbf{B}_k$ and $\mathbf{d}_k$ as constants, which mean that $\mathbf{A}_{k+i} = \mathbf{A}_k = \mathbf{A}$, $\mathbf{B}_{k+i} = \mathbf{B}_k = \mathbf{B}$ and $\mathbf{d}_{k+i} = \mathbf{d}_k = \mathbf{d}$ for $i = 1, ..., \mathrm{H_p}$. With this assumption the future states can be computed by applying $\mathrm{H_p}$ times the equation (15):

$$\mathbf{Y}_k = \begin{pmatrix} \mathbf{x}_{k+1} \\ \mathbf{x}_{k+2} \\ \vdots \\ \mathbf{x}_{k+\mathrm{H_p}} \end{pmatrix} = \underbrace{\begin{bmatrix} \mathbf{A} \\ \mathbf{A}^2 \\ \vdots \\ \mathbf{A}^{\mathrm{H_p}} \end{bmatrix}}_{\Psi} \mathbf{x}_k + \underbrace{\begin{bmatrix} \mathbf{B} \\ \mathbf{AB} + \mathbf{B} \\ \vdots \\ \sum_{i=0}^{\mathrm{H_p}-1} \mathbf{A}^i \mathbf{B} \end{bmatrix}}_{\gamma} \mathbf{u}_{k-1} + \underbrace{\begin{bmatrix} \mathbf{d} \\ \mathbf{Ad} + \mathbf{d} \\ \vdots \\ \sum_{i=0}^{\mathrm{H_p}-1} \mathbf{A}^i \mathbf{d} \end{bmatrix}}_{\mathbf{D}} + \ldots$$

$$\cdots + \underbrace{\begin{bmatrix} \mathbf{B} & 0 & \ldots & 0 \\ \mathbf{AB} + \mathbf{B} & \mathbf{B} & 0 \ldots & 0 \\ \vdots & & & \vdots \\ \sum_{i=0}^{\mathrm{H_p}-1} \mathbf{A}^i \mathbf{B} & \sum_{i=0}^{\mathrm{H_p}-2} \mathbf{A}^i \mathbf{B} & \ldots & \sum_{i=0}^{\mathrm{H_p}-\mathrm{H_u}} \mathbf{A}^i \mathbf{B} \end{bmatrix}}_{\Theta} \underbrace{\begin{bmatrix} \boldsymbol{\Delta}\mathbf{u}_k \\ \boldsymbol{\Delta}\mathbf{u}_{k+1} \\ \vdots \\ \boldsymbol{\Delta}\mathbf{u}_{k+\mathrm{H_u}-1} \end{bmatrix}}_{\boldsymbol{\Delta}\mathbf{U}} \quad (17)$$

To take into account the unilateral tensions in the cables, a constrained optimization problem has to be considered. As the control variables of the MPC controller are the torques of the actuators, the tension limits, $\boldsymbol{\tau}_{\min} = [\tau_{\min}, \tau_{\min}]^T$

and $\boldsymbol{\tau}_{\max} = [\tau_{\max}, \tau_{\max}]^T$, have to be converted into torques limits $\mathbf{u}_{\min} = [\mathrm{C}_{\min}, \mathrm{C}_{\min}]^T$ and $\mathbf{u}_{\max} = [\mathrm{C}_{\max}, \mathrm{C}_{\max}]^T$. The solution of (16) can be obtained by solving the constrained quadratic programming (QP) problem:

$$\min_{\boldsymbol{\Delta}\mathbf{U}} \boldsymbol{\Delta}\mathbf{U}^T\mathbf{H}\boldsymbol{\Delta}\mathbf{U} + \mathbf{G}^T\boldsymbol{\Delta}\mathbf{U} \tag{18}$$

$$\text{subject to}: \quad \mathbf{u}_{\min} \leq \mathbf{u} \leq \mathbf{u}_{\max} \tag{19}$$

with $\mathbf{H} = \Theta^T\mathbf{Q}\Theta + \mathbf{R}$ and $\mathbf{G} = 2\Theta^T\mathbf{Q}(\mathbf{T}_k - \Psi\mathbf{x}_k - \gamma\mathbf{u}_{k-1} - \mathbf{D})$. This problem has to be solved at each step k. Once the optimal solution $\boldsymbol{\Delta}\mathbf{U}$ is found, the first control output is applied and kept until the next time step.

3.2.1 Trajectory Tracking

The overall control approach is composed of an external MPC controller and an internal tension control loop (cf. Fig. 2). The control output $\mathbf{u}_k$ of the MPC is converted into a desired tension $\boldsymbol{\tau}^d$. This vector of tension is then compared to $\boldsymbol{\tau}^m$ that contains the real tensions in the cables. These tensions are measured using force sensors. A tension controller $\mathbf{C}_\tau$ is considered to control these tensions in the inner loop. The output of this tension controller is sent to the actuators as desired velocities $\dot{\boldsymbol{\alpha}}^d = [\dot{\alpha}_1^d,\ \dot{\alpha}_2^d]^T$.

During the trajectory tracking, the variables θ and $\dot{\theta}$ in the state vector $\mathbf{x}$ are to be controlled. To do so, the weight matrix $\mathbf{Q}$ has to be defined with high values for the components Q_θ and $\mathrm{Q}_{\dot{\theta}}$ and values near to zero for the other components $\mathrm{Q}_{\rho_1}, \mathrm{Q}_{\rho_2}, \mathrm{Q}_{\dot{\rho}_1}$. and $\mathrm{Q}_{\dot{\rho}_2}$:

$$\mathbf{Q} = \mathrm{diag}(\mathrm{Q}_{\rho_1}, \mathrm{Q}_{\rho_2}, \mathrm{Q}_\theta, \mathrm{Q}_{\dot{\rho}_1}, \mathrm{Q}_{\dot{\rho}_2}, \mathrm{Q}_{\dot{\theta}}) \tag{20}$$

3.2.2 Stiffness Modulation

The stiffness modulation is possible by changing the control parameter $\mathbf{Q}$. During the modulation of the angular stiffness K_θ, the mechanism has to conserve a static position θ^d. As K_θ depends on the cables tensions (5), an initial and a final desired tensions, respectively $\boldsymbol{\tau}^i = [\tau_1^i,\ \tau_2^i]^T$ and $\boldsymbol{\tau}^f = [\tau_1^f,\ \tau_2^f]^T$, have to be defined. These tensions are computed by solving (1). The solution has the form [18]:

$$\boldsymbol{\tau} = \mathbf{W}^+\Gamma + \mathbf{N}\lambda \tag{21}$$

where $\mathbf{W}^+$ denotes the Moore-Penrose pseudoinverse of $\mathbf{W}$, $\mathbf{N} = [-L\cos(\theta/2),\ -L\sin(\theta/2)]^T$ a basis of $\mathbf{W}$ null space and λ is a scalar to be defined.

Since the mechanisms is in an equilibrium position, the generated torque produced by the cables is equal to zero $\Gamma = 0$. The bounds of λ, defined as $[\lambda_{min}, \lambda_{max}]$, can therefore be obtained by solving the inequality:

$$\tau_{min} \leq \mathbf{N}\lambda \leq \tau_{max} \tag{22}$$

Once the bounds of λ are obtained, the tensions $\boldsymbol{\tau}^i$ and $\boldsymbol{\tau}^f$ could be chosen for given values of λ that satisfy (22). Then, using (2), the tensions $\boldsymbol{\tau}^i$ and

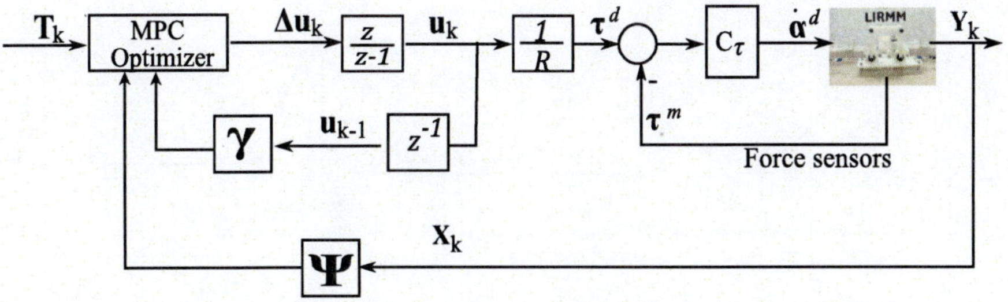

Fig. 2. Strategy of control proposed based on a MPC controller

$\boldsymbol{\tau}^f$ are converted into initial and final desired cables displacement, respectively $\boldsymbol{\rho}^i = [\rho_1^i, \rho_2^i]^T$ and $\boldsymbol{\rho}^f = [\rho_1^f, \rho_2^f]^T$. A desired trajectory, from $\boldsymbol{\rho}^i$ to $\boldsymbol{\rho}^f$, can therefore be defined in order to modulate K_θ.

Since the trajectory of interest is now in terms of the cables displacement, the weight matrix $\mathbf{Q}$ has to be selected with high values for $\mathrm{Q}_{\rho_1}, \mathrm{Q}_{\rho_2}$ and with small values for the rest of the components. A special attention has to be made with the component Q_θ, because its weight allows to ensure that the orientation of the mechanism will not change during the stiffness modulation.

4 Results

The control approach is evaluated on the developed prototype shown in Fig. 1. The mechanism is actuated using 2 DC motors. Polyethylene fibre cables are deployed for motion transmission. 2 force sensors are used to measure the cables tensions. Each sensor is placed between one motor and a spring. Moreover, an optical encoder, with a resolution of 0.225°, is used to measure the angle θ.

Trajectory tracking with low and relatively-high dynamic is first assessed. Three trajectories, going from 90° to 45°, are created using fifth-order polynomial equation. The duration of these trajectories are respectively 30 s, 5 s and 1 s. The cables tensions limits are set to $\tau_{min} = 4N$ and $\tau_{max} = 10N$ and the control parameters of $\mathbf{Q}$ are defined in function of the trajectory. For the first trajectory, $Q_\theta = 8e4$, $\mathrm{Q}_{\dot{\theta}} = 50$. For relatively-high dynamic trajectories, $Q_\theta = 2.1e5$, $\mathrm{Q}_{\dot{\theta}} = 8$. The control parameters $Q_{\rho_1} = Q_{\rho_2} = \mathrm{Q}_{\dot{\rho}_1} = \mathrm{Q}_{\dot{\rho}_2} = 10^{-3}$ and $\mathbf{R} = \mathrm{diag}(9e8, 9e8)$ are set equal for all the trajectories. H_p and H_u are set equal to 40 with a sampling time of 7 ms. The constrained MPC optimization problem (19) is solved using the Interior point method implemented in the Eigen quadprog solver. The results can be observed in Fig. 3.

The root mean square errors (RMSE) of the position tracking for the three trajectories are respectively 0.47°, 1.28° and 4.36°. The static errors is less than or equal to 0.225° which is the encoder resolution.

The obtained results are compared to the literature ones [10], where similar experiments have been conducted using a PID controller. Reductions of 51% and 64% in the RMSE for the two trajectories, 30 s and 5 s, are observed (cf. Table 1).

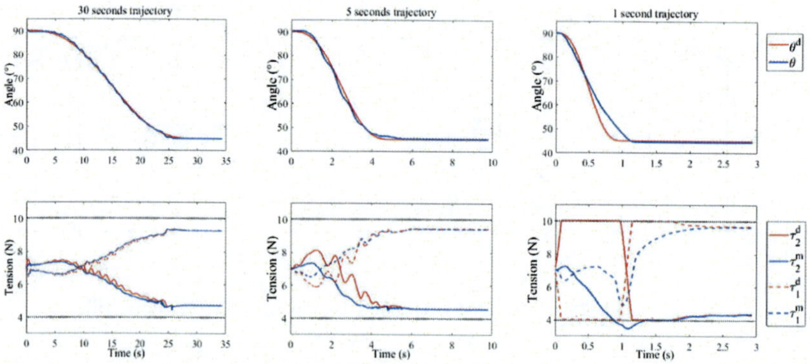

Fig. 3. Trajectory tracking with low and relatively high dynamic.

Table 1. Trajectory tracking results for PID controller [10] and MPC approach.

Trajectory	PID	MPC
30 s	0.97°	0.47°
5 s	3.6°	1.28°

Besides, the RMSE increases when the duration of trajectories decreases. This is mainly due the bandwidth of the tension inner loop, limited actually to 1 s of time response. A solution could be to reformulate the MPC problem by changing the control vector **u** from motor's torques to motor's angular velocities. Thus, this reformulation eliminates the tension inner loop, and would improve the time response of the system.

A final experiment is performed in order to validate the stiffness modulation of the system. This experiment is divided into two phases. In the first phase, trajectory tracking is performed during 30 s, going from 90° to 70°. Once the trajectory tracking phase is finished, the stiffness modulation is assessed during 10 s, while the position of the mechanism is kept at $\theta_f = 70°$. $\boldsymbol{\tau}^i$ is defined as desired tensions at the end of the trajectory tracking. $\boldsymbol{\tau}^f$ is defined by setting $\lambda = \lambda_{max}$. A linear displacement of the cables, going from $\boldsymbol{\rho}^i$ to $\boldsymbol{\rho}^f$, is therefore defined and sent to the MPC controller. The weights variables of $\mathbf{Q}$ are defined as $\mathrm{Q}_{\rho_1} = \mathrm{Q}_{\rho_2} = 7.2e5$, $Q_\theta = \mathrm{Q}_{\dot{\rho}_1} = \mathrm{Q}_{\dot{\rho}_2} = 1$, $Q_{\dot{\theta}} = 10^{-3}$ and $\mathbf{R} = \mathrm{diag}(7e7, 7e7)$. The results are depicted in Fig. 4. During the trajectory tracking, the angular stiffness is quite constant. During the stiffness modulation, K_θ varies linearly from $0.98 Nm/rad$ to $0.89 Nm/rad$. This variation is related to the cables tensions variation, as expected in (5) and observed in Fig. 4(b and c). The evolution of the angular stiffness K_θ is inversely proportional to the variation of $\boldsymbol{\tau}$.

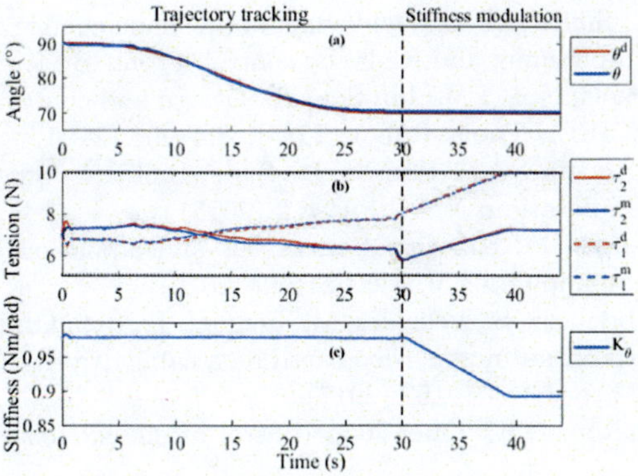

Fig. 4. Results of the trajectory tracking and stiffness modulation experiment

5 Conclusion

This work presents a linear MPC strategy for tensegrity mechanisms. The approach allows the position tracking and the stiffness modulation by only changing the values of the control parameters **Q** and **R**. A planar cable-driven tensegrity mechanism was used to validate the performance of the proposed controller. The results have shown a static error of less than 0.225° and good performance tracking relatively-high and low dynamics trajectories. A reduction of up to 64% on the RMSE during the trajectory tracking, with respect to literature results, was obtained. Furthermore, the MPC proposed shows a decoupled control of the position and the stiffness.

Future work will be the extension of this approach for spatial tensegrity mechanisms. An interesting perspective would be the proposal of a MPC with the motor's angular velocity as the output control vector, in order to avoid the use of an inner tension control loop.

Acknowledgements. This work was supported by French state funds managed by the ANR within the Investissements d'Avenir programme (Robotex ANR-10-EQPX-44, Labex CAMI - ANR- 11-LABX-0004).

References

1. Fuller, B.: Tensile-integrity structures. United States Patent No. 3,063,521 (1962)
2. Pugh, A.: An Introduction to Tensegrity, 1st edn. University of California Press, Berkeley and Los Angeles (1976)
3. Snelson, K.: Space Frame Structure Made by 3-D Weaving of Rod Members. US Patent 6739937.2004
4. Sultan, C.: Modeling, design, and control of tensegrity structures with applications. Purdue University (1999)
5. Skelton, R.E., De Oliveira, M.C.: Tensegrity Systems. Springer (2009)

6. Sychterz, A.C., Smith, I.F.C.: Deployment and shape change of a tensegrity structure using path-planning and feedback control. Front. Built Environ. (2018)
7. Paul, C., Valero-Cuevas, F.J., Lip-Son, H.: Design and control of tensegrity robots for locomotion. IEEE Trans. Robot. **22**(5), 944–957 (2006)
8. Sunspiral, V., Agogino, A., Atkinson, D.: Super Ball Bot - Structures for Planetary Landing and Exploration. NIAC Phase 2 (2015)
9. Furet, M., Wenger, P.: Kinetostatic analysis and actuation strategy of a planar tensegrity 2-X manipulator. J. Mech. Rob. (2019)
10. Boehler, Q., Abdelaziz, S., Vedrines, M., Poignet, P., Renaud, P.: From modeling to control of a variable stiffness device based on a cable-driven tensegrity mechanism. Mech. Mach. Theory **1–12**, 107 (2017)
11. Masic, M., Skelton, R.E.: Open-loop control of class-2 tensegrity towers, p. 298 (2004)
12. Iscen, A., Caluwaerts, K., et al.: Learning tensegrity locomotion using open-loop control signals and coevolutionary algorithms. Artif. Life **21**, 119–140 (2015)
13. Kanchanasaratool, N., Williamson, D.: Modelling and control of class NSP tensegrity structures. Int. J. Control **75**(2), 123–139 (2002)
14. Hustig-Schultz, D., SunSpiral, V., Teodorescu, M.: Morphological design for controlled tensegrity quadruped locomotion. In: IEEE/RSJ International Conference on Intelligent Robots and Systems (IROS), pp. 4714–4719 (2016)
15. Wroldsen, A.S., de Oliveira, M.C., Skelton, R.E.: Modelling and control of non-minimal non-linear realisations of tensegrity systems (2009)
16. Sabelhaus, A.P., et al.: Model-predictive control with reference input tracking for tensegrity spine robots. arXiv arXiv:1806.08868 (2019)
17. Maciejowski, J.: Predictive Control with Constraints. Prentice Hall (2002)
18. Gouttefarde, M., Lamaury, J., Reichert, C., Bruckmann, T.: A versatile tension distribution algorithm for n-DOF parallel robots driven by n+2 cables. IEEE Trans. Robot. **31**(6), 1444–1457 (2015)

Author Index

A
Abdelaziz, Salih, 15, 349
Altuzarra, Oscar, 327
Andreff, Nicolas, 49

B
Babič, Jan, 341
Bai, Shaoping, 82
Bandyopadhyay, Sandipan, 277
Barreto, Rodrigo Luis Pereira, 190
Baskar, Aravind, 303
Begey, Jérémy, 49
Birlescu, Iosif, 57
Blaysat, B., 117
Bogateanu, Remus, 57
Bongardt, Bertold, 250
Boudon, B., 117
Briot, Sébastien, 65, 150

C
Campa, Francisco J., 327
Carboni, Andrea Piga, 190
Caro, Stéphane, 209
Carricato, Marco, 65
Chablat, Damien, 226
Chanal, H., 117
Chen, Kevin, 242
Chorsi, Meysam T., 41
Claveau, Fabien, 209
Conconi, Michele, 109
Condurache, Daniel, 142
Corves, Burkhard, 158
Coste, Michel, 234
Cruz-Martinez, G. M., 15
Cui, Chuangchuang, 268

D
Di Gironimo, Giuseppe, 335
Dietrich, Franz, 33
Djintelbe, Nestor, 234
Donelan, Peter, 133
Doshi, Dhruvin, 33

F
Fabritius, Marc, 218
Furet, Matthieu, 319

G
Gherman, Bogdan, 57
Gonzalez, A. Vilchis, 15
Gorbatyuk, Vitaliy, 41
Gouttefarde, Marc, 199
Gracia, Alba Perez, 7
Grazioso, Stanislao, 335
Guyon, J.-B., 117

H
Hayes, M. John D., 90
Heidari, Omid, 7
Hüsing, Mathias, 158
Husty, Manfred, 57, 166
Husty, Manfred L., 90

I
Idà, Edoardo, 65
Ilies, Horea, 41

J. Lenarčič and B. Siciliano (Eds.): ARK 2020, SPAR 15, pp. 359–360, 2020.
https://doi.org/10.1007/978-3-030-50975-0

J
Jamšek, Marko, 341
Jilich, Michal, 311
Jurado Realpe, JR, 349

K
Kazerounian, Kazem, 41
Kecskemethy, Andres, 293
Klimmek, Stefan, 158
Kumar, Shivesh, 259

L
Lenarčič, Jadran, 1
Li, Ruiqin, 82

M
Martins, Daniel, 190
McCarthy, J. Michael, 242
Merlet, J-P., 98
Mihelj, Matjaž, 174
Müller, Andreas, 125, 259
Müller, Katharina, 293
Mundrane, Caitlyn, 41
Munih, Marko, 174

N
Nawratil, G., 182
Nayak, Abhilash, 150, 166

P
Parenti-Castelli, Vincenzo, 109
Patra, Bibekananda, 277
Peitsch, Katrin, 158
Petrič, Tadej, 341
Pfurner, Martin, 90, 166
Picard, Etienne, 209
Pisla, Doina, 57
Plecnik, Mark, 303
Plestan, Franck, 209
Podobnik, Janez, 174
Poignet, Philippe, 15, 349
Porta, Josep M., 285
Pott, Andreas, 218
Prasad, Prem Kumar, 277

R
Ramadoss, Vishal, 311
Ramos, Juan Mauricio Toro, 33
Renaud, Pierre, 49
Rosati, Luciano, 335
Rotzoll, Mirja, 90

S
Sagar, Keerthi, 311
Sancisi, Nicola, 109
Santos, João Cavalcanti, 199
Schütz, Daniel, 33
Selig, J. M., 73
Shahidi, Amir, 158
Shen, Huiping, 166, 268
Shende, Argaja Deepak, 277
Siciliano, Bruno, 335
Simoni, Roberto, 190
Stoeffler, Christoph, 259

T
Tavousi, Pouya, 41
Thomas, Federico, 285

U
Uicker, John J., 250
Ulinici, Ionut, 57

V
Vaida, Calin, 57
van der Wijk, Volkert, 25
Vedrines, Marc, 49
Venkateswaran, Swaminath, 226
Vilchis, J-C Avila, 15

W
Wenger, Philippe, 319
Wu, Guanglei, 268
Wu, Rui, 82

Z
Zanella, Alessandro, 311
Zhang, Ning, 268
Zhang, Xuping, 268
Zlatanov, Dimiter, 311
Zoppi, Matteo, 311